T0076762

Probability

This lively introduction to measure-theoretic probability theory covers laws of large numbers, central limit theorems, random walks, martingales, Markov chains, ergodic theorems, and Brownian motion. Concentrating on the results that are the most useful for applications, this comprehensive treatment is a rigorous, measure theory–based graduate text and reference. Operating under the philosophy that the best way to learn probability is to see it in action, the book contains many extended examples that apply the theory to concrete applications. Readers learn to recognize when a method works and, more important, when it does not.

This fifth edition contains a new chapter on multidimensional Brownian motion and its relationship to PDEs, an advanced topic that is finding new applications. Setting the foundation for this expansion, Chapter 7 now features a proof of Itôs formula. Key exercises that previously were simply proofs left to the reader have been directly inserted into the text as lemmas. The new edition also reinstates discussion about the central limit theorem for martingales and stationary sequences.

RICK DURRETT is a James B. Duke professor in the mathematics department of Duke University. He received his Ph.D. in Operations Research from Stanford University in 1976. After nine years at UCLA and twenty-five at Cornell University, he moved to Duke in 2010. He is the author of 8 books and more than 220 journal articles on a wide variety of topics, and has supervised more than 45 Ph.D. students. He is a member of National Academy of Science, American Academy of Arts and Sciences, and a fellow of the Institute of Mathematical Statistics and of the American Mathematical Society.

CAMBRIDGE SERIES IN STATISTICAL AND PROBABILISTIC MATHEMATICS

Editorial Board

Z. Ghahramani (Department of Engineering, University of Cambridge)

R. Gill (Mathematical Institute, Leiden University)

F. P. Kelly (Department of Pure Mathematics and Mathematical Statistics, University of Cambridge)

B. D. Ripley (Department of Statistics, University of Oxford)

S. Ross (Department of Industrial and Systems Engineering, University of Southern California)

M. Stein (Department of Statistics, University of Chicago)

This series of high-quality upper-division textbooks and expository monographs covers all aspects of stochastic applicable mathematics. The topics range from pure and applied statistics to probability theory, operations research, optimization, and mathematical programming. The books contain clear presentations of new developments in the field and also of the state of the art in classical methods. While emphasizing rigorous treatment of theoretical methods, the books also contain applications and discussions of new techniques made possible by advances in computational practice.

A complete list of books in the series can be found at www.cambridge.org/statistics.

Recent titles include the following:

22. *Saddlepoint Approximations with Applications,* by Ronald W. Butler
23. *Applied Asymptotics,* by A. R. Brazzale, A. C. Davison and N. Reid
24. *Random Networks for Communication,* by Massimo Franceschetti and Ronald Meester
25. *Design of Comparative Experiments,* by R. A. Bailey
26. *Symmetry Studies,* by Marlos A. G. Viana
27. *Model Selection and Model Averaging,* by Gerda Claeskens and Nils Lid Hjort
28. *Bayesian Nonparametrics,* edited by Nils Lid Hjort *et al.*
29. *From Finite Sample to Asymptotic Methods in Statistics,* by Pranab K. Sen, Julio M. Singer and Antonio C. Pedrosa de Lima
30. *Brownian Motion,* by Peter Mörters and Yuval Peres
31. *Probability (Fourth Edition),* by Rick Durrett
33. *Stochastic Processes,* by Richard F. Bass
34. *Regression for Categorical Data,* by Gerhard Tutz
35. *Exercises in Probability (Second Edition),* by Loïc Chaumont and Marc Yor
36. *Statistical Principles for the Design of Experiments,* by R. Mead, S. G. Gilmour and A. Mead
37. *Quantum Stochastics,* by Mou-Hsiung Chang
38. *Nonparametric Estimation under Shape Constraints*, by Piet Groeneboom and Geurt Jongbloed
39. *Large Sample Covariance Matrices and High-Dimensional Data Analysis,* by Jianfeng Yao, Shurong Zheng and Zhidong Bai
40. *Mathematical Foundations of Infinite-Dimensional Statistical Models,* by Evarist Giné and Richard Nickl

41. *Confidence, Likelihood, Probability,* by Tore Schweder and Nils Lid Hjort
42. *Probability on Trees and Networks*, by Russell Lyons and Yuval Peres
43. *Random Graphs and Complex Networks (Volume 1)*, by Remco van der Hofstad
44. *Fundamentals of Nonparametric Bayesian Inference*, by Subhashis Ghosal and Aad van der Vaart
45. *Long-Range Dependence and Self-Similarity*, by Vladas Pipiras and Murad S. Taqqu
46. *Predictive Statistics*, by Bertrand S. Clarke and Jennifer L. Clarke
47. *High-Dimensional Probability,* by Roman Vershynin
48. *High-Dimensional Statistics,* by Martin J. Wainwright
49. *Probability: Theory and Examples (Fifth Edition)*, by Rick Durrett

Probability

Theory and Examples

FIFTH EDITION

Rick Durrett
Duke University

CAMBRIDGE
UNIVERSITY PRESS

CAMBRIDGE
UNIVERSITY PRESS

University Printing House, Cambridge CB2 8BS, United Kingdom

One Liberty Plaza, 20th Floor, New York, NY 10006, USA

477 Williamstown Road, Port Melbourne, VIC 3207, Australia

314–321, 3rd Floor, Plot 3, Splendor Forum, Jasola District Centre, New Delhi – 110025, India

79 Anson Road, #06–04/06, Singapore 079906

Cambridge University Press is part of the University of Cambridge.

It furthers the University's mission by disseminating knowledge in the pursuit of education, learning, and research at the highest international levels of excellence.

www.cambridge.org
Information on this title: www.cambridge.org/9781108473682
DOI: 10.1017/9781108591034

© Rick Durrett 2019

This publication is in copyright. Subject to statutory exception and to the provisions of relevant collective licensing agreements, no reproduction of any part may take place without the written permission of Cambridge University Press.

First published 2019
3rd printing 2020

Printed in the United Kingdom by TJ International Ltd. Padstow Cornwall

A catalogue record for this publication is available from the British Library.

Library of Congress Cataloging-in-Publication Data
Names: Durrett, Richard, 1951– author.
Title: Probability : theory and examples / Rick Durrett (Duke University, Durham, North Carolina).
Description: Fifth edition. | Cambridge ; New York, NY : Cambridge University Press, 2019. | Series: Cambridge series in statistical and probabilistic mathematics ; 49 | Includes bibliographical references and index.
Identifiers: LCCN 2018047195 | ISBN 9781108473682 (hardback : alk. paper)
Subjects: LCSH: Probabilities. | Probabilities–Textbooks.
Classification: LCC QA273 .D865 2019 | DDC 519.2–dc23
LC record available at https://lccn.loc.gov/2018047195

ISBN 978-1-108-47368-2 Hardback

Cambridge University Press has no responsibility for the persistence or accuracy of URLs for external or third-party internet websites referred to in this publication and does not guarantee that any content on such websites is, or will remain, accurate or appropriate.

Contents

Preface

> Some times the lights are shining on me. Other times I can barely see.
> Lately it occurs to me what a long strange trip its been. (Grateful Dead)

In 1989, when the first edition of the book was completed, my sons David and Greg were 3 and 1, and the cover picture showed the Dow Jones at 2650. The last 29 years have brought many changes, but the song remains the same: "The title of the book indicates that as we develop the theory, we will focus our attention on examples. Hoping that the book would be a useful reference for people who apply probability in their work, we have tried to emphasize the results that are important for applications, and illustrated their use with roughly 200 examples. Probability is not a spectator sport, so the book contains almost 450 exercises to challenge the reader and to deepen their understanding."

The fifth edition has a number of changes:

- The exercises have been moved to the end of the section. The Examples, Theorems, and Lemmas are now numbered in one sequence to make it easier to find things.
- There is a new chapter on multidimensional Brownian motion and its relationship to PDEs. To make this possible, a proof of Itô's formula has been added to Chapter 7.
- The lengthy Brownian motion chapter has been split into two, with the second one focusing on Donsker's theorem, etc. The material on the central limit theorem for martingales and stationary sequences deleted from the fourth edition has been reinstated.
- The four sections of the random walk chapter have been relocated. Stopping times have been moved to the martingale chapter; recurrence of random walks and the arcsine laws to the Markov chain chapter; renewal theory has been moved to Chapter 2.
- Some of the exercises that were simply proofs left to the reader have been put into the text as lemmas. There are a few new exercises.

Typos The fourth edition contains a list of the people who made corrections to the first three editions. With apologies to those whose contributions I lost track of, this time I need to thank: Richard Arratia, Benson Au, Swee Hong Chan, Conrado Costa, Nate Eldredge, Steve Evans, Jason Farnon, Christina Goldschmidt, Eduardo Hota, Martin Hildebrand, Shlomo Leventhal, Jan Lieke, Kyle MacDonald, Ron Peled, Jonathan Peterson, Erfan Salavati, Byron Schmuland, Timo Seppalainen, Antonio Carlos de Azevedo Sodre, Shouda Wang, and Ruth Williams. I must confess that Christophe Leuridan pointed one out that I have not corrected. Lemma 3.4.19 incorrectly asserts that the distributions in its statement have mean 0, but their means do not exist. The conclusion remains valid, since they are differentiable at 0. A sixth edition is extremely unlikely, but you can e-mail me about typos and I will post them on my website.

Family update As the fourth edition was being completed, David had recently graduated from Ithaca College and Greg was in his last semester at MIT applying to graduate school in computer science. Now, eight years later, Greg has graduated from Berkeley University, and is an assistant professor in the Computer Science department at the University of Texas in Austin. Greg works in the field of machine learning, specifically natural language processing. No, I don't know what that means, but it seems to pay well. David got his degree in journalism. After an extensive job search process and some freelance work, David has settled into a steady job working for a company that produces newsletters for athletic directors and trainers.

In the summer of 2010, Susan and I moved to Durham. Since many people think that the move was about the weather, I will mention that during our first summer it was 104 degrees (and humid!) three days in a row. Yes, it almost never snows here, but when it does, three inches of snow (typically mixed with ice) will shut down the whole town for four days. It took some time for us to adjust to the Durham/Chapel Hill area, which has about 10 times as many people as Ithaca and is criss-crossed by freeways, but we live in a nice quiet neighborhood near the campus. Susan enjoys volunteering at the Sarah P. Duke gardens and listening to their talks about the plants of North Carolina and future plans for the gardens.

As I write this, it is the last week before school starts.

1

Measure Theory

In this chapter, we will recall some definitions and results from measure theory. Our purpose here is to provide an introduction to readers who have not seen these concepts before and to review that material for those who have. Harder proofs, especially those that do not contribute much to one's intuition, are hidden away in the Appendix. Readers with a solid background in measure theory can skip Sections 1.4, 1.5, and 1.7, which were previously part of the Appendix.

1.1 Probability Spaces

Here and throughout the book, terms being defined are set in **boldface**. We begin with the most basic quantity. A **probability space** is a triple $(\Omega, \mathcal{F}, P)$, where Ω is a set of "outcomes," $\mathcal{F}$ is a set of "events," and $P : \mathcal{F} \to [0,1]$ is a function that assigns probabilities to events. We assume that $\mathcal{F}$ is a σ**-field** (or σ**-algebra**), i.e., a (nonempty) collection of subsets of Ω that satisfy

(i) if $A \in \mathcal{F}$, then $A^c \in \mathcal{F}$, and

(ii) if $A_i \in \mathcal{F}$ is a countable sequence of sets, then $\cup_i A_i \in \mathcal{F}$.

Here and in what follows, **countable** means finite or countably infinite. Since $\cap_i A_i = (\cup_i A_i^c)^c$, it follows that a σ-field is closed under countable intersections. We omit the last property from the definition to make it easier to check.

Without P, $(\Omega, \mathcal{F})$ is called a **measurable space**, i.e., it is a space on which we can put a measure. A **measure** is a nonnegative countably additive set function; that is, a function $\mu : \mathcal{F} \to \mathbf{R}$ with

(i) $\mu(A) \geq \mu(\emptyset) = 0$ for all $A \in \mathcal{F}$, and

(ii) if $A_i \in \mathcal{F}$ is a countable sequence of disjoint sets, then

$$\mu(\cup_i A_i) = \sum_i \mu(A_i)$$

If $\mu(\Omega) = 1$, we call μ a **probability measure**. In this book, probability measures are usually denoted by P.

The next result gives some consequences of the definition of a measure that we will need later. In all cases, we assume that the sets we mention are in $\mathcal{F}$.

Theorem 1.1.1 *Let μ be a measure on $(\Omega, \mathcal{F})$*

(i) **monotonicity**. *If $A \subset B$, then $\mu(A) \le \mu(B)$.*

(ii) **subadditivity**. *If $A \subset \cup_{m=1}^{\infty} A_m$, then $\mu(A) \le \sum_{m=1}^{\infty} \mu(A_m)$.*

(iii) **continuity from below**. *If $A_i \uparrow A$ (i.e., $A_1 \subset A_2 \subset \ldots$ and $\cup_i A_i = A$), then $\mu(A_i) \uparrow \mu(A)$.*

(iv) **continuity from above**. *If $A_i \downarrow A$ (i.e., $A_1 \supset A_2 \supset \ldots$ and $\cap_i A_i = A$), with $\mu(A_1) < \infty$, then $\mu(A_i) \downarrow \mu(A)$.*

Proof (i) Let $B - A = B \cap A^c$ be the **difference** of the two sets. Using $+$ to denote disjoint union, $B = A + (B - A)$ so

$$\mu(B) = \mu(A) + \mu(B - A) \ge \mu(A).$$

(ii) Let $A'_n = A_n \cap A$, $B_1 = A'_1$ and for $n > 1$, $B_n = A'_n - \cup_{m=1}^{n-1} A'_m$. Since the B_n are disjoint and have union A, we have using (ii) of the definition of measure, $B_m \subset A_m$, and (i) of this theorem

$$\mu(A) = \sum_{m=1}^{\infty} \mu(B_m) \le \sum_{m=1}^{\infty} \mu(A_m)$$

(iii) Let $B_n = A_n - A_{n-1}$. Then the B_n are disjoint and have $\cup_{m=1}^{\infty} B_m = A$, $\cup_{m=1}^{n} B_m = A_n$ so

$$\mu(A) = \sum_{m=1}^{\infty} \mu(B_m) = \lim_{n\to\infty} \sum_{m=1}^{n} \mu(B_m) = \lim_{n\to\infty} \mu(A_n)$$

(iv) $A_1 - A_n \uparrow A_1 - A$ so (iii) implies $\mu(A_1 - A_n) \uparrow \mu(A_1 - A)$. Since $A_1 \supset A$, we have $\mu(A_1 - A) = \mu(A_1) - \mu(A)$ and it follows that $\mu(A_n) \downarrow \mu(A)$. □

The simplest setting, which should be familiar from undergraduate probability, is:

Example 1.1.2 (Discrete probability spaces) Let Ω = a countable set, i.e., finite or countably infinite. Let $\mathcal{F}$ = the set of all subsets of Ω. Let

$$P(A) = \sum_{\omega \in A} p(\omega), \quad \text{where } p(\omega) \ge 0 \text{ and } \sum_{\omega \in \Omega} p(\omega) = 1$$

A little thought reveals that this is the most general probability measure on this space. In many cases when Ω is a finite set, we have $p(\omega) = 1/|\Omega|$, where $|\Omega|$ = the number of points in Ω.

For a simple concrete example that requires this level of generality, consider the astragali, dice used in ancient Egypt made from the ankle bones of sheep. This die could come to rest on the top side of the bone for four points or on the bottom for three points. The side of the bone was slightly rounded. The die could come to rest on a flat and narrow piece for six points or somewhere on the rest of the side for one point. There is no reason to think that all four outcomes are equally likely, so we need probabilities p_1, p_3, p_4, and p_6 to describe P.

To prepare for our next definition, we need to note that it follows easily from the definition: If $\mathcal{F}_i$, $i \in I$ are σ-fields, then $\cap_{i\in I} \mathcal{F}_i$ is. Here $I \ne \emptyset$ is an arbitrary index set (i.e., possibly uncountable). From this it follows that if we are given a set Ω and a collection $\mathcal{A}$ of subsets of Ω, then there is a smallest σ-field containing $\mathcal{A}$. We will call this the **σ-field generated by** $\mathcal{A}$ and denote it by $\sigma(\mathcal{A})$.

Let $\mathbf{R}^d$ be the set of vectors $(x_1, \dots x_d)$ of real numbers and $\mathcal{R}^d$ be the **Borel sets,** the smallest σ-field containing the open sets. When $d = 1$, we drop the superscript.

Example 1.1.3 (Measures on the real line) Measures on $(\mathbf{R}, \mathcal{R})$ are defined by giving a **Stieltjes measure function** with the following properties:

(i) F is nondecreasing.

(ii) F is right continuous, i.e., $\lim_{y \downarrow x} F(y) = F(x)$.

Theorem 1.1.4 *Associated with each Stieltjes measure function F there is a unique measure μ on $(\mathbf{R}, \mathcal{R})$ with $\mu((a,b]) = F(b) - F(a)$*

$$\mu((a,b]) = F(b) - F(a) \tag{1.1.1}$$

When $F(x) = x$, the resulting measure is called **Lebesgue measure.**

The proof of Theorem 1.1.4 is a long and winding road, so we will content ourselves with describing the main ideas involved in this section and hiding the remaining details in the Appendix in Section A.1. The choice of "closed on the right" in $(a,b]$ is dictated by the fact that if $b_n \downarrow b$, then we have

$$\cap_n (a, b_n] = (a,b]$$

The next definition will explain the choice of "open on the left."

A collection $\mathcal{S}$ of sets is said to be a **semialgebra** if (i) it is closed under intersection, i.e., $S, T \in \mathcal{S}$ implies $S \cap T \in \mathcal{S}$, and (ii) if $S \in \mathcal{S}$, then S^c is a finite disjoint union of sets in $\mathcal{S}$. An important example of a semialgebra is

Example 1.1.5 $\mathcal{S}_d$ = the empty set plus all sets of the form

$$(a_1, b_1] \times \cdots \times (a_d, b_d] \subset \mathbf{R}^d \quad \text{where } -\infty \le a_i < b_i \le \infty$$

The definition in (1.1.1) gives the values of μ on the semialgebra $\mathcal{S}_1$. To go from semialgebra to σ-algebra, we use an intermediate step. A collection $\mathcal{A}$ of subsets of Ω is called an **algebra** (or **field**) if $A, B \in \mathcal{A}$ implies A^c and $A \cup B$ are in $\mathcal{A}$. Since $A \cap B = (A^c \cup B^c)^c$, it follows that $A \cap B \in \mathcal{A}$. Obviously, a σ-algebra is an algebra. An example in which the converse is false is:

Example 1.1.6 Let $\Omega = \mathbf{Z}$ = the integers. $\mathcal{A}$ = the collection of $A \subset \mathbf{Z}$ so that A or A^c is finite is an algebra.

Lemma 1.1.7 *If $\mathcal{S}$ is a semialgebra, then $\bar{\mathcal{S}}$ = {finite disjoint unions of sets in $\mathcal{S}$} is an algebra, called the* **algebra generated by** $\mathcal{S}$.

Proof Suppose $A = +_i S_i$ and $B = +_j T_j$, where $+$ denotes disjoint union and we assume the index sets are finite. Then $A \cap B = +_{i,j} S_i \cap T_j \in \bar{\mathcal{S}}$. As for complements, if $A = +_i S_i$ then $A^c = \cap_i S_i^c$. The definition of $\mathcal{S}$ implies $S_i^c \in \bar{\mathcal{S}}$. We have shown that $\bar{\mathcal{S}}$ is closed under intersection, so it follows by induction that $A^c \in \bar{\mathcal{S}}$. □

Example 1.1.8 Let $\Omega = \mathbf{R}$ and $\mathcal{S} = \mathcal{S}_1$ then $\bar{\mathcal{S}}_1$ = the empty set plus all sets of the form

$$\cup_{i=1}^{k}(a_i, b_i] \quad \text{where } -\infty \leq a_i < b_i \leq \infty$$

Given a set function μ on $\mathcal{S}$ we can extend it to $\bar{\mathcal{S}}$ by

$$\mu\left(+_{i=1}^{n} A_i\right) = \sum_{i=1}^{n} \mu(A_i)$$

By a **measure on an algebra** $\mathcal{A}$, we mean a set function μ with

(i) $\mu(A) \geq \mu(\emptyset) = 0$ for all $A \in \mathcal{A}$, and

(ii) if $A_i \in \mathcal{A}$ are disjoint *and their union is in* $\mathcal{A}$, then

$$\mu\left(\cup_{i=1}^{\infty} A_i\right) = \sum_{i=1}^{\infty} \mu(A_i)$$

μ is said to be σ**-finite** if there is a sequence of sets $A_n \in \mathcal{A}$ so that $\mu(A_n) < \infty$ and $\cup_n A_n = \Omega$. Letting $A_1' = A_1$ and for $n \geq 2$,

$$A_n' = \cup_{m=1}^{n} A_m \quad \text{or} \quad A_n' = A_n \cap \left(\cap_{m=1}^{n-1} A_m^c\right) \in \mathcal{A}$$

we can without loss of generality assume that $A_n \uparrow \Omega$ or the A_n are disjoint.

The next result helps us to extend a measure defined on a semialgebra $\mathcal{S}$ to the σ-algebra it generates, $\sigma(\mathcal{S})$

Theorem 1.1.9 *Let $\mathcal{S}$ be a semialgebra and let μ defined on $\mathcal{S}$ have $\mu(\emptyset) = 0$. Suppose (i) if $S \in \mathcal{S}$, is a finite disjoint union of sets $S_i \in \mathcal{S}$, then $\mu(S) = \sum_i \mu(S_i)$, and (ii) if $S_i, S \in \mathcal{S}$ with $S = +_{i\geq 1} S_i$, then $\mu(S) \leq \sum_{i\geq 1} \mu(S_i)$. Then μ has a unique extension $\bar{\mu}$ that is a measure on $\bar{\mathcal{S}}$, the algebra generated by $\mathcal{S}$. If $\bar{\mu}$ is sigma-finite, then there is a unique extension ν that is a measure on $\sigma(\mathcal{S})$.*

In (ii) above, and in what follows, $i \geq 1$ indicates a countable union, while a plain subscript i or j indicates a finite union. The proof of Theorems 1.1.9 is rather involved so it is given in Section A.1. To check condition (ii) in the theorem the following is useful.

Lemma 1.1.10 *Suppose only that (i) holds.*
(a) If $A, B_i \in \bar{\mathcal{S}}$ with $A = +_{i=1}^{n} B_i$, then $\bar{\mu}(A) = \sum_i \bar{\mu}(B_i)$.
(b) If $A, B_i \in \bar{\mathcal{S}}$ with $A \subset \cup_{i=1}^{n} B_i$, then $\bar{\mu}(A) \leq \sum_i \bar{\mu}(B_i)$.

Proof Observe that it follows from the definition that if $A = +_i B_i$ is a finite disjoint union of sets in $\bar{\mathcal{S}}$ and $B_i = +_j S_{i,j}$, then

$$\bar{\mu}(A) = \sum_{i,j} \mu(S_{i,j}) = \sum_i \bar{\mu}(B_i)$$

To prove (b), we begin with the case $n = 1$, $B_1 = B$. $B = A + (B \cap A^c)$ and $B \cap A^c \in \bar{\mathcal{S}}$, so

$$\bar{\mu}(A) \leq \bar{\mu}(A) + \bar{\mu}(B \cap A^c) = \bar{\mu}(B)$$

To handle $n > 1$ now, let $F_k = B_1^c \cap \ldots \cap B_{k-1}^c \cap B_k$ and note

$$\cup_i B_i = F_1 + \cdots + F_n$$
$$A = A \cap (\cup_i B_i) = (A \cap F_1) + \cdots + (A \cap F_n)$$

so using (a), (b) with $n = 1$, and (a) again

$$\bar{\mu}(A) = \sum_{k=1}^{n} \bar{\mu}(A \cap F_k) \le \sum_{k=1}^{n} \bar{\mu}(F_k) = \bar{\mu}(\cup_i B_i)$$ □

Proof of Theorem 1.1.4. Let $\mathcal{S}$ be the semialgebra of half-open intervals $(a,b]$ with $-\infty \le a < b \le \infty$. To define μ on $\mathcal{S}$, we begin by observing that

$$F(\infty) = \lim_{x \uparrow \infty} F(x) \quad \text{and} \quad F(-\infty) = \lim_{x \downarrow -\infty} F(x) \quad \text{exist}$$

and $\mu((a,b]) = F(b) - F(a)$ makes sense for all $-\infty \le a < b \le \infty$ since $F(\infty) > -\infty$ and $F(-\infty) < \infty$.

If $(a,b] = +_{i=1}^{n}(a_i, b_i]$, then after relabeling the intervals we must have $a_1 = a$, $b_n = b$, and $a_i = b_{i-1}$ for $2 \le i \le n$, so condition (i) in Theorem 1.1.9 holds. To check (ii), suppose first that $-\infty < a < b < \infty$, and $(a,b] \subset \cup_{i \ge 1}(a_i, b_i]$ where (without loss of generality) $-\infty < a_i < b_i < \infty$. Pick $\delta > 0$ so that $F(a+\delta) < F(a) + \epsilon$ and pick η_i so that

$$F(b_i + \eta_i) < F(b_i) + \epsilon 2^{-i}$$

The open intervals $(a_i, b_i + \eta_i)$ cover $[a+\delta, b]$, so there is a finite subcover (α_j, β_j), $1 \le j \le J$. Since $(a+\delta, b] \subset \cup_{j=1}^{J}(\alpha_j, \beta_j]$, (b) in Lemma 1.1.10 implies

$$F(b) - F(a+\delta) \le \sum_{j=1}^{J} F(\beta_j) - F(\alpha_j) \le \sum_{i=1}^{\infty} (F(b_i + \eta_i) - F(a_i))$$

So, by the choice of δ and η_i,

$$F(b) - F(a) \le 2\epsilon + \sum_{i=1}^{\infty} (F(b_i) - F(a_i))$$

and since ϵ is arbitrary, we have proved the result in the case $-\infty < a < b < \infty$. To remove the last restriction, observe that if $(a,b] \subset \cup_i (a_i, b_i]$ and $(A,B] \subset (a,b]$ has $-\infty < A < B < \infty$, then we have

$$F(B) - F(A) \le \sum_{i=1}^{\infty} (F(b_i) - F(a_i))$$

Since the last result holds for any finite $(A,B] \subset (a,b]$, the desired result follows. □

Measures on $\mathbf{R}^\mathbf{d}$

Our next goal is to prove a version of Theorem 1.1.4 for $\mathbf{R}^d$. The first step is to introduce the assumptions on the defining function F. By analogy with the case $d = 1$ it is natural to assume:

0	2/3	1
0	0	2/3
0	0	0

Figure 1.1 Picture of the counterexample.

(i) It is nondecreasing, i.e., if $x \le y$ (meaning $x_i \le y_i$ for all i), then $F(x) \le F(y)$.

(ii) F is right continuous, i.e., $\lim_{y \downarrow x} F(y) = F(x)$ (here $y \downarrow x$ means each $y_i \downarrow x_i$).

(iii) If $x_n \downarrow -\infty$, i.e., each coordinate does, then $F(x_n) \downarrow 0$. If $x_n \uparrow -\infty$, i.e., each coordinate does, then $F(x_n) \uparrow 1$.

However, this time it is not enough. Consider the following F

$$F(x_1,x_2) = \begin{cases} 1 & \text{if } x_1,x_2 \ge 1 \\ 2/3 & \text{if } x_1 \ge 1 \text{ and } 0 \le x_2 < 1 \\ 2/3 & \text{if } x_2 \ge 1 \text{ and } 0 \le x_1 < 1 \\ 0 & \text{otherwise} \end{cases}$$

See Figure 1.1 for a picture. A little thought shows that

$$\begin{aligned} \mu((a_1,b_1] \times (a_2,b_2]) &= \mu((-\infty,b_1] \times (-\infty,b_2]) - \mu((-\infty,a_1] \times (-\infty,b_2]) \\ &\quad - \mu((-\infty,b_1] \times (-\infty,a_2]) + \mu((-\infty,a_1] \times (-\infty,a_2]) \\ &= F(b_1,b_2) - F(a_1,b_2) - F(b_1,a_2) + F(a_1,a_2) \end{aligned}$$

Using this with $a_1 = a_2 = 1 - \epsilon$ and $b_1 = b_2 = 1$ and letting $\epsilon \to 0$ we see that

$$\mu(\{1,1\}) = 1 - 2/3 - 2/3 + 0 = -1/3$$

Similar reasoning shows that $\mu(\{1,0\}) = \mu(\{0,1\}) = 2/3$.

To formulate the third and final condition for F to define a measure, let

$$\begin{aligned} A &= (a_1,b_1] \times \cdots \times (a_d,b_d] \\ V &= \{a_1,b_1\} \times \cdots \times \{a_d,b_d\} \end{aligned}$$

where $-\infty < a_i < b_i < \infty$. To emphasize that ∞'s are not allowed, we will call A a finite rectangle. Then $V =$ the vertices of the rectangle A. If $v \in V$, let

$$\begin{aligned} \text{sgn}\,(v) &= (-1)^{\#\text{ of } a\text{'s in } v} \\ \Delta_A F &= \sum_{v \in V} \text{sgn}\,(v) F(v) \end{aligned}$$

We will let $\mu(A) = \Delta_A F$, so we must assume

(iv) $\Delta_A F \geq 0$ for all rectangles A.

Theorem 1.1.11 *Suppose* $F : \mathbf{R}^d \to [0,1]$ *satisfies (i)–(iv) given above. Then there is a unique probability measure* μ *on* $(\mathbf{R}^d, \mathcal{R}^d)$ *so that* $\mu(A) = \Delta_A F$ *for all finite rectangles.*

Example 1.1.12 Suppose $F(x) = \prod_{i=1}^{d} F_i(x)$, where the F_i satisfy (i) and (ii) of Theorem 1.1.4. In this case,

$$\Delta_A F = \prod_{i=1}^{d} (F_i(b_i) - F_i(a_i))$$

When $F_i(x) = x$ for all i, the resulting measure is Lebesgue measure on $\mathbf{R}^d$.

Proof We let $\mu(A) = \Delta_A F$ for all finite rectangles and then use monotonicity to extend the definition to $\mathcal{S}_d$. To check (i) of Theorem 1.1.9, call $A = +_k B_k$ a **regular subdivision** of A if there are sequences $a_i = \alpha_{i,0} < \alpha_{i,1} \ldots < \alpha_{i,n_i} = b_i$ so that each rectangle B_k has the form

$$(\alpha_{1,j_1-1}, \alpha_{1,j_1}] \times \cdots \times (\alpha_{d,j_d-1}, \alpha_{d,j_d}] \quad \text{where } 1 \leq j_i \leq n_i$$

It is easy to see that for regular subdivisions $\lambda(A) = \sum_k \lambda(B_k)$. (First consider the case in which all the endpoints are finite and then take limits to get the general case.) To extend this result to a general finite subdivision $A = +_j A_j$, subdivide further to get a regular one.

The proof of (ii) is almost identical to that in Theorem 1.1.4. To make things easier to write and to bring out the analogies with Theorem 1.1.4, we let

$$(x, y) = (x_1, y_1) \times \cdots \times (x_d, y_d)$$
$$(x, y] = (x_1, y_1] \times \cdots \times (x_d, y_d]$$
$$[x, y] = [x_1, y_1] \times \cdots \times [x_d, y_d]$$

for $x, y \in \mathbf{R}^d$. Suppose first that $-\infty < a < b < \infty$, where the inequalities mean that each component is finite, and suppose $(a,b] \subset \cup_{i \geq 1}(a^i, b^i]$, where (without loss of generality) $-\infty < a^i < b^i < \infty$. Let $\bar{1} = (1, \ldots, 1)$, pick $\delta > 0$ so that

$$\mu((a - \delta\bar{1}, b]) > \mu((a,b]) - \epsilon$$

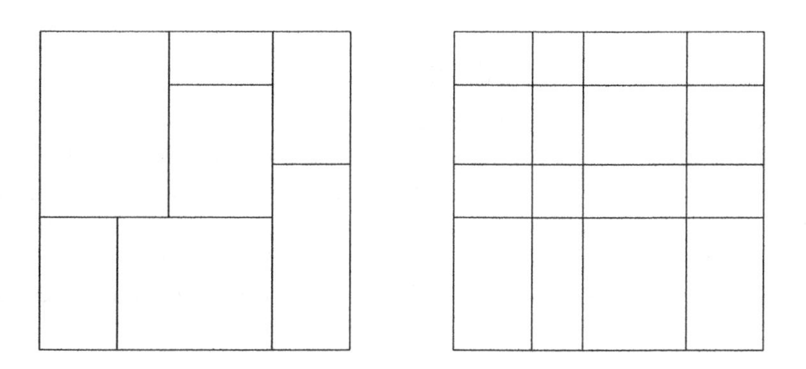

Figure 1.2 Conversion of a subdivision to a regular one.

and pick η_i so that

$$\mu((a, b^i + \eta_i \bar{1}]) < \mu((a^i, b^i]) + \epsilon 2^{-i}$$

The open rectangles $(a^i, b^i + \eta_i \bar{1})$ cover $[a + \delta\bar{1}, b]$, so there is a finite subcover (α^j, β^j), $1 \le j \le J$. Since $(a + \delta\bar{1}, b] \subset \cup_{j=1}^J (\alpha^j, \beta^j]$, (b) in Lemma 1.1.10 implies

$$\mu([a + \delta\bar{1}, b]) \le \sum_{j=1}^{J} \mu((\alpha^j, \beta^j]) \le \sum_{i=1}^{\infty} \mu((a^i, b^i + \eta_i \bar{1}])$$

So, by the choice of δ and η_i,

$$\mu((a, b]) \le 2\epsilon + \sum_{i=1}^{\infty} \mu((a^i, b^i])$$

and since ϵ is arbitrary, we have proved the result in the case $-\infty < a < b < \infty$. The proof can now be completed exactly as before. □

Exercises

1.1.1 Let $\Omega = \mathbf{R}$, $\mathcal{F}$ = all subsets so that A or A^c is countable, $P(A) = 0$ in the first case and $= 1$ in the second. Show that $(\Omega, \mathcal{F}, P)$ is a probability space.

1.1.2 Recall the definition of $\mathcal{S}_d$ from Example 1.1.5. Show that $\sigma(\mathcal{S}_d) = \mathcal{R}^d$, the Borel subsets of $\mathbf{R}^d$.

1.1.3 A σ-field $\mathcal{F}$ is said to be **countably generated** if there is a countable collection $\mathcal{C} \subset \mathcal{F}$ so that $\sigma(\mathcal{C}) = \mathcal{F}$. Show that $\mathcal{R}^d$ is countably generated.

1.1.4 (i) Show that if $\mathcal{F}_1 \subset \mathcal{F}_2 \subset \ldots$ are σ-algebras, then $\cup_i \mathcal{F}_i$ is an algebra. (ii) Give an example to show that $\cup_i \mathcal{F}_i$ need not be a σ-algebra.

1.1.5 A set $A \subset \{1, 2, \ldots\}$ is said to have **asymptotic density** θ if

$$\lim_{n \to \infty} |A \cap \{1, 2, \ldots, n\}|/n = \theta$$

Let $\mathcal{A}$ be the collection of sets for which the asymptotic density exists. Is $\mathcal{A}$ a σ-algebra? an algebra?

1.2 Distributions

Probability spaces become a little more interesting when we define random variables on them. A real-valued function X defined on Ω is said to be a **random variable** if for every Borel set $B \subset \mathbf{R}$ we have $X^{-1}(B) = \{\omega : X(\omega) \in B\} \in \mathcal{F}$. When we need to emphasize the σ-field, we will say that X is $\mathcal{F}$-**measurable** or write $X \in \mathcal{F}$. If Ω is a discrete probability space (see Example 1.1.2), then any function $X : \Omega \to \mathbf{R}$ is a random variable. A second trivial, but useful, type of example of a random variable is the **indicator function** of a set $A \in \mathcal{F}$:

$$1_A(\omega) = \begin{cases} 1 & \omega \in A \\ 0 & \omega \notin A \end{cases}$$

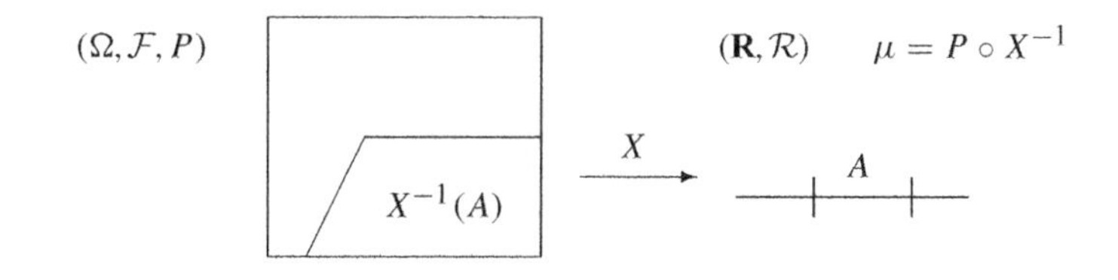

Figure 1.3 Definition of the distribution of X.

The notation is supposed to remind you that this function is 1 on A. Analysts call this object the characteristic function of A. In probability, that term is used for something quite different. (See Section 3.3.)

If X is a random variable, then X induces a probability measure on $\mathbf{R}$ called its **distribution** by setting $\mu(A) = P(X \in A)$ for Borel sets A. Using the notation introduced previously, the right-hand side can be written as $P(X^{-1}(A))$. In words, we pull $A \in \mathcal{R}$ back to $X^{-1}(A) \in \mathcal{F}$ and then take P of that set.

To check that μ is a probability measure we observe that if the A_i are disjoint, then using the definition of μ; the fact that X lands in the union if and only if it lands in one of the A_i; the fact that if the sets $A_i \in \mathcal{R}$ are disjoint, then the events $\{X \in A_i\}$ are disjoint; and the definition of μ again; we have:

$$\mu\left(\cup_i A_i\right) = P\left(X \in \cup_i A_i\right) = P\left(\cup_i \{X \in A_i\}\right) = \sum_i P(X \in A_i) = \sum_i \mu(A_i)$$

The distribution of a random variable X is usually described by giving its **distribution function**, $F(x) = P(X \le x)$.

Theorem 1.2.1 *Any distribution function F has the following properties:*

(i) F is nondecreasing.

(ii) $\lim_{x \to \infty} F(x) = 1$, $\lim_{x \to -\infty} F(x) = 0$.

(iii) F is right continuous, i.e., $\lim_{y \downarrow x} F(y) = F(x)$.

(iv) If $F(x-) = \lim_{y \uparrow x} F(y)$, *then* $F(x-) = P(X < x)$.

(v) $P(X = x) = F(x) - F(x-)$.

Proof To prove (i), note that if $x \le y$, then $\{X \le x\} \subset \{X \le y\}$, and then use (i) in Theorem 1.1.1 to conclude that $P(X \le x) \le P(X \le y)$.

To prove (ii), we observe that if $x \uparrow \infty$, then $\{X \le x\} \uparrow \Omega$, while if $x \downarrow -\infty$, then $\{X \le x\} \downarrow \emptyset$ and then use (iii) and (iv) of Theorem 1.1.1.

To prove (iii), we observe that if $y \downarrow x$, then $\{X \le y\} \downarrow \{X \le x\}$.

To prove (iv), we observe that if $y \uparrow x$, then $\{X \le y\} \uparrow \{X < x\}$.

For (v), note $P(X = x) = P(X \le x) - P(X < x)$ and use (iii) and (iv). □

The next result shows that we have found more than enough properties to characterize distribution functions.

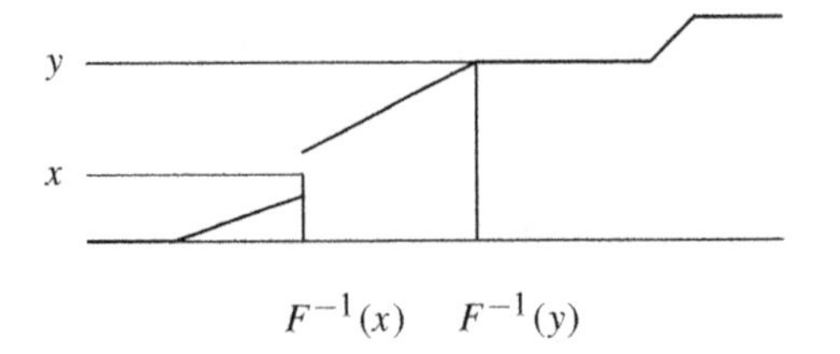

Figure 1.4 Picture of the inverse defined in the proof of Theorem 1.2.2.

Theorem 1.2.2 *If F satisfies (i), (ii), and (iii) in Theorem 1.2.1, then it is the distribution function of some random variable.*

Proof Let $\Omega = (0,1)$, $\mathcal{F}$ = the Borel sets, and P = Lebesgue measure. If $\omega \in (0,1)$, let

$$X(\omega) = \sup\{y : F(y) < \omega\}$$

Once we show that

$$(\star) \qquad \{\omega : X(\omega) \le x\} = \{\omega : \omega \le F(x)\}$$

the desired result follows immediately since $P(\omega : \omega \le F(x)) = F(x)$. (Recall P is Lebesgue measure.) To check ($\star$), we observe that if $\omega \le F(x)$, then $X(\omega) \le x$, since $x \notin \{y : F(y) < \omega\}$. On the other hand if $\omega > F(x)$, then since F is right continuous, there is an $\epsilon > 0$ so that $F(x+\epsilon) < \omega$ and $X(\omega) \ge x + \epsilon > x$. □

Even though F may not be 1-1 and onto we will call X the inverse of F and denote it by F^{-1}. The scheme in the proof of Theorem 1.2.2 is useful in generating random variables on a computer. Standard algorithms generate random variables U with a uniform distribution, then one applies the inverse of the distribution function defined in Theorem 1.2.2 to get a random variable $F^{-1}(U)$ with distribution function F.

If X and Y induce the same distribution μ on $(\mathbf{R}, \mathcal{R})$, we say X and Y are **equal in distribution**. In view of Theorem 1.1.4, this holds if and only if X and Y have the same distribution function, i.e., $P(X \le x) = P(Y \le x)$ for all x. When X and Y have the same distribution, we like to write

$$X \stackrel{d}{=} Y$$

but this is too tall to use in text, so for typographical reasons we will also use $X =_d Y$.

When the distribution function $F(x) = P(X \le x)$ has the form

$$F(x) = \int_{-\infty}^{x} f(y)\, dy \tag{1.2.1}$$

we say that X has **density function** f. In remembering formulas, it is often useful to think of $f(x)$ as being $P(X = x)$ although

$$P(X = x) = \lim_{\epsilon \to 0} \int_{x-\epsilon}^{x+\epsilon} f(y)\, dy = 0$$

By popular demand, we have ceased our previous practice of writing $P(X = x)$ for the density function. Instead we will use things like the lovely and informative $f_X(x)$.

We can start with f and use (1.2.1) to define a distribution function F. In order to end up with a distribution function, it is necessary and sufficient that $f(x) \geq 0$ and $\int f(x)\,dx = 1$. Three examples that will be important in what follows are:

Example 1.2.3 (Uniform distribution on (0,1)) $f(x) = 1$ for $x \in (0,1)$ and 0 otherwise. Distribution function:

$$F(x) = \begin{cases} 0 & x \leq 0 \\ x & 0 \leq x \leq 1 \\ 1 & x > 1 \end{cases}$$

Example 1.2.4 (Exponential distribution with rate λ) $f(x) = \lambda e^{-\lambda x}$ for $x \geq 0$ and 0 otherwise. Distribution function:

$$F(x) = \begin{cases} 0 & x \leq 0 \\ 1 - e^{-\lambda x} & x \geq 0 \end{cases}$$

Example 1.2.5 (Standard normal distribution)

$$f(x) = (2\pi)^{-1/2} \exp(-x^2/2)$$

In this case, there is no closed form expression for $F(x)$, but we have the following bounds that are useful for large x:

Theorem 1.2.6 *For $x > 0$,*

$$(x^{-1} - x^{-3}) \exp(-x^2/2) \leq \int_x^\infty \exp(-y^2/2) dy \leq x^{-1} \exp(-x^2/2)$$

Proof Changing variables $y = x + z$ and using $\exp(-z^2/2) \leq 1$ gives

$$\int_x^\infty \exp(-y^2/2)\,dy \leq \exp(-x^2/2) \int_0^\infty \exp(-xz)\,dz = x^{-1} \exp(-x^2/2)$$

For the other direction, we observe

$$\int_x^\infty (1 - 3y^{-4}) \exp(-y^2/2)\,dy = (x^{-1} - x^{-3}) \exp(-x^2/2)$$

□

A distribution function on **R** is said to be **absolutely continuous** if it has a density, and **singular** if the corresponding measure is singular w.r.t. Lebesgue measure. See Section A.4 for more on these notions. An example of a singular distribution is:

Example 1.2.7 (Uniform distribution on the Cantor set) The Cantor set C is defined by removing (1/3, 2/3) from [0,1] and then removing the middle third of each interval that remains. We define an associated distribution function by setting $F(x) = 0$ for $x \leq 0$, $F(x) = 1$ for $x \geq 1$, $F(x) = 1/2$ for $x \in [1/3, 2/3]$, $F(x) = 1/4$ for $x \in [1/9, 2/9]$, $F(x) = 3/4$ for $x \in [7/9, 8/9]$, ... Then extend F to all of $[0,1]$ using monotonicity. There is no f for which (1.2.1) holds because such an f would be equal to 0 on a set of measure 1. From the definition, it is immediate that the corresponding measure has $\mu(C^c) = 0$.

A probability measure P (or its associated distribution function) is said to be **discrete** if there is a countable set S with $P(S^c) = 0$. The simplest example of a discrete distribution is

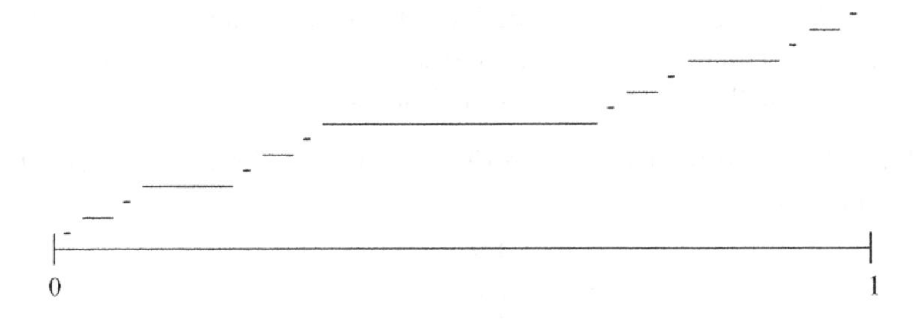

Figure 1.5 Cantor distribution function.

Example 1.2.8 (Point mass at 0) $F(x) = 1$ for $x \geq 0$, $F(x) = 0$ for $x < 0$.

In Section 1.6, we will see the Bernoulli, Poisson, and geometric distributions. The next example shows that the distribution function associated with a discrete probability measure can be quite wild.

Example 1.2.9 (Dense discontinuities) Let $q_1, q_2 \ldots$ be an enumeration of the rationals. Let $\alpha_i > 0$ have $\sum_{i=1}^{\infty} \alpha_1 = 1$ and let

$$F(x) = \sum_{i=1}^{\infty} \alpha_i 1_{[q_i,\infty)}$$

where $1_{[\theta,\infty)}(x) = 1$ if $x \in [\theta,\infty) = 0$ otherwise.

Exercises

1.2.1 Suppose X and Y are random variables on $(\Omega, \mathcal{F}, P)$ and let $A \in \mathcal{F}$. Show that if we let $Z(\omega) = X(\omega)$ for $\omega \in A$ and $Z(\omega) = Y(\omega)$ for $\omega \in A^c$, then Z is a random variable.

1.2.2 Let χ have the standard normal distribution. Use Theorem 1.2.6 to get upper and lower bounds on $P(\chi \geq 4)$.

1.2.3 Show that a distribution function has at most countably many discontinuities.

1.2.4 Show that if $F(x) = P(X \leq x)$ is continuous, then $Y = F(X)$ has a uniform distribution on (0,1), that is, if $y \in [0,1]$, $P(Y \leq y) = y$.

1.2.5 Suppose X has continuous density f, $P(\alpha \leq X \leq \beta) = 1$ and g is a function that is strictly increasing and differentiable on (α, β). Then $g(X)$ has density $f(g^{-1}(y))/g'(g^{-1}(y))$ for $y \in (g(\alpha), g(\beta))$ and 0 otherwise. When $g(x) = ax + b$ with $a > 0$, $g^{-1}(y) = (y - b)/a$ so the answer is $(1/a)f((y - b)/a)$.

1.2.6 Suppose X has a normal distribution. Use the previous exercise to compute the density of $\exp(X)$. (The answer is called the **lognormal distribution**.)

1.2.7 (i) Suppose X has density function f. Compute the distribution function of X^2 and then differentiate to find its density function. (ii) Work out the answer when X has a standard normal distribution to find the density of the **chi-square distribution**.

1.3 Random Variables

In this section, we will develop some results that will help us later to prove that quantities we define are random variables, i.e., they are measurable. Since most of what we have to say is true for random elements of an arbitrary measurable space $(S, \mathcal{S})$ and the proofs are the same (sometimes easier), we will develop our results in that generality. First we need a definition. A function $X : \Omega \to S$ is said to be a **measurable map** from $(\Omega, \mathcal{F})$ to $(S, \mathcal{S})$ if

$$X^{-1}(B) \equiv \{\omega : X(\omega) \in B\} \in \mathcal{F} \quad \text{for all } B \in \mathcal{S}$$

If $(S, \mathcal{S}) = (\mathbf{R}^d, \mathcal{R}^d)$ and $d > 1$, then X is called a **random vector**. Of course, if $d = 1$, X is called a **random variable**, or r.v. for short.

The next result is useful for proving that maps are measurable.

Theorem 1.3.1 *If* $\{\omega : X(\omega) \in A\} \in \mathcal{F}$ *for all* $A \in \mathcal{A}$ *and* $\mathcal{A}$ **generates** $\mathcal{S}$ *(i.e.,* $\mathcal{S}$ *is the smallest* σ*-field that contains* $\mathcal{A}$*), then* X *is measurable.*

Proof Writing $\{X \in B\}$ as shorthand for $\{\omega : X(\omega) \in B\}$, we have

$$\{X \in \cup_i B_i\} = \cup_i \{X \in B_i\}$$
$$\{X \in B^c\} = \{X \in B\}^c$$

So the class of sets $\mathcal{B} = \{B : \{X \in B\} \in \mathcal{F}\}$ is a σ-field. Since $\mathcal{B} \supset \mathcal{A}$ and $\mathcal{A}$ generates $\mathcal{S}$, $\mathcal{B} \supset \mathcal{S}$. □

It follows from the two equations displayed in the previous proof that if $\mathcal{S}$ is a σ-field, then $\{\{X \in B\} : B \in \mathcal{S}\}$ is a σ-field. It is the smallest σ-field on Ω that makes X a measurable map. It is called the **σ-field generated by** X and denoted $\sigma(X)$. For future reference we note that

$$\sigma(X) = \{\{X \in B\} : B \in \mathcal{S}\} \tag{1.3.1}$$

Example 1.3.2 If $(S, \mathcal{S}) = (\mathbf{R}, \mathcal{R})$, then possible choices of $\mathcal{A}$ in Theorem 1.3.1 are $\{(-\infty, x] : x \in \mathbf{R}\}$ or $\{(-\infty, x) : x \in \mathbf{Q}\}$ where $\mathbf{Q} =$ the rationals.

Example 1.3.3 If $(S, \mathcal{S}) = (\mathbf{R}^d, \mathcal{R}^d)$, a useful choice of $\mathcal{A}$ is

$$\{(a_1, b_1) \times \cdots \times (a_d, b_d) : -\infty < a_i < b_i < \infty\}$$

or occasionally the larger collection of open sets.

Theorem 1.3.4 *If* $X : (\Omega, \mathcal{F}) \to (S, \mathcal{S})$ *and* $f : (S, \mathcal{S}) \to (T, \mathcal{T})$ *are measurable maps, then* $f(X)$ *is a measurable map from* $(\Omega, \mathcal{F})$ *to* $(T, \mathcal{T})$

Proof Let $B \in \mathcal{T}$. $\{\omega : f(X(\omega)) \in B\} = \{\omega : X(\omega) \in f^{-1}(B)\} \in \mathcal{F}$, since by assumption $f^{-1}(B) \in \mathcal{S}$. □

From Theorem 1.3.4, it follows immediately that if X is a random variable, then so is cX for all $c \in \mathbf{R}$, X^2, $\sin(X)$, etc. The next result shows why we wanted to prove Theorem 1.3.4 for measurable maps.

Theorem 1.3.5 *If* $X_1, \ldots X_n$ *are random variables and* $f : (\mathbf{R}^n, \mathcal{R}^n) \to (\mathbf{R}, \mathcal{R})$ *is measurable, then* $f(X_1, \ldots, X_n)$ *is a random variable.*

Proof In view of Theorem 1.3.4, it suffices to show that $(X_1, \dots, X_n)$ is a random vector. To do this, we observe that if $A_1, \dots, A_n$ are Borel sets, then

$$\{(X_1, \dots, X_n) \in A_1 \times \cdots \times A_n\} = \cap_i \{X_i \in A_i\} \in \mathcal{F}$$

Since sets of the form $A_1 \times \cdots \times A_n$ generate $\mathcal{R}^n$, the desired result follows from Theorem 1.3.1. □

Theorem 1.3.6 *If $X_1, \dots, X_n$ are random variables, then $X_1 + \cdots + X_n$ is a random variable.*

Proof In view of Theorem 1.3.5 it suffices to show that $f(x_1, \dots, x_n) = x_1 + \cdots + x_n$ is measurable. To do this, we use Example 1.3.2 and note that $\{x : x_1 + \cdots + x_n < a\}$ is an open set and hence is in $\mathcal{R}^n$. □

Theorem 1.3.7 *If $X_1, X_2, \dots$ are random variables then so are*

$$\inf_n X_n \qquad \sup_n X_n \qquad \limsup_n X_n \qquad \liminf_n X_n$$

Proof Since the infimum of a sequence is $< a$ if and only if some term is $< a$ (if all terms are $\geq a$ then the infimum is), we have

$$\{\inf_n X_n < a\} = \cup_n \{X_n < a\} \in \mathcal{F}$$

A similar argument shows $\{\sup_n X_n > a\} = \cup_n \{X_n > a\} \in \mathcal{F}$. For the last two, we observe

$$\liminf_{n\to\infty} X_n = \sup_n \left(\inf_{m\geq n} X_m \right)$$

$$\limsup_{n\to\infty} X_n = \inf_n \left(\sup_{m\geq n} X_m \right)$$

To complete the proof in the first case, note that $Y_n = \inf_{m\geq n} X_m$ is a random variable for each n so $\sup_n Y_n$ is as well. □

From Theorem 1.3.7, we see that

$$\Omega_o \equiv \{\omega : \lim_{n\to\infty} X_n \text{ exists }\} = \{\omega : \limsup_{n\to\infty} X_n - \liminf_{n\to\infty} X_n = 0\}$$

is a measurable set. (Here $\equiv$ indicates that the first equality is a definition.) If $P(\Omega_o) = 1$, we say that X_n **converges almost surely**, or a.s. for short. This type of convergence is called **almost everywhere** in measure theory. To have a limit defined on the whole space, it is convenient to let

$$X_\infty = \limsup_{n\to\infty} X_n$$

but this random variable may take the value $+\infty$ or $-\infty$. To accommodate this and some other headaches, we will generalize the definition of random variable.

A function whose domain is a set $D \in \mathcal{F}$ and whose range is $\mathbf{R}^* \equiv [-\infty, \infty]$ is said to be a **random variable** if for all $B \in \mathcal{R}^*$ we have $X^{-1}(B) = \{\omega : X(\omega) \in B\} \in \mathcal{F}$. Here $\mathcal{R}^*$ = the Borel subsets of $\mathbf{R}^*$ with $\mathbf{R}^*$ given the usual topology, i.e., the one generated by intervals of the form $[-\infty, a)$, (a, b) and $(b, \infty]$ where $a, b \in \mathbf{R}$. The reader should note that the **extended real line** $(\mathbf{R}^*, \mathcal{R}^*)$ is a measurable space, so all the results above generalize immediately.

Exercises

1.3.1 Show that if $\mathcal{A}$ generates $\mathcal{S}$, then $X^{-1}(\mathcal{A}) \equiv \{\{X \in A\} : A \in \mathcal{A}\}$ generates $\sigma(X) = \{\{X \in B\} : B \in \mathcal{S}\}$.

1.3.2 Prove Theorem 1.3.6 when $n = 2$ by checking $\{X_1 + X_2 < x\} \in \mathcal{F}$.

1.3.3 Show that if f is continuous and $X_n \to X$ almost surely then $f(X_n) \to f(X)$ almost surely.

1.3.4 (i) Show that a continuous function from $\mathbf{R}^d \to \mathbf{R}$ is a measurable map from $(\mathbf{R}^d, \mathcal{R}^d)$ to $(\mathbf{R}, \mathcal{R})$. (ii) Show that $\mathcal{R}^d$ is the smallest σ-field that makes all the continuous functions measurable.

1.3.5 A function f is said to be **lower semicontinuous** or l.s.c. if

$$\liminf_{y \to x} f(y) \geq f(x)$$

and **upper semicontinuous** (u.s.c.) if $-f$ is l.s.c. Show that f is l.s.c. if and only if $\{x : f(x) \leq a\}$ is closed for each $a \in \mathbf{R}$ and conclude that semicontinuous functions are measurable.

1.3.6 Let $f : \mathbf{R}^d \to \mathbf{R}$ be an arbitrary function and let $f^\delta(x) = \sup\{f(y) : |y - x| < \delta\}$ and $f_\delta(x) = \inf\{f(y) : |y - x| < \delta\}$ where $|z| = (z_1^2 + \cdots + z_d^2)^{1/2}$. Show that f^δ is l.s.c. and f_δ is u.s.c. Let $f^0 = \lim_{\delta \downarrow 0} f^\delta$, $f_0 = \lim_{\delta \downarrow 0} f_\delta$, and conclude that the set of points at which f is discontinuous $= \{f^0 \neq f_0\}$ is measurable.

1.3.7 A function $\varphi : \Omega \to \mathbf{R}$ is said to be **simple** if

$$\varphi(\omega) = \sum_{m=1}^{n} c_m 1_{A_m}(\omega)$$

where the c_m are real numbers and $A_m \in \mathcal{F}$. Show that the class of $\mathcal{F}$ measurable functions is the smallest class containing the simple functions and closed under pointwise limits.

1.3.8 Use the previous exercise to conclude that Y is measurable with respect to $\sigma(X)$ if and only if $Y = f(X)$ where $f : \mathbf{R} \to \mathbf{R}$ is measurable.

1.3.9 To get a constructive proof of the last result, note that $\{\omega : m2^{-n} \leq Y < (m+1)2^{-n}\} = \{X \in B_{m,n}\}$ for some $B_{m,n} \in \mathcal{R}$ and set $f_n(x) = m2^{-n}$ for $x \in B_{m,n}$ and show that as $n \to \infty$ $f_n(x) \to f(x)$ and $Y = f(X)$.

1.4 Integration

Let μ be a σ-finite measure on $(\Omega, \mathcal{F})$. We will be primarily interested in the special case μ is a probability measure, but we will sometimes need to integrate with respect to infinite measure and it is no harder to develop the results in general.

In this section we will define $\int f \, d\mu$ for a class of measurable functions. This is a four-step procedure:

1. Simple functions
2. Bounded functions
3. Nonnegative functions
4. General functions

This sequence of four steps is also useful in proving integration formulas. See, for example, the proofs of Theorems 1.6.9 and 1.7.2.

Step 1. φ is said to be a **simple function** if $\varphi(\omega) = \sum_{i=1}^n a_i 1_{A_i}$ and A_i are disjoint sets with $\mu(A_i) < \infty$. If φ is a simple function, we let

$$\int \varphi \, d\mu = \sum_{i=1}^n a_i \mu(A_i)$$

The representation of φ is not unique since we have not supposed that the a_i are distinct. However, it is easy to see that the last definition does not contradict itself.

We will prove the next three conclusions four times, but before we can state them for the first time, we need a definition. $\varphi \geq \psi$ **μ-almost everywhere** (or $\varphi \geq \psi$ μ-a.e.) means $\mu(\{\omega : \varphi(\omega) < \psi(\omega)\}) = 0$. When there is no doubt about what measure we are referring to, we drop the μ.

Lemma 1.4.1 *Let φ and ψ be simple functions.*
(i) If $\varphi \geq 0$ a.e. then $\int \varphi \, d\mu \geq 0$.
(ii) For any $a \in \mathbf{R}$, $\int a\varphi \, d\mu = a \int \varphi \, d\mu$.
(iii) $\int \varphi + \psi \, d\mu = \int \varphi \, d\mu + \int \psi \, d\mu$.

Proof (i) and (ii) are immediate consequences of the definition. To prove (iii), suppose

$$\varphi = \sum_{i=1}^m a_i 1_{A_i} \quad \text{and} \quad \psi = \sum_{j=1}^n b_j 1_{B_j}$$

To make the supports of the two functions the same, we let $A_0 = \cup_i B_i - \cup_i A_i$, let $B_0 = \cup_i A_i - \cup_i B_i$, and let $a_0 = b_0 = 0$. Now

$$\varphi + \psi = \sum_{i=0}^m \sum_{j=0}^n (a_i + b_j) 1_{(A_i \cap B_j)}$$

and the $A_i \cap B_j$ are pairwise disjoint, so

$$\begin{aligned}
\int (\varphi + \psi) \, d\mu &= \sum_{i=0}^m \sum_{j=0}^n (a_i + b_j) \mu(A_i \cap B_j) \\
&= \sum_{i=0}^m \sum_{j=0}^n a_i \mu(A_i \cap B_j) + \sum_{j=0}^n \sum_{i=0}^m b_j \mu(A_i \cap B_j) \\
&= \sum_{i=0}^m a_i \, \mu(A_i) + \sum_{j=0}^n b_j \, \mu(B_j) = \int \varphi \, d\mu + \int \psi \, d\mu
\end{aligned}$$

In the next-to-last step, we used $A_i = +_j (A_i \cap B_j)$ and $B_j = +_i (A_i \cap B_j)$, where $+$ denotes a disjoint union. □

We will prove (i)–(iii) three more times as we generalize our integral. As a consequence of (i)–(iii), we get three more useful properties. To keep from repeating their proofs, which do not change, we will prove

Lemma 1.4.2 *If (i) and (iii) hold, then we have:*
(iv) If $\varphi \le \psi$ *a.e. then* $\int \varphi \, d\mu \le \int \psi \, d\mu$.
(v) If $\varphi = \psi$ *a.e. then* $\int \varphi \, d\mu = \int \psi \, d\mu$.
If, in addition, (ii) holds when $a = -1$, *we have*
(vi) $|\int \varphi \, d\mu| \le \int |\varphi| \, d\mu$

Proof By (iii), $\int \psi \, d\mu = \int \varphi \, d\mu + \int (\psi - \varphi) \, d\mu$ and the second integral is ≥ 0 by (i), so (iv) holds. $\varphi = \psi$ a.e. implies $\varphi \le \psi$ a.e. and $\psi \le \varphi$ a.e. so (v) follows from two applications of (iv). To prove (vi) now, notice that $\varphi \le |\varphi|$ so (iv) implies $\int \varphi \, d\mu \le \int |\varphi| \, d\mu$. $-\varphi \le |\varphi|$, so (iv) and (ii) imply $-\int \varphi \, d\mu \le \int |\varphi| \, d\mu$. Since $|y| = \max(y, -y)$, the result follows. □

Step 2. Let E be a set with $\mu(E) < \infty$ and let f be a bounded function that vanishes on E^c. To define the integral of f, we observe that if φ, ψ are simple functions that have $\varphi \le f \le \psi$, then we want to have

$$\int \varphi \, d\mu \le \int f \, d\mu \le \int \psi \, d\mu$$

so we let

$$\int f \, d\mu = \sup_{\varphi \le f} \int \varphi \, d\mu = \inf_{\psi \ge f} \int \psi \, d\mu \tag{1.4.1}$$

Here and for the rest of Step 2, we assume that φ and ψ vanish on E^c. To justify the definition, we have to prove that the sup and inf are equal. It follows from (iv) in Lemma 1.4.2 that

$$\sup_{\varphi \le f} \int \varphi \, d\mu \le \inf_{\psi \ge f} \int \psi \, d\mu$$

To prove the other inequality, suppose $|f| \le M$ and let

$$E_k = \left\{ x \in E : \frac{kM}{n} \ge f(x) > \frac{(k-1)M}{n} \right\} \quad \text{for } -n \le k \le n$$

$$\psi_n(x) = \sum_{k=-n}^{n} \frac{kM}{n} 1_{E_k} \qquad \varphi_n(x) = \sum_{k=-n}^{n} \frac{(k-1)M}{n} 1_{E_k}$$

By definition, $\psi_n(x) - \varphi_n(x) = (M/n) 1_E$, so

$$\int \psi_n(x) - \varphi_n(x) \, d\mu = \frac{M}{n} \mu(E)$$

Since $\varphi_n(x) \le f(x) \le \psi_n(x)$, it follows from (iii) in Lemma 1.4.1 that

$$\begin{aligned} \sup_{\varphi \le f} \int \varphi \, d\mu \ge \int \varphi_n \, d\mu &= -\frac{M}{n} \mu(E) + \int \psi_n \, d\mu \\ &\ge -\frac{M}{n} \mu(E) + \inf_{\psi \ge f} \int \psi \, d\mu \end{aligned}$$

The last inequality holds for all n, so the proof is complete. □

Lemma 1.4.3 *Let E be a set with $\mu(E) < \infty$. If f and g are bounded functions that vanish on E^c, then:*

(i) If $f \geq 0$ a.e. then $\int f \, d\mu \geq 0$.
(ii) For any $a \in \mathbf{R}$, $\int af \, d\mu = a \int f \, d\mu$.
(iii) $\int f + g \, d\mu = \int f \, d\mu + \int g \, d\mu$.
(iv) If $g \leq f$ a.e. then $\int g \, d\mu \leq \int f \, d\mu$.
(v) If $g = f$ a.e. then $\int g \, d\mu = \int f \, d\mu$.
(vi) $|\int f \, d\mu| \leq \int |f| \, d\mu$.

Proof Since we can take $\varphi \equiv 0$, (i) is clear from the definition. To prove (ii), we observe that if $a > 0$, then $a\varphi \leq af$ if and only if $\varphi \leq f$, so

$$\int af \, d\mu = \sup_{\varphi \leq f} \int a\varphi \, d\mu = \sup_{\varphi \leq f} a \int \varphi \, d\mu = a \sup_{\varphi \leq f} \int \varphi \, d\mu = a \int f \, d\mu$$

For $a < 0$, we observe that $a\varphi \leq af$ if and only if $\varphi \geq f$, so

$$\int af \, d\mu = \sup_{\varphi \geq f} \int a\varphi \, d\mu = \sup_{\varphi \geq f} a \int \varphi \, d\mu = a \inf_{\varphi \geq f} \int \varphi \, d\mu = a \int f \, d\mu$$

To prove (iii), we observe that if $\psi_1 \geq f$ and $\psi_2 > g$, then $\psi_1 + \psi_2 \geq f + g$ so

$$\inf_{\psi \geq f+g} \int \psi \, d\mu \leq \inf_{\psi_1 \geq f, \psi_2 \geq g} \int \psi_1 + \psi_2 \, d\mu$$

Using linearity for simple functions, it follows that

$$\begin{aligned} \int f + g \, d\mu &= \inf_{\psi \geq f+g} \int \psi \, d\mu \\ &\leq \inf_{\psi_1 \geq f, \psi_2 \geq g} \int \psi_1 \, d\mu + \int \psi_2 \, d\mu = \int f \, d\mu + \int g \, d\mu \end{aligned}$$

To prove the other inequality, observe that the last conclusion applied to $-f$ and $-g$ and (ii) imply

$$-\int f + g \, d\mu \leq -\int f \, d\mu - \int g \, d\mu$$

(iv)–(vi) follow from (i)–(iii) by Lemma 1.4.2. □

Notation. We define the integral of f over the set E:

$$\int_E f \, d\mu \equiv \int f \cdot 1_E \, d\mu$$

Step 3. If $f \geq 0$, then we let

$$\int f \, d\mu = \sup \left\{ \int h \, d\mu : 0 \leq h \leq f, h \text{ is bounded and } \mu(\{x : h(x) > 0\}) < \infty \right\}$$

The last definition is nice since it is clear that this is well defined. The next result will help us compute the value of the integral.

Lemma 1.4.4 *Let $E_n \uparrow \Omega$ have $\mu(E_n) < \infty$ and let $a \wedge b = \min(a,b)$. Then*

$$\int_{E_n} f \wedge n \, d\mu \uparrow \int f \, d\mu \quad \text{as } n \uparrow \infty$$

Proof It is clear from (iv) in Lemma 1.4.3 that the left-hand side increases as n does. Since $h = (f \wedge n)1_{E_n}$ is a possibility in the sup, each term is smaller than the integral on the right. To prove that the limit is $\int f \, d\mu$, observe that if $0 \le h \le f$, $h \le M$, and $\mu(\{x : h(x) > 0\}) < \infty$, then for $n \ge M$ using $h \le M$, (iv), and (iii),

$$\int_{E_n} f \wedge n \, d\mu \ge \int_{E_n} h \, d\mu = \int h \, d\mu - \int_{E_n^c} h \, d\mu$$

Now $0 \le \int_{E_n^c} h \, d\mu \le M\mu(E_n^c \cap \{x : h(x) > 0\}) \to 0$ as $n \to \infty$, so

$$\liminf_{n \to \infty} \int_{E_n} f \wedge n \, d\mu \ge \int h \, d\mu$$

which proves the desired result since h is an arbitrary member of the class that defines the integral of f. □

Lemma 1.4.5 *Suppose $f, g \ge 0$.*

(i) $\int f \, d\mu \ge 0$
(ii) If $a > 0$, then $\int af d\mu = a \int f \, d\mu$.
(iii) $\int f + g \, d\mu = \int f \, d\mu + \int g \, d\mu$
(iv) If $0 \le g \le f$ a.e. then $\int g \, d\mu \le \int f \, d\mu$.
(v) If $0 \le g = f$ a.e. then $\int g \, d\mu = \int f \, d\mu$.

Here we have dropped (vi) because it is trivial for $f \ge 0$.

Proof (i) is trivial from the definition. (ii) is clear, since when $a > 0$, $ah \le af$ if and only if $h \le f$ and we have $\int ah \, d\mu = a \int h \, du$ for h in the defining class. For (iii), we observe that if $f \ge h$ and $g \ge k$, then $f + g \ge h + k$ so taking the sup over h and k in the defining classes for f and g gives

$$\int f + g \, d\mu \ge \int f \, d\mu + \int g \, d\mu$$

To prove the other direction, we observe $(a + b) \wedge n \le (a \wedge n) + (b \wedge n)$ so (iv) from Lemma 1.4.3 and (iii) from Lemma 1.4.4 imply

$$\int_{E_n} (f + g) \wedge n \, d\mu \le \int_{E_n} f \wedge n \, d\mu + \int_{E_n} g \wedge n \, d\mu$$

Letting $n \to \infty$ and using Lemma 1.4.4 gives (iii). As before, (iv) and (v) follow from (i), (iii), and Lemma 1.4.2. □

Step 4. We say f is integrable if $\int |f| \, d\mu < \infty$. Let

$$f^+(x) = f(x) \vee 0 \quad \text{and} \quad f^-(x) = (-f(x)) \vee 0$$

where $a \vee b = \max(a,b)$. Clearly,

$$f(x) = f^+(x) - f^-(x) \quad \text{and} \quad |f(x)| = f^+(x) + f^-(x)$$

We define the integral of f by

$$\int f\,d\mu = \int f^+\,d\mu - \int f^-\,d\mu$$

The right-hand side is well defined since $f^+, f^- \le |f|$ and we have (iv) in Lemma 1.4.5. For the final time, we will prove our six properties. To do this, it is useful to know:

Lemma 1.4.6 *If $f = f_1 - f_2$ where $f_1, f_2 \ge 0$ and $\int f_i\,d\mu < \infty$, then*

$$\int f\,d\mu = \int f_1\,d\mu - \int f_2\,d\mu$$

Proof $f_1 + f^- = f_2 + f^+$ and all four functions are ≥ 0, so by (iii) of Lemma 1.4.5,

$$\int f_1\,d\mu + \int f^-\,d\mu = \int f_1 + f^-\,d\mu = \int f_2 + f^+\,d\mu = \int f_2\,d\mu + \int f^+\,d\mu$$

Rearranging gives the desired conclusion. □

Theorem 1.4.7 *Suppose f and g are integrable.*
(i) If $f \ge 0$ a.e. then $\int f\,d\mu \ge 0$.
(ii) For all $a \subset \mathbf{R}$, $\int af\,d\mu = a\int f\,d\mu$.
(iii) $\int f + g\,d\mu = \int f\,d\mu + \int g\,d\mu$
(iv) If $g \le f$ a.e. then $\int g\,d\mu \le \int f\,d\mu$.
(v) If $g = f$ a.e. then $\int g\,d\mu = \int f\,d\mu$.
(vi) $|\int f\,d\mu| \le \int |f|\,d\mu$.

Proof (i) is trivial. (ii) is clear since if $a > 0$, then $(af)^+ = a(f^+)$, and so on. To prove (iii), observe that $f+g = (f^+ + g^+) - (f^- + g^-)$, so using Lemma 1.4.6 and Lemma 1.4.5

$$\begin{aligned}\int f + g\,d\mu &= \int f^+ + g^+\,d\mu - \int f^- + g^-\,d\mu \\ &= \int f^+\,d\mu + \int g^+\,d\mu - \int f^-\,d\mu - \int g^-\,d\mu\end{aligned}$$

As usual, (iv)–(vi) follow from (i)–(iii) and Lemma 1.4.2. □

Notation for special cases:

(a) When $(\Omega, \mathcal{F}, \mu) = (\mathbf{R}^d, \mathcal{R}^d, \lambda)$, we write $\int f(x)\,dx$ for $\int f\,d\lambda$.

(b) When $(\Omega, \mathcal{F}, \mu) = (\mathbf{R}, \mathcal{R}, \lambda)$ and $E = [a,b]$, we write $\int_a^b f(x)\,dx$ for $\int_E f\,d\lambda$.

(c) When $(\Omega, \mathcal{F}, \mu) = (\mathbf{R}, \mathcal{R}, \mu)$ with $\mu((a,b]) = G(b) - G(a)$ for $a < b$, we write $\int f(x)\,dG(x)$ for $\int f\,d\mu$.

(d) When Ω is a countable set, $\mathcal{F} =$ all subsets of Ω, and μ is counting measure, we write $\sum_{i\in\Omega} f(i)$ for $\int f\,d\mu$.

We mention example (d) primarily to indicate that results for sums follow from those for integrals. The notation for the special case in which μ is a probability measure will be taken up in Section 1.6.

Exercises

1.4.1 Show that if $f \geq 0$ and $\int f\,d\mu = 0$, then $f = 0$ a.e.

1.4.2 Let $f \geq 0$ and $E_{n,m} = \{x : m/2^n \leq f(x) < (m+1)/2^n\}$. As $n \uparrow \infty$,

$$\sum_{m=1}^{\infty} \frac{m}{2^n} \mu(E_{n,m}) \uparrow \int f\,d\mu$$

1.4.3 Let g be an integrable function on $\mathbf{R}$ and $\epsilon > 0$. (i) Use the definition of the integral to conclude there is a simple function $\varphi = \sum_k b_k 1_{A_k}$ with $\int |g - \varphi|\,dx < \epsilon$. (ii) Use Exercise A.2.1 to approximate the A_k by finite unions of intervals to get a **step function**

$$q = \sum_{j=1}^{k} c_j 1_{(a_{j-1}, a_j)}$$

with $a_0 < a_1 < \ldots < a_k$, so that $\int |\varphi - q| < \epsilon$. (iii) Round the corners of q to get a continuous function r so that $\int |q - r|\,dx < \epsilon$.

(iv) To make a continuous function replace each $c_j 1_{(a_{j-1}, a_j)}$ by a function that is 0 $(a_{j-1}, a_j)^c$, c_j on $[a_{j-1} + \delta - j, a_j - \delta_j]$, and linear otherwise. If the δ_j are small enough and we let $r(x) = \sum_{j=1}^{k} r_j(x)$, then

$$\int |q(x) - r(x)|\,d\mu = \sum_{j=1}^{k} \delta_j c_j < \epsilon$$

1.4.4 Prove the **Riemann-Lebesgue lemma.** If g is integrable, then

$$\lim_{n \to \infty} \int g(x) \cos nx\,dx = 0$$

Hint: If g is a step function, this is easy. Now use the previous exercise.

1.5 Properties of the Integral

In this section, we will develop properties of the integral defined in the last section. Our first result generalizes (vi) from Theorem 1.4.7.

Theorem 1.5.1 (Jensen's inequality) *Suppose φ is convex, that is,*

$$\lambda\varphi(x) + (1 - \lambda)\varphi(y) \geq \varphi(\lambda x + (1 - \lambda) y)$$

for all $\lambda \in (0,1)$ and $x, y \in \mathbf{R}$. If μ is a probability measure, and f and $\varphi(f)$ are integrable, then

$$\varphi\left(\int f\,d\mu\right) \leq \int \varphi(f)\,d\mu$$

Proof Let $c = \int f\,d\mu$ and let $\ell(x) = ax + b$ be a linear function that has $\ell(c) = \varphi(c)$ and $\varphi(x) \geq \ell(x)$. To see that such a function exists, recall that convexity implies

$$\lim_{h \downarrow 0} \frac{\varphi(c) - \varphi(c - h)}{h} \leq \lim_{h \downarrow 0} \frac{\varphi(c + h) - \varphi(c)}{h}$$

(The limits exist since the sequences are monotone.) If we let a be any number between the two limits and let $\ell(x) = a(x-c) + \varphi(c)$, then ℓ has the desired properties. With the existence of ℓ established, the rest is easy. (iv) in Theorem 1.4.7 implies

$$\int \varphi(f)\,d\mu \geq \int (af+b)\,d\mu = a\int f\,d\mu + b = \ell\left(\int f\,d\mu\right) = \varphi\left(\int f\,d\mu\right)$$

since $c = \int f\,d\mu$ and $\ell(c) = \varphi(c)$. □

Let $\|f\|_p = (\int |f|^p\,d\mu)^{1/p}$ for $1 \leq p < \infty$, and notice $\|cf\|_p = |c| \cdot \|f\|_p$ for any real number c.

Theorem 1.5.2 (Hölder's inequality) *If $p, q \in (1,\infty)$ with $1/p + 1/q = 1$, then*

$$\int |fg|\,d\mu \leq \|f\|_p\|g\|_q$$

Proof If $\|f\|_p$ or $\|g\|_q = 0$, then $|fg| = 0$ a.e., so it suffices to prove the result when $\|f\|_p$ and $\|g\|_q > 0$ or by dividing both sides by $\|f\|_p\|g\|_q$, when $\|f\|_p = \|g\|_q = 1$. Fix $y \geq 0$ and let

$$\varphi(x) = x^p/p + y^q/q - xy \quad \text{for } x \geq 0$$

$$\varphi'(x) = x^{p-1} - y \quad \text{and} \quad \varphi''(x) = (p-1)x^{p-2}$$

so φ has a minimum at $x_o = y^{1/(p-1)}$. $q = p/(p-1)$ and $x_o^p = y^{p/(p-1)} = y^q$ so

$$\varphi(x_o) = y^q(1/p + 1/q) - y^{1/(p-1)}y = 0$$

Since x_o is the minimum, it follows that $xy \leq x^p/p + y^q/q$. Letting $x = |f|$, $y = |g|$, and integrating

$$\int |fg|\,d\mu \leq \frac{1}{p} + \frac{1}{q} = 1 = \|f\|_p\|g\|_q$$ □

Remark The special case $p = q = 2$ is called the **Cauchy-Schwarz inequality**. One can give a direct proof of the result in this case by observing that for any θ,

$$0 \leq \int (f+\theta g)^2\,d\mu = \int f^2\,d\mu + \theta\left(2\int fg\,d\mu\right) + \theta^2\left(\int g^2\,d\mu\right)$$

so the quadratic $a\theta^2 + b\theta + c$ on the right-hand side has at most one real root. Recalling the formula for the roots of a quadratic

$$\frac{-b \pm \sqrt{b^2 - 4ac}}{2a}$$

we see $b^2 - 4ac \leq 0$, which is the desired result.

Our next goal is to give conditions that guarantee

$$\lim_{n\to\infty}\int f_n\,d\mu = \int \left(\lim_{n\to\infty} f_n\right)d\mu$$

First, we need a definition. We say that $f_n \to f$ **in measure**, i.e., for any $\epsilon > 0$, $\mu(\{x : |f_n(x) - f(x)| > \epsilon\}) \to 0$ as $n \to \infty$. On a space of finite measure, this is a weaker assumption than $f_n \to f$ a.e., but the next result is easier to prove in the greater generality.

Theorem 1.5.3 (Bounded convergence theorem) *Let E be a set with $\mu(E) < \infty$. Suppose f_n vanishes on E^c, $|f_n(x)| \leq M$, and $f_n \to f$ in measure. Then*

$$\int f \, d\mu = \lim_{n\to\infty} \int f_n \, d\mu$$

Example 1.5.4 Consider the real line **R** equipped with the Borel sets $\mathcal{R}$ and Lebesgue measure λ. The functions $f_n(x) = 1/n$ on $[0, n]$ and 0 otherwise on show that the conclusion of Theorem 1.5.3 does not hold when $\mu(E) = \infty$.

Proof Let $\epsilon > 0$, $G_n = \{x : |f_n(x) - f(x)| < \epsilon\}$ and $B_n = E - G_n$. Using (iii) and (vi) from Theorem 1.4.7,

$$\begin{aligned}\left|\int f \, d\mu - \int f_n \, d\mu\right| &= \left|\int (f - f_n) \, d\mu\right| \leq \int |f - f_n| \, d\mu \\ &= \int_{G_n} |f - f_n| \, d\mu + \int_{B_n} |f - f_n| \, d\mu \\ &\leq \epsilon\mu(E) + 2M\mu(B_n)\end{aligned}$$

$f_n \to f$ in measure implies $\mu(B_n) \to 0$. $\epsilon > 0$ is arbitrary and $\mu(E) < \infty$, so the proof is complete. □

Theorem 1.5.5 (Fatou's lemma) *If $f_n \geq 0$, then*

$$\liminf_{n\to\infty} \int f_n \, d\mu \geq \int \left(\liminf_{n\to\infty} f_n\right) d\mu$$

Example 1.5.6 Example 1.5.4 shows that we may have strict inequality in Theorem 1.5.5. The functions $f_n(x) = n1_{(0,1/n]}(x)$ on (0,1) equipped with the Borel sets and Lebesgue measure show that this can happen on a space of finite measure.

Proof Let $g_n(x) = \inf_{m\geq n} f_m(x)$. $f_n(x) \geq g_n(x)$ and as $n \uparrow \infty$,

$$g_n(x) \uparrow g(x) = \liminf_{n\to\infty} f_n(x)$$

Since $\int f_n \, d\mu \geq \int g_n \, d\mu$, it suffices then to show that

$$\liminf_{n\to\infty} \int g_n \, d\mu \geq \int g \, d\mu$$

Let $E_m \uparrow \Omega$ be sets of finite measure. Since $g_n \geq 0$ and for fixed m

$$(g_n \wedge m) \cdot 1_{E_m} \to (g \wedge m) \cdot 1_{E_m} \quad \text{a.e.}$$

the bounded convergence theorem, 1.5.3, implies

$$\liminf_{n\to\infty} \int g_n \, d\mu \geq \int_{E_m} g_n \wedge m \, d\mu \to \int_{E_m} g \wedge m \, d\mu$$

Taking the sup over m and using Theorem 1.4.4 gives the desired result. □

Theorem 1.5.7 (Monotone convergence theorem) *If $f_n \geq 0$ and $f_n \uparrow f$, then*

$$\int f_n \, d\mu \uparrow \int f \, d\mu$$

Proof Fatou's lemma, Theorem 1.5.5, implies $\liminf \int f_n \, d\mu \geq \int f \, d\mu$. On the other hand, $f_n \leq f$ implies $\limsup \int f_n \, d\mu \leq \int f \, d\mu$. □

Theorem 1.5.8 (Dominated convergence theorem) *If $f_n \to f$ a.e., $|f_n| \leq g$ for all n, and g is integrable, then $\int f_n \, d\mu \to \int f \, d\mu$.*

Proof $f_n + g \geq 0$ so Fatou's lemma implies

$$\liminf_{n\to\infty} \int f_n + g \, d\mu \geq \int f + g \, d\mu$$

Subtracting $\int g \, d\mu$ from both sides gives

$$\liminf_{n\to\infty} \int f_n \, d\mu \geq \int f \, d\mu$$

Applying the last result to $-f_n$, we get

$$\limsup_{n\to\infty} \int f_n \, d\mu \leq \int f \, d\mu$$

and the proof is complete. □

Exercises

1.5.1 Let $\|f\|_\infty = \inf\{M : \mu(\{x : |f(x)| > M\}) = 0\}$. Prove that

$$\int |fg| d\mu \leq \|f\|_1 \|g\|_\infty$$

1.5.2 Show that if μ is a probability measure, then

$$\|f\|_\infty = \lim_{p\to\infty} \|f\|_p$$

1.5.3 **Minkowski's inequality.** (i) Suppose $p \in (1,\infty)$. The inequality $|f + g|^p \leq 2^p(|f|^p + |g|^p)$ shows that if $\|f\|_p$ and $\|g\|_p$ are $< \infty$, then $\|f + g\|_p < \infty$. Apply Hölder's inequality to $|f||f + g|^{p-1}$ and $|g||f + g|^{p-1}$ to show $\|f + g\|_p \leq \|f\|_p + \|g\|_p$. (ii) Show that the last result remains true when $p = 1$ or $p = \infty$.

1.5.4 If f is integrable and E_m are disjoint sets with union E, then

$$\sum_{m=0}^{\infty} \int_{E_m} f \, d\mu = \int_E f \, d\mu$$

So if $f \geq 0$, then $\nu(E) = \int_E f \, d\mu$ defines a measure.

1.5.5 If $g_n \uparrow g$ and $\int g_1^- \, d\mu < \infty$, then $\int g_n \, d\mu \uparrow \int g \, d\mu$.

1.5.6 If $g_m \geq 0$, then $\int \sum_{m=0}^{\infty} g_m \, d\mu = \sum_{m=0}^{\infty} \int g_m \, d\mu$.

1.5.7 Let $f \geq 0$. (i) Show that $\int f \wedge n \, d\mu \uparrow \int f \, d\mu$ as $n \to \infty$. (ii) Use (i) to conclude that if g is integrable and $\epsilon > 0$, then we can pick $\delta > 0$ so that $\mu(A) < \delta$ implies $\int_A |g| d\mu < \epsilon$.

1.5.8 Show that if f is integrable on $[a,b]$, $g(x) = \int_{[a,x]} f(y) \, dy$ is continuous on (a,b).

1.5.9 Show that if f has $\|f\|_p = (\int |f|^p d\mu)^{1/p} < \infty$, then there are simple functions φ_n so that $\|\varphi_n - f\|_p \to 0$.

1.5.10 Show that if $\sum_n \int |f_n| d\mu < \infty$, then $\sum_n \int f_n \, d\mu = \int \sum_n f_n \, d\mu$.

1.6 Expected Value

We now specialize to integration with respect to a probability measure P. If $X \geq 0$ is a random variable on $(\Omega, \mathcal{F}, P)$, then we define its **expected value** to be $EX = \int X \, dP$, which always makes sense, but may be ∞. To reduce the general case to the nonnegative case, let $x^+ = \max\{x, 0\}$ be the **positive part** and let $x^- = \max\{-x, 0\}$ be the **negative part** of x. We declare that EX **exists** and set $EX = EX^+ - EX^-$ whenever the subtraction makes sense, i.e., $EX^+ < \infty$ or $EX^- < \infty$.

EX is often called the **mean** of X and denoted by μ. EX is defined by integrating X, so it has all the properties that integrals do. From Theorems 1.4.5 and 1.4.7 and the trivial observation that $E(b) = b$ for any real number b, we get the following:

Theorem 1.6.1 *Suppose $X, Y \geq 0$ or $E|X|, E|Y| < \infty$.*

(a) $E(X+Y) = EX + EY$.

(b) $E(aX+b) = aE(X) + b$ for any real numbers a, b.

(c) If $X \geq Y$, then $EX \geq EY$.

In this section, we will restate some properties of the integral derived in the last section in terms of expected value and prove some new ones. To organize things, we will divide the developments into three subsections.

1.6.1 Inequalities

For probability measures, Theorem 1.5.1 becomes:

Theorem 1.6.2 (Jensen's inequality) *Suppose φ is* **convex**, *that is,*

$$\lambda\varphi(x) + (1-\lambda)\varphi(y) \geq \varphi(\lambda x + (1-\lambda)y)$$

for all $\lambda \in (0,1)$ and $x, y \in \mathbf{R}$. Then

$$E(\varphi(X)) \geq \varphi(EX)$$

provided both expectations exist, i.e., $E|X|$ and $E|\varphi(X)| < \infty$.

To recall the direction in which the inequality goes note that if $P(X = x) = \lambda$ and $P(X = y) = 1 - \lambda$, then

$$E\varphi(X) = \lambda\varphi(x) + (1-\lambda)\varphi(y) \geq \varphi(\lambda x + (1-\lambda)y) = \varphi(EX)$$

Two useful special cases are $|EX| \leq E|X|$ and $(EX)^2 \leq E(X^2)$.

Theorem 1.6.3 (Hölder's inequality) *If $p, q \in [1, \infty]$ with $1/p + 1/q = 1$, then*

$$E|XY| \leq \|X\|_p \|Y\|_q$$

Here $\|X\|_r = (E|X|^r)^{1/r}$ for $r \in [1, \infty)$; $\|X\|_\infty = \inf\{M : P(|X| > M) = 0\}$.

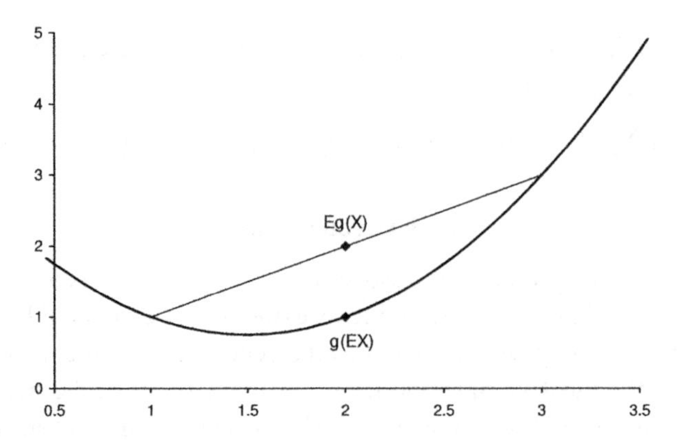

Figure 1.6 Jensen's inequality for $g(x) = x^2 - 3x + 3$, $P(X = 1) = P(X = 3) = 1/2$.

To state our next result, we need some notation. If we only integrate over $A \subset \Omega$, we write

$$E(X; A) = \int_A X \, dP$$

Theorem 1.6.4 (Chebyshev's inequality) *Suppose* $\varphi : \mathbf{R} \to \mathbf{R}$ *has* $\varphi \geq 0$, *let* $A \in \mathcal{R}$ *and let* $i_A = \inf\{\varphi(y) : y \in A\}$.

$$i_A P(X \in A) \leq E(\varphi(X); X \in A) \leq E\varphi(X)$$

Proof The definition of i_A and the fact that $\varphi \geq 0$ imply that

$$i_A 1_{(X \in A)} \leq \varphi(X) 1_{(X \in A)} \leq \varphi(X)$$

So taking expected values and using part (c) of Theorem 1.6.1 gives the desired result. □

Remark Some authors call this result **Markov's inequality** and use the name Chebyshev's inequality for the special case in which $\varphi(x) = x^2$ and $A = \{x : |x| \geq a\}$:

$$a^2 P(|X| \geq a) \leq EX^2 \tag{1.6.1}$$

1.6.2 Integration to the Limit

Our next step is to restate the three classic results from the previous section about what happens when we interchange limits and integrals.

Theorem 1.6.5 (Fatou's lemma) *If* $X_n \geq 0$, *then*

$$\liminf_{n \to \infty} EX_n \geq E(\liminf_{n \to \infty} X_n)$$

Theorem 1.6.6 (Monotone convergence theorem) *If* $0 \leq X_n \uparrow X$, *then* $EX_n \uparrow EX$.

Theorem 1.6.7 (Dominated convergence theorem) *If* $X_n \to X$ *a.s.*, $|X_n| \leq Y$ *for all* n, *and* $EY < \infty$, *then* $EX_n \to EX$.

The special case of Theorem 1.6.7 in which Y is constant is called the **bounded convergence theorem**.

In the developments that follow, we will need another result on integration to the limit. Perhaps the most important special case of this result occurs when $g(x) = |x|^p$ with $p > 1$ and $h(x) = x$.

Theorem 1.6.8 *Suppose $X_n \to X$ a.s. Let g, h be continuous functions with*

(i) $g \geq 0$ and $g(x) \to \infty$ as $|x| \to \infty$,

(ii) $|h(x)|/g(x) \to 0$ as $|x| \to \infty$,

and (iii) $Eg(X_n) \leq K < \infty$ for all n.

Then $Eh(X_n) \to Eh(X)$.

Proof By subtracting a constant from h, we can suppose without loss of generality that $h(0) = 0$. Pick M large so that $P(|X| = M) = 0$ and $g(x) > 0$ when $|x| \geq M$. Given a random variable Y, let $\bar{Y} = Y1_{(|Y| \leq M)}$. Since $P(|X| = M) = 0$, $\bar{X}_n \to \bar{X}$ a.s. Since $h(\bar{X}_n)$ is bounded and h is continuous, it follows from the bounded convergence theorem that

(a) $$Eh(\bar{X}_n) \to Eh(\bar{X})$$

To control the effect of the truncation, we use the following:

(b) $$|Eh(\bar{Y}) - Eh(Y)| \leq E|h(\bar{Y}) - h(Y)| \leq E(|h(Y)|; |Y| > M) \leq \epsilon_M Eg(Y)$$

where $\epsilon_M = \sup\{|h(x)|/g(x) : |x| \geq M\}$. To check the second inequality, note that when $|Y| \leq M$, $\bar{Y} = Y$, and we have supposed $h(0) = 0$. The third inequality follows from the definition of ϵ_M.

Taking $Y = X_n$ in (b) and using (iii), it follows that

(c) $$|Eh(\bar{X}_n) - Eh(X_n)| \leq K\epsilon_M$$

To estimate $|Eh(\bar{X}) - Eh(X)|$, we observe that $g \geq 0$ and g is continuous, so Fatou's lemma implies

$$Eg(X) \leq \liminf_{n\to\infty} Eg(X_n) \leq K$$

Taking $Y = X$ in (b) gives

(d) $$|Eh(\bar{X}) - Eh(X)| \leq K\epsilon_M$$

The triangle inequality implies

$$|Eh(X_n) - Eh(X)| \leq |Eh(X_n) - Eh(\bar{X}_n)| + |Eh(\bar{X}_n) - Eh(\bar{X})| + |Eh(\bar{X}) - Eh(X)|$$

Taking limits and using (a), (c), (d), we have

$$\limsup_{n\to\infty} |Eh(X_n) - Eh(X)| \leq 2K\epsilon_M$$

which proves the desired result since $K < \infty$ and $\epsilon_M \to 0$ as $M \to \infty$. □

1.6.3 Computing Expected Values

Integrating over $(\Omega, \mathcal{F}, P)$ is nice in theory, but to do computations we have to shift to a space on which we can do calculus. In most cases, we will apply the next result with $S = \mathbf{R}^d$.

Theorem 1.6.9 (Change of variables formula) *Let X be a random element of $(S, \mathcal{S})$ with distribution μ, i.e., $\mu(A) = P(X \in A)$. If f is a measurable function from $(S, \mathcal{S})$ to $(\mathbf{R}, \mathcal{R})$ so that $f \geq 0$ or $E|f(X)| < \infty$, then*

$$Ef(X) = \int_S f(y)\, \mu(dy)$$

Remark To explain the name, write h for X and $P \circ h^{-1}$ for μ to get

$$\int_\Omega f(h(\omega))\, dP = \int_S f(y)\, d(P \circ h^{-1})$$

Proof We will prove this result by verifying it in four increasingly more general special cases that parallel the way that the integral was defined in Section 1.4. The reader should note the method employed, since it will be used several times.

Case 1: Indicator functions. If $B \in \mathcal{S}$ and $f = 1_B$, then recalling the relevant definitions shows

$$E 1_B(X) = P(X \in B) = \mu(B) = \int_S 1_B(y)\, \mu(dy)$$

Case 2: Simple functions. Let $f(x) = \sum_{m=1}^n c_m 1_{B_m}$ where $c_m \in \mathbf{R}$, $B_m \in \mathcal{S}$. The linearity of expected value, the result of Case 1, and the linearity of integration imply

$$\begin{aligned} Ef(X) &= \sum_{m=1}^n c_m E 1_{B_m}(X) \\ &= \sum_{m=1}^n c_m \int_S 1_{B_m}(y)\, \mu(dy) = \int_S f(y)\, \mu(dy) \end{aligned}$$

Case 3: Nonnegative functions. Now if $f \geq 0$ and we let

$$f_n(x) = ([2^n f(x)]/2^n) \wedge n$$

where $[x]$ = the largest integer $\leq x$ and $a \wedge b = \min\{a, b\}$, then the f_n are simple and $f_n \uparrow f$, so using the result for simple functions and the monotone convergence theorem:

$$Ef(X) = \lim_n Ef_n(X) = \lim_n \int_S f_n(y)\, \mu(dy) = \int_S f(y)\, \mu(dy)$$

Case 4: Integrable functions. The general case now follows by writing $f(x) = f(x)^+ - f(x)^-$. The condition $E|f(X)| < \infty$ guarantees that $Ef(X)^+$ and $Ef(X)^-$ are finite. So using the result for nonnegative functions and linearity of expected value and integration:

$$Ef(X) = Ef(X)^+ - Ef(X)^- = \int_S f(y)^+ \mu(dy) - \int_S f(y)^- \mu(dy)$$
$$= \int_S f(y)\, \mu(dy)$$

which completes the proof. □

A consequence of Theorem 1.6.9 is that we can compute expected values of functions of random variables by performing integrals on the real line. Before we can treat some examples, we need to introduce the terminology for what we are about to compute. If k is a positive integer, then EX^k is called the **kth moment** of X. The first moment EX is usually called the **mean** and denoted by μ. If $EX^2 < \infty$, then the **variance** of X is defined to be $\text{var}\,(X) = E(X - \mu)^2$. To compute the variance the following formula is useful:

$$\begin{aligned} \text{var}\,(X) &= E(X - \mu)^2 \\ &= EX^2 - 2\mu EX + \mu^2 = EX^2 - \mu^2 \end{aligned} \tag{1.6.2}$$

From this it is immediate that

$$\text{var}\,(X) \le EX^2 \tag{1.6.3}$$

Here EX^2 is the expected value of X^2. When we want the square of EX, we will write $(EX)^2$. Since $E(aX+b) = aEX+b$ by (b) of Theorem 1.6.1, it follows easily from the definition that

$$\begin{aligned} \text{var}\,(aX + b) &= E(aX + b - E(aX + b))^2 \\ &= a^2 E(X - EX)^2 = a^2\, \text{var}\,(X) \end{aligned} \tag{1.6.4}$$

We turn now to concrete examples and leave the calculus in the first two examples to the reader. (Integrate by parts.)

Example 1.6.10 If X has an **exponential distribution** with rate 1, then

$$EX^k = \int_0^\infty x^k e^{-x} dx = k!$$

So the mean of X is 1 and variance is $EX^2 - (EX)^2 = 2 - 1^2 = 1$. If we let $Y = X/\lambda$, then by Exercise 1.2.5, Y has density $\lambda e^{-\lambda y}$ for $y \ge 0$, the **exponential density** with parameter λ. From (b) of Theorem 1.6.1 and (1.6.4), it follows that Y has mean $1/\lambda$ and variance $1/\lambda^2$.

Example 1.6.11 If X has a **standard normal distribution**,

$$EX = \int x(2\pi)^{-1/2} \exp(-x^2/2)\, dx = 0 \quad \text{(by symmetry)}$$
$$\text{var}\,(X) = EX^2 = \int x^2 (2\pi)^{-1/2} \exp(-x^2/2)\, dx = 1$$

If we let $\sigma > 0$, $\mu \in \mathbf{R}$, and $Y = \sigma X + \mu$, then (b) of Theorem 1.6.1 and (1.6.4), imply $EY = \mu$ and $\text{var}\,(Y) = \sigma^2$. By Exercise 1.2.5, Y has density

$$(2\pi\sigma^2)^{-1/2} \exp(-(y - \mu)^2/2\sigma^2)$$

the **normal distribution** with mean μ and variance σ^2.

We will next consider some discrete distributions. The first is very simple, but will be useful several times, so we record it here.

Example 1.6.12 We say that X has a **Bernoulli distribution** with parameter p if $P(X = 1) = p$ and $P(X = 0) = 1 - p$. Clearly,

$$EX = p \cdot 1 + (1 - p) \cdot 0 = p$$

Since $X^2 = X$, we have $EX^2 = EX = p$ and

$$\text{var}\,(X) = EX^2 - (EX)^2 = p - p^2 = p(1 - p)$$

Example 1.6.13 We say that X has a **Poisson distribution** with parameter λ if

$$P(X = k) = e^{-\lambda}\lambda^k/k! \text{ for } k = 0, 1, 2, \ldots$$

To evaluate the moments of the Poisson random variable, we use a little inspiration to observe that for $k \geq 1$

$$\begin{aligned} E(X(X-1)\cdots(X-k+1)) &= \sum_{j=k}^{\infty} j(j-1)\cdots(j-k+1)e^{-\lambda}\frac{\lambda^j}{j!} \\ &= \lambda^k \sum_{j=k}^{\infty} e^{-\lambda}\frac{\lambda^{j-k}}{(j-k)!} = \lambda^k \end{aligned}$$

where the equalities follow from (i) the facts that $j(j-1)\cdots(j-k+1) = 0$ when $j < k$, (ii) cancelling part of the factorial, and (iii) the fact that Poisson distribution has total mass 1. Using the last formula, it follows that $EX = \lambda$ while

$$\text{var}\,(X) = EX^2 - (EX)^2 = E(X(X-1)) + EX - \lambda^2 = \lambda$$

Example 1.6.14 N is said to have a **geometric distribution** with success probability $p \in (0, 1)$ if

$$P(N = k) = p(1 - p)^{k-1} \quad \text{for } k = 1, 2, \ldots$$

N is the number of independent trials needed to observe an event with probability p. Differentiating the identity

$$\sum_{k=0}^{\infty} (1 - p)^k = 1/p$$

and referring to Example A.5.6 for the justification gives

$$\begin{aligned} -\sum_{k=1}^{\infty} k(1 - p)^{k-1} &= -1/p^2 \\ \sum_{k=2}^{\infty} k(k-1)(1 - p)^{k-2} &= 2/p^3 \end{aligned}$$

From this it follows that

$$EN = \sum_{k=1}^{\infty} kp(1-p)^{k-1} = 1/p$$

$$EN(N-1) = \sum_{k=1}^{\infty} k(k-1)p(1-p)^{k-1} = 2(1-p)/p^2$$

$$\begin{aligned}\text{var}(N) &= EN^2 - (EN)^2 = EN(N-1) + EN - (EN)^2 \\ &= \frac{2(1-p)}{p^2} + \frac{p}{p^2} - \frac{1}{p^2} = \frac{1-p}{p^2}\end{aligned}$$

Exercises

1.6.1 Suppose φ is strictly convex, i.e., $>$ holds for $\lambda \in (0,1)$. Show that, under the assumptions of Theorem 1.6.2, $\varphi(EX) = E\varphi(X)$ implies $X = EX$ a.s.

1.6.2 Suppose $\varphi : \mathbf{R}^n \to \mathbf{R}$ is convex. Imitate the proof of Theorem 1.5.1 to show

$$E\varphi(X_1, \dots, X_n) \geq \varphi(EX_1, \dots, EX_n)$$

provided $E|\varphi(X_1, \dots, X_n)| < \infty$ and $E|X_i| < \infty$ for all i.

1.6.3 **Chebyshev's inequality is and is not sharp.** (i) Show that Theorem 1.6.4 is sharp by showing that if $0 < b \leq a$ are fixed there is an X with $EX^2 = b^2$ for which $P(|X| \geq a) = b^2/a^2$. (ii) Show that Theorem 1.6.4 is not sharp by showing that if X has $0 < EX^2 < \infty$, then

$$\lim_{a\to\infty} a^2 P(|X| \geq a)/EX^2 = 0$$

1.6.4 **One-sided Chebyshev bound.** (i) Let $a > b > 0$, $0 < p < 1$, and let X have $P(X = a) = p$ and $P(X = -b) = 1 - p$. Apply Theorem 1.6.4 to $\varphi(x) = (x + b)^2$ and conclude that if Y is any random variable with $EY = EX$ and $\text{var}(Y) = \text{var}(X)$, then $P(Y \geq a) \leq p$ and equality holds when $Y = X$.
(ii) Suppose $EY = 0$, $\text{var}(Y) = \sigma^2$, and $a > 0$. Show that $P(Y \geq a) \leq \sigma^2/(a^2 + \sigma^2)$, and there is a Y for which equality holds.

1.6.5 **Two nonexistent lower bounds.**
Show that: (i) if $\epsilon > 0$, $\inf\{P(|X| > \epsilon) : EX = 0, \text{var}(X) = 1\} = 0$.
(ii) if $y \geq 1$, $\sigma^2 \in (0,\infty)$, $\inf\{P(|X| > y) : EX = 1, \text{var}(X) = \sigma^2\} = 0$.

1.6.6 **A useful lower bound.** Let $Y \geq 0$ with $EY^2 < \infty$. Apply the Cauchy-Schwarz inequality to $Y1_{(Y>0)}$ and conclude

$$P(Y > 0) \geq (EY)^2/EY^2$$

1.6.7 Let $\Omega = (0,1)$ equipped with the Borel sets and Lebesgue measure. Let $\alpha \in (1,2)$ and $X_n = n^\alpha 1_{(1/(n+1),1/n)} \to 0$ a.s. Show that Theorem 1.6.8 can be applied with $h(x) = x$ and $g(x) = |x|^{2/\alpha}$, but the X_n are not dominated by an integrable function.

1.6.8 Suppose that the probability measure μ has $\mu(A) = \int_A f(x)\,dx$ for all $A \in \mathcal{R}$. Use the proof technique of Theorem 1.6.9 to show that for any g with $g \ge 0$ or $\int |g(x)|\,\mu(dx) < \infty$ we have

$$\int g(x)\,\mu(dx) = \int g(x) f(x)\,dx$$

1.6.9 **Inclusion-exclusion formula**. Let $A_1, A_2, \dots A_n$ be events and $A = \cup_{i=1}^n A_i$. Prove that $1_A = 1 - \prod_{i=1}^n (1 - 1_{A_i})$. Expand out the right-hand side, then take expected value to conclude

$$P\left(\cup_{i=1}^n A_i\right) = \sum_{i=1}^n P(A_i) - \sum_{i<j} P(A_i \cap A_j) + \sum_{i<j<k} P(A_i \cap A_j \cap A_k) - \cdots + (-1)^{n-1} P(\cap_{i=1}^n A_i)$$

1.6.10 **Bonferroni inequalities.** Let $A_1, A_2, \dots A_n$ be events and $A = \cup_{i=1}^n A_i$. Show that $1_A \le \sum_{i=1}^n 1_{A_i}$, etc. and then take expected values to conclude

$$P\left(\cup_{i=1}^n A_i\right) \le \sum_{i=1}^n P(A_i)$$

$$P\left(\cup_{i=1}^n A_i\right) \ge \sum_{i=1}^n P(A_i) - \sum_{i<j} P(A_i \cap A_j)$$

$$P\left(\cup_{i=1}^n A_i\right) \le \sum_{i=1}^n P(A_i) - \sum_{i<j} P(A_i \cap A_j) + \sum_{i<j<k} P(A_i \cap A_j \cap A_k)$$

In general, if we stop the inclusion–exclusion formula after an even (odd) number of sums, we get a(n) lower (upper) bound.

1.6.11 If $E|X|^k < \infty$, then for $0 < j < k$, $E|X|^j < \infty$, and furthermore

$$E|X|^j \le (E|X|^k)^{j/k}$$

1.6.12 Apply Jensen's inequality with $\varphi(x) = e^x$ and $P(X = \log\, y_m) = p(m)$ to conclude that if $\sum_{m=1}^n p(m) = 1$ and $p(m), y_m > 0$, then

$$\sum_{m=1}^n p(m) y_m \ge \prod_{m=1}^n y_m^{p(m)}$$

When $p(m) = 1/n$, this says the arithmetic mean exceeds the geometric mean.

1.6.13 If $EX_1^- < \infty$ and $X_n \uparrow X$, then $EX_n \uparrow EX$.

1.6.14 Let $X \ge 0$ but do NOT assume $E(1/X) < \infty$. Show

$$\lim_{y\to\infty} yE(1/X; X > y) = 0, \qquad \lim_{y\downarrow 0} yE(1/X; X > y) = 0.$$

1.6.15 If $X_n \ge 0$, then $E(\sum_{n=0}^\infty X_n) = \sum_{n=0}^\infty EX_n$.

1.6.16 If X is integrable and A_n are disjoint sets with union A, then

$$\sum_{n=0}^{\infty} E(X; A_n) = E(X; A)$$

i.e., the sum converges absolutely and has the value on the right.

1.7 Product Measures, Fubini's Theorem

Let $(X, \mathcal{A}, \mu_1)$ and $(Y, \mathcal{B}, \mu_2)$ be two σ-finite measure spaces. Let

$$\Omega = X \times Y = \{(x, y) : x \in X, y \in Y\}$$
$$\mathcal{S} = \{A \times B : A \in \mathcal{A}, B \in \mathcal{B}\}$$

Sets in $\mathcal{S}$ are called **rectangles**. It is easy to see that $\mathcal{S}$ is a semi-algebra:

$$(A \times B) \cap (C \times D) = (A \cap C) \times (B \cap D)$$
$$(A \times B)^c = (A^c \times B) \cup (A \times B^c) \cup (A^c \times B^c)$$

Let $\mathcal{F} = \mathcal{A} \times \mathcal{B}$ be the σ-algebra generated by $\mathcal{S}$.

Theorem 1.7.1 *There is a unique measure μ on $\mathcal{F}$ with*

$$\mu(A \times B) = \mu_1(A)\mu_2(B)$$

Notation μ is often denoted by $\mu_1 \times \mu_2$.

Proof By Theorem 1.1.9 it is enough to show that if $A \times B = +_i(A_i \times B_i)$ is a finite or countable disjoint union, then

$$\mu(A \times B) = \sum_i \mu(A_i \times B_i)$$

For each $x \in A$, let $I(x) = \{i : x \in A_i\}$. $B = +_{i \in I(x)} B_i$, so

$$1_A(x)\mu_2(B) = \sum_i 1_{A_i}(x)\mu_2(B_i)$$

Integrating with respect to μ_1 and using Exercise 1.5.6 gives

$$\mu_1(A)\mu_2(B) = \sum_i \mu_1(A_i)\mu_2(B_i)$$

which proves the result. □

Using Theorem 1.7.1 and induction, it follows that if $(\Omega_i, \mathcal{F}_i, \mu_i)$, $i = 1, \ldots, n$, are σ-finite measure spaces and $\Omega = \Omega_1 \times \cdots \times \Omega_n$, there is a unique measure μ on the σ-algebra $\mathcal{F}$ generated by sets of the form $A_1 \times \cdots \times A_n$, $A_i \in \mathcal{F}_i$, which has

$$\mu(A_1 \times \cdots \times A_n) = \prod_{m=1}^{n} \mu_m(A_m)$$

When $(\Omega_i, \mathcal{F}_i, \mu_i) = (\mathbf{R}, \mathcal{R}, \lambda)$ for all i, the result is Lebesgue measure on the Borel subsets of n dimensional Euclidean space $\mathbf{R}^n$.

Returning to the case in which $(\Omega, \mathcal{F}, \mu)$ is the product of two measure spaces, $(X, \mathcal{A}, \mu)$ and $(Y, \mathcal{B}, \nu)$, our next goal is to prove:

Theorem 1.7.2 (Fubini's theorem) *If $f \geq 0$ or $\int |f|\, d\mu < \infty$, then*

$$(*) \qquad \int_X \int_Y f(x,y)\, \mu_2(dy)\, \mu_1(dx) = \int_{X \times Y} f\, d\mu = \int_Y \int_X f(x,y)\, \mu_1(dx)\, \mu_2(dy)$$

Proof We will prove only the first equality, since the second follows by symmetry. Two technical things that need to be proved before we can assert that the first integral makes sense are:

When x is fixed, $y \to f(x,y)$ is $\mathcal{B}$ measurable.

$x \to \int_Y f(x,y)\mu_2(dy)$ is $\mathcal{A}$ measurable.

We begin with the case $f = 1_E$. Let $E_x = \{y : (x,y) \in E\}$ be the **cross-section** at x.

Lemma 1.7.3 *If $E \in \mathcal{F}$, then $E_x \in \mathcal{B}$.*

Proof $(E^c)_x = (E_x)^c$ and $(\cup_i E_i)_x = \cup_i (E_i)_x$, so if $\mathcal{E}$ is the collection of sets E for which $E_x \in \mathcal{B}$, then $\mathcal{E}$ is a σ-algebra. Since $\mathcal{E}$ contains the rectangles, the result follows. □

Lemma 1.7.4 *If $E \in \mathcal{F}$, then $g(x) \equiv \mu_2(E_x)$ is $\mathcal{A}$ measurable and*

$$\int_X g\, d\mu_1 = \mu(E)$$

Notice that it is not obvious that the collection of sets for which the conclusion is true is a σ-algebra since $\mu(E_1 \cup E_2) = \mu(E_1) + \mu(E_2) - \mu(E_1 \cap E_2)$. Dynkin's $\pi - \lambda$ Theorem (A.1.4) was tailor-made for situations like this.

Proof If conclusions hold for E_n and $E_n \uparrow E$, then Theorem 1.3.7 and the monotone convergence theorem imply that they hold for E. Since μ_1 and μ_2 are σ-finite, it is enough then to prove the result for $E \subset F \times G$ with $\mu_1(F) < \infty$ and $\mu_2(G) < \infty$, or taking $\Omega = F \times G$ we can suppose without loss of generality that $\mu(\Omega) < \infty$. Let $\mathcal{L}$ be the collection of sets E for which the conclusions hold. We will now check that $\mathcal{L}$ is a λ-system. Property (i) of a λ-system is trivial. (iii) follows from the first sentence in the proof. To check (ii) we observe that

$$\mu_2((A - B)_x) = \mu_2(A_x - B_x) = \mu_2(A_x) - \mu_2(B_x)$$

and integrating over x gives the second conclusion. Since $\mathcal{L}$ contains the rectangles, a π-system that generates $\mathcal{F}$, the desired result follows from the $\pi - \lambda$ theorem. □

We are now ready to prove Theorem 1.7.2 by verifying it in four increasingly more general special cases.

Case 1. If $E \in \mathcal{F}$ and $f = 1_E$, then $(*)$ follows from Lemma 1.7.4.

Case 2. Since each integral is linear in f, it follows that $(*)$ holds for simple functions.

Case 3. Now if $f \geq 0$ and we let $f_n(x) = ([2^n f(x)]/2^n) \wedge n$, where $[x] =$ the largest integer $\leq x$, then the f_n are simple and $f_n \uparrow f$, so it follows from the monotone convergence theorem that $(*)$ holds for all $f \geq 0$.

Case 4. The general case now follows by writing $f(x) = f(x)^+ - f(x)^-$ and applying Case 3 to f^+, f^-, and $|f|$. □

To illustrate why the various hypotheses of Theorem 1.7.2 are needed, we will now give some examples where the conclusion fails.

Example 1.7.5 Let $X = Y = \{1,2,\dots\}$ with $\mathcal{A} = \mathcal{B} =$ all subsets and $\mu_1 = \mu_2 =$ counting measure. For $m \geq 1$, let $f(m,m) = 1$ and $f(m+1,m) = -1$, and let $f(m,n) = 0$ otherwise. We claim that

$$\sum_m \sum_n f(m,n) = 1 \quad \text{but} \quad \sum_n \sum_m f(m,n) = 0$$

A picture is worth several dozen words:

$$\begin{array}{cccccc} & \vdots & \vdots & \vdots & \vdots & \\ & 0 & 0 & 0 & 1 & \dots \\ \uparrow & 0 & 0 & 1 & -1 & \dots \\ n & 0 & 1 & -1 & 0 & \dots \\ & 1 & -1 & 0 & 0 & \dots \\ & & m & \to & & \end{array}$$

In words, if we sum the columns first, the first one gives us a 1 and the others 0, while if we sum the rows each one gives us a 0.

Example 1.7.6 Let $X = (0,1)$, $Y = (1,\infty)$, both equipped with the Borel sets and Lebesgue measure. Let $f(x,y) = e^{-xy} - 2e^{-2xy}$.

$$\int_0^1 \int_1^\infty f(x,y)\,dy\,dx = \int_0^1 x^{-1}(e^{-x} - e^{-2x})\,dx > 0$$

$$\int_1^\infty \int_0^1 f(x,y)\,dx\,dy = \int_1^\infty y^{-1}(e^{-2y} - e^{-y})\,dy < 0$$

The next example indicates why μ_1 and μ_2 must be σ-finite.

Example 1.7.7 Let $X = (0,1)$ with $\mathcal{A} =$ the Borel sets and $\mu_1 =$ Lebesgue measure. Let $Y = (0,1)$ with $\mathcal{B} =$ all subsets and $\mu_2 =$ counting measure. Let $f(x,y) = 1$ if $x = y$ and 0 otherwise

$$\int_Y f(x,y)\,\mu_2(dy) = 1 \quad \text{for all } x \text{ so} \quad \int_X \int_Y f(x,y)\,\mu_2(dy)\,\mu_1(dx) = 1$$

$$\int_X f(x,y)\,\mu_1(dx) = 0 \quad \text{for all } y \text{ so} \quad \int_Y \int_X f(x,y)\,\mu_1(dy)\,\mu_2(dx) = 0$$

Our last example shows that measurability is important or maybe that some of the axioms of set theory are not as innocent as they seem.

Example 1.7.8 By the axiom of choice and the continuum hypothesis one can define an order relation $<'$ on (0,1) so that $\{x : x <' y\}$ is countable for each y. Let $X = Y = (0,1)$, let

$\mathcal{A} = \mathcal{B} =$ the Borel sets and $\mu_1 = \mu_2 =$ Lebesgue measure. Let $f(x,y) = 1$ if $x <' y$, $= 0$ otherwise. Since $\{x : x <' y\}$ and $\{y : x <' y\}^c$ are countable,

$$\int_X f(x,y)\,\mu_1(dx) = 0 \quad \text{for all } y$$
$$\int_Y f(x,y)\,\mu_2(dy) = 1 \quad \text{for all } x$$

Exercises

1.7.1 If $\int_X \int_Y |f(x,y)|\mu_2(dy)\mu_1(dx) < \infty$, then

$$\int_X \int_Y f(x,y)\mu_2(dy)\mu_1(dx) = \int_{X\times Y} f\,d(\mu_1 \times \mu_2) = \int_Y \int_X f(x,y)\mu_1(dx)\mu_2(dy)$$

Corollary Let $X = \{1,2,\ldots\}$, $\mathcal{A} =$ all subsets of X, and $\mu_1 =$ counting measure. If $\sum_n \int |f_n| d\mu < \infty$, then $\sum_n \int f_n\,d\mu = \int \sum_n f_n\,d\mu$.

1.7.2 Let $g \geq 0$ be a measurable function on $(X, \mathcal{A}, \mu)$. Use Theorem 1.7.2 to conclude that

$$\int_X g\,d\mu = (\mu \times \lambda)(\{(x,y) : 0 \leq y < g(x)\}) = \int_0^\infty \mu(\{x : g(x) > y\})\,dy$$

1.7.3 Let F, G be Stieltjes measure functions and let μ, ν be the corresponding measures on $(\mathbf{R}, \mathcal{R})$. Show that

(i) $\int_{(a,b]}\{F(y) - F(a)\}dG(y) = (\mu \times \nu)(\{(x,y) : a < x \leq y \leq b\})$

(ii) $\int_{(a,b]} F(y)\,dG(y) + \int_{(a,b]} G(y)\,dF(y)$

$$= F(b)G(b) - F(a)G(a) + \sum_{x\in(a,b]} \mu(\{x\})\nu(\{x\})$$

(iii) If $F = G$ is continuous, then $\int_{(a,b]} 2F(y)dF(y) = F^2(b) - F^2(a)$.

To see the second term in (ii) is needed, let $F(x) = G(x) = 1_{[0,\infty)}(x)$ and $a < 0 < b$.

1.7.4 Let μ be a finite measure on $\mathbf{R}$ and $F(x) = \mu((-\infty, x])$. Show that

$$\int (F(x+c) - F(x))\,dx = c\mu(\mathbf{R})$$

1.7.5 Show that $e^{-xy} \sin x$ is integrable in the strip $0 < x < a$, $0 < y$. Perform the double integral in the two orders to get:

$$\int_0^a \frac{\sin x}{x}\,dx = \arctan(a) - (\cos a)\int_0^\infty \frac{e^{-ay}}{1+y^2}\,dy - (\sin a)\int_0^\infty \frac{ye^{-ay}}{1+y^2}\,dy$$

and replace $1 + y^2$ by 1 to conclude $\left|\int_0^a (\sin x)/x\,dx - \arctan(a)\right| \leq 2/a$ for $a \geq 1$.

2

Laws of Large Numbers

2.1 Independence

Measure theory ends and probability begins with the definition of independence. We begin with what is hopefully a familiar definition and then work our way up to a definition that is appropriate for our current setting.

Two events A and B are **independent** if $P(A \cap B) = P(A)P(B)$.

Two random variables X and Y are **independent** if for all $C, D \in \mathcal{R}$,

$$P(X \in C, Y \in D) = P(X \in C)P(Y \in D)$$

i.e., the events $A = \{X \in C\}$ and $B = \{Y \in D\}$ are independent.

Two σ-fields $\mathcal{F}$ and $\mathcal{G}$ are **independent** if for all $A \in \mathcal{F}$ and $B \in \mathcal{G}$ the events A and B are independent.

As the next result shows, the second definition is a special case of the third.

Theorem 2.1.1 *(i) If X and Y are independent then $\sigma(X)$ and $\sigma(Y)$ are. (ii) Conversely, if $\mathcal{F}$ and $\mathcal{G}$ are independent, $X \in \mathcal{F}$, and $Y \in \mathcal{G}$, then X and Y are independent.*

Proof (i) If $A \in \sigma(X)$, then it follows from the definition of $\sigma(X)$ that $A = \{X \in C\}$ for some $C \in \mathcal{R}$. Likewise, if $B \in \sigma(Y)$, then $B = \{Y \in D\}$ for some $D \in \mathcal{R}$, so using these facts and the independence of X and Y,

$$P(A \cap B) = P(X \in C, Y \in D) = P(X \in C)P(Y \in D) = P(A)P(B)$$

(ii) Conversely, if $X \in \mathcal{F}$, $Y \in \mathcal{G}$, and $C, D \in \mathcal{R}$, it follows from the definition of measurability that $\{X \in C\} \in \mathcal{F}$, $\{Y \in D\} \in \mathcal{G}$. Since $\mathcal{F}$ and $\mathcal{G}$ are independent, it follows that $P(X \in C, Y \in D) = P(X \in C)P(Y \in D)$. □

The first definition is, in turn, a special case of the second.

Theorem 2.1.2 *(i) If A and B are independent, then so are A^c and B, A and B^c, and A^c and B^c. (ii) Conversely events A and B are independent if and only if their indicator random variables 1_A and 1_B are independent.*

Proof (i) Subtracting $P(A \cap B) = P(A)P(B)$ from $P(B) = P(B)$ shows $P(A^c \cap B) = P(A^c)P(B)$. The second and third conclusions follow by applying the first one to the pairs of independent events (B, A) and (A, B^c).

(ii) If $C, D \in \mathcal{R}$, then $\{1_A \in C\} \in \{\emptyset, A, A^c, \Omega\}$ and $\{1_B \in D\} \in \{\emptyset, B, B^c, \Omega\}$, so there are 16 things to check. When either set involved is $\emptyset$ or Ω, the equality holds, so there are only four cases to worry about and they are all covered by (i). □

In view of the fact that the first definition is a special case of the second, which is a special case of the third, we take things in the opposite order when we say what it means for several things to be independent. We begin by reducing to the case of finitely many objects. An infinite collection of objects (σ-fields, random variables, or sets) is said to be independent if every finite subcollection is.

σ-fields $\mathcal{F}_1, \mathcal{F}_2, \ldots, \mathcal{F}_n$ are **independent** if whenever $A_i \in \mathcal{F}_i$ for $i = 1, \ldots, n$, we have

$$P\left(\cap_{i=1}^n A_i\right) = \prod_{i=1}^n P(A_i)$$

Random variables $X_1, \ldots, X_n$ are **independent** if whenever $B_i \in \mathcal{R}$ for $i = 1, \ldots, n$ we have

$$P\left(\cap_{i=1}^n \{X_i \in B_i\}\right) = \prod_{i=1}^n P(X_i \in B_i)$$

Sets $A_1, \ldots, A_n$ are **independent** if whenever $I \subset \{1, \ldots n\}$ we have

$$P\left(\cap_{i \in I} A_i\right) = \prod_{i \in I} P(A_i)$$

At first glance, it might seem that the last definition does not match the other two. However, if you think about it for a minute, you will see that if the indicator variables 1_{A_i}, $1 \le i \le n$ are independent and we take $B_i = \{1\}$ for $i \in I$, and $B_i = \mathbf{R}$ for $i \notin I$, then the condition in the definition results. Conversely:

Theorem 2.1.3 *Let $A_1, A_2, \ldots, A_n$ be independent. (i) $A_1^c, A_2, \ldots, A_n$ are independent; (ii) $1_{A_1}, \ldots, 1_{A_n}$ are independent.*

Proof (i) Let $B_1 = A_1^c$ and $B_i = A_i$ for $i > 1$. If $I \subset \{1, \ldots, n\}$ does not contain 1, it is clear that $P(\cap_{i \in I} B_i) = \prod_{i \in I} P(B_i)$. Suppose now that $1 \in I$ and let $J = I - \{1\}$. Subtracting $P(\cap_{i \in I} A_i) = \prod_{i \in I} P(A_i)$ from $P(\cap_{i \in J} A_i) = \prod_{i \in J} P(A_i)$ gives $P(A_1^c \cap \cap_{i \in J} A_i) = P(A_1^c) \prod_{i \in J} P(A_i)$.

(ii) Iterating (i) we see that if $B_i \in \{A_i, A_i^c\}$, then $B_1, \ldots, B_n$ are independent. Thus if $C_i \in \{A_i, A_i^c, \Omega\}$ $P(\cap_{i=1}^n C_i) = \prod_{i=1}^n P(C_i)$. The last equality holds trivially if some $C_i = \emptyset$, so noting $1_{A_i} \in \{\emptyset, A_i, A_i^c, \Omega\}$ the desired result follows. □

One of the first things to understand about the definition of independent events is that it is not enough to assume $P(A_i \cap A_j) = P(A_i)P(A_j)$ for all $i \neq j$. A sequence of events $A_1, \ldots, A_n$ with the last property is called **pairwise independent**. It is clear that independent events are pairwise independent. The next example shows that the converse is not true.

Example 2.1.4 Let X_1, X_2, X_3 be independent random variables with

$$P(X_i = 0) = P(X_i = 1) = 1/2$$

Let $A_1 = \{X_2 = X_3\}$, $A_2 = \{X_3 = X_1\}$, and $A_3 = \{X_1 = X_2\}$. These events are pairwise independent since if $i \neq j$, then

$$P(A_i \cap A_j) = P(X_1 = X_2 = X_3) = 1/4 = P(A_i)P(A_j)$$

but they are not independent since

$$P(A_1 \cap A_2 \cap A_3) = 1/4 \neq 1/8 = P(A_1)P(A_2)P(A_3)$$

In order to show that random variables X and Y are independent, we have to check that $P(X \in A, Y \in B) = P(X \in A)P(Y \in B)$ for all Borel sets A and B. Since there are a lot of Borel sets, our next topic is:

2.1.1 Sufficient Conditions for Independence

Our main result is Theorem 2.1.7. To state that result, we need a definition that generalizes all our earlier definitions.

Collections of sets $\mathcal{A}_1, \mathcal{A}_2, \ldots, \mathcal{A}_n \subset \mathcal{F}$ are said to be **independent** if whenever $A_i \in \mathcal{A}_i$ and $I \subset \{1, \ldots, n\}$, we have $P(\cap_{i \in I} A_i) = \prod_{i \in I} P(A_i)$.

If each collection is a single set, i.e., $\mathcal{A}_i = \{A_i\}$, this definition reduces to the one for sets.

Lemma 2.1.5 *Without loss of generality we can suppose each $\mathcal{A}_i$ contains Ω. In this case, the condition is equivalent to*

$$P\left(\cap_{i=1}^n A_i\right) = \prod_{i=1}^n P(A_i) \quad \text{whenever } A_i \in \mathcal{A}_i$$

since we can set $A_i = \Omega$ for $i \notin I$.

Proof If $\mathcal{A}_1, \mathcal{A}_2, \ldots, \mathcal{A}_n$ are independent and $\bar{\mathcal{A}}_i = \mathcal{A}_i \cup \{\Omega\}$ then $\bar{\mathcal{A}}_1, \bar{\mathcal{A}}_2, \ldots, \bar{\mathcal{A}}_n$ are independent, since if $A_i \in \bar{\mathcal{A}}_i$ and $I = \{j : A_j = \Omega\}$ $\cap_i A_i = \cap_{i \in I} A_i$. □

The proof of Theorem 2.1.7 is based on Dynkin's $\pi - \lambda$ theorem. To state this result, we need two definitions. We say that $\mathcal{A}$ is a **π-system** if it is closed under intersection, i.e., if $A, B \in \mathcal{A}$, then $A \cap B \in \mathcal{A}$. We say that $\mathcal{L}$ is a **λ-system** if: (i) $\Omega \in \mathcal{L}$. (ii) If $A, B \in \mathcal{L}$ and $A \subset B$, then $B - A \in \mathcal{L}$. (iii) If $A_n \in \mathcal{L}$ and $A_n \uparrow A$, then $A \in \mathcal{L}$.

Theorem 2.1.6 ($\pi - \lambda$ Theorem) *If $\mathcal{P}$ is a π-system and $\mathcal{L}$ is a λ-system that contains $\mathcal{P}$, then $\sigma(\mathcal{P}) \subset \mathcal{L}$.*

The proof is hidden away in Section A.1 of the Appendix.

Theorem 2.1.7 *Suppose $\mathcal{A}_1, \mathcal{A}_2, \ldots, \mathcal{A}_n$ are independent and each $\mathcal{A}_i$ is a π-system. Then $\sigma(\mathcal{A}_1), \sigma(\mathcal{A}_2), \ldots, \sigma(\mathcal{A}_n)$ are independent.*

Proof Let $A_2, \ldots, A_n$ be sets with $A_i \in \mathcal{A}_i$, let $F = A_2 \cap \cdots \cap A_n$ and let $\mathcal{L} = \{A : P(A \cap F) = P(A)P(F)\}$. Since $P(\Omega \cap F) = P(\Omega)P(F)$, $\Omega \in \mathcal{L}$. To check (ii) of the definition of a λ-system, we note that if $A, B \in \mathcal{L}$ with $A \subset B$ then $(B - A) \cap F = (B \cap F) - (A \cap F)$. So using (i) in Theorem 1.1.1, $A, B \in \mathcal{L}$ and then (i) in Theorem 1.1.1 again:

$$P((B-A)\cap F) = P(B\cap F) - P(A\cap F) = P(B)P(F) - P(A)P(F)$$
$$= \{P(B) - P(A)\}P(F) = P(B-A)P(F)$$

and we have $B - A \in \mathcal{L}$. To check (iii) let $B_k \in \mathcal{L}$ with $B_k \uparrow B$ and note that $(B_k \cap F) \uparrow (B \cap F)$ so using (iii) in Theorem 1.1.1, the fact that $B_k \in \mathcal{L}$, and then (iii) in Theorem 1.1.1 again:

$$P(B\cap F) = \lim_k P(B_k \cap F) = \lim_k P(B_k)P(F) = P(B)P(F)$$

Applying the $\pi - \lambda$ theorem now gives $\mathcal{L} \supset \sigma(\mathcal{A}_1)$. It follows that if $A_1 \in \sigma(\mathcal{A}_1)$ and $A_i \in \mathcal{A}_i$ for $2 \le i \le n$, then

$$P(\cap_{i=1}^n A_i) = P(A_1)P(\cap_{i=2}^n A_i) = \prod_{i=1}^n P(A_i)$$

Using Lemma 2.1.5 now, we have:

$(*)$ If $\mathcal{A}_1, \mathcal{A}_2, \ldots, \mathcal{A}_n$ are independent then $\sigma(\mathcal{A}_1), \mathcal{A}_2, \ldots, \mathcal{A}_n$ are independent.

Applying $(*)$ to $\mathcal{A}_2, \ldots, \mathcal{A}_n, \sigma(\mathcal{A}_1)$ (which are independent since the definition is unchanged by permuting the order) shows that $\sigma(\mathcal{A}_2), \mathcal{A}_3, \ldots, \mathcal{A}_n, \sigma(\mathcal{A}_1)$ are independent, and after n iterations we have the desired result. □

Remark The reader should note that it is not easy to show that if $A, B \in \mathcal{L}$, then $A \cap B \in \mathcal{L}$, or $A \cup B \in \mathcal{L}$, but it is easy to check that if $A, B \in \mathcal{L}$ with $A \subset B$, then $B - A \in \mathcal{L}$.

Having worked to establish Theorem 2.1.7, we get several corollaries.

Theorem 2.1.8 *In order for $X_1, \ldots, X_n$ to be independent, it is sufficient that for all $x_1, \ldots, x_n \in (-\infty, \infty]$*

$$P(X_1 \le x_1, \ldots, X_n \le x_n) = \prod_{i=1}^n P(X_i \le x_i)$$

Proof Let $\mathcal{A}_i =$ the sets of the form $\{X_i \le x_i\}$. Since

$$\{X_i \le x\} \cap \{X_i \le y\} = \{X_i \le x \wedge y\},$$

where $(x \wedge y)_i = x_i \wedge y_i = \min\{x_i, y_i\}$. $\mathcal{A}_i$ is a π-system. Since we have allowed $x_i = \infty$, $\Omega \in \mathcal{A}_i$. Exercise 1.3.1 implies $\sigma(\mathcal{A}_i) = \sigma(X_i)$, so the result follows from Theorem 2.1.7. □

The last result expresses independence of random variables in terms of their distribution functions. Exercises 2.1.1 and 2.1.2 treat density functions and discrete random variables.

Our next goal is to prove that functions of disjoint collections of independent random variables are independent. See Theorem 2.1.10 for the precise statement. First, we will prove an analogous result for σ-fields.

Theorem 2.1.9 *Suppose $\mathcal{F}_{i,j}, 1 \le i \le n, 1 \le j \le m(i)$ are independent and let $\mathcal{G}_i = \sigma(\cup_j \mathcal{F}_{i,j})$. Then $\mathcal{G}_1, \ldots, \mathcal{G}_n$ are independent.*

Proof Let $\mathcal{A}_i$ be the collection of sets of the form $\cap_j A_{i,j}$ where $A_{i,j} \in \mathcal{F}_{i,j}$. $\mathcal{A}_i$ is a π-system that contains Ω and contains $\cup_j \mathcal{F}_{i,j}$ so Theorem 2.1.7 implies $\sigma(\mathcal{A}_i) = \mathcal{G}_i$ are independent. □

Theorem 2.1.10 *If for $1 \leq i \leq n$, $1 \leq j \leq m(i)$, $X_{i,j}$ are independent and $f_i : \mathbf{R}^{m(i)} \to \mathbf{R}$ are measurable, then $f_i(X_{i,1}, \ldots, X_{i,m(i)})$ are independent.*

Proof Let $\mathcal{F}_{i,j} = \sigma(X_{i,j})$ and $\mathcal{G}_i = \sigma(\cup_j \mathcal{F}_{i,j})$. Since $f_i(X_{i,1}, \ldots, X_{i,m(i)}) \in \mathcal{G}_i$, the desired result follows from Theorem 2.1.9 and Exercise 2.1.1. □

A concrete special case of Theorem 2.1.10 that we will use shortly is: if $X_1, \ldots, X_n$ are independent, then $X = X_1$ and $Y = X_2 \cdots X_n$ are independent. Later, when we study sums $S_m = X_1 + \cdots + X_m$ of independent random variables $X_1, \ldots, X_n$, we will use Theorem 2.1.10 to conclude that if $m < n$, then $S_n - S_m$ is independent of the indicator function of the event $\{\max_{1 \leq k \leq m} S_k > x\}$.

2.1.2 Independence, Distribution, and Expectation

Our next goal is to obtain formulas for the distribution and expectation of independent random variables.

Theorem 2.1.11 *Suppose $X_1, \ldots, X_n$ are independent random variables and X_i has distribution μ_i, then $(X_1, \ldots, X_n)$ has distribution $\mu_1 \times \cdots \times \mu_n$.*

Proof Using the definitions of (i) $A_1 \times \cdots \times A_n$, (ii) independence, (iii) μ_i, and (iv) $\mu_1 \times \cdots \times \mu_n$

$$P((X_1, \ldots, X_n) \in A_1 \times \cdots \times A_n) = P(X_1 \in A_1, \ldots, X_n \in A_n)$$
$$= \prod_{i=1}^n P(X_i \in A_i) = \prod_{i=1}^n \mu_i(A_i) = \mu_1 \times \cdots \times \mu_n(A_1 \times \cdots \times A_n)$$

The last formula shows that the distribution of $(X_1, \ldots, X_n)$ and the measure $\mu_1 \times \cdots \times \mu_n$ agree on sets of the form $A_1 \times \cdots \times A_n$, a π-system that generates $\mathcal{R}^n$. So Theorem 2.1.6 implies they must agree. □

Theorem 2.1.12 *Suppose X and Y are independent and have distributions μ and ν. If $h : \mathbf{R}^2 \to \mathbf{R}$ is a measurable function with $h \geq 0$ or $E|h(X,Y)| < \infty$, then*

$$Eh(X,Y) = \iint h(x,y)\, \mu(dx)\, \nu(dy)$$

In particular, if $h(x,y) = f(x)g(y)$ where $f, g : \mathbf{R} \to \mathbf{R}$ are measurable functions with $f, g \geq 0$ or $E|f(X)|$ and $E|g(Y)| < \infty$, then

$$Ef(X)g(Y) = Ef(X) \cdot Eg(Y)$$

Proof Using Theorem 1.6.9 and then Fubini's theorem (Theorem 1.7.2), we have

$$Eh(X,Y) = \int_{\mathbf{R}^2} h\, d(\mu \times \nu) = \iint h(x,y)\, \mu(dx)\, \nu(dy)$$

To prove the second result, we start with the result when $f, g \geq 0$. In this case, using the first result, the fact that $g(y)$ does not depend on x and then Theorem 1.6.9 twice we get

$$Ef(X)g(Y) = \iint f(x)g(y)\,\mu(dx)\,\nu(dy) = \int g(y) \int f(x)\,\mu(dx)\,\nu(dy)$$
$$= \int Ef(X)g(y)\,\nu(dy) = Ef(X)Eg(Y)$$

Applying the result for nonnegative f and g to $|f|$ and $|g|$ shows $E|f(X)g(Y)| = E|f(X)|E|g(Y)| < \infty$, and we can repeat the last argument to prove the desired result. □

From Theorem 2.1.12, it is only a small step to:

Theorem 2.1.13 *If $X_1, \ldots, X_n$ are independent and have (a) $X_i \geq 0$ for all i, or (b) $E|X_i| < \infty$ for all i, then*

$$E\left(\prod_{i=1}^{n} X_i\right) = \prod_{i=1}^{n} EX_i$$

i.e., the expectation on the left exists and has the value given on the right.

Proof $X = X_1$ and $Y = X_2 \cdots X_n$ are independent by Theorem 2.1.10, so taking $f(x) = |x|$ and $g(y) = |y|$, we have $E|X_1 \cdots X_n| = E|X_1|E|X_2 \cdots X_n|$, and it follows by induction that if $1 \leq m \leq n$

$$E|X_m \cdots X_n| = \prod_{i=m}^{n} E|X_k|$$

If the $X_i \geq 0$, then $|X_i| = X_i$ and the desired result follows from the special case $m = 1$. To prove the result in general, note that the special case $m = 2$ implies $E|Y| = E|X_2 \cdots X_n| < \infty$, so using Theorem 2.1.12 with $f(x) = x$ and $g(y) = y$ shows $E(X_1 \cdots X_n) = EX_1 \cdot E(X_2 \cdots X_n)$, and the desired result follows by induction. □

Example 2.1.14 It can happen that $E(XY) = EX \cdot EY$ without the variables being independent. Suppose the joint distribution of X and Y is given by the following table

			Y	
		1	0	−1
	1	0	a	0
X	0	b	c	b
	−1	0	a	0

where $a, b > 0$, $c \geq 0$, and $2a + 2b + c = 1$. Things are arranged so that $XY \equiv 0$. Symmetry implies $EX = 0$ and $EY = 0$, so $E(XY) = 0 = EXEY$. The random variables are not independent since

$$P(X = 1, Y = 1) = 0 < ab = P(X = 1)P(Y = 1)$$

Two random variables X and Y with $EX^2, EY^2 < \infty$ that have $EXY = EXEY$ are said to be **uncorrelated**. The finite second moments are needed so that we know $E|XY| < \infty$ by the Cauchy-Schwarz inequality.

2.1.3 Sums of Independent Random Variables

Theorem 2.1.15 *If X and Y are independent, $F(x) = P(X \le x)$, and $G(y) = P(Y \le y)$, then*

$$P(X+Y \le z) = \int F(z-y)\, dG(y)$$

The integral on the right-hand side is called the **convolution** of F and G and is denoted $F * G(z)$. The meaning of $dG(y)$ will be explained in the proof.

Proof Let $h(x,y) = 1_{(x+y \le z)}$. Let μ and ν be the probability measures with distribution functions F and G. Since for fixed y

$$\int h(x,y)\, \mu(dx) = \int 1_{(-\infty, z-y]}(x)\, \mu(dx) = F(z-y)$$

using Theorem 2.1.12 gives

$$\begin{aligned} P(X+Y \le z) &= \int\int 1_{(x+y \le z)}\, \mu(dx)\, \nu(dy) \\ &= \int F(z-y)\, \nu(dy) = \int F(z-y)\, dG(y) \end{aligned}$$

The last equality is just a change of notation: We regard $dG(y)$ as a shorthand for "integrate with respect to the measure ν with distribution function G." □

To treat concrete examples, we need a special case of Theorem 2.1.15.

Theorem 2.1.16 *Suppose that X with density f and Y with distribution function G are independent. Then $X+Y$ has density*

$$h(x) = \int f(x-y)\, dG(y)$$

When Y has density g, the last formula can be written as

$$h(x) = \int f(x-y)\, g(y)\, dy$$

Proof From Theorem 2.1.15, the definition of density function, and Fubini's theorem (Theorem 1.7.2), which is justified since everything is nonnegative, we get

$$\begin{aligned} P(X+Y \le z) &= \int F(z-y)\, dG(y) = \int \int_{-\infty}^{z} f(x-y)\, dx\, dG(y) \\ &= \int_{-\infty}^{z} \int f(x-y)\, dG(y)\, dx \end{aligned}$$

The last equation says that $X+Y$ has density $h(x) = \int f(x-y) dG(y)$. The second formula follows from the first when we recall the meaning of $dG(y)$ given in the proof of Theorem 2.1.15 and use Exercise 1.6.8. □

Theorem 2.1.16 plus some ugly calculus allows us to treat two standard examples. These facts should be familiar from undergraduate probability.

Example 2.1.17 The **gamma density** with parameters α and λ is given by

$$f(x) = \begin{cases} \lambda^\alpha x^{\alpha-1} e^{-\lambda x}/\Gamma(\alpha) & \text{for } x \geq 0 \\ 0 & \text{for } x < 0 \end{cases}$$

where $\Gamma(\alpha) = \int_0^\infty x^{\alpha-1} e^{-x}\, dx$.

Theorem 2.1.18 *If $X = \text{gamma}(\alpha,\lambda)$ and $Y = \text{gamma}(\beta,\lambda)$ are independent then $X + Y$ is gamma$(\alpha+\beta,\lambda)$. Consequently, if $X_1, \ldots X_n$ are independent exponential(λ) r.v.'s, then $X_1 + \cdots + X_n$ has a gamma(n,λ) distribution.*

Proof Writing $f_{X+Y}(z)$ for the density function of $X + Y$ and using Theorem 2.1.16

$$\begin{aligned} f_{X+Y}(x) &= \int_0^x \frac{\lambda^\alpha (x-y)^{\alpha-1}}{\Gamma(\alpha)} e^{-\lambda(x-y)} \frac{\lambda^\beta y^{\beta-1}}{\Gamma(\beta)} e^{-\lambda y}\, dy \\ &= \frac{\lambda^{\alpha+\beta} e^{-\lambda x}}{\Gamma(\alpha)\Gamma(\beta)} \int_0^x (x-y)^{\alpha-1} y^{\beta-1}\, dy \end{aligned}$$

so it suffices to show the integral is $x^{\alpha+\beta-1}\Gamma(\alpha)\Gamma(\beta)/\Gamma(\alpha+\beta)$. To do this, we begin by changing variables $y = xu$, $dy = x\, du$ to get

$$x^{\alpha+\beta-1} \int_0^1 (1-u)^{\alpha-1} u^{\beta-1}\, du = \int_0^x (x-y)^{\alpha-1} y^{\beta-1}\, dy \tag{2.1.1}$$

There are two ways to complete the proof at this point. The soft solution is to note that we have shown that the density $f_{X+Y}(x) = c_{\alpha,\beta} e^{-\lambda x} \lambda^{\alpha+\beta} x^{\alpha+\beta-1}$ where

$$c_{\alpha,\beta} = \frac{1}{\Gamma(\alpha)\Gamma(\beta)} \int_0^1 (1-u)^{\alpha-1} u^{\beta-1}\, du$$

There is only one norming constant $c_{\alpha,\beta}$ that makes this a probability distribution, so recalling the definition of the beta distribution, we must have $c_{\alpha,\beta} = 1/\Gamma(\alpha+\beta)$.

The less elegant approach for those of us who cannot remember the definition of the beta is to prove the last equality by calculus. Multiplying each side of the last equation by e^{-x}, integrating from 0 to ∞, and then using Fubini's theorem on the right we have

$$\begin{aligned} &\Gamma(\alpha+\beta) \int_0^1 (1-u)^{\alpha-1} u^{\beta-1}\, du \\ &\quad = \int_0^\infty \int_0^x y^{\beta-1} e^{-y} (x-y)^{\alpha-1} e^{-(x-y)}\, dy\, dx \\ &\quad = \int_0^\infty y^{\beta-1} e^{-y} \int_y^\infty (x-y)^{\alpha-1} e^{-(x-y)}\, dx\, dy = \Gamma(\alpha)\Gamma(\beta) \end{aligned}$$

which gives the first result. The second follows from the fact that a gamma$(1,\lambda)$ is an exponential with parameter λ and induction. □

Example 2.1.19 Normal distribution. In Example 1.6.11, we introduced the normal density with mean μ and variance a,

$$(2\pi a)^{-1/2} \exp(-(x-\mu)^2/2a).$$

Theorem 2.1.20 *If $X = \text{normal}(\mu, a)$ and $Y = \text{normal}(\nu, b)$ are independent, then $X + Y = \text{normal}(\mu + \nu, a + b)$.*

Proof It is enough to prove the result for $\mu = \nu = 0$. Suppose $Y_1 = \text{normal}(0, a)$ and $Y_2 = \text{normal}(0, b)$. Then Theorem 2.1.16 implies

$$f_{Y_1+Y_2}(z) = \frac{1}{2\pi\sqrt{ab}} \int e^{-x^2/2a} e^{-(z-x)^2/2b}\, dx$$

Dropping the constant in front, the integral can be rewritten as

$$\begin{aligned}
&\int \exp\left(-\frac{bx^2 + ax^2 - 2axz + az^2}{2ab}\right) dx \\
&= \int \exp\left(-\frac{a+b}{2ab}\left\{x^2 - \frac{2a}{a+b}xz + \frac{a}{a+b}z^2\right\}\right) dx \\
&= \int \exp\left(-\frac{a+b}{2ab}\left\{\left(x - \frac{a}{a+b}z\right)^2 + \frac{ab}{(a+b)^2}z^2\right\}\right) dx
\end{aligned}$$

since $-\{a/(a+b)\}^2 + \{a/(a+b)\} = ab/(a+b)^2$. Factoring out the term that does not depend on x, the last integral

$$\begin{aligned}
&= \exp\left(-\frac{z^2}{2(a+b)}\right) \int \exp\left(-\frac{a+b}{2ab}\left(x - \frac{a}{a+b}z\right)^2\right) dx \\
&= \exp\left(-\frac{z^2}{2(a+b)}\right) \sqrt{2\pi ab/(a+b)}
\end{aligned}$$

since the last integral is the normal density with parameters $\mu = az/(a+b)$ and $\sigma^2 = ab/(a+b)$ without its proper normalizing constant. Reintroducing the constant we dropped at the beginning,

$$f_{Y_1+Y_2}(z) = \frac{1}{2\pi\sqrt{ab}} \sqrt{2\pi ab/(a+b)} \exp\left(-\frac{z^2}{2(a+b)}\right) \qquad \square$$

2.1.4 Constructing Independent Random Variables

The last question that we have to address before we can study independent random variables is: Do they exist? (If they don't exist, then there is no point in studying them!) If we are given a finite number of distribution functions $F_i, 1 \le i \le n$, it is easy to construct independent random variables $X_1, \ldots, X_n$ with $P(X_i \le x) = F_i(x)$. Let $\Omega = \mathbf{R}^n$, $\mathcal{F} = \mathcal{R}^n$, $X_i(\omega_1, \ldots, \omega_n) = \omega_i$ (the ith coordinate of $\omega \in \mathbf{R}^n$), and let P be the measure on $\mathcal{R}^n$ that has

$$P((a_1, b_1] \times \cdots \times (a_n, b_n]) = (F_1(b_1) - F_1(a_1)) \cdots (F_n(b_n) - F_n(a_n))$$

If μ_i is the measure with distribution function F_i, then $P = \mu_1 \times \cdots \times \mu_n$.

To construct an infinite sequence $X_1, X_2, \ldots$ of independent random variables with given distribution functions, we want to perform the last construction on the infinite product space

$$\mathbf{R}^{\mathbf{N}} = \{(\omega_1, \omega_2, \ldots) : \omega_i \in \mathbf{R}\} = \{\text{functions } \omega : \mathbf{N} \to \mathbf{R}\}$$

where $\mathbf{N} = \{1, 2, \ldots\}$ and $\mathbf{N}$ stands for **natural numbers**. We define $X_i(\omega) = \omega_i$ and we equip $\mathbf{R}^{\mathbf{N}}$ with the product σ-field $\mathcal{R}^{\mathbf{N}}$, which is generated by the **finite dimensional sets** = sets of the form $\{\omega : \omega_i \in B_i, 1 \le i \le n\}$, where $B_i \in \mathcal{R}$. It is clear how we want to define P for finite dimensional sets. To assert the existence of a unique extension to $\mathcal{R}^{\mathbf{N}}$, we use Theorem A.3.1:

Theorem 2.1.21 (Kolmogorov's extension theorem) *Suppose we are given probability measures μ_n on $(\mathbf{R}^n, \mathcal{R}^n)$ that are consistent, that is,*

$$\mu_{n+1}((a_1, b_1] \times \cdots \times (a_n, b_n] \times \mathbf{R}) = \mu_n((a_1, b_1] \times \cdots \times (a_n, b_n])$$

Then there is a unique probability measure P on $(\mathbf{R}^{\mathbf{N}}, \mathcal{R}^{\mathbf{N}})$ with

$$P(\omega : \omega_i \in (a_i, b_i], 1 \le i \le n) = \mu_n((a_1, b_1] \times \cdots \times (a_n, b_n])$$

In what follows we will need to construct sequences of random variables that take values in other measurable spaces $(S, \mathcal{S})$. Unfortunately, Theorem 2.1.21 is not valid for arbitrary measurable spaces. The first example (on an infinite product of different spaces $\Omega_1 \times \Omega_2 \times \ldots$) was constructed by Andersen and Jessen (1948). (See Halmos (1950) p. 214 or Neveu (1965) p. 84.) For an example in which all the spaces Ω_i are the same see Wegner (1973). Fortunately, there is a class of spaces that is adequate for all of our results and for which the generalization of Kolmogorov's theorem is trivial.

$(S, \mathcal{S})$ is said to be **nice** if there is a 1-1 map φ from S into $\mathbf{R}$ so that φ and φ^{-1} are both measurable.

Such spaces are often called **standard Borel spaces**, but we already have too many things named after Borel. The next result shows that most spaces arising in applications are nice.

Theorem 2.1.22 *If S is a Borel subset of a complete separable metric space M, and $\mathcal{S}$ is the collection of Borel subsets of S, then $(S, \mathcal{S})$ is nice.*

Proof We begin with the special case $S = [0, 1)^{\mathbf{N}}$ with metric

$$\rho(x, y) = \sum_{n=1}^{\infty} |x_n - y_n| / 2^n$$

If $x = (x^1, x^2, x^3, \ldots)$, expand each component in binary $x^j = .x_1^j x_2^j x_3^j \ldots$ (taking the expansion with an infinite number of 0's). Let

$$\varphi_o(x) = .x_1^1 x_2^1 x_1^2 x_3^1 x_2^2 x_1^3 x_4^1 x_3^2 x_2^3 x_1^4 \ldots$$

To treat the general case, we observe that by letting

$$d(x, y) = \rho(x, y)/(1 + \rho(x, y))$$

(for more details, see Exercise 2.1.3), we can suppose that the metric has $d(x, y) < 1$ for all x, y. Let $q_1, q_2, \ldots$ be a countable dense set in S. Let

$$\psi(x) = (d(x, q_1), d(x, q_2), \ldots).$$

$\psi : S \to [0, 1)^N$ is continuous and 1-1. $\varphi_o \circ \psi$ gives the desired mapping. □

Caveat emptor. The proof here is somewhat light when it comes to details. For a more comprehensive discussion, see section 13.1 of Dudley (1989). An interesting consequence of the analysis there is that for Borel subsets of a complete separable metric space the continuum hypothesis is true: i.e., all sets are either finite, countably infinite, or have the cardinality of the real numbers.

Exercises

2.1.1 Suppose $(X_1, \dots, X_n)$ has density $f(x_1, x_2, \dots, x_n)$, that is

$$P((X_1, X_2, \dots, X_n) \in A) = \int_A f(x)\,dx \text{ for } A \in \mathcal{R}^n$$

If $f(x)$ can be written as $g_1(x_1) \cdots g_n(x_n)$ where the $g_m \geq 0$ are measurable, then $X_1, X_2, \dots, X_n$ are independent. Note that the g_m are not assumed to be probability densities.

2.1.2 Suppose $X_1, \dots, X_n$ are random variables that take values in countable sets $S_1, \dots, S_n$. Then in order for $X_1, \dots, X_n$ to be independent, it is sufficient that whenever $x_i \in S_i$

$$P(X_1 = x_1, \dots, X_n = x_n) = \prod_{i=1}^{n} P(X_i = x_i)$$

2.1.3 Let $\rho(x, y)$ be a metric. (i) Suppose h is differentiable with $h(0) = 0$, $h'(x) > 0$ for $x > 0$, and $h'(x)$ decreasing on $[0, \infty)$. Then $h(\rho(x, y))$ is a metric. (ii) $h(x) = x/(x + 1)$ satisfies the hypotheses in (i).

2.1.4 Let $\Omega = (0, 1)$, $\mathcal{F}$ = Borel sets, P = Lebesgue measure. $X_n(\omega) = \sin(2\pi n\omega)$, $n = 1, 2, \dots$ are uncorrelated but not independent.

2.1.5 (i) Show that if X and Y are independent with distributions μ and ν, then

$$P(X + Y = 0) = \sum_y \mu(\{-y\})\nu(\{y\})$$

(ii) Conclude that if X has continuous distribution $P(X = Y) = 0$.

2.1.6 Prove directly from the definition that if X and Y are independent and f and g are measurable functions, then $f(X)$ and $g(Y)$ are independent.

2.1.7 Let $K \geq 3$ be a prime and let X and Y be independent random variables that are uniformly distributed on $\{0, 1, \dots, K - 1\}$. For $0 \leq n < K$, let $Z_n = X + nY \bmod K$. Show that $Z_0, Z_1, \dots, Z_{K-1}$ are **pairwise independent**, i.e., each pair is independent. They are not independent because if we know the values of two of the variables then we know the values of all the variables.

2.1.8 Find four random variables taking values in $\{-1, 1\}$ so that any three are independent but all four are not. Hint: Consider products of independent random variables.

2.1.9 Let $\Omega = \{1, 2, 3, 4\}$, $\mathcal{F}$ = all subsets of Ω, and $P(\{i\}) = 1/4$. Give an example of two collections of sets $\mathcal{A}_1$ and $\mathcal{A}_2$ that are independent but whose generated σ-fields are not.

2.1.10 Show that if X and Y are independent, integer-valued random variables, then

$$P(X+Y=n)=\sum_m P(X=m)P(Y=n-m)$$

2.1.11 In Example 1.6.13, we introduced the Poisson distribution with parameter λ, which is given by $P(Z=k)=e^{-\lambda}\lambda^k/k!$ for $k=0,1,2,\ldots$ Use the previous exercise to show that if $X=\text{Poisson}(\lambda)$ and $Y=\text{Poisson}(\mu)$ are independent, then $X+Y=\text{Poisson}(\lambda+\mu)$.

2.1.12 X is said to have a Binomial(n,p) distribution if

$$P(X=m)=\binom{n}{m}p^m(1-p)^{n-m}$$

(i) Show that if $X=\text{Binomial}(n,p)$ and $Y=\text{Binomial}(m,p)$ are independent then $X+Y=\text{Binomial}(n+m,p)$. (ii) Look at Example 1.6.12 and use induction to conclude that the sum of n independent Bernoulli(p) random variables is Binomial(n,p).

2.1.13 It should not be surprising that the distribution of $X+Y$ can be $F*G$ without the random variables being independent. Suppose $X,Y\in\{0,1,2\}$ and take each value with probability 1/3. (a) Find the distribution of $X+Y$ assuming X and Y are independent. (b) Find all the joint distributions (X,Y) so that the distribution of $X+Y$ is the same as the answer to (a).

2.1.14 Let $X,Y\geq 0$ be independent with distribution functions F and G. Find the distribution function of XY.

2.1.15 If we want an infinite sequence of coin tossings, we do not have to use Kolmogorov's theorem. Let Ω be the unit interval (0,1) equipped with the Borel sets $\mathcal{F}$ and Lebesgue measure P. Let $Y_n(\omega)=1$ if $[2^n\omega]$ is odd and 0 if $[2^n\omega]$ is even. Show that $Y_1,Y_2,\ldots$ are independent with $P(Y_k=0)=P(Y_k=1)=1/2$.

2.2 Weak Laws of Large Numbers

In this section, we will prove several "weak laws of large numbers." The first order of business is to define the mode of convergence that appears in the conclusions of the theorems. We say that Y_n converges to Y **in probability** if for all $\epsilon>0$, $P(|Y_n-Y|>\epsilon)\to 0$ as $n\to\infty$.

2.2.1 *L^2 Weak Laws*

Our first set of weak laws come from computing variances and using Chebyshev's inequality. Extending a definition given in Example 2.1.14 for two random variables, a family of random variables X_i, $i\in I$ with $EX_i^2<\infty$ is said to be **uncorrelated** if we have

$$E(X_iX_j)=EX_iEX_j \quad \text{whenever } i\neq j$$

The key to our weak law for uncorrelated random variables, Theorem 2.2.3, is:

Theorem 2.2.1 *Let $X_1, \dots, X_n$ have $E(X_i^2) < \infty$ and be uncorrelated. Then*

$$var(X_1 + \cdots + X_n) = var(X_1) + \cdots + var(X_n)$$

where $var(Y)$ = *the variance of* Y.

Proof Let $\mu_i = EX_i$ and $S_n = \sum_{i=1}^n X_i$. Since $ES_n = \sum_{i=1}^n \mu_i$, using the definition of the variance, writing the square of the sum as the product of two copies of the sum, and then expanding, we have

$$\begin{aligned} \text{var}(S_n) &= E(S_n - ES_n)^2 = E\left(\sum_{i=1}^n (X_i - \mu_i)\right)^2 \\ &= E\left(\sum_{i=1}^n \sum_{j=1}^n (X_i - \mu_i)(X_j - \mu_j)\right) \\ &= \sum_{i=1}^n E(X_i - \mu_i)^2 + 2\sum_{i=1}^n \sum_{j=1}^{i-1} E((X_i - \mu_i)(X_j - \mu_j)) \end{aligned}$$

where in the last equality we have separated out the diagonal terms $i = j$ and used the fact that the sum over $1 \le i < j \le n$ is the same as the sum over $1 \le j < i \le n$.

The first sum is $\text{var}(X_1) + \cdots + \text{var}(X_n)$ so we want to show that the second sum is zero. To do this, we observe

$$\begin{aligned} E((X_i - \mu_i)(X_j - \mu_j)) &= EX_iX_j - \mu_i EX_j - \mu_j EX_i + \mu_i\mu_j \\ &= EX_iX_j - \mu_i\mu_j = 0 \end{aligned}$$

since X_i and X_j are uncorrelated. □

In words, Theorem 2.2.1 says that for uncorrelated random variables the variance of the sum is the sum of the variances. The second ingredient in our proof of Theorem 2.2.3 is the following consequence of (1.6.4):

$$\text{var}(cY) = c^2\,\text{var}(Y)$$

The third and final ingredient is

Lemma 2.2.2 *If $p > 0$ and $E|Z_n|^p \to 0$, then $Z_n \to 0$ in probability.*

Proof Chebyshev's inequality, Theorem 1.6.4, with $\varphi(x) = x^p$ and $X = |Z_n|$ implies that if $\epsilon > 0$ then $P(|Z_n| \ge \epsilon) \le \epsilon^{-p} E|Z_n|^p \to 0$. □

We can now easily prove:

Theorem 2.2.3 (L^2 weak law) *Let $X_1, X_2, \dots$ be uncorrelated random variables with $EX_i = \mu$ and $var(X_i) \le C < \infty$. If $S_n = X_1 + \cdots + X_n$, then as $n \to \infty$, $S_n/n \to \mu$ in L^2 and in probability.*

Proof To prove L^2 convergence, observe that $E(S_n/n) = \mu$, so

$$E(S_n/n - \mu)^2 = \text{var}(S_n/n) = \frac{1}{n^2}(\text{var}(X_1) + \cdots + \text{var}(X_n)) \le \frac{Cn}{n^2} \to 0$$

To conclude there is also convergence in probability, we apply the Lemma 2.2.2 to $Z_n = S_n/n - \mu$. □

The most important special case of Theorem 2.2.3 occurs when $X_1, X_2, \ldots$ are independent random variables that all have the same distribution. In the jargon, they are **independent and identically distributed** or **i.i.d.** for short. Theorem 2.2.3 tells us in this case that if $EX_i^2 < \infty$ then S_n/n converges to $\mu = EX_i$ in probability as $n \to \infty$. In Theorem 2.2.14, we will see that $E|X_i| < \infty$ is sufficient for the last conclusion, but for the moment we will concern ourselves with consequences of the weaker result.

Our first application is to a situation that on the surface has nothing to do with randomness.

Example 2.2.4 (Polynomial approximation) Let f be a continuous function on [0,1], and let

$$f_n(x) = \sum_{m=0}^{n} \binom{n}{m} x^m (1-x)^{n-m} f(m/n) \quad \text{where } \binom{n}{m} = \frac{n!}{m!\,(n-m)!}$$

be the **Bernstein polynomial of degree** n associated with f. Then as $n \to \infty$

$$\sup_{x \in [0,1]} |f_n(x) - f(x)| \to 0$$

Proof First observe that if S_n is the sum of n independent random variables with $P(X_i = 1) = p$ and $P(X_i = 0) = 1 - p$, then $EX_i = p$, $\text{var}\,(X_i) = p(1-p)$ and

$$P(S_n = m) = \binom{n}{m} p^m (1-p)^{n-m}$$

so $Ef(S_n/n) = f_n(p)$. Theorem 2.2.3 tells us that as $n \to \infty$, $S_n/n \to p$ in probability. The last two observations motivate the definition of $f_n(p)$, but to prove the desired conclusion we have to use the proof of Theorem 2.2.3 rather than the result itself.

Combining the proof of Theorem 2.2.3 with our formula for the variance of X_i and the fact that $p(1-p) \le 1/4$ when $p \in [0,1]$, we have

$$P(|S_n/n - p| > \delta) \le \frac{\text{var}\,(S_n/n)}{\delta^2} = \frac{p(1-p)}{n\delta^2} \le \frac{1}{4n\delta^2}$$

To conclude now that $Ef(S_n/n) \to f(p)$, let $M = \sup_{x \in [0,1]} |f(x)|$, let $\epsilon > 0$, and pick $\delta > 0$ so that if $|x - y| < \delta$, then $|f(x) - f(y)| < \epsilon$. (This is possible since a continuous function is uniformly continuous on each bounded interval.) Now, using Jensen's inequality, Theorem 1.6.2, gives

$$|Ef(S_n/n) - f(p)| \le E|f(S_n/n) - f(p)| \le \epsilon + 2MP(|S_n/n - p| > \delta)$$

Letting $n \to \infty$, we have $\limsup_{n\to\infty} |Ef(S_n/n) - f(p)| \le \epsilon$, but ϵ is arbitrary, so this gives the desired result. □

Our next result is for comic relief.

Example 2.2.5 (A high-dimensional cube is almost the boundary of a ball) Let $X_1, X_2, \ldots$ be independent and uniformly distributed on $(-1,1)$. Let $Y_i = X_i^2$, which are independent since they are functions of independent random variables. $EY_i = 1/3$ and $\text{var}\,(Y_i) \le EY_i^2 \le 1$, so Theorem 2.2.3 implies

$$(X_1^2 + \cdots + X_n^2)/n \to 1/3 \quad \text{in probability as } n \to \infty$$

Let $A_{n,\epsilon} = \{x \in \mathbf{R}^n : (1-\epsilon)\sqrt{n/3} < |x| < (1+\epsilon)\sqrt{n/3}\}$, where $|x| = (x_1^2 + \cdots + x_n^2)^{1/2}$. If we let $|S|$ denote the Lebesgue measure of S, then the last conclusion implies that for any $\epsilon > 0$, $|A_{n,\epsilon} \cap (-1,1)^n|/2^n \to 1$, or, in words, most of the volume of the cube $(-1,1)^n$ comes from $A_{n,\epsilon}$, which is almost the boundary of the ball of radius $\sqrt{n/3}$.

2.2.2 Triangular Arrays

Many classical limit theorems in probability concern arrays $X_{n,k}$, $1 \le k \le n$ of random variables and investigate the limiting behavior of their row sums $S_n = X_{n,1} + \cdots + X_{n,n}$. In most cases, we assume that the random variables on each row are independent, but for the next trivial (but useful) result we do not need that assumption. Indeed, here S_n can be any sequence of random variables.

Theorem 2.2.6 *Let $\mu_n = ES_n$, $\sigma_n^2 = \text{var}\,(S_n)$. If $\sigma_n^2/b_n^2 \to 0$ then*

$$\frac{S_n - \mu_n}{b_n} \to 0 \qquad \textit{in probability}$$

Proof Our assumptions imply $E((S_n - \mu_n)/b_n)^2 = b_n^{-2}\,\text{var}\,(S_n) \to 0$, so the desired conclusion follows from Lemma 2.2.2. □

We will now give three applications of Theorem 2.2.6. For these three examples, the following calculation is useful:

$$\sum_{m=1}^{n} \frac{1}{m} \ge \int_1^n \frac{dx}{x} \ge \sum_{m=2}^{n} \frac{1}{m}$$

$$\log n \le \sum_{m=1}^{n} \frac{1}{m} \le 1 + \log n \tag{2.2.1}$$

Example 2.2.7 (Coupon collector's problem) Let $X_1, X_2, \ldots$ be i.i.d. uniform on $\{1,2,\ldots,n\}$. To motivate the name, think of collecting baseball cards (or coupons). Suppose that the ith item we collect is chosen at random from the set of possibilities and is independent of the previous choices. Let $\tau_k^n = \inf\{m : |\{X_1,\ldots,X_m\}| = k\}$ be the first time we have k different items. In this problem, we are interested in the asymptotic behavior of $T_n = \tau_n^n$, the time to collect a complete set. It is easy to see that $\tau_1^n = 1$. To make later formulas work out nicely, we will set $\tau_0^n = 0$. For $1 \le k \le n$, $X_{n,k} \equiv \tau_k^n - \tau_{k-1}^n$ represents the time to get a choice different from our first $k-1$, so $X_{n,k}$ has a geometric distribution with parameter $1 - (k-1)/n$ and is independent of the earlier waiting times $X_{n,j}$, $1 \le j < k$. Example 1.6.14 tells us that if X has a geometric distribution with parameter p then $EX = 1/p$ and $\text{var}\,(X) \le 1/p^2$. Using the linearity of expected value, bounds on $\sum_{m=1}^n 1/m$ in (2.2.1), and Theorem 2.2.1 we see that

$$ET_n = \sum_{k=1}^{n}\left(1 - \frac{k-1}{n}\right)^{-1} = n\sum_{m=1}^{n} m^{-1} \sim n\log n$$

$$\text{var}\,(T_n) \le \sum_{k=1}^{n}\left(1 - \frac{k-1}{n}\right)^{-2} = n^2\sum_{m=1}^{n} m^{-2} \le n^2\sum_{m=1}^{\infty} m^{-2}$$

Taking $b_n = n \log n$ and using Theorem 2.2.6, it follows that

$$\frac{T_n - n\sum_{m=1}^{n} m^{-1}}{n \log n} \to 0 \quad \text{in probability}$$

and hence $T_n/(n \log n) \to 1$ in probability.

For a concrete example, take $n = 365$, i.e., we are interested in the number of people we need to meet until we have seen someone with every birthday. In this case, the limit theorem says it will take about $365 \log 365 = 2,153.46$ tries to get a complete set. Note that the number of trials is 5.89 times the number of birthdays.

Example 2.2.8 (Random permutations) Let Ω_n consist of the $n!$ permutations (i.e., one-to-one mappings from $\{1,\ldots,n\}$ onto $\{1,\ldots,n\}$) and make this into a probability space by assuming all the permutations are equally likely. This application of the weak law concerns the cycle structure of a random permutation π, so we begin by describing the decomposition of a permutation into cycles. Consider the sequence $1, \pi(1), \pi(\pi(1)), \ldots$ Eventually, $\pi^k(1) = 1$. When it does, we say the first cycle is completed and has length k. To start the second cycle, we pick the smallest integer i not in the first cycle and look at $i, \pi(i), \pi(\pi(i)), \ldots$ until we come back to i. We repeat the construction until all the elements are accounted for. For example, if the permutation is

i	1	2	3	4	5	6	7	8	9
$\pi(i)$	3	9	6	8	2	1	5	4	7

then the cycle decomposition is (136) (2975) (48).

Let $X_{n,k} = 1$ if a right parenthesis occurs after the kth number in the decomposition, $X_{n,k} = 0$ otherwise, and let $S_n = X_{n,1} + \cdots + X_{n,n} =$ the number of cycles. (In the example, $X_{9,3} = X_{9,7} = X_{9,9} = 1$, and the other $X_{9,m} = 0$.) I claim that

Lemma 2.2.9 *$X_{n,1}, \ldots, X_{n,n}$ are independent and $P(X_{n,j} = 1) = \frac{1}{n-j+1}$.*

Intuitively, this is true since, independent of what has happened so far, there are $n - j + 1$ values that have not appeared in the range, and only one of them will complete the cycle.

Proof To prove this, it is useful to generate the permutation in a special way. Let $i_1 = 1$. Pick j_1 at random from $\{1,\ldots,n\}$ and let $\pi(i_1) = j_1$. If $j_1 \neq 1$, let $i_2 = j_1$. If $j_1 = 1$, let $i_2 = 2$. In either case, pick j_2 at random from $\{1,\ldots,n\} - \{j_1\}$. In general, if $i_1, j_1, \ldots, i_{k-1}, j_{k-1}$ have been selected and we have set $\pi(i_\ell) = j_\ell$ for $1 \le \ell < k$, then (a) if $j_{k-1} \in \{i_1, \ldots, i_{k-1}\}$ so a cycle has just been completed, we let $i_k = \inf(\{1,\ldots,n\} - \{i_1,\ldots,i_{k-1}\})$ and (b) if $j_{k-1} \notin \{i_1,\ldots,i_{k-1}\}$, we let $i_k = j_{k-1}$. In either case, we pick j_k at random from $\{1,\ldots,n\} - \{j_1,\ldots,j_{k-1}\}$ and let $\pi(i_k) = j_k$.

The construction here is tedious to write out, or to read, but now I can claim with a clear conscience that $X_{n,1}, \ldots, X_{n,n}$ are independent and $P(X_{n,k} = 1) = 1/(n - k + 1)$ since when we pick j_k there are $n - k + 1$ values in $\{1,\ldots,n\} - \{j_1,\ldots,j_{k-1}\}$ and only one of them will complete the cycle. □

To check the conditions of Theorem 2.2.6, now note

$$ES_n = 1/n + 1/(n-1) + \cdots + 1/2 + 1$$

$$\text{var}\,(S_n) = \sum_{k=1}^{n} \text{var}\,(X_{n,k}) \le \sum_{k=1}^{n} E(X_{n,k}^2) = \sum_{k=1}^{n} E(X_{n,k}) = ES_n$$

where the results on the second line follow from Theorem 2.2.1, the fact that $\text{var}(Y) \le EY^2$, and $X_{n,k}^2 = X_{n,k}$. Now $ES_n \sim \log n$, so if $b_n = (\log n)^{.5+\epsilon}$ with $\epsilon > 0$, the conditions of Theorem 2.2.6 are satisfied and it follows that

$$\frac{S_n - \sum_{m=1}^n m^{-1}}{(\log n)^{.5+\epsilon}} \to 0 \quad \text{in probability} \tag{$*$}$$

Taking $\epsilon = 0.5$, we have that $S_n/\log n \to 1$ in probability, but $(*)$ says more. We will see in Example 3.4.11 that $(*)$ is false if $\epsilon = 0$.

Example 2.2.10 (An occupancy problem) Suppose we put r balls at random in n boxes, i.e., all n^r assignments of balls to boxes have equal probability. Let A_i be the event that the ith box is empty and $N_n =$ the number of empty boxes. It is easy to see that

$$P(A_i) = (1 - 1/n)^r \qquad \text{and} \qquad EN_n = n(1 - 1/n)^r$$

A little calculus (take logarithms) shows that if $r/n \to c$, $EN_n/n \to e^{-c}$. (For a proof, see Lemma 3.1.1.) To compute the variance of N_n, we observe that

$$\begin{aligned}
EN_n^2 &= E\left(\sum_{m=1}^n 1_{A_m}\right)^2 = \sum_{1\le k,m\le n} P(A_k \cap A_m) \\
\text{var}(N_n) &= EN_n^2 - (EN_n)^2 = \sum_{1\le k,m\le n} P(A_k \cap A_m) - P(A_k)P(A_m) \\
&= n(n-1)\{(1-2/n)^r - (1-1/n)^{2r}\} + n\{(1-1/n)^r - (1-1/n)^{2r}\}
\end{aligned}$$

The first term comes from $k \ne m$ and the second from $k = m$. Since $(1-2/n)^r \to e^{-2c}$ and $(1-1/n)^r \to e^{-c}$, it follows easily from the last formula that $\text{var}(N_n/n) = \text{var}(N_n)/n^2 \to 0$. Taking $b_n = n$ in Theorem 2.2.6, now we have

$$N_n/n \to e^{-c} \quad \text{in probability}$$

2.2.3 Truncation

To truncate a random variable X at level M means considering

$$\bar{X} = X1_{(|X|\le M)} = \begin{cases} X & \text{if } |X| \le M \\ 0 & \text{if } |X| > M \end{cases}$$

To extend the weak law to random variables without a finite second moment, we will truncate and then use Chebyshev's inequality. We begin with a very general but also very useful result. Its proof is easy because we have assumed what we need for the proof. Later we will have to work a little to verify the assumptions in special cases, but the general result serves to identify the essential ingredients in the proof.

Theorem 2.2.11 (Weak law for triangular arrays) *For each n let $X_{n,k}$, $1 \le k \le n$, be independent. Let $b_n > 0$ with $b_n \to \infty$, and let $\bar{X}_{n,k} = X_{n,k}1_{(|X_{n,k}|\le b_n)}$. Suppose that as $n \to \infty$*

(i) $\sum_{k=1}^{n} P(|X_{n,k}| > b_n) \to 0$, *and*

(ii) $b_n^{-2} \sum_{k=1}^{n} E\bar{X}_{n,k}^2 \to 0$.

If we let $S_n = X_{n,1} + \cdots + X_{n,n}$ *and put* $a_n = \sum_{k=1}^{n} E\bar{X}_{n,k}$, *then*

$$(S_n - a_n)/b_n \to 0 \text{ in probability}$$

Proof Let $\bar{S}_n = \bar{X}_{n,1} + \cdots + \bar{X}_{n,n}$. Clearly,

$$P\left(\left|\frac{S_n - a_n}{b_n}\right| > \epsilon\right) \le P(S_n \ne \bar{S}_n) + P\left(\left|\frac{\bar{S}_n - a_n}{b_n}\right| > \epsilon\right)$$

To estimate the first term, we note that

$$P(S_n \ne \bar{S}_n) \le P\left(\cup_{k=1}^{n}\{\bar{X}_{n,k} \ne X_{n,k}\}\right) \le \sum_{k=1}^{n} P(|X_{n,k}| > b_n) \to 0$$

by (i). For the second term, we note that Chebyshev's inequality, $a_n = E\bar{S}_n$, Theorem 2.2.1, and $\text{var}(X) \le EX^2$ imply

$$P\left(\left|\frac{\bar{S}_n - a_n}{b_n}\right| > \epsilon\right) \le \epsilon^{-2} E\left|\frac{\bar{S}_n - a_n}{b_n}\right|^2 = \epsilon^{-2} b_n^{-2}\, \text{var}(\bar{S}_n)$$
$$= (b_n\epsilon)^{-2} \sum_{k=1}^{n} \text{var}(\bar{X}_{n,k}) \le (b_n\epsilon)^{-2} \sum_{k=1}^{n} E(\bar{X}_{n,k})^2 \to 0$$

by (ii), and the proof is complete. □

From Theorem 2.2.11, we get the following result for a single sequence.

Theorem 2.2.12 (Weak law of large numbers) *Let* $X_1, X_2, \ldots$ *be i.i.d. with*

$$xP(|X_i| > x) \to 0 \quad \text{as } x \to \infty$$

Let $S_n = X_1 + \cdots + X_n$ *and let* $\mu_n = E(X_1 1_{(|X_1| \le n)})$. *Then* $S_n/n - \mu_n \to 0$ *in probability.*

Remark The assumption in the theorem is necessary for the existence of constants a_n so that $S_n/n - a_n \to 0$. See Feller (1971) pp. 234–236 for a proof.

Proof We will apply Theorem 2.2.11 with $X_{n,k} = X_k$ and $b_n = n$. To check (i), we note

$$\sum_{k=1}^{n} P(|X_{n,k}| > n) = nP(|X_i| > n) \to 0$$

by assumption. To check (ii), we need to show $n^{-2} \cdot nE\bar{X}_{n,1}^2 \to 0$. To do this, we need the following result, which will be useful several times later.

Lemma 2.2.13 *If* $Y \ge 0$ *and* $p > 0$ *then* $E(Y^p) = \int_0^\infty py^{p-1} P(Y > y)\, dy$.

Proof Using the definition of expected value, Fubini's theorem (for nonnegative random variables), and then calculating the resulting integrals gives

$$\begin{aligned}\int_0^\infty py^{p-1}P(Y>y)\,dy &= \int_0^\infty \int_\Omega py^{p-1}1_{(Y>y)}\,dP\,dy \\ &= \int_\Omega \int_0^\infty py^{p-1}1_{(Y>y)}\,dy\,dP \\ &= \int_\Omega \int_0^Y py^{p-1}\,dy\,dP = \int_\Omega Y^p\,dP = EY^p\end{aligned}$$

which is the desired result. □

Returning to the proof of Theorem 2.2.12, we observe that Lemma 2.2.13 and the fact that $\bar{X}_{n,1} = X_1 1_{(|X_1|\le n)}$ imply

$$E(\bar{X}_{n,1}^2) = \int_0^\infty 2yP(|\bar{X}_{n,1}| > y)\,dy \le \int_0^n 2yP(|X_1| > y)\,dy$$

since $P(|\bar{X}_{n,1}| > y) = 0$ for $y \ge n$ and $= P(|X_1| > y) - P(|X_1| > n)$ for $y \le n$. We claim that $yP(|X_1| > y) \to 0$ implies

$$E(\bar{X}_{n,1}^2)/n = \frac{1}{n}\int_0^n 2yP(|X_1| > y)\,dy \to 0$$

as $n \to \infty$. Intuitively, this holds since the right-hand side is the average of $g(y) = 2yP(|X_1| > y)$ over $[0,n]$ and $g(y) \to 0$ as $y \to \infty$. To spell out the details, note that $0 \le g(y) \le 2y$ and $g(y) \to 0$ as $y \to \infty$, so we must have $M = \sup g(y) < \infty$. Let $g_n(y) = g(ny)$. Since g_n is bounded and $g_n(y) \to 0$ a.s.,

$$(1/n)\int_0^n g(y)\,dy = \int_0^1 g_n(x)\,dx \to 0$$

which completes the proof. □

Remark Applying Lemma 2.2.13 with $p = 1-\epsilon$ and $\epsilon > 0$, we see that $xP(|X_1| > x) \to 0$ implies $E|X_1|^{1-\epsilon} < \infty$, so the assumption in Theorem 2.2.12 is not much weaker than finite mean.

Finally, we have the weak law in its most familiar form.

Theorem 2.2.14 *Let $X_1, X_2, \ldots$ be i.i.d. with $E|X_i| < \infty$. Let $S_n = X_1 + \cdots + X_n$ and let $\mu = EX_1$. Then $S_n/n \to \mu$ in probability.*

Proof Two applications of the dominated convergence theorem imply

$$\begin{aligned} xP(|X_1| > x) &\le E(|X_1|1_{(|X_1|>x)}) \to 0 \quad \text{as } x \to \infty \\ \mu_n &= E(X_1 1_{(|X_1|\le n)}) \to E(X_1) = \mu \quad \text{as } n \to \infty\end{aligned}$$

Using Theorem 2.2.12, we see that if $\epsilon > 0$ then $P(|S_n/n - \mu_n| > \epsilon/2) \to 0$. Since $\mu_n \to \mu$, it follows that $P(|S_n/n - \mu| > \epsilon) \to 0$. □

Example 2.2.15 For an example where the weak law does not hold, suppose $X_1, X_2, \ldots$ are independent and have a **Cauchy distribution**:

$$P(X_i \le x) = \int_{-\infty}^x \frac{dt}{\pi(1+t^2)}$$

As $x \to \infty$,

$$P(|X_1| > x) = 2\int_x^\infty \frac{dt}{\pi(1+t^2)} \sim \frac{2}{\pi}\int_x^\infty t^{-2}dt = \frac{2}{\pi}x^{-1}$$

From the necessity of this condition, we can conclude that there is no sequence of constants μ_n so that $S_n/n - \mu_n \to 0$. We will see later that S_n/n always has the same distribution as X_1 (See Exercise 3.3.6.).

As the next example shows, we can have a weak law in some situations in which $E|X| = \infty$.

Example 2.2.16 (The "St. Petersburg paradox.") Let $X_1, X_2, \dots$ be independent random variables with

$$P(X_i = 2^j) = 2^{-j} \quad \text{for } j \geq 1$$

In words, you win 2^j dollars if it takes j tosses to get a heads. The paradox here is that $EX_1 = \infty$, but you clearly wouldn't pay an infinite amount to play this game. An application of Theorem 2.2.11 will tell us how much we should pay to play the game n times.

In this example, $X_{n,k} = X_k$. To apply Theorem 2.2.11, we have to pick b_n. To do this, we are guided by the principle that in checking (ii) we want to take b_n as small as we can and have (i) hold. With this in mind, we observe that if m is an integer

$$P(X_1 \geq 2^m) = \sum_{j=m}^\infty 2^{-j} = 2^{-m+1}$$

Let $m(n) = \log_2 n + K(n)$ where $K(n) \to \infty$ and is chosen so that $m(n)$ is an integer (and hence the displayed formula is valid). Letting $b_n = 2^{m(n)}$, we have

$$nP(X_1 \geq b_n) = n2^{-m(n)+1} = 2^{-K(n)+1} \to 0$$

proving (i). To check (ii), we observe that if $\bar{X}_{n,k} = X_k 1_{(|X_k| \leq b_n)}$, then

$$E\bar{X}_{n,k}^2 = \sum_{j=1}^{m(n)} 2^{2j} \cdot 2^{-j} \leq 2^{m(n)} \sum_{k=0}^\infty 2^{-k} = 2b_n$$

So the expression in (ii) is smaller than $2n/b_n$, which $\to 0$ since

$$b_n = 2^{m(n)} = n2^{K(n)} \text{ and } K(n) \to \infty$$

The last two steps are to evaluate a_n and to apply Theorem 2.2.11.

$$E\bar{X}_{n,k} = \sum_{j=1}^{m(n)} 2^j 2^{-j} = m(n)$$

so $a_n = nm(n)$. We have $m(n) = \log n + K(n)$ (here and until the end of the example all logs are base 2), so if we pick $K(n)/\log n \to 0$ then $a_n/n\log n \to 1$ as $n \to \infty$. Using Theorem 2.2.11 now, we have

$$\frac{S_n - a_n}{n2^{K(n)}} \to 0 \quad \text{in probability}$$

If we suppose that $K(n) \le \log\log n$ for large n, then the last conclusion holds with the denominator replaced by $n \log n$, and it follows that $S_n/(n \log n) \to 1$ in probability.

Returning to our original question, we see that a fair price for playing n times is \$ $\log_2 n$ per play. When $n = 1024$, this is \$10 per play. Nicolas Bernoulli wrote in 1713, "There ought not to exist any even halfway sensible person who would not sell the right of playing the game for 40 ducats (per play)." If the wager were 1 ducat, one would need $2^{40} \approx 10^{12}$ plays to start to break even.

Exercises

2.2.1 Let $X_1, X_2, \ldots$ be uncorrelated with $EX_i = \mu_i$ and $\text{var}(X_i)/i \to 0$ as $i \to \infty$. Let $S_n = X_1 + \cdots + X_n$ and $\nu_n = ES_n/n$ then as $n \to \infty$, $S_n/n - \nu_n \to 0$ in L^2 and in probability.

2.2.2 The L^2 weak law generalizes immediately to certain dependent sequences. Suppose $EX_n = 0$ and $EX_nX_m \le r(n-m)$ for $m \le n$ (no absolute value on the left-hand side!) with $r(k) \to 0$ as $k \to \infty$. Show that $(X_1 + \cdots + X_n)/n \to 0$ in probability.

2.2.3 **Monte Carlo integration.** (i) Let f be a measurable function on $[0,1]$ with $\int_0^1 |f(x)|dx < \infty$. Let $U_1, U_2, \ldots$ be independent and uniformly distributed on $[0,1]$, and let

$$I_n = n^{-1}(f(U_1) + \cdots + f(U_n))$$

Show that $I_n \to I \equiv \int_0^1 f\, dx$ in probability. (ii) Suppose $\int_0^1 |f(x)|^2\, dx < \infty$. Use Chebyshev's inequality to estimate $P(|I_n - I| > a/n^{1/2})$.

2.2.4 Let $X_1, X_2, \ldots$ be i.i.d. with $P(X_i = (-1)^k k) = C/k^2 \log k$ for $k \ge 2$ where C is chosen to make the sum of the probabilities $= 1$. Show that $E|X_i| = \infty$, but there is a finite constant μ so that $S_n/n \to \mu$ in probability.

2.2.5 Let $X_1, X_2, \ldots$ be i.i.d. with $P(X_i > x) = e/x \log x$ for $x \ge e$. Show that $E|X_i| = \infty$, but there is a sequence of constants $\mu_n \to \infty$ so that $S_n/n - \mu_n \to 0$ in probability.

2.2.6 (i) Show that if $X \ge 0$ is integer valued $EX = \sum_{n\ge1} P(X \ge n)$. (ii) Find a similar expression for EX^2.

2.2.7 Generalize Lemma 2.2.13 to conclude that if $H(x) = \int_{(-\infty,x]} h(y)\, dy$ with $h(y) \ge 0$, then

$$E\, H(X) = \int_{-\infty}^{\infty} h(y) P(X \ge y)\, dy$$

An important special case is $H(x) = \exp(\theta x)$ with $\theta > 0$.

2.2.8 **An unfair "fair game."** Let $p_k = 1/2^k k(k+1)$, $k = 1, 2, \ldots$ and $p_0 = 1 - \sum_{k\ge1} p_k$.

$$\sum_{k=1}^{\infty} 2^k p_k = (1 - \frac{1}{2}) + (\frac{1}{2} - \frac{1}{3}) + \cdots = 1$$

so if we let $X_1, X_2, \ldots$ be i.i.d. with $P(X_n = -1) = p_0$ and

$$P(X_n = 2^k - 1) = p_k \quad \text{for } k \geq 1$$

then $EX_n = 0$. Let $S_n = X_1 + \cdots + X_n$. Use Theorem 2.2.11 with $b_n = 2^{m(n)}$, where $m(n) = \min\{m : 2^{-m} m^{-3/2} \leq n^{-1}\}$ to conclude that

$$S_n/(n/\log_2 n) \to -1 \text{ in probability}$$

2.2.9 **Weak law for positive variables.** Suppose $X_1, X_2, \ldots$ are i.i.d., $P(0 \leq X_i < \infty) = 1$ and $P(X_i > x) > 0$ for all x. Let $\mu(s) = \int_0^s x\, dF(x)$ and $\nu(s) = \mu(s)/s(1 - F(s))$. It is known that there exist constants a_n so that $S_n/a_n \to 1$ in probability, if and only if $\nu(s) \to \infty$ as $s \to \infty$. Pick $b_n \geq 1$ so that $n\mu(b_n) = b_n$ (this works for large n), and use Theorem 2.2.11 to prove that the condition is sufficient.

2.3 Borel-Cantelli Lemmas

If A_n is a sequence of subsets of Ω, we let

$$\limsup A_n = \lim_{m\to\infty} \cup_{n=m}^{\infty} A_n = \{\omega \text{ that are in infinitely many } A_n\}$$

(the limit exists since the sequence is decreasing in m) and let

$$\liminf A_n = \lim_{m\to\infty} \cap_{n=m}^{\infty} A_n = \{\omega \text{ that are in all but finitely many } A_n\}$$

(the limit exists since the sequence is increasing in m). The names lim sup and lim inf can be explained by noting that

$$\limsup_{n\to\infty} 1_{A_n} = 1_{(\limsup A_n)} \qquad \liminf_{n\to\infty} 1_{A_n} = 1_{(\liminf A_n)}$$

It is common to write $\limsup A_n = \{\omega : \omega \in A_n \text{ i.o.}\}$, where i.o. stands for infinitely often. An example that illustrates the use of this notation is: "$X_n \to 0$ a.s. if and only if for all $\epsilon > 0$, $P(|X_n| > \epsilon \text{ i.o.}) = 0$." The reader will see many other examples later.

Exercise 2.3.1 Prove that $P(\limsup A_n) \geq \limsup P(A_n)$ and $P(\liminf A_n) \leq \liminf P(A_n)$.

The next result should be familiar from measure theory, even though its name may not be.

Theorem 2.3.1 (Borel-Cantelli lemma) *If $\sum_{n=1}^{\infty} P(A_n) < \infty$, then*

$$P(A_n \text{ i.o.}) = 0.$$

Proof Let $N = \sum_k 1_{A_k}$ be the number of events that occur. Fubini's theorem implies $EN = \sum_k P(A_k) < \infty$, so we must have $N < \infty$ a.s. □

The next result is a typical application of the Borel-Cantelli lemma.

Theorem 2.3.2 *$X_n \to X$ in probability if and only if for every subsequence $X_{n(m)}$ there is a further subsequence $X_{n(m_k)}$ that converges almost surely to X.*

Proof Let ϵ_k be a sequence of positive numbers that $\downarrow 0$. For each k, there is an $n(m_k) > n(m_{k-1})$ so that $P(|X_{n(m_k)} - X| > \epsilon_k) \le 2^{-k}$. Since

$$\sum_{k=1}^{\infty} P(|X_{n(m_k)} - X| > \epsilon_k) < \infty$$

the Borel-Cantelli lemma implies $P(|X_{n(m_k)} - X| > \epsilon_k \text{ i.o.}) = 0$, i.e., $X_{n(m_k)} \to X$ a.s. To prove the second conclusion, we note that if for every subsequence $X_{n(m)}$ there is a further subsequence $X_{n(m_k)}$ that converges almost surely to X, then we can apply the next lemma to the sequence of numbers $y_n = P(|X_n - X| > \delta)$ for any $\delta > 0$ to get the desired result. □

Theorem 2.3.3 *Let y_n be a sequence of elements of a topological space. If every subsequence $y_{n(m)}$ has a further subsequence $y_{n(m_k)}$ that converges to y, then $y_n \to y$.*

Proof If $y_n \not\to y$, then there is an open set G containing y and a subsequence $y_{n(m)}$ with $y_{n(m)} \notin G$ for all m, but clearly no subsequence of $y_{n(m)}$ converges to y. □

Remark Since there is a sequence of random variables that converges in probability but not a.s. (for an example, see Exercises 2.3.14 or 2.3.11), it follows from Theorem 2.3.3 that a.s. convergence does not come from a metric, or even from a topology. Exercise 2.3.6 will give a metric for convergence in probability, and Exercise 2.3.7 will show that the space of random variables is a complete space under this metric.

Theorem 2.3.2 allows us to upgrade convergence in probability to convergence almost surely. An example of the usefulness of this is:

Theorem 2.3.4 *If f is continuous and $X_n \to X$ in probability, then $f(X_n) \to f(X)$ in probability. If, in addition, f is bounded, then $Ef(X_n) \to Ef(X)$.*

Proof If $X_{n(m)}$ is a subsequence, then Theorem 2.3.2 implies there is a further subsequence $X_{n(m_k)} \to X$ almost surely. Since f is continuous, Exercise 1.3.3 implies $f(X_{n(m_k)}) \to f(X)$ almost surely and Theorem 2.3.2 implies $f(X_n) \to f(X)$ in probability. If f is bounded, then the bounded convergence theorem implies $Ef(X_{n(m_k)}) \to Ef(X)$, and applying Theorem 2.3.3 to $y_n = Ef(X_n)$ gives the desired result. □

Exercise 2.3.2 Prove the first result in Theorem 2.3.4 directly from the definition.

As our second application of the Borel-Cantelli lemma, we get our first strong law of large numbers:

Theorem 2.3.5 *Let $X_1, X_2, \ldots$ be i.i.d. with $EX_i = \mu$ and $EX_i^4 < \infty$. If $S_n = X_1 + \cdots + X_n$, then $S_n/n \to \mu$ a.s.*

Proof By letting $X_i' = X_i - \mu$, we can suppose without loss of generality that $\mu = 0$. Now

$$ES_n^4 = E\left(\sum_{i=1}^{n} X_i\right)^4 = E \sum_{1 \le i,j,k,\ell \le n} X_i X_j X_k X_\ell$$

Terms in the sum of the form $E(X_i^3 X_j)$, $E(X_i^2 X_j X_k)$, and $E(X_i X_j X_k X_\ell)$ are 0 (if i, j, k, ℓ are distinct) since the expectation of the product is the product of the expectations, and in each case one of the terms has expectation 0. The only terms that do not vanish are those of the form EX_i^4 and $EX_i^2X_j^2 = (EX_i^2)^2$. There are n and $3n(n-1)$ of these terms, respectively. (In the second case, we can pick the two indices in $n(n-1)/2$ ways, and with the indices fixed, the term can arise in a total of six ways.) The last observation implies

$$ES_n^4 = nEX_1^4 + 3(n^2 - n)(EX_1^2)^2 \le Cn^2$$

where $C < \infty$. Chebyshev's inequality gives us

$$P(|S_n| > n\epsilon) \le E(S_n^4)/(n\epsilon)^4 \le C/(n^2\epsilon^4)$$

Summing on n and using the Borel-Cantelli lemma gives $P(|S_n| > n\epsilon \text{ i.o.}) = 0$. Since ϵ is arbitrary, the proof is complete. □

The converse of the Borel-Cantelli lemma is trivially false.

Example 2.3.6 Let $\Omega = (0,1)$, $\mathcal{F}$ = Borel sets, P = Lebesgue measure. If $A_n = (0, a_n)$ where $a_n \to 0$ as $n \to \infty$, then $\limsup A_n = \emptyset$, but if $a_n \ge 1/n$, we have $\sum a_n = \infty$.

The example just given suggests that for general sets we cannot say much more than the result in Exercise 2.3.1.

For independent events, however, the necessary condition for $P(\limsup A_n) > 0$ is sufficient for $P(\limsup A_n) = 1$.

Theorem 2.3.7 (The second Borel-Cantelli lemma) *If the events A_n are independent, then $\sum P(A_n) = \infty$ implies $P(A_n \text{ i.o.}) = 1$.*

Proof Let $M < N < \infty$. Independence and $1 - x \le e^{-x}$ imply

$$P\left(\cap_{n=M}^N A_n^c\right) = \prod_{n=M}^N (1 - P(A_n)) \le \prod_{n=M}^N \exp(-P(A_n))$$

$$= \exp\left(-\sum_{n=M}^N P(A_n)\right) \to 0 \quad \text{as } N \to \infty$$

So $P(\cup_{n=M}^\infty A_n) = 1$ for all M, and since $\cup_{n=M}^\infty A_n \downarrow \limsup A_n$ it follows that $P(\limsup A_n) = 1$. □

A typical application of the second Borel-Cantelli lemma is:

Theorem 2.3.8 *If $X_1, X_2, \ldots$ are i.i.d. with $E|X_i| = \infty$, then $P(|X_n| \ge n \text{ i.o.}) = 1$. So if $S_n = X_1 + \cdots + X_n$, then $P(\lim S_n/n \text{ exists} \in (-\infty, \infty)) = 0$.*

Proof From Lemma 2.2.13, we get

$$E|X_1| = \int_0^\infty P(|X_1| > x)\, dx \le \sum_{n=0}^\infty P(|X_1| > n)$$

Since $E|X_1| = \infty$ and $X_1, X_2, \ldots$ are i.i.d., it follows from the second Borel-Cantelli lemma that $P(|X_n| \ge n \text{ i.o.}) = 1$. To prove the second claim, observe that

$$\frac{S_n}{n} - \frac{S_{n+1}}{n+1} = \frac{S_n}{n(n+1)} - \frac{X_{n+1}}{n+1}$$

and on $C \equiv \{\omega : \lim_{n\to\infty} S_n/n \text{ exists} \in (-\infty,\infty)\}$, $S_n/(n(n+1)) \to 0$. So, on $C \cap \{\omega : |X_n| \geq n \text{ i.o.}\}$, we have

$$\left|\frac{S_n}{n} - \frac{S_{n+1}}{n+1}\right| > 2/3 \quad \text{i.o.}$$

contradicting the fact that $\omega \in C$. From the last observation, we conclude that

$$\{\omega : |X_n| \geq n \text{ i.o.}\} \cap C = \emptyset$$

and since $P(|X_n| \geq n \text{ i.o.}) = 1$, it follows that $P(C) = 0$. □

Theorem 2.3.8 shows that $E|X_i| < \infty$ is necessary for the strong law of large numbers. The reader will have to wait until Theorem 2.4.1 to see that condition is also sufficient. The next result extends the second Borel-Cantelli lemma and sharpens its conclusion.

Theorem 2.3.9 *If $A_1, A_2, \ldots$ are pairwise independent and $\sum_{n=1}^\infty P(A_n) = \infty$, then as $n \to \infty$*

$$\sum_{m=1}^n 1_{A_m} \Big/ \sum_{m=1}^n P(A_m) \to 1 \quad a.s.$$

Proof Let $X_m = 1_{A_m}$ and let $S_n = X_1 + \cdots + X_n$. Since the A_m are pairwise independent, the X_m are uncorrelated, and hence Theorem 2.2.1 implies

$$\text{var}(S_n) = \text{var}(X_1) + \cdots + \text{var}(X_n)$$

$\text{var}(X_m) \leq E(X_m^2) = E(X_m)$, since $X_m \in \{0,1\}$, so $\text{var}(S_n) \leq E(S_n)$. Chebyshev's inequality implies

$$(*) \qquad P(|S_n - ES_n| > \delta ES_n) \leq \text{var}(S_n)/(\delta ES_n)^2 \leq 1/(\delta^2 ES_n) \to 0$$

as $n \to \infty$. (Since we have assumed $ES_n \to \infty$.)

The last computation shows that $S_n/ES_n \to 1$ in probability. To get almost sure convergence, we have to take subsequences. Let $n_k = \inf\{n : ES_n \geq k^2\}$. Let $T_k = S_{n_k}$ and note that the definition and $EX_m \leq 1$ imply $k^2 \leq ET_k \leq k^2 + 1$. Replacing n by n_k in (∗) and using $ET_k \geq k^2$ shows

$$P(|T_k - ET_k| > \delta ET_k) \leq 1/(\delta^2 k^2)$$

So $\sum_{k=1}^\infty P(|T_k - ET_k| > \delta ET_k) < \infty$, and the Borel-Cantelli lemma implies $P(|T_k - ET_k| > \delta ET_k \text{ i.o.}) = 0$. Since δ is arbitrary, it follows that $T_k/ET_k \to 1$ a.s. To show $S_n/ES_n \to 1$ a.s., pick an ω so that $T_k(\omega)/ET_k \to 1$ and observe that if $n_k \leq n < n_{k+1}$ then

$$\frac{T_k(\omega)}{ET_{k+1}} \leq \frac{S_n(\omega)}{ES_n} \leq \frac{T_{k+1}(\omega)}{ET_k}$$

To show that the terms at the left and right ends $\to 1$, we rewrite the last inequalities as

$$\frac{ET_k}{ET_{k+1}} \cdot \frac{T_k(\omega)}{ET_k} \leq \frac{S_n(\omega)}{ES_n} \leq \frac{T_{k+1}(\omega)}{ET_{k+1}} \cdot \frac{ET_{k+1}}{ET_k}$$

From this, we see it is enough to show $ET_{k+1}/ET_k \to 1$, but this follows from

$$k^2 \leq ET_k \leq ET_{k+1} \leq (k+1)^2 + 1$$

and the fact that $\{(k+1)^2 + 1\}/k^2 = 1 + 2/k + 2/k^2 \to 1$. □

The moral of the proof of Theorem 2.3.9 is that if you want to show that $X_n/c_n \to 1$ a.s. for sequences $c_n, X_n \geq 0$ that are increasing, it is enough to prove the result for a subsequence $n(k)$ that has $c_{n(k+1)}/c_{n(k)} \to 1$.

Example 2.3.10 (Record values) Let $X_1, X_2 \ldots$ be a sequence of random variables and think of X_k as the distance for an individual's kth high jump or shot-put toss so that $A_k = \{X_k > \sup_{j<k} X_j\}$ is the event that a record occurs at time k. Ignoring the fact that an athlete's performance may get better with more experience or that injuries may occur, we will suppose that $X_1, X_2 \ldots$ are i.i.d. with a distribution $F(x)$ that is continuous. Even though it may seem that the occurrence of a record at time k will make it less likely that one will occur at time $k+1$, we

Claim. The A_k are independent with $P(A_k) = 1/k$.

To prove this, we start by observing that since F is continuous $P(X_j = X_k) = 0$ for any $j \neq k$ (see Exercise 2.1.5), so we can let $Y_1^n > Y_2^n > \cdots > Y_n^n$ be the random variables $X_1, \ldots, X_n$ put into decreasing order and define a random permutation of $\{1, \ldots, n\}$ by $\pi_n(i) = j$ if $X_i = Y_j^n$, i.e., if the ith random variable has rank j. Since the distribution of $(X_1, \ldots, X_n)$ is not affected by changing the order of the random variables, it is easy to see:

(a) The permutation π_n is uniformly distributed over the set of $n!$ possibilities.

Proof of (a) This is "obvious" by symmetry, but if one wants to hear more, we can argue as follows. Let π_n be the permutation induced by $(X_1, \ldots, X_n)$, and let σ_n be a randomly chosen permutation of $\{1, \ldots, n\}$ independent of the X sequence. Then we can say two things about the permutation induced by $(X_{\sigma(1)}, \ldots, X_{\sigma(n)})$: (i) it is $\pi_n \circ \sigma_n$, and (ii) it has the same distribution as π_n. The desired result follows now by noting that if π is any permutation, $\pi \circ \sigma_n$, is uniform over the $n!$ possibilities. □

Once you believe (a), the rest is easy:

(b) $P(A_n) = P(\pi_n(n) = 1) = 1/n$.

(c) If $m < n$ and $i_{m+1}, \ldots i_n$ are distinct elements of $\{1, \ldots, n\}$, then

$$P(A_m | \pi_n(j) = i_j \text{ for } m+1 \leq j \leq n) = 1/m$$

Intuitively, this is true since if we condition on the ranks of $X_{m+1}, \ldots, X_n$, then this determines the set of ranks available for $X_1, \ldots, X_m$, but all possible orderings of the ranks are equally likely, and hence there is probability $1/m$ that the smallest rank will end up at m.

Proof of (c) If we let σ_m be a randomly chosen permutation of $\{1, \ldots, m\}$, then (i) $\pi_n \circ \sigma_m$ has the same distribution as π_n, and (ii) since the application of σ_m randomly rearranges $\pi_n(1), \ldots, \pi_n(m)$ the desired result follows. □

If we let $m_1 < m_2 \ldots < m_k$, then it follows from (c) that

$$P(A_{m_1} | A_{m_2} \cap \ldots \cap A_{m_k}) = P(A_{m_1})$$

and the claim follows by induction.

Using Theorem 2.3.9 and the by now familiar fact that $\sum_{m=1}^{n} 1/m \sim \log n$, we have:

Theorem 2.3.11 *If $R_n = \sum_{m=1}^n 1_{A_m}$ is the number of records at time n. then as $n \to \infty$,*

$$R_n / \log n \to 1 \quad a.s.$$

The reader should note that the last result is independent of the distribution F (as long as it is continuous).

Remark Let $X_1, X_2, \dots$ be i.i.d. with a distribution that is continuous. Let Y_i be the number of $j \le i$ with $X_j > X_i$. It follows from (a) that Y_i are independent random variables with $P(Y_i = j) = 1/i$ for $0 \le j < i - 1$.

Comic relief Let $X_0, X_1, \dots$ be i.i.d. and imagine they are the offers you get for a car you are going to sell. Let $N = \inf\{n \ge 1 : X_n > X_0\}$. Symmetry implies $P(N > n) \ge 1/(n+1)$. (When the distribution is continuous, this probability is exactly $1/(n+1)$, but our distribution now is general and ties go to the first person who calls.) Using Exercise 2.2.7 now:

$$EN = \sum_{n=0}^{\infty} P(N > n) \ge \sum_{n=0}^{\infty} \frac{1}{n+1} = \infty$$

so the expected time you have to wait until you get an offer better than the first one is ∞. To avoid lawsuits, let me hasten to add that I am not suggesting that you should take the first offer you get!

Example 2.3.12 (Head runs) Let X_n, $n \in \mathbf{Z}$, be i.i.d. with $P(X_n = 1) = P(X_n = -1) = 1/2$. Let $\ell_n = \max\{m : X_{n-m+1} = \dots = X_n = 1\}$ be the length of the run of $+1$'s at time n, and let $L_n = \max_{1 \le m \le n} \ell_m$ be the longest run at time n. We use a two-sided sequence so that for all n, $P(\ell_n = k) = (1/2)^{k+1}$ for $k \ge 0$. Since $\ell_1 < \infty$, the result we are going to prove

$$L_n / \log_2 n \to 1 \quad \text{a.s.} \tag{2.3.1}$$

is also true for a one-sided sequence. To prove (2.3.1), we begin by observing

$$P(\ell_n \ge (1+\epsilon)\log_2 n) \le n^{-(1+\epsilon)}$$

for any $\epsilon > 0$, so it follows from the Borel-Cantelli lemma that $\ell_n \le (1+\epsilon)\log_2 n$ for $n \ge N_\epsilon$. Since ϵ is arbitrary, it follows that

$$\limsup_{n\to\infty} L_n / \log_2 n \le 1 \quad \text{a.s.}$$

To get a result in the other direction, we break the first n trials into disjoint blocks of length $[(1-\epsilon)\log_2 n] + 1$, on which the variables are all 1 with probability

$$2^{-[(1-\epsilon)\log_2 n]-1} \ge n^{-(1-\epsilon)}/2,$$

to conclude that if n is large enough so that $[n/\{[(1-\epsilon)\log_2 n]+1\}] \ge n/\log_2 n$

$$P(L_n \le (1-\epsilon)\log_2 n) \le (1 - n^{-(1-\epsilon)}/2)^{n/(\log_2 n)} \le \exp(-n^{\epsilon}/2\log_2 n)$$

which is summable, so the Borel-Cantelli lemma implies

$$\liminf_{n\to\infty} L_n / \log_2 n \ge 1 \quad \text{a.s.}$$

Exercise 2.3.3 Let ℓ_n be the length of the head run at time. See Example 8.3.1 for the precise definition. Show that $\limsup_{n\to\infty} \ell_n/\log_2 n = 1$, $\liminf_{n\to\infty} \ell_n = 0$ a.s.

Exercises

2.3.4 **Fatou's lemma.** Suppose $X_n \geq 0$ and $X_n \to X$ in probability. Show that $\liminf_{n\to\infty} EX_n \geq EX$.

2.3.5 **Dominated convergence.** Suppose $X_n \to X$ in probability and (a) $|X_n| \leq Y$ with $EY < \infty$ or (b) there is a continuous function g with $g(x) > 0$ for large x with $|x|/g(x) \to 0$ as $|x| \to \infty$ so that $Eg(X_n) \leq C < \infty$ for all n. Show that $EX_n \to EX$.

2.3.6 **Metric for convergence in probability.** Show (a) that $d(X,Y) = E(|X-Y|/(1+|X-Y|))$ defines a metric on the set of random variables, i.e., (i) $d(X,Y) = 0$ if and only if $X = Y$ a.s., (ii) $d(X,Y) = d(Y,X)$, (iii) $d(X,Z) \leq d(X,Y) + d(Y,Z)$ and (b) that $d(X_n, X) \to 0$ as $n \to \infty$ if and only if $X_n \to X$ in probability.

2.3.7 Show that random variables are a complete space under the metric defined in the previous exercise, i.e., if $d(X_m, X_n) \to 0$ whenever $m, n \to \infty$, then there is a r.v. X_∞ so that $X_n \to X_\infty$ in probability.

2.3.8 Let A_n be a sequence of independent events with $P(A_n) < 1$ for all n. Show that $P(\cup A_n) = 1$ implies $\sum_n P(A_n) = \infty$ and hence $P(A_n \text{ i.o.}) = 1$.

2.3.9 (i) If $P(A_n) \to 0$ and $\sum_{n=1}^\infty P(A_n^c \cap A_{n+1}) < \infty$, then $P(A_n \text{ i.o.}) = 0$. (ii) Find an example of a sequence A_n to which the result in (i) can be applied but the Borel-Cantelli lemma cannot.

2.3.10 **Kochen-Stone lemma.** Suppose $\sum P(A_k) = \infty$. Use Exercises 1.6.6 and 2.3.1 to show that if

$$\limsup_{n\to\infty} \left(\sum_{k=1}^n P(A_k)\right)^2 \Bigg/ \left(\sum_{1\leq j,k\leq n} P(A_j \cap A_k)\right) = \alpha > 0$$

then $P(A_n \text{ i.o.}) \geq \alpha$. The case $\alpha = 1$ contains Theorem 2.3.7.

2.3.11 Let $X_1, X_2, \ldots$ be independent with $P(X_n = 1) = p_n$ and $P(X_n = 0) = 1 - p_n$. Show that (i) $X_n \to 0$ in probability if and only if $p_n \to 0$, and (ii) $X_n \to 0$ a.s. if and only if $\sum p_n < \infty$.

2.3.12 Let $X_1, X_2, \ldots$ be a sequence of r.v.'s on $(\Omega, \mathcal{F}, P)$, where Ω is a countable set and $\mathcal{F}$ consists of all subsets of Ω. Show that $X_n \to X$ in probability implies $X_n \to X$ a.s.

2.3.13 If X_n is any sequence of random variables, there are constants $c_n \to \infty$ so that $X_n/c_n \to 0$ a.s.

2.3.14 Let $X_1, X_2, \ldots$ be independent. Show that $\sup X_n < \infty$ a.s. if and only if $\sum_n P(X_n > A) < \infty$ for some A.

2.3.15 Let $X_1, X_2, \ldots$ be i.i.d. with $P(X_i > x) = e^{-x}$, let $M_n = \max_{1 \le m \le n} X_m$. Show that (i) $\limsup_{n\to\infty} X_n/\log n = 1$ a.s. and (ii) $M_n/\log n \to 1$ a.s.

2.3.16 Let $X_1, X_2, \ldots$ be i.i.d. with distribution F, let $\lambda_n \uparrow \infty$, and let $A_n = \{\max_{1\le m\le n} X_m > \lambda_n\}$. Show that $P(A_n \text{ i.o.}) = 0$ or 1 according as $\sum_{n\ge 1}(1 - F(\lambda_n)) < \infty$ or $= \infty$.

2.3.17 Let $Y_1, Y_2, \ldots$ be i.i.d. Find necessary and sufficient conditions for (i) $Y_n/n \to 0$ almost surely, (ii) $(\max_{m\le n} Y_m)/n \to 0$ almost surely, (iii) $(\max_{m\le n} Y_m)/n \to 0$ in probability, and (iv) $Y_n/n \to 0$ in probability.

2.3.18 Let $0 \le X_1 \le X_2 \ldots$ be random variables with $EX_n \sim an^\alpha$ with $a, \alpha > 0$, and $\text{var}(X_n) \le Bn^\beta$ with $\beta < 2\alpha$. Show that $X_n/n^\alpha \to a$ a.s.

2.3.19 Let X_n be independent Poisson r.v.'s with $EX_n = \lambda_n$, and let $S_n = X_1 + \cdots + X_n$. Show that if $\sum \lambda_n = \infty$, then $S_n/ES_n \to 1$ a.s.

2.3.20 Show that if X_n is the outcome of the nth play of the St. Petersburg game (Example 2.2.16), then $\limsup_{n\to\infty} X_n/(n \log_2 n) = \infty$ a.s. and hence the same result holds for S_n. This shows that the convergence $S_n/(n \log_2 n) \to 1$ in probability proved in Section 2.2 does not occur a.s.

2.4 Strong Law of Large Numbers

We are now ready to give Etemadi's (1981) proof of:

Theorem 2.4.1 (Strong law of large numbers) *Let $X_1, X_2, \ldots$ be pairwise independent identically distributed random variables with $E|X_i| < \infty$. Let $EX_i = \mu$ and $S_n = X_1 + \cdots + X_n$. Then $S_n/n \to \mu$ a.s. as $n \to \infty$.*

Proof As in the proof of the weak law of large numbers, we begin by truncating.

Lemma 2.4.2 *Let $Y_k = X_k 1_{(|X_k|\le k)}$ and $T_n = Y_1 + \cdots + Y_n$. It is sufficient to prove that $T_n/n \to \mu$ a.s.*

Proof $\sum_{k=1}^\infty P(|X_k| > k) \le \int_0^\infty P(|X_1| > t)\,dt = E|X_1| < \infty$ so $P(X_k \ne Y_k \text{ i.o.}) = 0$. This shows that $|S_n(\omega) - T_n(\omega)| \le R(\omega) < \infty$ a.s. for all n, from which the desired result follows. □

The second step is not so intuitive, but it is an important part of this proof and the one given in Section 2.5.

Lemma 2.4.3 $\sum_{k=1}^\infty$ *var* $(Y_k)/k^2 \le 4E|X_1| < \infty$.

Proof To bound the sum, we observe

$$\text{var}(Y_k) \le E(Y_k^2) = \int_0^\infty 2yP(|Y_k| > y)\,dy \le \int_0^k 2yP(|X_1| > y)\,dy$$

so using Fubini's theorem (since everything is ≥ 0 and the sum is just an integral with respect to counting measure on $\{1,2,\ldots\}$)

$$\sum_{k=1}^{\infty} E(Y_k^2)/k^2 \leq \sum_{k=1}^{\infty} k^{-2} \int_0^{\infty} 1_{(y<k)}\, 2y\, P(|X_1| > y)\, dy$$
$$= \int_0^{\infty} \left\{ \sum_{k=1}^{\infty} k^{-2} 1_{(y<k)} \right\} 2y P(|X_1| > y)\, dy$$

Since $E|X_1| = \int_0^\infty P(|X_1| > y)\, dy$, we can complete the proof by showing:

Lemma 2.4.4 *If* $y \geq 0$ *then* $2y \sum_{k>y} k^{-2} \leq 4$.

Proof We begin with the observation that if $m \geq 2$, then

$$\sum_{k \geq m} k^{-2} \leq \int_{m-1}^{\infty} x^{-2} dx = (m-1)^{-1}$$

When $y \geq 1$, the sum starts with $k = [y] + 1 \geq 2$, so

$$2y \sum_{k>y} k^{-2} \leq 2y/[y] \leq 4$$

since $y/[y] \leq 2$ for $y \geq 1$ (the worst case being y close to 2). To cover $0 \leq y < 1$, we note that in this case

$$2y \sum_{k>y} k^{-2} \leq 2 \left(1 + \sum_{k=2}^{\infty} k^{-2} \right) \leq 4$$

This establishes Lemma 2.4.4, which completes the proof of Lemma 2.4.3 and of the theorem. □

The first two steps, Lemmas 2.4.2 and 2.4.3, are standard. Etemadi's inspiration was that since $X_n^+, n \geq 1$, and $X_n^-, n \geq 1$ satisfy the assumptions of the theorem and $X_n = X_n^+ - X_n^-$, we can without loss of generality suppose $X_n \geq 0$. As in the proof of Theorem 2.3.9, we will prove the result first for a subsequence and then use monotonicity to control the values in between. This time, however, we let $\alpha > 1$ and $k(n) = [\alpha^n]$. Chebyshev's inequality implies that if $\epsilon > 0$

$$\sum_{n=1}^{\infty} P(|T_{k(n)} - ET_{k(n)}| > \epsilon k(n)) \leq \epsilon^{-2} \sum_{n=1}^{\infty} \text{var}\,(T_{k(n)})/k(n)^2$$
$$= \epsilon^{-2} \sum_{n=1}^{\infty} k(n)^{-2} \sum_{m=1}^{k(n)} \text{var}\,(Y_m) = \epsilon^{-2} \sum_{m=1}^{\infty} \text{var}\,(Y_m) \sum_{n:k(n) \geq m} k(n)^{-2}$$

where we have used Fubini's theorem to interchange the two summations of nonnegative terms. Now $k(n) = [\alpha^n]$ and $[\alpha^n] \geq \alpha^n/2$ for $n \geq 1$, so summing the geometric series and noting that the first term is $\leq m^{-2}$:

$$\sum_{n:\alpha^n \geq m} [\alpha^n]^{-2} \leq 4 \sum_{n:\alpha^n \geq m} \alpha^{-2n} \leq 4(1 - \alpha^{-2})^{-1} m^{-2}$$

Combining our computations shows

$$\sum_{n=1}^{\infty} P(|T_{k(n)} - ET_{k(n)}| > \epsilon k(n)) \leq 4(1-\alpha^{-2})^{-1}\epsilon^{-2} \sum_{m=1}^{\infty} E(Y_m^2)m^{-2} < \infty$$

by Lemma 2.4.3. Since ϵ is arbitrary $(T_{k(n)} - ET_{k(n)})/k(n) \to 0$ a.s. The dominated convergence theorem implies $EY_k \to EX_1$ as $k \to \infty$, so $ET_{k(n)}/k(n) \to EX_1$ and we have shown $T_{k(n)}/k(n) \to EX_1$ a.s. To handle the intermediate values, we observe that if $k(n) \leq m < k(n+1)$

$$\frac{T_{k(n)}}{k(n+1)} \leq \frac{T_m}{m} \leq \frac{T_{k(n+1)}}{k(n)}$$

(here we use $Y_i \geq 0$), so recalling $k(n) = [\alpha^n]$, we have $k(n+1)/k(n) \to \alpha$ and

$$\frac{1}{\alpha}EX_1 \leq \liminf_{n\to\infty} T_m/m \leq \limsup_{m\to\infty} T_m/m \leq \alpha EX_1$$

Since $\alpha > 1$ is arbitrary, the proof is complete. □

The next result shows that the strong law holds whenever EX_i exists.

Theorem 2.4.5 *Let $X_1, X_2, \ldots$ be i.i.d. with $EX_i^+ = \infty$ and $EX_i^- < \infty$. If $S_n = X_1 + \cdots + X_n$, then $S_n/n \to \infty$ a.s.*

Proof Let $M > 0$ and $X_i^M = X_i \wedge M$. The X_i^M are i.i.d. with $E|X_i^M| < \infty$, so if $S_n^M = X_1^M + \cdots + X_n^M$, then Theorem 2.4.1 implies $S_n^M/n \to EX_i^M$. Since $X_i \geq X_i^M$, it follows that

$$\liminf_{n\to\infty} S_n/n \geq \lim_{n\to\infty} S_n^M/n = EX_i^M$$

The monotone convergence theorem implies $E(X_i^M)^+ \uparrow EX_i^+ = \infty$ as $M \uparrow \infty$, so $EX_i^M = E(X_i^M)^+ - E(X_i^M)^- \uparrow \infty$, and we have $\liminf_{n\to\infty} S_n/n \geq \infty$, which implies the desired result. □

The rest of this section is devoted to applications of the strong law of large numbers.

Example 2.4.6 (Renewal theory) Let $X_1, X_2, \ldots$ be i.i.d. with $0 < X_i < \infty$. Let $T_n = X_1 + \cdots + X_n$ and think of T_n as the time of nth occurrence of some event. For a concrete situation, consider a diligent janitor who replaces a light bulb the instant it burns out. Suppose the first bulb is put in at time 0 and let X_i be the lifetime of the ith light bulb. In this interpretation, T_n is the time the nth light bulb burns out and $N_t = \sup\{n : T_n \leq t\}$ is the number of light bulbs that have burnt out by time t.

Theorem 2.4.7 *If $EX_1 = \mu \leq \infty$, then as $t \to \infty$,*

$$N_t/t \to 1/\mu \text{ a.s.} \quad (1/\infty = 0).$$

Proof By Theorems 2.4.1 and 2.4.5, $T_n/n \to \mu$ a.s. From the definition of N_t, it follows that $T(N_t) \leq t < T(N_t + 1)$, so dividing through by N_t gives

$$\frac{T(N_t)}{N_t} \leq \frac{t}{N_t} \leq \frac{T(N_t+1)}{N_t+1} \cdot \frac{N_t+1}{N_t}$$

To take the limit, we note that since $T_n < \infty$ for all n, we have $N_t \uparrow \infty$ as $t \to \infty$. The strong law of large numbers implies that for $\omega \in \Omega_0$ with $P(\Omega_0) = 1$, we have $T_n(\omega)/n \to \mu$, $N_t(\omega) \uparrow \infty$, and hence

$$\frac{T_{N_t(\omega)}(\omega)}{N_t(\omega)} \to \mu \qquad \frac{N_t(\omega)+1}{N_t(\omega)} \to 1$$

From this it follows that for $\omega \in \Omega_0$ that $t/N_t(\omega) \to \mu$ a.s. □

The last argument shows that if $X_n \to X_\infty$ a.s. and $N(n) \to \infty$ a.s., then $X_{N(n)} \to X_\infty$ a.s. We have written this out with care because the analogous result for convergence in probability is false. If $X_n \in \{0,1\}$ are independent with $P(X = 1) = a_n \to 0$ and $\sum_n a_n = \infty$, then $X_n \to 0$ in probability, but if we let $N(n) = \inf\{m \geq n : X_m = 1\}$, then $X_{N(n)} = 1$ a.s.

Example 2.4.8 (Empirical distribution functions) Let $X_1, X_2, \ldots$ be i.i.d. with distribution F and let

$$F_n(x) = n^{-1} \sum_{m=1}^{n} 1_{(X_m \leq x)}$$

$F_n(x)$ = the observed frequency of values that are $\leq x$, hence the name given here. The next result shows that F_n converges uniformly to F as $n \to \infty$.

Theorem 2.4.9 (The Glivenko-Cantelli theorem) *As $n \to \infty$,*

$$\sup_x |F_n(x) - F(x)| \to 0 \quad a.s.$$

Proof Fix x and let $Y_n = 1_{(X_n \leq x)}$. Since the Y_n are i.i.d. with $EY_n = P(X_n \leq x) = F(x)$, the strong law of large numbers implies that $F_n(x) = n^{-1}\sum_{m=1}^n Y_m \to F(x)$ a.s. In general, if F_n is a sequence of nondecreasing functions that converges pointwise to a bounded and continuous limit F, then $\sup_x |F_n(x) - F(x)| \to 0$. However, the distribution function $F(x)$ may have jumps, so we have to work a little harder.

Again, fix x and let $Z_n = 1_{(X_n < x)}$. Since the Z_n are i.i.d. with $EZ_n = P(X_n < x) = F(x-) = \lim_{y \uparrow x} F(y)$, the strong law of large numbers implies that $F_n(x-) = n^{-1}\sum_{m=1}^n Z_m \to F(x-)$ a.s. For $1 \leq j \leq k-1$ let $x_{j,k} = \inf\{y : F(y) \geq j/k\}$. The pointwise convergence of $F_n(x)$ and $F_n(x-)$ imply that we can pick $N_k(\omega)$ so that if $n \geq N_k(\omega)$ then

$$|F_n(x_{j,k}) - F(x_{j,k})| < k^{-1} \quad \text{and} \quad |F_n(x_{j,k}-) - F(x_{j,k}-)| < k^{-1}$$

for $1 \leq j \leq k-1$. If we let $x_{0,k} = -\infty$ and $x_{k,k} = \infty$, then the last two inequalities hold for $j = 0$ or k. If $x \in (x_{j-1,k}, x_{j,k})$ with $1 \leq j \leq k$ and $n \geq N_k(\omega)$, then using the monotonicity of F_n and F, and $F(x_{j,k}-) - F(x_{j-1,k}) \leq k^{-1}$, we have

$$F_n(x) \leq F_n(x_{j,k}-) \leq F(x_{j,k}-) + k^{-1} \leq F(x_{j-1,k}) + 2k^{-1} \leq F(x) + 2k^{-1}$$
$$F_n(x) \geq F_n(x_{j-1,k}) \geq F(x_{j-1,k}) - k^{-1} \geq F(x_{j,k}-) - 2k^{-1} \geq F(x) - 2k^{-1}$$

so $\sup_x |F_n(x) - F(x)| \leq 2k^{-1}$, and we have proved the result. □

Example 2.4.10 (Shannon's theorem) Let $X_1, X_2, \ldots \in \{1, \ldots, r\}$ be independent with $P(X_i = k) = p(k) > 0$ for $1 \le k \le r$. Here we are thinking of $1, \ldots, r$ as the letters of an alphabet, and $X_1, X_2, \ldots$ are the successive letters produced by an information source. In this i.i.d. case, it is the proverbial monkey at a typewriter. Let $\pi_n(\omega) = p(X_1(\omega)) \cdots p(X_n(\omega))$ be the probability of the realization we observed in the first n trials. Since $\log \pi_n(\omega)$ is a sum of independent random variables, it follows from the strong law of large numbers that

$$-n^{-1} \log \pi_n(\omega) \to H \equiv -\sum_{k=1}^{r} p(k) \log p(k) \text{ a.s.}$$

The constant H is called the **entropy** of the source and is a measure of how random it is. The last result is the **asymptotic equipartition property**: If $\epsilon > 0$, then as $n \to \infty$

$$P\{\exp(-n(H + \epsilon)) \le \pi_n(\omega) \le \exp(-n(H - \epsilon)\} \to 1$$

Exercises

2.4.1 **Lazy janitor.** Suppose the ith light bulb burns for an amount of time X_i and then remains burned out for time Y_i before being replaced. Suppose the X_i, Y_i are positive and independent with the X's having distribution F and the Y's having distribution G, both of which have finite mean. Let R_t be the amount of time in $[0, t]$ that we have a working light bulb. Show that $R_t/t \to EX_i/(EX_i + EY_i)$ almost surely.

2.4.2 Let $X_0 = (1, 0)$ and define $X_n \in \mathbf{R}^2$ inductively by declaring that X_{n+1} is chosen at random from the ball of radius $|X_n|$ centered at the origin, i.e., $X_{n+1}/|X_n|$ is uniformly distributed on the ball of radius 1 and independent of $X_1, \ldots, X_n$. Prove that $n^{-1} \log |X_n| \to c$ a.s. and compute c.

2.4.3 **Investment problem.** We assume that at the beginning of each year you can buy bonds for \$1 that are worth \$ a at the end of the year or stocks that are worth a random amount $V \ge 0$. If you always invest a fixed proportion p of your wealth in bonds, then your wealth at the end of year $n + 1$ is $W_{n+1} = (ap + (1 - p)V_n)W_n$. Suppose $V_1, V_2, \ldots$ are i.i.d. with $EV_n^2 < \infty$ and $E(V_n^{-2}) < \infty$. (i) Show that $n^{-1} \log W_n \to c(p)$ a.s. (ii) Show that $c(p)$ is concave. [Use Theorem A.5.1 in the Appendix to justify differentiating under the expected value.] (iii) By investigating $c'(0)$ and $c'(1)$, give conditions on V that guarantee that the optimal choice of p is in (0,1). (iv) Suppose $P(V = 1) = P(V = 4) = 1/2$. Find the optimal p as a function of a.

2.5 Convergence of Random Series*

In this section, we will pursue a second approach to the strong law of large numbers based on the convergence of random series. This approach has the advantage that it leads to estimates on the rate of convergence under moment assumptions, Theorems 2.5.11 and 2.5.12, and to a negative result for the infinite mean case, Theorem 2.5.13, which is stronger than the one in Theorem 2.3.8. The first two results in this section are of considerable interest in their own right, although we will see more general versions in Lemma 3.1.1 and Theorem 3.4.2.

To state the first result, we need some notation. Let $\mathcal{F}'_n = \sigma(X_n, X_{n+1}, \ldots)$ = the future after time n = the smallest σ-field with respect to which all the X_m, $m \geq n$ are measurable. Let $\mathcal{T} = \cap_n \mathcal{F}'_n$ = the remote future, or **tail σ-field.** Intuitively, $A \in \mathcal{T}$ if and only if changing a finite number of values does not affect the occurrence of the event. As usual, we turn to examples to help explain the definition.

Example 2.5.1 If $B_n \in \mathcal{R}$, then $\{X_n \in B_n \text{ i.o.}\} \in \mathcal{T}$. If we let $X_n = 1_{A_n}$ and $B_n = \{1\}$, this example becomes $\{A_n \text{ i.o.}\}$.

Example 2.5.2 Let $S_n = X_1 + \cdots + X_n$. It is easy to check that
$\{\lim_{n\to\infty} S_n \text{ exists }\} \in \mathcal{T}$,
$\{\limsup_{n\to\infty} S_n > 0\} \notin \mathcal{T}$,
$\{\limsup_{n\to\infty} S_n/c_n > x\} \in \mathcal{T}$ if $c_n \to \infty$.

The next result shows that all examples are trivial.

Theorem 2.5.3 (Kolmogorov's 0-1 law) *If $X_1, X_2, \ldots$ are independent and $A \in \mathcal{T}$, then $P(A) = 0$ or 1.*

Proof We will show that A is independent of itself, that is, $P(A \cap A) = P(A)P(A)$, so $P(A) = P(A)^2$, and hence $P(A) = 0$ or 1. We will sneak up on this conclusion in two steps:

(a) $A \in \sigma(X_1, \ldots, X_k)$ and $B \in \sigma(X_{k+1}, X_{k+2}, \ldots)$ are independent.

Proof of (a). If $B \in \sigma(X_{k+1}, \ldots, X_{k+j})$ for some j, this follows from Theorem 2.1.9. Since $\sigma(X_1, \ldots, X_k)$ and $\cup_j \sigma(X_{k+1}, \ldots, X_{k+j})$ are π-systems that contain Ω (a) follows from Theorem 2.1.7.

(b) $A \in \sigma(X_1, X_2, \ldots)$ and $B \in \mathcal{T}$ are independent.

Proof of (b). Since $\mathcal{T} \subset \sigma(X_{k+1}, X_{k+2}, \ldots)$, if $A \in \sigma(X_1, \ldots, X_k)$ for some k, this follows from (a). $\cup_k \sigma(X_1, \ldots, X_k)$ and $\mathcal{T}$ are π-systems that contain Ω, so (b) follows from Theorem 2.1.7.

Since $\mathcal{T} \subset \sigma(X_1, X_2, \ldots)$, (b) implies an $A \in \mathcal{T}$ is independent of itself and Theorem 2.5.3 follows. □

Before taking up our main topic, we will prove a 0-1 law that, in the i.i.d. case, generalizes Kolmogorov's. To state the new 0-1 law, we need two definitions. A **finite permutation** of $\mathbf{N} = \{1, 2, \ldots\}$ is a map π from $\mathbf{N}$ onto $\mathbf{N}$ so that $\pi(i) \neq i$ for only finitely many i. If π is a finite permutation of $\mathbf{N}$ and $\omega \in S^{\mathbf{N}}$, we define $(\pi\omega)_i = \omega_{\pi(i)}$. In words, the coordinates of ω are rearranged according to π. Since $X_i(\omega) = \omega_i$ this is the same as rearranging the random variables. An event A is **permutable** if $\pi^{-1}A \equiv \{\omega : \pi\omega \in A\}$ is equal to A for any finite permutation π, or, in other words, if its occurrence is not affected by rearranging finitely many of the random variables. The collection of permutable events is a σ-field. It is called the **exchangeable** σ-field and denoted by $\mathcal{E}$.

To see the reason for interest in permutable events, suppose $S = \mathbf{R}$ and let $S_n(\omega) = X_1(\omega) + \cdots + X_n(\omega)$. Two examples of permutable events are

(i) $\{\omega : S_n(\omega) \in B \text{ i.o.}\}$

(ii) $\{\omega : \limsup_{n\to\infty} S_n(\omega)/c_n \geq 1\}$

In each case, the event is permutable because $S_n(\omega) = S_n(\pi\omega)$ for large n. The list of examples can be enlarged considerably by observing:

(iii) All events in the tail σ-field $\mathcal{T}$ are permutable.

To see this, observe that if $A \in \sigma(X_{n+1}, X_{n+2}, \ldots)$, then the occurrence of A is unaffected by a permutation of $X_1, \ldots, X_n$. (i) shows that the converse of (iii) is false. The next result shows that for an i.i.d. sequence there is no difference between $\mathcal{E}$ and $\mathcal{T}$. They are both trivial.

Theorem 2.5.4 (Hewitt-Savage 0-1 law) *If $X_1, X_2, \ldots$ are i.i.d. and $A \in \mathcal{E}$ then $P(A) \in \{0, 1\}$.*

Proof Let $A \in \mathcal{E}$. As in the proof of Kolmogorov's 0-1 law, we will show A is independent of itself, i.e., $P(A) = P(A \cap A) = P(A)P(A)$ so $P(A) \in \{0, 1\}$. Let $A_n \in \sigma(X_1, \ldots, X_n)$ so that

(a)
$$P(A_n \Delta A) \to 0$$

Here $A \Delta B = (A - B) \cup (B - A)$ is the symmetric difference. The existence of the A_n's is proved in part ii of Lemma A.2.1. A_n can be written as $\{\omega : (\omega_1, \ldots, \omega_n) \in B_n\}$ with $B_n \in \mathcal{S}^n$. Let

$$\pi(j) = \begin{cases} j + n & \text{if } 1 \le j \le n \\ j - n & \text{if } n + 1 \le j \le 2n \\ j & \text{if } j \ge 2n + 1 \end{cases}$$

Observing that π^2 is the identity (so we don't have to worry about whether to write π or π^{-1}) and the coordinates are i.i.d. (so the permuted coordinates are) gives

(b)
$$P(\omega : \omega \in A_n \Delta A) = P(\omega : \pi\omega \in A_n \Delta A)$$

Now $\{\omega : \pi\omega \in A\} = \{\omega : \omega \in A\}$, since A is permutable, and

$$\{\omega : \pi\omega \in A_n\} = \{\omega : (\omega_{n+1}, \ldots, \omega_{2n}) \in B_n\}$$

If we use A'_n to denote the last event, then we have

(c)
$$\{\omega : \pi\omega \in A_n \Delta A\} = \{\omega : \omega \in A'_n \Delta A\}$$

Combining (b) and (c) gives

(d)
$$P(A_n \Delta A) = P(A'_n \Delta A)$$

It is easy to see that

$$|P(B) - P(C)| \le |P(B \Delta C|$$

so (d) implies $P(A_n), P(A'_n) \to P(A)$. Now $A - C \subset (A - B) \cup (B - C)$ and with a similar inequality for $C - A$ implies $A \Delta C \subset (A \Delta B) \cup (B \Delta C)$. The last inequality, (d), and (a) imply

$$P(A_n \Delta A'_n) \le P(A_n \Delta A) + P(A \Delta A'_n) \to 0$$

The last result implies

$$\begin{aligned} 0 &\le P(A_n) - P(A_n \cap A'_n) \\ &\le P(A_n \cup A'_n) - P(A_n \cap A'_n) = P(A_n \Delta A'_n) \to 0 \end{aligned}$$

so $P(A_n \cap A'_n) \to P(A)$. But A_n and A'_n are independent, so

$$P(A_n \cap A'_n) = P(A_n)P(A'_n) \to P(A)^2$$

This shows $P(A) = P(A)^2$, and proves Theorem 2.5.4. □

If $A_1, A_2, \ldots$ are independent, then Theorem 2.5.3 implies $P(A_n \text{ i.o.}) = 0$ or 1. Applying Theorem 2.5.3 to Example 2.5.2 gives $P(\lim_{n\to\infty} S_n \text{ exists}) = 0$ or 1. The next result will help us prove the probability is 1 in certain situations.

Theorem 2.5.5 (Kolmogorov's maximal inequality) *Suppose $X_1, \ldots, X_n$ are independent with $EX_i = 0$ and $var(X_i) < \infty$. If $S_n = X_1 + \cdots + X_n$, then*

$$P\left(\max_{1\le k\le n} |S_k| \ge x\right) \le x^{-2}\, var(S_n)$$

Remark Under the same hypotheses, Chebyshev's inequality (Theorem 1.6.4) gives only

$$P(|S_n| \ge x) \le x^{-2} \operatorname{var}(S_n)$$

Proof Let $A_k = \{|S_k| \ge x \text{ but } |S_j| < x \text{ for } j < k\}$, i.e., we break things down according to the time that $|S_k|$ first exceeds x. Since the A_k are disjoint and $(S_n - S_k)^2 \ge 0$,

$$\begin{aligned} ES_n^2 &\ge \sum_{k=1}^n \int_{A_k} S_n^2\, dP = \sum_{k=1}^n \int_{A_k} S_k^2 + 2S_k(S_n - S_k) + (S_n - S_k)^2\, dP \\ &\ge \sum_{k=1}^n \int_{A_k} S_k^2\, dP + \sum_{k=1}^n \int 2S_k 1_{A_k} \cdot (S_n - S_k)\, dP \end{aligned}$$

$S_k 1_{A_k} \in \sigma(X_1, \ldots, X_k)$ and $S_n - S_k \in \sigma(X_{k+1}, \ldots, X_n)$ are independent by Theorem 2.1.10, so using Theorem 2.1.13 and $E(S_n - S_k) = 0$ shows

$$\int 2S_k 1_{A_k} \cdot (S_n - S_k)\, dP = E(2S_k 1_{A_k}) \cdot E(S_n - S_k) = 0$$

Using now the fact that $|S_k| \ge x$ on A_k and the A_k are disjoint,

$$ES_n^2 \ge \sum_{k=1}^n \int_{A_k} S_k^2\, dP \ge \sum_{k=1}^n x^2 P(A_k) = x^2 P\left(\max_{1\le k\le n} |S_k| \ge x\right)$$ □

We turn now to our results on convergence of series. To state them, we need a definition. We say that $\sum_{n=1}^\infty a_n$ converges if $\lim_{N\to\infty} \sum_{n=1}^N a_n$ exists.

Theorem 2.5.6 *Suppose $X_1, X_2, \ldots$ are independent and have $EX_n = 0$. If*

$$\sum_{n=1}^\infty var(X_n) < \infty$$

then with probability one $\sum_{n=1}^\infty X_n(\omega)$ converges.

Proof Let $S_N = \sum_{n=1}^{N} X_n$. From Theorem 2.5.5, we get

$$P\left(\max_{M \le m \le N} |S_m - S_M| > \epsilon\right) \le \epsilon^{-2} \text{var}\,(S_N - S_M) = \epsilon^{-2} \sum_{n=M+1}^{N} \text{var}\,(X_n)$$

Letting $N \to \infty$ in the last result, we get

$$P\left(\sup_{m \ge M} |S_m - S_M| > \epsilon\right) \le \epsilon^{-2} \sum_{n=M+1}^{\infty} \text{var}\,(X_n) \to 0 \quad \text{as } M \to \infty$$

If we let $w_M = \sup_{m,n \ge M} |S_m - S_n|$, then $w_M \downarrow$ as $M \uparrow$ and

$$P(w_M > 2\epsilon) \le P\left(\sup_{m \ge M} |S_m - S_M| > \epsilon\right) \to 0$$

as $M \to \infty$ so $w_M \downarrow 0$ almost surely. But $w_M(\omega) \downarrow 0$ implies $S_n(\omega)$ is a Cauchy sequence and hence $\lim_{n\to\infty} S_n(\omega)$ exists, so the proof is complete. □

Example 2.5.7 Let $X_1, X_2, \ldots$ be independent with

$$P(X_n = n^{-\alpha}) = P(X_n = -n^{-\alpha}) = 1/2$$

$EX_n = 0$ and $\text{var}\,(X_n) = n^{-2\alpha}$ so if $\alpha > 1/2$, it follows from Theorem 2.5.6 that $\sum X_n$ converges. Theorem 2.5.8 shows that $\alpha > 1/2$ is also necessary for this conclusion. Notice that there is absolute convergence, i.e., $\sum |X_n| < \infty$, if and only if $\alpha > 1$.

Theorem 2.5.6 is sufficient for all of our applications, but our treatment would not be complete if we did not mention the last word on convergence of random series.

Theorem 2.5.8 (Kolmogorov's three-series theorem) *Let $X_1, X_2, \ldots$ be independent. Let $A > 0$ and let $Y_i = X_i 1_{(|X_i| \le A)}$. In order that $\sum_{n=1}^{\infty} X_n$ converges a.s., it is necessary and sufficient that*

$$\text{(i)} \sum_{n=1}^{\infty} P(|X_n| > A) < \infty, \text{ (ii)} \sum_{n=1}^{\infty} EY_n \text{ converges, and (iii)} \sum_{n=1}^{\infty} \text{var}\,(Y_n) < \infty$$

Proof We will prove the necessity in Example 3.4.12 as an application of the central limit theorem. To prove the sufficiency, let $\mu_n = EY_n$. (iii) and Theorem 2.5.6 imply that $\sum_{n=1}^{\infty}(Y_n - \mu_n)$ converges a.s. Using (ii) now gives that $\sum_{n=1}^{\infty} Y_n$ converges a.s. (i) and the Borel-Cantelli lemma imply $P(X_n \ne Y_n \text{ i.o.}) = 0$, so $\sum_{n=1}^{\infty} X_n$ converges a.s. □

The link between convergence of series and the strong law of large numbers is provided by

Theorem 2.5.9 (Kronecker's lemma) *If $a_n \uparrow \infty$ and $\sum_{n=1}^{\infty} x_n/a_n$ converges, then*

$$a_n^{-1} \sum_{m=1}^{n} x_m \to 0$$

Proof Let $a_0 = 0$, $b_0 = 0$, and for $m \geq 1$, let $b_m = \sum_{k=1}^{m} x_k/a_k$. Then $x_m = a_m(b_m - b_{m-1})$ and so

$$\begin{aligned} a_n^{-1}\sum_{m=1}^{n} x_m &= a_n^{-1}\left\{\sum_{m=1}^{n} a_m b_m - \sum_{m=1}^{n} a_m b_{m-1}\right\} \\ &= a_n^{-1}\left\{a_n b_n + \sum_{m=2}^{n} a_{m-1}b_{m-1} - \sum_{m=1}^{n} a_m b_{m-1}\right\} \\ &= b_n - \sum_{m=1}^{n} \frac{(a_m - a_{m-1})}{a_n} b_{m-1} \end{aligned}$$

(Recall $a_0 = 0$.) By hypothesis, $b_n \to b_\infty$ as $n \to \infty$. Since $a_m - a_{m-1} \geq 0$, the last sum is an average of $b_0, \ldots, b_n$. Intuitively, if $\epsilon > 0$ and $M < \infty$ are fixed and n is large, the average assigns mass $\geq 1 - \epsilon$ to the b_m with $m \geq M$, so

$$\sum_{m=1}^{n} \frac{(a_m - a_{m-1})}{a_n} b_{m-1} \to b_\infty$$

To argue formally, let $B = \sup |b_n|$, pick M so that $|b_m - b_\infty| < \epsilon/2$ for $m > M$, then pick N so that $a_M/a_n < \epsilon/4B$ for $n \geq N$. Now if $n \geq N$, we have

$$\begin{aligned} \left|\sum_{m=1}^{n} \frac{(a_m - a_{m-1})}{a_n} b_{m-1} - b_\infty\right| &\leq \sum_{m=1}^{n} \frac{(a_m - a_{m-1})}{a_n} |b_{m-1} - b_\infty| \\ &\leq \frac{a_M}{a_n} \cdot 2B + \frac{a_n - a_M}{a_n} \cdot \frac{\epsilon}{2} < \epsilon \end{aligned}$$

proving the desired result since ϵ is arbitrary. □

Theorem 2.5.10 (The strong law of large numbers) *Let $X_1, X_2 \ldots$ be i.i.d. random variables with $E|X_i| < \infty$. Let $EX_i = \mu$ and $S_n = X_1 + \cdots + X_n$. Then $S_n/n \to \mu$ a.s. as $n \to \infty$.*

Proof Let $Y_k = X_k 1_{(|X_k| \leq k)}$ and $T_n = Y_1 + \cdots + Y_n$. By (a) in the proof of Theorem 2.4.1 it suffices to show that $T_n/n \to \mu$. Let $Z_k = Y_k - EY_k$, so $EZ_k = 0$. Now $\text{var}(Z_k) = \text{var}(Y_k) \leq EY_k^2$ and (b) in the proof of Theorem 2.4.1 imply

$$\sum_{k=1}^{\infty} \text{var}(Z_k)/k^2 \leq \sum_{k=1}^{\infty} EY_k^2/k^2 < \infty$$

Applying Theorem 2.5.6 now, we conclude that $\sum_{k=1}^{\infty} Z_k/k$ converges a.s. so Theorem 2.5.9 implies

$$n^{-1}\sum_{k=1}^{n}(Y_k - EY_k) \to 0 \quad \text{and hence} \quad \frac{T_n}{n} - n^{-1}\sum_{k=1}^{n} EY_k \to 0 \text{ a.s.}$$

The dominated convergence theorem implies $EY_k \to \mu$ as $k \to \infty$. From this, it follows easily that $n^{-1}\sum_{k=1}^{n} EY_k \to \mu$, and hence $T_n/n \to \mu$. □

2.5.1 Rates of Convergence

As mentioned earlier, one of the advantages of the random series proof is that it provides estimates on the rate of convergence of $S_n/n \to \mu$. By subtracting μ from each random variable, we can and will suppose without loss of generality that $\mu = 0$.

Theorem 2.5.11 *Let $X_1, X_2, \ldots$ be i.i.d. random variables with $EX_i = 0$ and $EX_i^2 = \sigma^2 < \infty$. Let $S_n = X_1 + \cdots + X_n$. If $\epsilon > 0$ then*

$$S_n/n^{1/2}(\log n)^{1/2+\epsilon} \to 0 \quad a.s.$$

Remark The law of the iterated logarithm, Theorem 8.5.2 will show that

$$\limsup_{n\to\infty} S_n/n^{1/2}(\log\log n)^{1/2} = \sigma\sqrt{2} \quad \text{a.s.}$$

so the last result is not far from the best possible.

Proof Let $a_n = n^{1/2}(\log n)^{1/2+\epsilon}$ for $n \ge 2$ and $a_1 > 0$.

$$\sum_{n=1}^{\infty} \text{var}\,(X_n/a_n) = \sigma^2 \left(\frac{1}{a_1^2} + \sum_{n=2}^{\infty} \frac{1}{n(\log n)^{1+2\epsilon}} \right) < \infty$$

so applying Theorem 2.5.6, we get $\sum_{n=1}^{\infty} X_n/a_n$ converges a.s. and the indicated result follows from Theorem 2.5.9. □

The next result due to Marcinkiewicz and Zygmund treats the situation in which $EX_i^2 = \infty$ but $E|X_i|^p < \infty$ for some $1 < p < 2$.

Theorem 2.5.12 *Let $X_1, X_2, \ldots$ be i.i.d. with $EX_1 = 0$ and $E|X_1|^p < \infty$ where $1 < p < 2$. If $S_n = X_1 + \cdots + X_n$, then $S_n/n^{1/p} \to 0$ a.s.*

Proof Let $Y_k = X_k 1_{(|X_k| \le k^{1/p})}$ and $T_n = Y_1 + \cdots + Y_n$.

$$\sum_{k=1}^{\infty} P(Y_k \neq X_k) = \sum_{k=1}^{\infty} P(|X_k|^p > k) \le E|X_k|^p < \infty$$

so the Borel-Cantelli lemma implies $P(Y_k \neq X_k \text{ i.o.}) = 0$, and it suffices to show $T_n/n^{1/p} \to 0$. Using $\text{var}\,(Y_m) \le E(Y_m^2)$, Lemma 2.2.13 with $p = 2$, $P(|Y_m| > y) \le P(|X_1| > y)$, and Fubini's theorem (everything is ≥ 0), we have

$$\begin{aligned}
\sum_{m=1}^{\infty} \text{var}\,(Y_m/m^{1/p}) &\le \sum_{m-1}^{\infty} EY_m^2/m^{2/p} \\
&\le \sum_{m=1}^{\infty} \sum_{n=1}^{m} \int_{(n-1)^{1/p}}^{n^{1/p}} \frac{2y}{m^{2/p}} P(|X_1| > y)\, dy \\
&= \sum_{n=1}^{\infty} \int_{(n-1)^{1/p}}^{n^{1/p}} \sum_{m=n}^{\infty} \frac{2y}{m^{2/p}} P(|X_1| > y)\, dy
\end{aligned}$$

To bound the integral, we note that for $n \geq 2$ comparing the sum with the integral of $x^{-2/p}$

$$\sum_{m=n}^{\infty} m^{-2/p} \leq \frac{p}{2-p}(n-1)^{(p-2)/p} \leq Cy^{p-2}$$

when $y \in [(n-1)^{1/p}, n^{1/p}]$. Since $E|X_i|^p = \int_0^\infty px^{p-1} P(|X_i| > x)\,dx < \infty$, it follows that

$$\sum_{m=1}^{\infty} \text{var}\,(Y_m/m^{1/p}) < \infty$$

If we let $\mu_m = EY_m$ and apply Theorems 2.5.6 and 2.5.9, it follows that

$$n^{-1/p} \sum_{m=1}^{n} (Y_m - \mu_m) \to 0 \quad \text{a.s.}$$

To estimate μ_m, we note that since $EX_m = 0$, $\mu_m = -E(X_i; |X_i| > m^{1/p})$, so

$$\begin{aligned} |\mu_m| &\leq E(|X|; |X_i| > m^{1/p}) = m^{1/p} E(|X|/m^{1/p}; |X_i| > m^{1/p}) \\ &\leq m^{1/p} E((|X|/m^{1/p})^p; |X_i| > m^{1/p}) \\ &\leq m^{-1+1/p} p^{-1} E(|X_i|^p; |X_i| > m^{1/p}) \end{aligned}$$

Now $\sum_{m=1}^{n} m^{-1+1/p} \leq Cn^{1/p}$ and $E(|X_i|^p; |X_i| > m^{1/p}) \to 0$ as $m \to \infty$, so $n^{-1/p} \sum_{m=1}^{n} \mu_m \to 0$ and the desired result follows. □

2.5.2 *Infinite Mean*

The St. Petersburg game, discussed in Example 2.2.16 and Exercise 2.3.20, is a situation in which $EX_i = \infty$, $S_n/n \log_2 n \to 1$ in probability but

$$\limsup_{n\to\infty} S_n/(n \log_2 n) = \infty \text{ a.s.}$$

The next result, due to Feller (1946), shows that when $E|X_1| = \infty$, S_n/a_n cannot converge almost surely to a nonzero limit. In Theorem 2.3.8 we considered the special case $a_n = n$.

Theorem 2.5.13 *Let $X_1, X_2, \ldots$ be i.i.d. with $E|X_1| = \infty$ and let $S_n = X_1 + \cdots + X_n$. Let a_n be a sequence of positive numbers with a_n/n increasing. Then* $\limsup_{n\to\infty} |S_n|/a_n = 0$ *or ∞ according as $\sum_n P(|X_1| \geq a_n) < \infty$ or $= \infty$.*

Proof Since $a_n/n \uparrow$, $a_{kn} \geq ka_n$ for any integer k. Using this and $a_n \uparrow$,

$$\sum_{n=1}^{\infty} P(|X_1| \geq ka_n) \geq \sum_{n=1}^{\infty} P(|X_1| \geq a_{kn}) \geq \frac{1}{k} \sum_{m=k}^{\infty} P(|X_1| \geq a_m)$$

The last observation shows that if the sum is infinite, $\limsup_{n\to\infty} |X_n|/a_n = \infty$. Since $\max\{|S_{n-1}|, |S_n|\} \geq |X_n|/2$, it follows that $\limsup_{n\to\infty} |S_n|/a_n = \infty$.

To prove the other half, we begin with the identity

$$(*) \qquad \sum_{m=1}^{\infty} mP(a_{m-1} \leq |X_i| < a_m) = \sum_{n=1}^{\infty} P(|X_i| \geq a_{n-1})$$

To see this, write $m = \sum_{n=1}^{m} 1$ and then use Fubini's theorem. We now let $Y_n = X_n 1_{(|X_n|<a_n)}$, and $T_n = Y_1 + \cdots + Y_n$. When the sum is finite, $P(Y_n \neq X_n \text{ i.o.}) = 0$, and it suffices to investigate the behavior of the T_n. To do this, we let $a_0 = 0$ and compute

$$\begin{aligned}\sum_{n=1}^{\infty} \text{var}\,(Y_n/a_n) &\le \sum_{n=1}^{\infty} EY_n^2/a_n^2 \\ &= \sum_{n=1}^{\infty} a_n^{-2} \sum_{m=1}^{n} \int_{[a_{m-1},a_m)} y^2\, dF(y) \\ &= \sum_{m=1}^{\infty} \int_{[a_{m-1},a_m)} y^2\, dF(y) \sum_{n=m}^{\infty} a_n^{-2}\end{aligned}$$

Since $a_n \ge n a_m/m$, we have $\sum_{n=m}^{\infty} a_n^{-2} \le (m^2/a_m^2) \sum_{n=m}^{\infty} n^{-2} \le C m a_m^{-2}$, so

$$\le C \sum_{m=1}^{\infty} m \int_{[a_{m-1},a_m)} dF(y)$$

Using $(*)$ now, we conclude $\sum_{n=1}^{\infty} \text{var}\,(Y_n/a_n) < \infty$.

The last step is to show $ET_n/a_n \to 0$. To begin, we note that if $E|X_i| = \infty$, $\sum_{n=1}^{\infty} P(|X_i| > a_n) < \infty$, and $a_n/n \uparrow$, we must have $a_n/n \uparrow \infty$. To estimate ET_n/a_n now, we observe that

$$\begin{aligned}\left| a_n^{-1} \sum_{m=1}^{n} EY_m \right| &\le a_n^{-1} n \sum_{m=1}^{n} E(|X_m|; |X_m| < a_m) \\ &\le \frac{n a_N}{a_n} + \frac{n}{a_n} E(|X_i|; a_N \le |X_i| < a_n)\end{aligned}$$

where the last inequality holds for any fixed N. Since $a_n/n \to \infty$, the first term converges to 0. Since $m/a_m \downarrow$, the second is

$$\begin{aligned}&\le \sum_{m=N+1}^{n} \frac{m}{a_m} E(|X_i|; a_{m-1} \le |X_i| < a_m) \\ &\le \sum_{m=N+1}^{\infty} m P(a_{m-1} \le |X_i| < a_m)\end{aligned}$$

$(*)$ shows that the sum is finite, so it is small if N is large and the desired result follows. □

Exercises

2.5.1 Suppose $X_1, X_2, \ldots$ are i.i.d. with $EX_i = 0$, $\text{var}\,(X_i) = C < \infty$. Use Theorem 2.5.5 with $n = m^{\alpha}$ where $\alpha(2p - 1) > 1$ to conclude that if $S_n = X_1 + \cdots + X_n$ and $p > 1/2$, then $S_n/n^p \to 0$ almost surely.

2.5.2 The converse of Theorem 2.5.12 is much easier. Let $p > 0$. If $S_n/n^{1/p} \to 0$ a.s., then $E|X_1|^p < \infty$.

2.5.3 Let $X_1, X_2, \dots$ be i.i.d. standard normals. Show that for any t

$$\sum_{n=1}^{\infty} X_n \cdot \frac{\sin(n\pi t)}{n} \quad \text{converges a.s.}$$

2.5.4 Let $X_1, X_2, \dots$ be independent with $EX_n = 0$, $\text{var}(X_n) = \sigma_n^2$. (i) Show that if $\sum_n \sigma_n^2/n^2 < \infty$, then $\sum_n X_n/n$ converges a.s., and hence $n^{-1}\sum_{m=1}^n X_m \to 0$ a.s. (ii) Suppose $\sum \sigma_n^2/n^2 = \infty$ and without loss of generality that $\sigma_n^2 \le n^2$ for all n. Show that there are independent random variables X_n with $EX_n = 0$ and $\text{var}(X_n) \le \sigma_n^2$ so that X_n/n, and hence $n^{-1}\sum_{m\le n} X_m$ does not converge to 0 a.s.

2.5.5 Let $X_n \ge 0$ be independent for $n \ge 1$. The following are equivalent:
(i) $\sum_{n=1}^\infty X_n < \infty$ a.s. (ii) $\sum_{n=1}^\infty [P(X_n > 1) + E(X_n 1_{(X_n \le 1)})] < \infty$
(iii) $\sum_{n=1}^\infty E(X_n/(1+X_n)) < \infty$.

2.5.6 Let $\psi(x) = x^2$ when $|x| \le 1$ and $= |x|$ when $|x| \ge 1$. Show that if $X_1, X_2, \dots$ are independent with $EX_n = 0$ and $\sum_{n=1}^\infty E\psi(X_n) < \infty$, then $\sum_{n=1}^\infty X_n$ converges a.s.

2.5.7 Let X_n be independent. Suppose $\sum_{n=1}^\infty E|X_n|^{p(n)} < \infty$, where $0 < p(n) \le 2$ for all n and $EX_n = 0$ when $p(n) > 1$. Show that $\sum_{n=1}^\infty X_n$ converges a.s.

2.5.8 Let $X_1, X_2, \dots$ be i.i.d. and not $\equiv 0$. Then the radius of convergence of the power series $\sum_{n\ge 1} X_n(\omega) z^n$ (i.e., $r(\omega) = \sup\{c : \sum |X_n(\omega)| c^n < \infty\}$) is 1 a.s. or 0 a.s., according as $E \log^+ |X_1| < \infty$ or $= \infty$, where $\log^+ x = \max(\log x, 0)$.

2.5.9 Let $X_1, X_2, \dots$ be independent and let $S_{m,n} = X_{m+1} + \dots + X_n$. Then

$$(\star) \qquad P\left(\max_{m<j\le n} |S_{m,j}| > 2a\right) \min_{m<k\le n} P(|S_{k,n}| \le a) \le P(|S_{m,n}| > a)$$

2.5.10 Use $(\star)$ to prove a theorem of P. Lévy: Let $X_1, X_2, \dots$ be independent and let $S_n = X_1 + \dots + X_n$. If $\lim_{n\to\infty} S_n$ exists in probability, then it also exists a.s.

2.5.11 Let $X_1, X_2, \dots$ be i.i.d. and $S_n = X_1 + \dots + X_n$. Use $(\star)$ to conclude that if $S_n/n \to 0$ in probability, then $(\max_{1\le m\le n} S_m)/n \to 0$ in probability.

2.5.12 Let $X_1, X_2, \dots$ be i.i.d. and $S_n = X_1 + \dots + X_n$. Suppose $a_n \uparrow \infty$ and $a(2^n)/a(2^{n-1})$ is bounded. (i) Use $(\star)$ to show that if $S_n/a(n) \to 0$ in probability and $S_{2^n}/a(2^n) \to 0$ a.s., then $S_n/a(n) \to 0$ a.s. (ii) Suppose in addition that $EX_1 = 0$ and $EX_1^2 < \infty$. Use the previous exercise and Chebyshev's inequality to conclude that $S_n/n^{1/2}(\log_2 n)^{1/2+\epsilon} \to 0$ a.s.

2.6 Renewal Theory*

Let $\xi_1, \xi_2, \dots$ be i.i.d. positive random variables (i.e., $P(\xi_i > 0) = 1$) with distribution F, and define a sequence of times by $T_0 = 0$, and $T_k = T_{k-1} + \xi_k$ for $k \ge 1$. As explained in Example 2.4.6, we think of ξ_i as the lifetime of the ith light bulb, and T_k is the time the kth bulb burns out. A second interpretation from the discussion of Poisson process in Section 3.6 is that T_k is the time of arrival of the kth customer. A third interpretation from Chapter 5 is that T_k is the time of the kth visit to state k.

To have a neutral terminology, we will refer to the T_k as **renewals**. The term renewal refers to the fact that the process "starts afresh" at T_k, i.e., $\{T_{k+j} - T_k, j \geq 1\}$ has the same distribution as $\{T_j, j \geq 1\}$.

Figure 2.1 Renewal sequence.

Departing slightly from the notation in Example 2.4.6, we let $N_t = \inf\{k : T_k > t\}$. N_t is the number of renewals in $[0,t]$, counting the renewal at time 0. The advantage of this definition is that N_t is a stopping time, i.e., $\{N_t = k\}$ is measurable with respect to $\mathcal{F}_k$.

In Theorem 2.4.7, we showed that:

Theorem 2.6.1 *As $t \to \infty$, $N_t/t \to 1/\mu$ a.s., where $\mu = E\xi_i \in (0,\infty]$ and $1/\infty = 0$.*

Our next result concerns the asymptotic behavior of $U(t) = EN_t$. To derive the result we need:

Theorem 2.6.2 (Wald's equation) *Let $X_1, X_2, \ldots$ be i.i.d. with $E|X_i| < \infty$. If N is a stopping time with $EN < \infty$, then $ES_N = EX_1 EN$.*

Proof First suppose the $X_i \geq 0$.

$$ES_N = \int S_N dP = \sum_{n=1}^{\infty} \int S_n 1_{\{N=n\}} dP = \sum_{n=1}^{\infty} \sum_{m=1}^{n} \int X_m 1_{\{N=n\}} dP$$

Since the $X_i \geq 0$, we can interchange the order of summation (i.e., use Fubini's theorem) to conclude that the last expression

$$= \sum_{m=1}^{\infty} \sum_{n=m}^{\infty} \int X_m 1_{\{N=n\}} dP = \sum_{m=1}^{\infty} \int X_m 1_{\{N \geq m\}} dP$$

Now $\{N \geq m\} = \{N \leq m-1\}^c \in \mathcal{F}_{m-1}$ and is independent of X_m, so the last expression

$$= \sum_{m=1}^{\infty} EX_m P(N \geq m) = EX_1 EN$$

To prove the result in general, we run the last argument backward. If we have $EN < \infty$, then

$$\infty > \sum_{m=1}^{\infty} E|X_m| P(N \geq m) = \sum_{m=1}^{\infty} \sum_{n=m}^{\infty} \int |X_m| 1_{\{N=n\}}\, dP$$

The last formula shows that the double sum converges absolutely in one order, so Fubini's theorem gives

$$\sum_{m=1}^{\infty} \sum_{n=m}^{\infty} \int X_m 1_{\{N=n\}} dP = \sum_{n=1}^{\infty} \sum_{m=1}^{n} \int X_m 1_{\{N=n\}} dP$$

Using the independence of $\{N \geq m\} \in \mathcal{F}_{m-1}$ and X_m, and rewriting the last identity, it follows that

$$\sum_{m=1}^{\infty} EX_m P(N \geq m) = ES_N$$

Since the left-hand side is $ENEX_1$, the proof is complete. □

Theorem 2.6.3 *As $t \to \infty$, $U(t)/t \to 1/\mu$.*

Proof We will apply Wald's equation to the stopping time N_t. The first step is to show that $EN_t < \infty$. To do this, pick $\delta > 0$ so that $P(\xi_i > \delta) = \epsilon > 0$ and pick K so that $K\delta \geq t$. Since K consecutive $\xi_i' s$ that are $> \delta$ will make $T_n > t$, we have

$$P(N_t > mK) \leq (1 - \epsilon^K)^m$$

and $EN_t < \infty$. If $\mu < \infty$, applying Wald's equation now gives

$$\mu EN_t = ET_{N_t} \geq t$$

so $U(t) \geq t/\mu$. The last inequality is trivial when $\mu = \infty$ so it holds in general.

Turning to the upper bound, we observe that if $P(\xi_i \leq c) = 1$, then repeating the last argument shows $\mu EN_t = ES_{N_t} \leq t + c$, and the result holds for bounded distributions. If we let $\bar{\xi}_i = \xi_i \wedge c$ and define $\bar{T}_n$ and $\bar{N}_t$ in the obvious way, then

$$EN_t \leq E\bar{N}_t \leq (t + c)/E(\bar{\xi}_i)$$

Letting $t \to \infty$ and then $c \to \infty$ gives $\limsup_{t\to\infty} EN_t/t \leq 1/\mu$, and the proof is complete. □

To take a closer look at when the renewals occur, we let

$$U(A) = \sum_{n=0}^{\infty} P(T_n \in A)$$

U is called the **renewal measure**. We absorb the old definition, $U(t) = EN_t$, into the new one by regarding $U(t)$ as shorthand for $U([0,t])$. This should not cause problems since $U(t)$ is the distribution function for the renewal measure. The asymptotic behavior of $U(t)$ depends upon whether the distribution F is **arithmetic**, i.e., concentrated on $\{\delta, 2\delta, 3\delta, \ldots\}$ for some $\delta > 0$, or **nonarithmetic**, i.e., not arithmetic. We will treat the first case in Chapter 5 as an application of Markov chains, so we will restrict our attention to the second case here.

Theorem 2.6.4 (Blackwell's renewal theorem) *If F is nonarithmetic, then*

$$U([t, t + h]) \to h/\mu \quad \text{as } t \to \infty.$$

We will prove the result in the case $\mu < \infty$ by "coupling" following Lindvall (1977) and Athreya, McDonald, and Ney (1978). To set the stage for the proof, we need a definition and some preliminary computations. If $T_0 \geq 0$ is independent of $\xi_1, \xi_2, \ldots$ and has distribution G, then $T_k = T_{k-1} + \xi_k$, $k \geq 1$ defines a **delayed renewal process**, and G is the **delay**

distribution. If we let $N_t = \inf\{k : T_k > t\}$ as before and set $V(t) = EN_t$, then breaking things down according to the value of T_0 gives

$$V(t) = \int_0^t U(t-s)\, dG(s) \tag{2.6.1}$$

The last integral, and all similar expressions here, is intended to include the contribution of any mass G has at 0. If we let $U(r) = 0$ for $r < 0$, then the last equation can be written as $V = U * G$, where $*$ denotes convolution.

Applying similar reasoning to U gives

$$U(t) = 1 + \int_0^t U(t-s)\, dF(s) \tag{2.6.2}$$

or, introducing convolution notation,

$$U = 1_{[0,\infty)}(t) + U * F.$$

Convolving each side with G (and recalling $G * U = U * G$) gives

$$V = G * U = G + V * F \tag{2.6.3}$$

We know $U(t) \sim t/\mu$. Our next step is to find a G so that $V(t) = t/\mu$. Plugging what we want into (2.6.3) gives

$$t/\mu = G(t) + \int_0^t \frac{t-y}{\mu}\, dF(y)$$

$$\text{so} \qquad G(t) = t/\mu - \int_0^t \frac{t-y}{\mu}\, dF(y)$$

The integration-by-parts formula is

$$\int_0^t K(y)\, dH(y) = H(t)K(t) - H(0)K(0) - \int_0^t H(y)\, dK(y)$$

If we let $H(y) = (y-t)/\mu$ and $K(y) = 1 - F(y)$, then

$$\frac{1}{\mu}\int_0^t 1 - F(y)\, dy = \frac{t}{\mu} - \int_0^t \frac{t-y}{\mu}\, dF(y)$$

so we have

$$G(t) = \frac{1}{\mu}\int_0^t 1 - F(y)\, dy \tag{2.6.4}$$

It is comforting to note that $\mu = \int_{[0,\infty)} 1 - F(y)\, dy$, so the last formula defines a probability distribution. When the delay distribution G is the one given in (2.6.4), we call the result the **stationary renewal process**. Something very special happens when $F(t) = 1 - \exp(-\lambda t)$, $t \geq 0$, where $\lambda > 0$ (i.e., the renewal process is a rate λ Poisson process). In this case, $\mu = 1/\lambda$ so $G(t) = F(t)$.

Proof of Theorem 2.6.4 for $\mu < \infty$. Let T_n be a renewal process (with $T_0 = 0$) and T'_n be an independent stationary renewal process. Our first goal is to find J and K so that $|T_J - T'_K| < \epsilon$ and the increments $\{T_{J+i} - T_J, i \geq 1\}$ and $\{T'_{K+i} - T'_K, i \geq 1\}$ are i.i.d. sequences independent of what has come before.

Let $\eta_1, \eta_2, \ldots$ and $\eta'_1, \eta'_2, \ldots$ be i.i.d. independent of T_n and T'_n and take the values 0 and 1 with probability 1/2 each. Let $\nu_n = \eta_1 + \cdots + \eta_n$ and $\nu'_n = 1 + \eta'_1 + \cdots + \eta'_n$, $S_n = T_{\nu_n}$ and $S'_n = T'_{\nu'_n}$. The increments of $S_n - S'_n$ are 0 with probability at least 1/4, and the support of their distribution is symmetric and contains the support of the ξ_k so if the distribution of the ξ_k is nonarithmetic, the random walk $S_n - S'_n$ is irreducible. Since the increments of $S_n - S'_n$ have mean 0, $N = \inf\{n : |S_n - S'_n| < \epsilon\}$ has $P(N < \infty) = 1$, and we can let $J = \nu_N$ and $K = \nu'_N$. Let

$$T''_n = \begin{cases} T_n & \text{if } J \geq n \\ T_J + T'_{K+(n-J)} - T'_K & \text{if } J < n \end{cases}$$

In other words, the increments $T''_{J+i} - T''_J$ are the same as $T'_{K+i} - T'_K$ for $i \geq 1$.

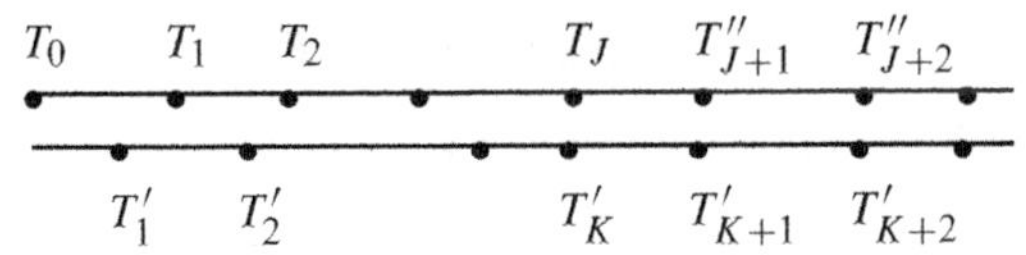

Figure 2.2 Coupling of renewal processes.

It is easy to see from the construction that T_n and T''_n have the same distribution. If we let

$$N'[s,t] = |\{n : T'_n \in [s,t]\}| \quad \text{and} \quad N''[s,t] = |\{n : T''_n \in [s,t]\}|$$

be the number of renewals in $[s,t]$ in the two processes, then on $\{T_J \leq t\}$

$$N''[t,t+h] = N'[t + T'_K - T_J, t + h + T'_K - T_J] \begin{cases} \geq N'[t+\epsilon, t+h-\epsilon] \\ \leq N'[t-\epsilon, t+h+\epsilon] \end{cases}$$

To relate the expected number of renewals in the two processes, we observe that even if we condition on the location of all the renewals in $[0,s]$, the expected number of renewals in $[s,s+t]$ is at most $U(t)$, since the worst thing that could happen is to have a renewal at time s. Combining the last two observations, we see that if $\epsilon < h/2$ (so $[t+\epsilon, t+h-\epsilon]$ has positive length)

$$\begin{aligned} U([t,t+h]) &= EN''[t,t+h] \geq E(N'[t+\epsilon, t+h-\epsilon]; T_J \leq t) \\ &\geq \frac{h-2\epsilon}{\mu} - P(T_J > t)U(h) \end{aligned}$$

since $EN'[t+\epsilon, t+h-\epsilon] = (h-2\epsilon)/\mu$ and $\{T_J > t\}$ is determined by the renewals of T in $[0,t]$ and the renewals of T' in $[0, t+\epsilon]$. For the other direction, we observe

$$\begin{aligned} U([t,t+h]) &\leq E(N'[t-\epsilon, t+h+\epsilon]; T_J \leq t) + E(N''[t,t+h]; T_J > t) \\ &\leq \frac{h+2\epsilon}{\mu} + P(T_J > t)U(h) \end{aligned}$$

The desired result now follows from the fact that $P(T_J > t) \to 0$ and $\epsilon < h/2$ is arbitrary. □

Proof of Theorem 2.6.4 for $\mu = \infty$. In this case, there is no stationary renewal process, so we have to resort to other methods. Let

$$\beta = \limsup_{t\to\infty} U(t,t+1] = \lim_{k\to\infty} U(t_k,t_k+1]$$

for some sequence $t_k \to \infty$. We want to prove that $\beta = 0$, for then by addition the previous conclusion holds with 1 replaced by any integer n and, by monotonicity, with n replaced by any $h < n$, and this gives us the result in Theorem 2.6.4. Fix i and let

$$a_{k,j} = \int_{(j-1,j]} U(t_k - y, t_k + 1 - y]\, dF^{i*}(y)$$

By considering the location of T_i we get

(a) $$\lim_{k\to\infty} \sum_{j=1}^{\infty} a_{k,j} = \lim_{k\to\infty} \int U(t_k - y, t_k + 1 - y]\, dF^{i*}(y) = \beta$$

Since β is the lim sup, we must have

(b) $$\limsup_{k\to\infty} a_{k,j} \le \beta \cdot P(T_i \in (j-1,j])$$

We want to conclude from (a) and (b) that

(c) $$\liminf_{k\to\infty} a_{k,j} \ge \beta \cdot P(T_i \in (j-1,j])$$

To do this, we observe that by considering the location of the first renewal in $(j-1,j]$

(d) $$0 \le a_{k,j} \le U(1) P(T_i \in (j-1,j])$$

(c) is trivial when $\beta = 0$ so we can suppose $\beta > 0$. To argue by contradiction, suppose there exist j_0 and $\epsilon > 0$ so that

$$\liminf_{k\to\infty} a_{k,j_0} \le \beta \cdot \{P(T_i \in (j_0 - 1, j_0]) - \epsilon\}$$

Pick $k_n \to \infty$ so that

$$a_{k_n,j_0} \to \beta \cdot \{P(T_i \in (j_0 - 1, j_0]) - \epsilon\}$$

Using (d), we can pick $J \ge j_0$ so that

$$\limsup_{n\to\infty} \sum_{j=J+1}^{\infty} a_{k_n,j} \le U(1) \sum_{j=J+1}^{\infty} P(T_i \in (j-1,j]) \le \beta\epsilon/2$$

Now an easy argument shows

$$\limsup_{n\to\infty} \sum_{j=1}^{J} a_{k_n,j} \le \sum_{j=1}^{J} \limsup_{n\to\infty} a_{k_n,j} \le \beta \left(\sum_{j=1}^{J} P(T_i \in (j-1,j]) - \epsilon \right)$$

by (b) and our assumption. Adding the last two results shows

$$\limsup_{n\to\infty} \sum_{j=1}^{\infty} a_{k_n,j} \le \beta(1 - \epsilon/2)$$

which contradicts (a), and proves (c).

Now, if $j - 1 < y \le j$, we have

$$U(t_k - y, t_k + 1 - y] \le U(t_k - j, t_k + 2 - j]$$

so using (c), it follows that for j with $P(T_i \in (j-1, j]) > 0$, we must have

$$\liminf_{k\to\infty} U(t_k - j, t_k + 2 - j] \ge \beta$$

Summing over i, we see that the last conclusion is true when $U(j-1, j] > 0$.

The support of U is closed under addition. (If x is in the support of F^{m*} and y is in the support of F^{n*}, then $x + y$ is in the support of $F^{(m+n)*}$.) We have assumed F is nonarithmetic, so $U(j-1, j] > 0$ for $j \ge j_0$. Letting $r_k = t_k - j_0$ and considering the location of the last renewal in $[0, r_k]$ and the index of the T_i gives

$$\begin{aligned}1 = \sum_{i=0}^{\infty} \int_0^{r_k} (1 - F(r_k - y))\, dF^{i*}(y) &= \int_0^{r_k} (1 - F(r_k - y))\, dU(y) \\ &\ge \sum_{n=1}^{\infty} (1 - F(2n))\, U(r_k - 2n, r_k + 2 - 2n]\end{aligned}$$

Since $\liminf_{k\to\infty} U(r_k - 2n, r_k + 2 - 2n] \ge \beta$ and

$$\sum_{n=0}^{\infty} (1 - F(2n)) \ge \mu/2 = \infty$$

β must be 0, and the proof is complete. □

Remark Following Lindvall (1977), we have based the proof for $\mu = \infty$ on part of Feller's (1961) proof of the discrete renewal theorem (i.e., for arithmetic distributions). See Freedman (1971b) pages 22–25 for an account of Feller's proof. Purists can find a proof that does everything by coupling in Thorisson (1987).

Our next topic is the **renewal equation**: $H = h + H * F$. Two cases we have seen in (2.6.2) and (2.6.3) are:

Example 2.6.5 $h \equiv 1$: $U(t) = 1 + \int_0^t U(t-s)\, dF(s)$

Example 2.6.6 $h(t) = G(t)$: $V(t) = G(t) + \int_0^t V(t-s)\, dF(s)$

The last equation is valid for an arbitrary delay distribution. If we let G be the distribution in (2.6.4) and subtract the last two equations, we get:

Example 2.6.7 $H(t) = U(t) - t/\mu$ satisfies the renewal equation with $h(t) = \frac{1}{\mu}\int_t^\infty 1 - F(s)\, ds$.

Last but not least, we have an example that is a typical application of the renewal equation.

Example 2.6.8 Let $x > 0$ be fixed, and let $H(t) = P(T_{N(t)} - t > x)$. By considering the value of T_1, we get

$$H(t) = (1 - F(t + x)) + \int_0^t H(t-s)\, dF(s)$$

The examples here should provide motivation for:

Theorem 2.6.9 *If h is bounded then the function*

$$H(t) = \int_0^t h(t-s)\, dU(s)$$

is the unique solution of the renewal equation that is bounded on bounded intervals.

Proof Let $U_n(A) = \sum_{m=0}^{n} P(T_m \in A)$ and

$$H_n(t) = \int_0^t h(t-s)\, dU_n(s) = \sum_{m=0}^{n} \left(h * F^{m*}\right)(t)$$

Here, F^{m*} is the distribution of T_m, and we have extended the definition of h by setting $h(r) = 0$ for $r < 0$. From the last expression, it should be clear that

$$H_{n+1} = h + H_n * F$$

The fact that $U(t) < \infty$ implies $U(t) - U_n(t) \to 0$. Since h is bounded,

$$|H_n(t) - H(t)| \leq \|h\|_\infty |U(t) - U_n(t)|$$

and $H_n(t) \to H(t)$ uniformly on bounded intervals. To estimate the convolution, we note that

$$\begin{aligned} |H_n * F(t) - H * F(t)| &\leq \sup_{s \leq t} |H_n(s) - H(s)| \\ &\leq \|h\|_\infty |U(t) - U_n(t)| \end{aligned}$$

since $U - U_n = \sum_{m=n+1}^{\infty} F^{m*}$ is increasing in t. Letting $n \to \infty$ in $H_{n+1} = h + H_n * F$, we see that H is a solution of the renewal equation that is bounded on bounded intervals.

To prove uniqueness, we observe that if H_1 and H_2 are two solutions, then $K = H_1 - H_2$ satisfies $K = K * F$. If K is bounded on bounded intervals, iterating gives $K = K * F^{n*} \to 0$ as $n \to \infty$, so $H_1 = H_2$. □

The proof of Theorem 2.6.9 is valid when $F(\infty) = P(\xi_i < \infty) < 1$. In this case, we have a **terminating renewal process**. After a geometric number of trials with mean $1/(1 - F(\infty))$, $T_n = \infty$. This "trivial case" has some interesting applications.

Example 2.6.10 (Pedestrian delay) A chicken wants to cross a road (we won't ask why) on which the traffic is a Poisson process with rate λ. She needs one unit of time with no arrival to safely cross the road. Let $M = \inf\{t \geq 0 : \text{there are no arrivals in } (t, t+1]\}$ be the waiting time until she starts to cross the road. By considering the time of the first arrival, we see that $H(t) = P(M \leq t)$ satisfies

$$H(t) = e^{-\lambda} + \int_0^1 H(t-y)\, \lambda e^{-\lambda y}\, dy$$

Comparing with Example 2.6.5 and using Theorem 2.6.9, we see that

$$H(t) = e^{-\lambda} \sum_{n=0}^{\infty} F^{n*}(t)$$

We could have gotten this answer without renewal theory by noting

$$P(M \le t) = \sum_{n=0}^{\infty} P(T_n \le t, T_{n+1} = \infty)$$

The last representation allows us to compute the mean of M. Let μ be the mean of the interarrival time given that it is < 1, and note that the lack of memory property of the exponential distribution implies

$$\mu = \int_0^1 x\lambda e^{-\lambda x}\, dx = \int_0^\infty - \int_1^\infty = \frac{1}{\lambda} - \left(1 + \frac{1}{\lambda}\right) e^{-\lambda}$$

Then, by considering the number of renewals in our terminating renewal process,

$$EM = \sum_{n=0}^{\infty} e^{-\lambda}(1 - e^{-\lambda})^n n\mu = (e^\lambda - 1)\mu$$

since if X is a geometric with success probability $e^{-\lambda}$ then $EM = \mu E(X - 1)$.

Example 2.6.11 (Cramér's estimates of ruin) Consider an insurance company that collects money at rate c and experiences i.i.d. claims at the arrival times of a Poisson process N_t with rate 1. If its initial capital is x, its wealth at time t is

$$W_x(t) = x + ct - \sum_{m=1}^{Nt} Y_i$$

Here $Y_1, Y_2 \ldots$ are i.i.d. with distribution G and mean μ. Let

$$R(x) = P(W_x(t) \ge 0 \text{ for all } t)$$

be the probability of never going bankrupt starting with capital x. By considering the time and size of the first claim:

$$\text{(a)} \qquad R(x) = \int_0^\infty e^{-s} \int_0^{x+cs} R(x + cs - y)\, dG(y)\, ds$$

This does not look much like a renewal equation, but with some ingenuity it can be transformed into one. Changing variables $t = x + cs$

$$R(x)e^{-x/c} = \int_x^\infty e^{-t/c} \int_0^t R(t - y)\, dG(y)\, \frac{dt}{c}$$

Differentiating w.r.t. x and then multiplying by $e^{x/c}$,

$$R'(x) = \frac{1}{c} R(x) - \int_0^x R(x - y)\, dG(y) \cdot \frac{1}{c}$$

Integrating x from 0 to w

$$\text{(b)} \qquad R(w) - R(0) = \frac{1}{c}\int_0^w R(x)\, dx - \frac{1}{c}\int_0^w \int_0^x R(x - y)\, dG(y)\, dx$$

Interchanging the order of integration in the double integral, letting

$$S(w) = \int_0^w R(x)\,dx$$

using $dG = -d(1-G)$, and then integrating by parts

$$\begin{aligned}-\frac{1}{c}\int_0^w \int_y^w R(x-y)\,dx\,dG(y) &= -\frac{1}{c}\int_0^w S(w-y)\,dG(y) \\ &= \frac{1}{c}\int_0^w S(w-y)\,d(1-G)(y) \\ &= \frac{1}{c}\left\{-S(w) + \int_0^w (1-G(y))R(w-y)\,dy\right\}\end{aligned}$$

Plugging this into (b), we finally have a renewal equation:

$$\text{(c)} \qquad R(w) = R(0) + \int_0^w R(w-y)\frac{1-G(y)}{c}\,dy$$

It took some cleverness to arrive at the last equation, but it is straightforward to analyze. First, we dismiss a trivial case. If $\mu > c$,

$$\frac{1}{t}\left(ct - \sum_{m=1}^{Nt} Y_i\right) \to c - \mu < 0 \quad \text{a.s.}$$

so $R(x) \equiv 0$. When $\mu < c$,

$$F(x) = \int_0^x \frac{1-G(y)}{c}\,dy$$

is a defective probability distribution with $F(\infty) = \mu/c$. Our renewal equation can be written as

$$\text{(d)} \qquad R = R(0) + R * F$$

so comparing with Example 2.6.5 and using Theorem 2.6.9 tells us $R(w) = R(0)U(w)$. To complete the solution, we have to compute the constant $R(0)$. Letting $w \to \infty$ and noticing $R(w) \to 1$, $U(w) \to (1 - F(\infty))^{-1} = (1 - \mu/c)^{-1}$, we have $R(0) = 1 - \mu/c$.

The basic fact about solutions of the renewal equation (in the nonterminating case) is:

Theorem 2.6.12 (The renewal theorem) *If F is nonarithmetic and h is directly Riemann integrable, then as $t \to \infty$*

$$H(t) \to \frac{1}{\mu}\int_0^\infty h(s)\,ds$$

Intuitively, this holds since Theorem 2.6.9 implies

$$H(t) = \int_0^t h(t-s)\,dU(s)$$

and Theorem 2.6.4 implies $dU(s) \to ds/\mu$ as $s \to \infty$. We will define directly Riemann integrable shortly. We will start doing the proof and then figure out what we need to assume.

Proof Suppose

$$h(s) = \sum_{k=0}^{\infty} a_k 1_{[k\delta,(k+1)\delta)}(s)$$

where $\sum_{k=0}^{\infty} |a_k| < \infty$. Since $U([t,t+\delta]) \le U([0,\delta]) < \infty$, it follows easily from Theorem 2.6.4 that

$$\int_0^t h(t-s)dU(s) = \sum_{k=0}^{\infty} a_k U((t-(k+1)\delta, t-k\delta]) \to \frac{1}{\mu}\sum_{k=0}^{\infty} a_k\delta$$

(Pick K so that $\sum_{k\ge K} |a_k| \le \epsilon/2U([0,\delta])$ and then T so that

$$|a_k| \cdot |U((t-(k+1)\delta, t-k\delta]) - \delta/\mu| \le \frac{\epsilon}{2K}$$

for $t \ge T$ and $0 \le k < K$.) If h is an arbitrary function on $[0,\infty)$, we let

$$I^\delta = \sum_{k=0}^{\infty} \delta \sup\{h(x) : x \in [k\delta,(k+1)\delta)\}$$

$$I_\delta = \sum_{k=0}^{\infty} \delta \inf\{h(x) : x \in [k\delta,(k+1)\delta)\}$$

be upper and lower Riemann sums approximating the integral of h over $[0,\infty)$. Comparing h with the obvious upper and lower bounds that are constant on $[k\delta,(k+1)\delta)$ and using the result for the special case,

$$\frac{I_\delta}{\mu} \le \liminf_{t\to\infty} \int_0^t h(t-s)\,dU(s) \le \limsup_{t\to\infty} \int_0^t h(t-s)\,dU(s) \le \frac{I^\delta}{\mu}$$

If I^δ and I_δ both approach the same finite limit I as $\delta \to 0$, then h is said to be **directly Riemann integrable**, and it follows that

$$\int_0^t h(t-s)\,dU(y) \to I/\mu$$

□

Remark The word "direct" in the name refers to the fact that while the Riemann integral over $[0,\infty)$ is usually defined as the limit of integrals over $[0,a]$, we are approximating the integral over $[0,\infty)$ directly.

In checking the new hypothesis in Theorem 2.6.12, the following result is useful.

Lemma 2.6.13 *If $h(x) \ge 0$ is decreasing with $h(0) < \infty$ and $\int_0^\infty h(x)\,dx < \infty$, then h is directly Riemann integrable.*

Proof Because h is decreasing, $I^\delta = \sum_{k=0}^{\infty} \delta h(k\delta)$ and $I_\delta = \sum_{k=0}^{\infty} \delta h((k+1)\delta)$. So

$$I^\delta \ge \int_0^\infty h(x)\,dx \ge I_\delta = I^\delta - h(0)\delta$$

proving the desired result. □

Returning now to our examples, we skip the first two because, in those cases, $h(t) \to 1$ as $t \to \infty$, so h is not integrable in any sense.

Example 2.6.14 (Continuation of Example 2.6.7) $h(t) = \frac{1}{\mu}\int_{[t,\infty)} 1 - F(s)\,ds$. h is decreasing, $h(0) = 1$, and

$$\begin{aligned}\mu \int_0^\infty h(t)\,dt &= \int_0^\infty \int_t^\infty 1 - F(s)\,ds\,dt \\ &= \int_0^\infty \int_0^s 1 - F(s)\,dt\,ds = \int_0^\infty s(1 - F(s))\,ds = E(\xi_i^2/2)\end{aligned}$$

So, if $\nu \equiv E(\xi_i^2) < \infty$, it follows from Lemma 2.6.13, Theorem 2.6.12, and the formula in Example 2.6.7 that

$$0 \le U(t) - t/\mu \to \nu/2\mu^2 \quad \text{as } t \to \infty$$

When the renewal process is a rate λ Poisson process, i.e., $P(\xi_i > t) = e^{-\lambda t}$, $N(t) - 1$ has a Poisson distribution with mean λt, so $U(t) = 1 + \lambda t$. According to Feller, Vol. II (1971), p. 385, if the ξ_i are uniform on (0,1), then

$$U(t) = \sum_{k=0}^{n} (-1)^k e^{t-k}(t-k)^k/k! \quad \text{for } n \le t \le n+1$$

As he says, the exact expression "reveals little about the nature of U. The asymptotic formula $0 \le U(t) - 2t \to 2/3$ is much more interesting."

Example 2.6.15 (Continuation of Example 2.6.8) $h(t) = 1 - F(t + x)$. Again, h is decreasing, but this time $h(0) \le 1$ and the integral of h is finite when $\mu = E(\xi_i) < \infty$. Applying Lemma 2.6.13 and Theorem 2.6.12 now gives

$$P(T_{N(t)} - t > x) \to \frac{1}{\mu}\int_0^\infty h(s)\,ds = \frac{1}{\mu}\int_x^\infty 1 - F(t)\,dt$$

so (when $\mu < \infty$) the distribution of the **residual waiting time** $T_{N(t)} - t$ converges to the delay distribution that produces the stationary renewal process. This fact also follows from our proof of 2.6.4.

Exercises

2.6.1 Show that $t/E(\xi_i \wedge t) \le U(t) \le 2t/E(\xi_i \wedge t)$.

2.6.2 Deduce Theorem 2.6.3 from Theorem 2.6.1 by showing

$$\limsup_{t\to\infty} E(N_t/t)^2 < \infty.$$

Hint: Use a comparison like the one in the proof of Theorem 2.6.3.

2.6.3 Customers arrive at times of a Poisson process with rate 1. If the server is occupied, they leave. (Think of a public telephone or prostitute.) If not, they enter service and require a service time with a distribution F that has mean μ. Show that the times at which customers enter service are a renewal process with mean $\mu + 1$, and use Theorem 2.6.1 to conclude that the asymptotic fraction of customers served is $1/(\mu+1)$.

In the remaining problems we assume that F is nonarithmetic, and in problems where the mean appears we assume it is finite.

2.6.4 Let $A_t = t - T_{N(t)-1}$ be the "age" at time t, i.e., the amount of time since the last renewal. If we fix $x > 0$, then $H(t) = P(A_t > x)$ satisfies the renewal equation

$$H(t) = (1 - F(t)) \cdot 1_{(x,\infty)}(t) + \int_0^t H(t-s)\,dF(s)$$

so $P(A_t > x) \to \frac{1}{\mu}\int_{(x,\infty)}(1-F(t))dt$, which is the limit distribution for the residual lifetime $B_t = T_{N(t)} - t$.

2.6.5 Use the renewal equation in the last problem and Theorem 2.6.9 to conclude that if T is a rate λ Poisson process, A_t has the same distribution as $\xi_i \wedge t$.

2.6.6 Let $A_t = t - T_{N(t)-1}$ and $B_t = T_{N(t)} - t$. Show that

$$P(A_t > x, B_t > y) \to \frac{1}{\mu}\int_{x+y}^{\infty} (1 - F(t))\,dt$$

2.6.7 **Alternating renewal process.** Let $\xi_1, \xi_2, \ldots > 0$ be i.i.d. with distribution F_1 and let $\eta_1, \eta_2, \ldots > 0$ be i.i.d. with distribution F_2. Let $T_0 = 0$ and for $k \geq 1$ let $S_k = T_{k-1} + \xi_k$ and $T_k = S_k + \eta_k$. In words, we have a machine that works for an amount of time ξ_k, breaks down, and then requires η_k units of time to be repaired. Let $F = F_1 * F_2$ and let $H(t)$ be the probability the machine is working at time t. Show that if F is nonarithmetic, then as $t \to \infty$

$$H(t) \to \mu_1/(\mu_1 + \mu_2)$$

where μ_i is the mean of F_i.

2.6.8 Write a renewal equation for $H(t) = P($ number of renewals in $[0,t]$ is odd$)$ and use the renewal theorem to show that $H(t) \to 1/2$. Note: This is a special case of the previous exercise.

2.6.9 **Renewal densities.** Show that if $F(t)$ has a directly Riemann integrable density function $f(t)$, then the $V = U - 1_{[0,\infty)}$ has a density v that satisfies

$$v(t) = f(t) + \int_0^t v(t-s)\,dF(s)$$

Use the renewal theorem to conclude that if f is directly Riemann integrable, then $v(t) \to 1/\mu$ as $t \to \infty$.

2.7 Large Deviations*

Let $X_1, X_2, \ldots$ be i.i.d. and let $S_n = X_1 + \cdots + X_n$. In this section, we will investigate the rate at which $P(S_n > na) \to 0$ for $a > \mu = EX_i$. We will ultimately conclude that if the

moment-generating function $\varphi(\theta) = E\exp(\theta X_i) < \infty$ for some $\theta > 0$, $P(S_n \geq na) \to 0$ exponentially rapidly and we will identify

$$\gamma(a) = \lim_{n\to\infty} \frac{1}{n} \log P(S_n \geq na)$$

Our first step is to prove that the limit exists. This is based on an observation that will be useful several times later. Let $\pi_n = P(S_n \geq na)$.

$$\pi_{m+n} \geq P(S_m \geq ma, S_{n+m} - S_m \geq na) = \pi_m \pi_n$$

since S_m and $S_{n+m} - S_m$ are independent. Letting $\gamma_n = \log \pi_n$ transforms multiplication into addition.

Lemma 2.7.1 *If $\gamma_{m+n} \geq \gamma_m + \gamma_n$, then as $n \to \infty$, $\gamma_n/n \to \sup_m \gamma_m/m$.*

Proof Clearly, $\limsup \gamma_n/n \leq \sup \gamma_m/m$. To complete the proof, it suffices to prove that for any m $\liminf \gamma_n/n \geq \gamma_m/m$. Writing $n = km + \ell$ with $0 \leq \ell < m$ and making repeated use of the hypothesis gives $\gamma_n \geq k\gamma_m + \gamma_\ell$. Dividing by $n = km + \ell$ gives

$$\frac{\gamma(n)}{n} \geq \left(\frac{km}{km+\ell}\right)\frac{\gamma(m)}{m} + \frac{\gamma(\ell)}{n}$$

Letting $n \to \infty$ and recalling $n = km + \ell$ with $0 \leq \ell < m$ gives the desired result. □

Lemma 2.7.1 implies that $\lim_{n\to\infty} \frac{1}{n} \log P(S_n \geq na) = \gamma(a)$ exists ≤ 0. It follows from the formula for the limit that

$$P(S_n \geq na) \leq e^{n\gamma(a)} \tag{2.7.1}$$

The last conclusion is valid for any distribution but it is not very useful if $\gamma(a0 = 0$. For the rest of this section, we will suppose:

(H1) $\varphi(\theta) = E\exp(\theta X_i) < \infty$ for some $\theta > 0$

Let $\theta_+ = \sup\{\theta : \phi(\theta) < \infty\}$, $\theta_- = \inf\{\theta : \phi(\theta) < \infty\}$ and note that $\phi(\theta) < \infty$ for $\theta \in (\theta_-, \theta_+)$. (H1) implies that $EX_i^+ < \infty$ so $\mu = EX^+ - EX^- \in [-\infty, \infty)$. If $\theta > 0$ Chebyshev's inequality implies

$$e^{\theta na} P(S_n \geq na) \leq E\exp(\theta S_n) = \varphi(\theta)^n$$

or letting $\kappa(\theta) = \log \varphi(\theta)$

$$P(S_n \geq na) \leq \exp(-n\{a\theta - \kappa(\theta)\}) \tag{2.7.2}$$

Our first goal is to show:

Lemma 2.7.2 *If $a > \mu$ and $\theta > 0$ is small, then $a\theta - \kappa(\theta) > 0$.*

Proof $\kappa(0) = \log \varphi(0) = 0$, so it suffices to show that (i) κ is continuous at 0, (ii) differentiable on $(0, \theta_+)$, and (iii) $\kappa'(\theta) \to \mu$ as $\theta \to 0$. For then

$$a\theta - \kappa(\theta) = \int_0^\theta a - \kappa'(x)\, dx > 0$$

for small θ.

Let $F(x) = P(X_i \le x)$. To prove (i) we note that if $0 < \theta < \theta_0 < \theta_-$

$$e^{\theta x} \le 1 + e^{\theta_0 x} \qquad (*)$$

so by the dominated convergence theorem as $\theta \to 0$

$$\int e^{\theta x}\, dF \to \int 1\, dF = 1$$

To prove (ii) we note that if $|h| < h_0$, then

$$|e^{hx} - 1| = \left| \int_0^{hx} e^y\, dy \right| \le |hx| e^{h_0 x}$$

so an application of the dominated convergence theorem shows that

$$\begin{aligned}
\varphi'(\theta) &= \lim_{h \to 0} \frac{\varphi(\theta + h) - \varphi(\theta)}{h} \\
&= \lim_{h \to 0} \int \frac{e^{hx} - 1}{h} e^{\theta x}\, dF(x) \\
&= \int x e^{\theta x} dF(x) \quad \text{for } \theta \in (0, \theta_+)
\end{aligned}$$

From the last equation, it follows that $\kappa(\theta) = \log \phi(\theta)$ has $\kappa'(\theta) = \phi'(\theta)/\phi(\theta)$. Using (*) and the dominated convergence theorem gives (iii) and the proof is complete. □

Having found an upper bound on $P(S_n \ge na)$, it is natural to optimize it by finding the maximum of $\theta a - \kappa(\theta)$:

$$\frac{d}{d\theta}\{\theta a - \log \varphi(\theta)\} = a - \varphi'(\theta)/\varphi(\theta)$$

so (assuming things are nice) the maximum occurs when $a = \varphi'(\theta)/\varphi(\theta)$. To turn the parenthetical clause into a mathematical hypothesis we begin by defining

$$F_\theta(x) = \frac{1}{\varphi(\theta)} \int_{-\infty}^{x} e^{\theta y}\, dF(y)$$

whenever $\phi(\theta) < \infty$. It follows from the proof of Lemma 2.7.2 that if $\theta \in (\theta_-, \theta_+)$, F_θ is a distribution function with mean

$$\int x\, dF_\theta(x) = \frac{1}{\varphi(\theta)} \int_{-\infty}^{\infty} x e^{\theta x}\, dF(x) = \frac{\varphi'(\theta)}{\varphi(\theta)}$$

Repeating the proof in Lemma 2.7.2, it is easy to see that if $\theta \in (\theta_-, \theta_+)$, then

$$\phi''(\theta) = \int_{-\infty}^{\infty} x^2 e^{\theta x}\, dF(x)$$

So we have

$$\frac{d}{d\theta} \frac{\varphi'(\theta)}{\varphi(\theta)} = \frac{\varphi''(\theta)}{\varphi(\theta)} - \left(\frac{\varphi'(\theta)}{\varphi(\theta)} \right)^2 = \int x^2\, dF_\theta(x) - \left(\int x\, dF_\theta(x) \right)^2 \ge 0$$

since the last expression is the variance of F_θ. If we assume

(H2) the distribution F is not a point mass at μ

then $\varphi'(\theta)/\varphi(\theta)$ is strictly increasing and $a\theta - \log\phi(\theta)$ is concave. Since we have $\varphi'(0)/\varphi(0) = \mu$, this shows that for each $a > \mu$ there is at most one $\theta_a \geq 0$ that solves $a = \varphi'(\theta_a)/\varphi(\theta_a)$, and this value of θ maximizes $a\theta - \log\varphi(\theta)$. Before discussing the existence of θ_a, we will consider some examples.

Example 2.7.3 (Normal distribution)

$$\int e^{\theta x}(2\pi)^{-1/2}\exp(-x^2/2)\,dx = \exp(\theta^2/2)\int (2\pi)^{-1/2}\exp(-(x-\theta)^2/2)\,dx$$

The integrand in the last integral is the density of a normal distribution with mean θ and variance 1, so $\varphi(\theta) = \exp(\theta^2/2)$, $\theta \in (-\infty,\infty)$. In this case, $\varphi'(\theta)/\varphi(\theta) = \theta$ and

$$F_\theta(x) = e^{-\theta^2/2}\int_{-\infty}^{x} e^{\theta y}(2\pi)^{-1/2}e^{-y^2/2}\,dy$$

is a normal distribution with mean θ and variance 1.

Example 2.7.4 (Exponential distribution with parameter λ) If $\theta < \lambda$

$$\int_0^\infty e^{\theta x}\lambda e^{-\lambda x}\,dx = \lambda/(\lambda - \theta)$$

$\varphi'(\theta)\varphi(\theta) = 1/(\lambda - \theta)$ and

$$F_\theta(x) = \frac{\lambda}{\lambda - \theta}\int_0^x e^{\theta y}\lambda e^{-\lambda y}\,dy$$

is an exponential distribution with parameter $\lambda - \theta$, and hence mean $1/(\lambda - \theta)$.

Example 2.7.5 (Coin flips) $P(X_i = 1) = P(X_i = -1) = 1/2$

$$\varphi(\theta) = (e^\theta + e^{-\theta})/2$$
$$\varphi'(\theta)/\varphi(\theta) = (e^\theta - e^{-\theta})/(e^\theta + e^{-\theta})$$

$F_\theta(\{x\})/F(\{x\}) = e^{\theta x}/\phi(\theta)$ so

$$F_\theta(\{1\}) = e^\theta/(e^\theta + e^{-\theta}) \quad \text{and} \quad F_\theta(\{-1\}) = e^{-\theta}/(e^\theta + e^{-\theta})$$

Example 2.7.6 (Perverted exponential) Let $g(x) = Cx^{-3}e^{-x}$ for $x \geq 1$, $g(x) = 0$ otherwise, and choose C so that g is a probability density. In this case,

$$\varphi(\theta) = \int e^{\theta x} g(x)dx < \infty$$

if and only if $\theta \leq 1$, and when $\theta \leq 1$, we have

$$\frac{\varphi'(\theta)}{\varphi(\theta)} \leq \frac{\varphi'(1)}{\varphi(1)} = \int_1^\infty Cx^{-2}\,dx \bigg/ \int_1^\infty Cx^{-3}dx = 2$$

Recall $\theta_+ = \sup\{\theta : \varphi(\theta) < \infty\}$. In Examples 2.7.3 and 2.7.4, we have $\phi'(\theta)/\phi(\theta) \uparrow \infty$ as $\theta \uparrow \theta_+$ so we can solve $a = \phi'(\theta)/\phi(\theta)$ for any $a > \mu$. In Example 2.7.5, $\phi'(\theta)/\phi(\theta) \uparrow 1$ as $\theta \to \infty$, but we cannot hope for much more since F, and hence F_θ is supported on $\{-1, 1\}$. Example 2.7.6 presents a problem since we cannot solve $a = \varphi'(\theta)/\varphi(\theta)$ when

$a > 2$. Theorem 2.7.10 will cover this problem case, but first we will treat the cases in which we can solve the equation.

Theorem 2.7.7 *Suppose in addition to (H1) and (H2) that there is a $\theta_a \in (0, \theta_+)$ so that $a = \varphi'(\theta_a)/\varphi(\theta_a)$. Then, as $n \to \infty$,*

$$n^{-1} \log P(S_n \geq na) \to -a\theta_a + \log \varphi(\theta_a)$$

Proof The fact that the limsup of the left-hand side $\leq$ the right-hand side follows from (2.7.2). To prove the other inequality, pick $\lambda \in (\theta_a, \theta_+)$, let $X_1^\lambda, X_2^\lambda, \ldots$ be i.i.d. with distribution F_λ and let $S_n^\lambda = X_1^\lambda + \cdots + X_n^\lambda$. Writing dF/dF_λ for the Radon-Nikodym derivative of the associated measures, it is immediate from the definition that $dF/dF_\lambda = e^{-\lambda x}\varphi(\lambda)$. If we let F_λ^n and F^n denote the distributions of S_n^λ and S_n, then

Lemma 2.7.8 $\dfrac{dF^n}{dF_\lambda^n} = e^{-\lambda x}\varphi(\lambda)^n$.

Proof We will prove this by induction. The result holds when $n = 1$. For $n > 1$, we note that

$$\begin{aligned}
F^n = F^{n-1} * F(z) &= \int_{-\infty}^{\infty} dF^{n-1}(x) \int_{-\infty}^{z-x} dF(y) \\
&= \int dF_\lambda^{n-1}(x) \int dF_\lambda(y)\, 1_{(x+y \leq z)} e^{-\lambda(x+y)} \varphi(\lambda)^n \\
&= E\left(1_{(S_{n-1}^\lambda + X_n^\lambda \leq z)} e^{-\lambda(S_{n-1}^\lambda + X_n^\lambda)} \varphi(\lambda)^n\right) \\
&= \int_{-\infty}^{z} dF_\lambda^n(u) e^{-\lambda u} \varphi(\lambda)^n
\end{aligned}$$

where in the last two equalities we have used Theorem 1.6.9 for $(S_{n-1}^\lambda, X_n^\lambda)$ and S_n^λ. □

If $\nu > a$, then the lemma and monotonicity imply

$$(*) \qquad P(S_n \geq na) \geq \int_{na}^{n\nu} e^{-\lambda x}\varphi(\lambda)^n dF_\lambda^n(x) \geq \varphi(\lambda)^n e^{-\lambda n \nu}(F_\lambda^n(n\nu) - F_\lambda^n(na))$$

F_λ has mean $\varphi'(\lambda)/\varphi(\lambda)$, so if we have $a < \varphi'(\lambda)/\varphi(\lambda) < \nu$, then the weak law of large numbers implies

$$F_\lambda^n(n\nu) - F_\lambda^n(na) \to 1 \text{ as } n \to \infty$$

From the last conclusion and $(*)$ it follows that

$$\liminf_{n\to\infty} n^{-1} \log P(S_n > na) \geq -\lambda\nu + \log \phi(\lambda)$$

Since $\lambda > \theta_a$ and $\nu > a$ are arbitrary, the proof is complete. □

To get a feel for what the answers look like, we consider our examples. To prepare for the computations, we recall some important information:

$$\begin{aligned}
&\kappa(\theta) = \log \phi(\theta) \quad \kappa'(\theta) = \phi'(\theta)/\phi(\theta) \quad \theta_a \text{ solves } \kappa'(\theta_a) = a \\
&\gamma(a) = \lim_{n\to\infty} (1/n) \log P(S_n \geq na) = -a\theta_a + \kappa(\theta_a)
\end{aligned}$$

Normal distribution (Example 2.7.3)

$$\kappa(\theta) = \theta^2/2 \qquad \kappa'(\theta) = \theta \qquad \theta_a = a$$
$$\gamma(a) = -a\theta_a + \kappa(\theta_a) = -a^2/2$$

One can check the last result by observing that S_n has a normal distribution with mean 0 and variance n, and then using Theorem 1.2.6.

Exponential distribution (Example 2.7.4) with $\lambda = 1$

$$\kappa(\theta) = -\log(1-\theta) \qquad \kappa'(\theta) = 1/(1-\theta) \qquad \theta_a = 1 - 1/a$$
$$\gamma(a) = -a\theta_a + \kappa(\theta_a) = -a + 1 + \log a$$

With these two examples as models, the reader should be able to do

Coin flips (Example 2.7.5). Here we take a different approach. To find the θ that makes the mean of $F_\theta = a$, we set $F_\theta(\{1\}) = e^\theta/(e^\theta + e^{-\theta}) = (1+a)/2$. Letting $x = e^\theta$ gives

$$2x = (1+a)(x + x^{-1}) \qquad (a-1)x^2 + (1+a) = 0$$

So $x = \sqrt{(1+a)/(1-a)}$ and $\theta_a = \log x = \{\log(1+a) - \log(1-a)\}/2$.

$$\phi(\theta_a) = \frac{e^{\theta_a} + e^{-\theta_a}}{2} = \frac{e^{\theta_a}}{1+a} = \frac{1}{\sqrt{(1+a)(1-a)}}$$
$$\gamma(a) = -a\theta_a + \kappa(\theta_a) = -\{(1+a)\log(1+a) + (1-a)\log(1-a)\}/2$$

In Exercise 3.1.3, this result will be proved by a direct computation. Since the formula for $\gamma(a)$ is rather ugly, the simpler bound in Exercise 2.7.4 is useful.

Turning now to the problematic values for which we cannot solve $a = \phi'(\theta_a)/\phi(\theta_a)$:

Theorem 2.7.9 *Suppose $x_o = \sup\{x : F(x) < 1\} < \infty$ and F is not a point mass at x_0. $\phi(\theta) < \infty$ for all $\theta > 0$ and $\phi'(\theta)/\phi(\theta) \to x_o$ as $\theta \uparrow \infty$.*

Proof Since $P(X \le x_o) = 1$, $Ee^{\theta X} < \infty$ for all $\theta > 0$. Since F_θ is concentrated on $(-\infty, x_o]$ it is clear that $\mu_\theta = \phi'(\theta)/\phi(\theta) \le x_o$. On the other hand, if $\delta > 0$, then $P(X \ge x_o - \delta) = c_\delta > 0$, $Ee^{\theta X} \ge c_\delta e^{\theta(x_o - \delta)}$, and hence

$$F_\theta(x_o - 2\delta) = \frac{1}{\phi(\theta)} \int_{-\infty}^{x_o - 2\delta} e^{\theta x} dF(x) \le \frac{e^{x_o - 2\delta)\theta}}{c_\delta e^{(x_o - \delta)\theta}} = e^{-\theta\delta}/c_\delta \to 0 \tag{2.7.3}$$

Since $\delta > 0$ is arbitrary, it follows that $\mu_\theta \to x_o$. □

The result for $a = x_o$ is trivial:

$$\frac{1}{n} \log P(S_n \ge n x_o) = \log P(X_i = x_o) \quad \text{for all } n$$

We leave it to the reader to show that as $a \uparrow x_o$, $\gamma(a) \downarrow \log P(X_i = x_o)$.

When $x_o = \infty$ and $\theta_+ = \infty$, the computation in (2.7.3) implies $\phi'(\theta)/\phi(\theta) \uparrow \infty$ as $\theta \uparrow \infty$, so the only case that remains is covered by:

Theorem 2.7.10 *Suppose $x_o = \infty$, $\theta_+ < \infty$, and $\varphi'(\theta)/\varphi(\theta)$ increases to a finite limit a_0 as $\theta \uparrow \theta_+$. If $a_0 \le a < \infty$*

$$n^{-1} \log P(S_n \ge na) \to -a\theta_+ + \log \varphi(\theta_+)$$

i.e., $\gamma(a)$ is linear for $a \ge a_0$.

Proof Since $(\log \varphi(\theta))' = \varphi'(\theta)/\varphi(\theta)$, integrating from 0 to θ_+ shows that $\log(\varphi(\theta_+)) < \infty$. Letting $\theta = \theta_+$ in (2.7.2) shows that the limsup of the left-hand side $\le$ the right-hand side. To get the other direction, we will use the transformed distribution F_λ, for $\lambda = \theta_+$. Letting $\theta \uparrow \theta_+$ and using the dominated convergence theorem for $x \le 0$ and the monotone convergence theorem for $x \ge 0$, we see that F_λ has mean a_0. From $(*)$ in the proof of Theorem 2.7.7, we see that if $a_0 \le a < \nu = a + 3\epsilon$

$$P(S_n \ge na) \ge \varphi(\lambda)^n e^{-n\lambda\nu}(F_\lambda^n(n\nu) - F_\lambda^n(na))$$

and hence

$$\frac{1}{n} \log P(S_n \ge na) \ge \log \varphi(\lambda) - \lambda\nu + \frac{1}{n} \log P(S_n^\lambda \in (na, n\nu])$$

Letting $X_1^\lambda, X_2^\lambda, \ldots$ be i.i.d. with distribution F_λ and $S_n^\lambda = X_1^\lambda + \cdots + X_n^\lambda$, we have

$$\begin{aligned} P(S_n^\lambda \in (na, n\nu]) &\ge P\{S_{n-1}^\lambda \in ((a_0 - \epsilon)n, (a_0 + \epsilon)n]\} \\ &\quad \cdot P\{X_n^\lambda \in ((a - a_0 + \epsilon)n, (a - a_0 + 2\epsilon)n]\} \\ &\ge \frac{1}{2} P\{X_n^\lambda \in ((a - a_0 + \epsilon)n, (a - a_0 + \epsilon)(n+1)]\} \end{aligned}$$

for large n by the weak law of large numbers. To get a lower bound on the right-hand side of the last equation, we observe that

$$\limsup_{n\to\infty} \frac{1}{n} \log P(X_1^\lambda \in ((a - a_0 + \epsilon)n, (a - a_0 + \epsilon)(n+1)]) = 0$$

for if the lim sup was < 0, we would have $E\exp(\eta X_1^\lambda) < \infty$ for some $\eta > 0$ and hence $E\exp((\lambda + \eta)X_1) < \infty$, contradicting the definition of $\lambda = \theta_+$. To finish the argument now, we recall that Theorem 2.7.1 implies that

$$\lim_{n\to\infty} \frac{1}{n} \log P(S_n \ge na) = \gamma(a)$$

exists, so our lower bound on the lim sup is good enough. □

By adapting the proof of the last result, you can show that (H1) is necessary for exponential convergence:

Exercises

2.7.1 Consider $\gamma(a)$ defined in (2.7.1). The following are equivalent: (a) $\gamma(a) = -\infty$, (b) $P(X_1 \ge a) = 0$, and (c) $P(S_n \ge na) = 0$ for all n.

2.7.2 Use the definition to conclude that if $\lambda \in [0,1]$ is rational, then $\gamma(\lambda a + (1-\lambda)b) \geq \lambda\gamma(a) + (1-\lambda)\gamma(b)$. Use monotonicity to conclude that the last relationship holds for all $\lambda \in [0,1]$ so γ is concave, and hence Lipschitz continuous on compact subsets of $\gamma(a) > -\infty$.

2.7.3 Let $X_1, X_2, \ldots$ be i.i.d. Poisson with mean 1, and let $S_n = X_1 + \cdots + X_n$. Find $\lim_{n\to\infty}(1/n)\log P(S_n \geq na)$ for $a > 1$. The answer and another proof can be found in Exercise 3.1.4.

2.7.4 Show that for coin flips $\varphi(\theta) \leq \exp(\varphi(\theta) - 1) \leq \exp(\beta\theta^2)$ for $\theta \leq 1$, where $\beta = \sum_{n=1}^{\infty} 1/(2n)! \approx 0.586$, and use (2.7.2) to conclude that $P(S_n \geq an) \leq \exp(-na^2/4\beta)$ for all $a \in [0,1]$. It is customary to simplify this further by using $\beta \leq \sum_{n=1}^{\infty} 2^{-n} = 1$.

2.7.5 Suppose $EX_i = 0$ and $E\exp(\theta X_i) = \infty$ for all $\theta > 0$. Then

$$\frac{1}{n}\log P(S_n \geq na) \to 0 \text{ for all } a > 0$$

2.7.6 Suppose $EX_i = 0$. Show that if $\epsilon > 0$, then

$$\liminf_{n\to\infty} P(S_n \geq na)/nP(X_1 \geq n(a+\epsilon)) \geq 1$$

Hint: Let $F_n = \{X_i \geq n(a+\epsilon) \text{ for exactly one } i \leq n\}$.

3

Central Limit Theorems

The first four sections of this chapter develop the central limit theorem. The last five treat various extensions and complements. We begin this chapter by considering special cases of these results that can be treated by elementary computations.

3.1 The De Moivre-Laplace Theorem

Let $X_1, X_2, \dots$ be i.i.d. with $P(X_1 = 1) = P(X_1 = -1) = 1/2$ and let $S_n = X_1 + \cdots + X_n$. In words, we are betting \$1 on the flipping of a fair coin and S_n is our winnings at time n. If n and k are integers

$$P(S_{2n} = 2k) = \binom{2n}{n+k} 2^{-2n}$$

since $S_{2n} = 2k$ if and only if there are $n+k$ flips that are $+1$ and $n-k$ flips that are -1 in the first $2n$. The first factor gives the number of such outcomes and the second the probability of each one. **Stirling's formula** (see Feller (1968), p. 52) tells us

$$n! \sim n^n e^{-n} \sqrt{2\pi n} \quad \text{as } n \to \infty \tag{3.1.1}$$

where $a_n \sim b_n$ means $a_n/b_n \to 1$ as $n \to \infty$, so

$$\begin{aligned} \binom{2n}{n+k} &= \frac{(2n)!}{(n+k)!\,(n-k)!} \\ &\sim \frac{(2n)^{2n}}{(n+k)^{n+k}(n-k)^{n-k}} \cdot \frac{(2\pi(2n))^{1/2}}{(2\pi(n+k))^{1/2}(2\pi(n-k))^{1/2}} \end{aligned}$$

and we have

$$\begin{aligned} \binom{2n}{n+k} 2^{-2n} \sim &\left(1 + \frac{k}{n}\right)^{-n-k} \cdot \left(1 - \frac{k}{n}\right)^{-n+k} \\ &\cdot (\pi n)^{-1/2} \cdot \left(1 + \frac{k}{n}\right)^{-1/2} \cdot \left(1 - \frac{k}{n}\right)^{-1/2} \end{aligned} \tag{3.1.2}$$

The first two terms on the right are

$$= \left(1 - \frac{k^2}{n^2}\right)^{-n} \cdot \left(1 + \frac{k}{n}\right)^{-k} \cdot \left(1 - \frac{k}{n}\right)^{k}$$

A little calculus shows that:

Lemma 3.1.1 *If $c_j \to 0$, $a_j \to \infty$ and $a_j c_j \to \lambda$ then $(1 + c_j)^{a_j} \to e^\lambda$.*

Proof As $x \to 0$, $\log(1 + x)/x \to 1$, so $a_j \log(1 + c_j) \to \lambda$ and the desired result follows. □

Using Lemma 3.1.1 now, we see that if $2k = x\sqrt{2n}$, i.e., $k = x\sqrt{n/2}$, then

$$\left(1 - \frac{k^2}{n^2}\right)^{-n} = \left(1 - x^2/2n\right)^{-n} \to e^{x^2/2}$$
$$\left(1 + \frac{k}{n}\right)^{-k} = \left(1 + x/\sqrt{2n}\right)^{-x\sqrt{n/2}} \to e^{-x^2/2}$$
$$\left(1 - \frac{k}{n}\right)^{k} = \left(1 - x/\sqrt{2n}\right)^{x\sqrt{n/2}} \to e^{-x^2/2}$$

For this choice of k, $k/n \to 0$, so

$$\left(1 + \frac{k}{n}\right)^{-1/2} \cdot \left(1 - \frac{k}{n}\right)^{-1/2} \to 1$$

and putting things together gives:

Theorem 3.1.2 *If $2k/\sqrt{2n} \to x$ then $P(S_{2n} = 2k) \sim (\pi n)^{-1/2} e^{-x^2/2}$.*

Our next step is to compute

$$P(a\sqrt{2n} \le S_{2n} \le b\sqrt{2n}) = \sum_{m \in [a\sqrt{2n}, b\sqrt{2n}] \cap 2\mathbf{Z}} P(S_{2n} = m)$$

Changing variables $m = x\sqrt{2n}$, we have that this is

$$\approx \sum_{x \in [a,b] \cap (2\mathbf{Z}/\sqrt{2n})} (2\pi)^{-1/2} e^{-x^2/2} \cdot (2/n)^{1/2}$$

where $2\mathbf{Z}/\sqrt{2n} = \{2z/\sqrt{2n} : z \in \mathbf{Z}\}$. We have multiplied and divided by $\sqrt{2}$ since the space between points in the sum is $(2/n)^{1/2}$, so if n is large, the sum just shown is

$$\approx \int_a^b (2\pi)^{-1/2} e^{-x^2/2} dx$$

The integrand is the density of the (standard) normal distribution, so changing notation we can write the last quantity as $P(a \le \chi \le b)$, where χ is a random variable with that distribution.

It is not hard to fill in the details to get:

Theorem 3.1.3 (The De Moivre-Laplace Theorem) *If $a < b$, then as $m \to \infty$*

$$P(a \le S_m/\sqrt{m} \le b) \to \int_a^b (2\pi)^{-1/2} e^{-x^2/2} dx$$

(To remove the restriction to even integers, observe $S_{2n+1} = S_{2n} \pm 1$.) The last result is a special case of the central limit theorem given in Section 3.4, so further details are left to the reader.

Exercises

3.1.1 Generalize the proof of Lemma 3.1.1 to conclude that if $\max_{1\le j\le n}|c_{j,n}| \to 0$, $\sum_{j=1}^{n} c_{j,n} \to \lambda$, and $\sup_n \sum_{j=1}^{n} |c_{j,n}| < \infty$, then $\prod_{j=1}^{n}(1+c_{j,n}) \to e^{\lambda}$.

The next three exercises illustrate the use of Stirling's formula. In them, $X_1, X_2, \ldots$ are i.i.d. and $S_n = X_1 + \cdots + X_n$.

3.1.2 If the X_i have a Poisson distribution with mean 1, then S_n has a Poisson distribution with mean n, i.e., $P(S_n = k) = e^{-n}n^k/k!$ Use Stirling's formula to show that if $(k-n)/\sqrt{n} \to x$, then

$$\sqrt{2\pi n}\, P(S_n = k) \to \exp(-x^2/2)$$

As in the case of coin flips it follows that

$$P(a \le (S_n - n)/\sqrt{n} \le b) \to \int_a^b (2\pi)^{-1/2} e^{-x^2/2}\, dx$$

but proving the last conclusion is not part of the exercise.

In the next two examples you should begin by considering $P(S_n = k)$ when $k/n \to a$ and then relate $P(S_n = j+1)$ to $P(S_n = j)$ to show $P(S_n \ge k) \le C P(S_n = k)$.

3.1.3 Suppose $P(X_i = 1) = P(X_i = -1) = 1/2$. Show that if $a \in (0,1)$

$$\frac{1}{2n} \log P(S_{2n} \ge 2na) \to -\gamma(a)$$

where $\gamma(a) = \frac{1}{2}\{(1+a)\log(1+a) + (1-a)\log(1-a)\}$.

3.1.4 Suppose $P(X_i = k) = e^{-1}/k!$ for $k = 0, 1, \ldots$ Show that if $a > 1$

$$\frac{1}{n} \log P(S_n \ge na) \to a - 1 - a \log a$$

3.2 Weak Convergence

In this section, we will define the type of convergence that appears in the central limit theorem and explore some of its properties. A sequence of distribution functions is said to **converge weakly** to a limit F (written $F_n \Rightarrow F$) if $F_n(y) \to F(y)$ for all y that are continuity points of F. A sequence of random variables X_n is said to **converge weakly** or **converge in distribution** to a limit X_∞ (written $X_n \Rightarrow X_\infty$) if their distribution functions $F_n(x) = P(X_n \le x)$ converge weakly. To see that convergence at continuity points is enough to identify the limit, observe that F is right continuous and by Exercise 1.2.3, the discontinuities of F are at most a countable set.

3.2.1 Examples

Two examples of weak convergence that we have seen earlier are:

Example 3.2.1 Let $X_1, X_2, \ldots$ be i.i.d. with $P(X_i = 1) = P(X_i = -1) = 1/2$ and let $S_n = X_1 + \cdots + X_n$. Then Theorem 3.1.3 implies

$$F_n(y) = P(S_n/\sqrt{n} \le y) \to \int_{-\infty}^{y} (2\pi)^{-1/2} e^{-x^2/2}\, dx$$

Example 3.2.2 Let $X_1, X_2, \ldots$ be i.i.d. with distribution F. The Glivenko-Cantelli theorem (Theorem 8.4.1) implies that for almost every ω,

$$F_n(y) = n^{-1} \sum_{m=1}^{n} 1_{(X_m(\omega) \le y)} \to F(y) \text{ for all } y$$

In the last two examples convergence occurred for all y, even though in the second case the distribution function could have discontinuities. The next example shows why we restrict our attention to continuity points.

Example 3.2.3 Let X have distribution F. Then $X + 1/n$ has distribution

$$F_n(x) = P(X + 1/n \le x) = F(x - 1/n)$$

As $n \to \infty$, $F_n(x) \to F(x-) = \lim_{y \uparrow x} F(y)$ so convergence only occurs at continuity points.

Example 3.2.4 (Waiting for rare events) Let X_p be the number of trials needed to get a success in a sequence of independent trials with success probability p. Then $P(X_p \ge n) = (1-p)^{n-1}$ for $n = 1, 2, 3, \ldots$ and it follows from Lemma 3.1.1 that as $p \to 0$,

$$P(pX_p > x) \to e^{-x} \quad \text{for all } x \ge 0$$

In words, pX_p converges weakly to an exponential distribution.

Example 3.2.5 (Birthday problem) Let $X_1, X_2, \ldots$ be independent and uniformly distributed on $\{1, \ldots, N\}$, and let $T_N = \min\{n : X_n = X_m \text{ for some } m < n\}$.

$$P(T_N > n) = \prod_{m=2}^{n} \left(1 - \frac{m-1}{N}\right)$$

When $N = 365$, this is the probability that two people in a group of size n do not have the same birthday (assuming all birthdays are equally likely). Using Exercise 3.1.1, it is easy to see that

$$P(T_N / N^{1/2} > x) \to \exp(-x^2/2) \text{ for all } x \ge 0$$

Taking $N = 365$ and noting $22/\sqrt{365} = 1.1515$ and $(1.1515)^2/2 = 0.6630$, this says that

$$P(T_{365} > 22) \approx e^{-0.6630} \approx 0.515$$

This answer is 2% smaller than the true probability 0.524.

Before giving our sixth example, we need a simple result called **Scheffé's Theorem**. Suppose we have probability densities f_n, $1 \le n \le \infty$, and $f_n \to f_\infty$ pointwise as $n \to \infty$. Then for all Borel sets B

$$\begin{aligned}\left| \int_B f_n(x)dx - \int_B f_\infty(x)dx \right| &\le \int |f_n(x) - f_\infty(x)|dx \\ &= 2\int (f_\infty(x) - f_n(x))^+ \, dx \to 0\end{aligned}$$

by the dominated convergence theorem, the equality following from the fact that the $f_n \geq 0$ and have integral $= 1$. Writing μ_n for the corresponding measures, we have shown that the **total variation norm**

$$\|\mu_n - \mu_\infty\| \equiv \sup_B |\mu_n(B) - \mu_\infty(B)| \to 0$$

a conclusion stronger than weak convergence. (Take $B = (-\infty, x]$.) The example μ_n = a point mass at $1/n$ (with $1/\infty = 0$) shows that we may have $\mu_n \Rightarrow \mu_\infty$ with $\|\mu_n - \mu_\infty\| = 1$ for all n.

Example 3.2.6 (Central order statistic) Put $(2n + 1)$ points at random in (0,1), i.e., with locations that are independent and uniformly distributed. Let V_{n+1} be the $(n + 1)$th largest point. It is easy to see that

Lemma 3.2.7 *V_{n+1} has density function*

$$f_{V_{n+1}}(x) = (2n+1)\binom{2n}{n} x^n (1-x)^n$$

Proof There are $2n + 1$ ways to pick the observation that falls at x, then we have to pick n indices for observations $< x$, which can be done in $\binom{2n}{n}$ ways. Once we have decided on the indices that will land $< x$ and $> x$, the probability the corresponding random variables will do what we want is $x^n(1 - x)^n$, and the probability density that the remaining one will land at x is 1. If you don't like the previous sentence, compute the probability $X_1 < x - \epsilon, \ldots, X_n < x - \epsilon, x - \epsilon < X_{n+1} < x + \epsilon$, $X_{n+2} > x + \epsilon, \ldots X_{2n+1} > x + \epsilon$ then let $\epsilon \to 0$. □

To compute the density function of $Y_n = 2(V_{n+1} - 1/2)\sqrt{2n}$, we use Exercise 1.2.5, or simply change variables $x = 1/2 + y/2\sqrt{2n}$, $dx = dy/2\sqrt{2n}$ to get

$$\begin{aligned} f_{Y_n}(y) &= (2n+1)\binom{2n}{n}\left(\frac{1}{2} + \frac{y}{2\sqrt{2n}}\right)^n \left(\frac{1}{2} - \frac{y}{2\sqrt{2n}}\right)^n \frac{1}{2\sqrt{2n}} \\ &= \binom{2n}{n} 2^{-2n} \cdot (1 - y^2/2n)^n \cdot \frac{2n+1}{2n} \cdot \sqrt{\frac{n}{2}} \end{aligned}$$

The first factor is $P(S_{2n} = 0)$ for a simple random walk, so Theorem 3.1.2 and Lemma 3.1.1 imply that

$$f_{Y_n}(y) \to (2\pi)^{-1/2} \exp(-y^2/2) \text{ as } n \to \infty$$

Here and in what follows we write $P(Y_n = y)$ for the density function of Y_n. Using Scheffé's theorem now, we conclude that Y_n converges weakly to a standard normal distribution.

3.2.2 Theory

The next result is useful for proving things about weak convergence.

Theorem 3.2.8 *If $F_n \Rightarrow F_\infty$, then there are random variables Y_n, $1 \leq n \leq \infty$, with distribution F_n so that $Y_n \to Y_\infty$ a.s.*

Proof Let $\Omega = (0,1)$, $\mathcal{F}$ = Borel sets, P = Lebesgue measure, and let $Y_n(x) = \sup\{y : F_n(y) < x\}$. By Theorem 1.2.2, Y_n has distribution F_n. We will now show that $Y_n(x) \to Y_\infty(x)$ for all but a countable number of x. To do this, it is convenient to write $Y_n(x)$ as $F_n^{-1}(x)$ and drop the subscript when $n = \infty$. We begin by identifying the exceptional set. Let $a_x = \sup\{y : F(y) < x\}$, $b_x = \inf\{y : F(y) > x\}$, and $\Omega_0 = \{x : (a_x, b_x) = \emptyset\}$, where (a_x, b_x) is the open interval with the indicated endpoints. $\Omega - \Omega_0$ is countable since the (a_x, b_x) are disjoint and each nonempty interval contains a different rational number. If $x \in \Omega_0$, then $F(y) < x$ for $y < F^{-1}(x)$ and $F(z) > x$ for $z > F^{-1}(x)$. To prove that $F_n^{-1}(x) \to F^{-1}(x)$ for $x \in \Omega_0$, there are two things to show:

(a) $\liminf_{n\to\infty} F_n^{-1}(x) \ge F^{-1}(x)$

Proof of (a). Let $y < F^{-1}(x)$ be such that F is continuous at y. Since $x \in \Omega_0$, $F(y) < x$ and if n is sufficiently large, $F_n(y) < x$, i.e., $F_n^{-1}(x) \ge y$. Since this holds for all y satisfying the indicated restrictions, the result follows.

(b) $\limsup_{n\to\infty} F_n^{-1}(x) \le F^{-1}(x)$

Proof of (b). Let $y > F^{-1}(x)$ be such that F is continuous at y. Since $x \in \Omega_0$, $F(y) > x$ and if n is sufficiently large, $F_n(y) > x$, i.e., $F_n^{-1}(x) \le y$. Since this holds for all y satisfying the indicated restrictions, the result follows and we have completed the proof. □

Theorem 3.2.8 allows us to immediately generalize some of our earlier results.

The next result illustrates the usefulness of Theorem 3.2.8 and gives an equivalent definition of weak convergence that makes sense in any topological space.

Theorem 3.2.9 *$X_n \Rightarrow X_\infty$ if and only if for every bounded continuous function g we have $Eg(X_n) \to Eg(X_\infty)$.*

Proof Let Y_n have the same distribution as X_n and converge a.s. Since g is continuous $g(Y_n) \to g(Y_\infty)$, a.s. and the bounded convergence theorem implies

$$Eg(X_n) = Eg(Y_n) \to Eg(Y_\infty) = Eg(X_\infty)$$

To prove the converse let

$$g_{x,\epsilon}(y) = \begin{cases} 1 & y \le x \\ 0 & y \ge x + \epsilon \\ \text{linear} & x \le y \le x + \epsilon \end{cases}$$

Since $g_{x,\epsilon}(y) = 1$ for $y \le x$, $g_{x,\epsilon}$ is continuous, and $g_{x,\epsilon}(y) = 0$ for $y > x + \epsilon$,

$$\limsup_{n\to\infty} P(X_n \le x) \le \limsup_{n\to\infty} Eg_{x,\epsilon}(X_n) = Eg_{x,\epsilon}(X_\infty) \le P(X_\infty \le x + \epsilon)$$

Letting $\epsilon \to 0$ gives $\limsup_{n\to\infty} P(X_n \le x) \le P(X_\infty \le x)$. The last conclusion is valid for any x. To get the other direction, we observe

$$\liminf_{n\to\infty} P(X_n \le x) \ge \liminf_{n\to\infty} Eg_{x-\epsilon,\epsilon}(X_n) = Eg_{x-\epsilon,\epsilon}(X_\infty) \ge P(X_\infty \le x - \epsilon)$$

Letting $\epsilon \to 0$ gives $\liminf_{n\to\infty} P(X_n \le x) \ge P(X_\infty < x) = P(X_\infty \le x)$ if x is a continuity point. The results for the lim sup and the lim inf combine to give the desired result. □

The next result is a trivial but useful generalization of Theorem 3.2.9.

Theorem 3.2.10 (Continuous mapping theorem) *Let g be a measurable function and $D_g = \{x : g \text{ is discontinuous at } x\}$. If $X_n \Rightarrow X_\infty$ and $P(X_\infty \in D_g) = 0$, then $g(X_n) \Rightarrow g(X)$. If in addition g is bounded, then $Eg(X_n) \to Eg(X_\infty)$.*

Remark D_g is always a Borel set. See Exercise 1.3.6.

Proof Let $Y_n =_d X_n$ with $Y_n \to Y_\infty$ a.s. If f is continuous, then $D_{f \circ g} \subset D_g$, so $P(Y_\infty \in D_{f \circ g}) = 0$ and it follows that $f(g(Y_n)) \to f(g(Y_\infty))$ a.s. If, in addition, f is bounded then the bounded convergence theorem implies $Ef(g(Y_n)) \to Ef(g(Y_\infty))$. Since this holds for all bounded continuous functions, it follows from Theorem 3.2.9 that $g(X_n) \Rightarrow g(X_\infty)$.

The second conclusion is easier. Since $P(Y_\infty \in D_g) = 0$, $g(Y_n) \to g(Y_\infty)$ a.s., and the desired result follows from the bounded convergence theorem. □

The next result provides a number of useful alternative definitions of weak convergence.

Theorem 3.2.11 *The following statements are equivalent: (i) $X_n \Rightarrow X_\infty$*
(ii) For all open sets G, $\liminf_{n\to\infty} P(X_n \in G) \ge P(X_\infty \in G)$.
(iii) For all closed sets K, $\limsup_{n\to\infty} P(X_n \in K) \le P(X_\infty \in K)$.
(iv) For all Borel sets A with $P(X_\infty \in \partial A) = 0$, $\lim_{n\to\infty} P(X_n \in A) = P(X_\infty \in A)$.

Remark To help remember the directions of the inequalities in (ii) and (iii), consider the special case in which $P(X_n = x_n) = 1$. In this case, if $x_n \in G$ and $x_n \to x_\infty \in \partial G$, then $P(X_n \in G) = 1$ for all n but $P(X_\infty \in G) = 0$. Letting $K = G^c$ gives an example for (iii).

Proof We will prove four things and leave it to the reader to check that we have proved the result given here.

(i) implies (ii): Let Y_n have the same distribution as X_n and $Y_n \to Y_\infty$ a.s. Since G is open

$$\liminf_{n\to\infty} 1_G(Y_n) \ge 1_G(Y_\infty)$$

so Fatou's Lemma implies

$$\liminf_{n\to\infty} P(Y_n \in G) \ge P(Y_\infty \in G)$$

(ii) is equivalent to (iii): This follows easily from: A is open if and only if A^c is closed and $P(A) + P(A^c) = 1$.

(ii) and (iii) imply (iv): Let $K = \bar{A}$ and $G = A^o$ be the closure and interior of A, respectively. The boundary of A, $\partial A = \bar{A} - A^o$ and $P(X_\infty \in \partial A) = 0$, so

$$P(X_\infty \in K) = P(X_\infty \in A) = P(X_\infty \in G)$$

Using (ii) and (iii) now

$$\limsup_{n\to\infty} P(X_n \in A) \le \limsup_{n\to\infty} P(X_n \in K) \le P(X_\infty \in K) = P(X_\infty \in A)$$
$$\liminf_{n\to\infty} P(X_n \in A) \ge \liminf_{n\to\infty} P(X_n \in G) \ge P(X_\infty \in G) = P(X_\infty \in A)$$

(iv) implies (i): Let x be such that $P(X_\infty = x) = 0$, i.e., x is a continuity point of F, and let $A = (-\infty, x]$. □

The next result is useful in studying limits of sequences of distributions:

Theorem 3.2.12 (Helly's selection theorem) *For every sequence F_n of distribution functions, there is a subsequence $F_{n(k)}$ and a right continuous nondecreasing function F so that $\lim_{k\to\infty} F_{n(k)}(y) = F(y)$ at all continuity points y of F.*

Remark The limit may not be a distribution function. For example, if $a + b + c = 1$ and $F_n(x) = a\, 1_{(x \geq n)} + b\, 1_{(x \geq -n)} + c\, G(x)$, where G is a distribution function, then $F_n(x) \to F(x) = b + cG(x)$,

$$\lim_{x \downarrow -\infty} F(x) = b \quad \text{and} \quad \lim_{x \uparrow \infty} F(x) = b + c = 1 - a$$

In words, an amount of mass a escapes to $+\infty$, and mass b escapes to $-\infty$. The type of convergence that occurs in Theorem 3.2.12 is sometimes called **vague convergence**, and will be denoted here by $\Rightarrow_v$.

Proof The first step is a diagonal argument. Let $q_1, q_2, \ldots$ be an enumeration of the rationals. Since for each k, $F_m(q_k) \in [0, 1]$ for all m, there is a sequence $m_k(i) \to \infty$ that is a subsequence of $m_{k-1}(j)$ (let $m_0(j) \equiv j$) so that

$$F_{m_k(i)}(q_k) \text{ converges to } G(q_k) \text{ as } i \to \infty$$

Let $F_{n(k)} = F_{m_k(k)}$. By construction, $F_{n(k)}(q) \to G(q)$ for all rational q. The function G may not be right continuous but $F(x) = \inf\{G(q) : q \in \mathbf{Q}, q > x\}$ is since

$$\begin{aligned}\lim_{x_n \downarrow x} F(x_n) &= \inf\{G(q) : q \in \mathbf{Q}, q > x_n \text{ for some } n\} \\ &= \inf\{G(q) : q \in \mathbf{Q}, q > x\} = F(x)\end{aligned}$$

To complete the proof, let x be a continuity point of F. Pick rationals r_1, r_2, s with $r_1 < r_2 < x < s$ so that

$$F(x) - \epsilon < F(r_1) \leq F(r_2) \leq F(x) \leq F(s) < F(x) + \epsilon$$

Since $F_{n(k)}(r_2) \to G(r_2) \geq F(r_1)$, and $F_{n(k)}(s) \to G(s) \leq F(s)$, it follows that if k is large

$$F(x) - \epsilon < F_{n(k)}(r_2) \leq F_{n(k)}(x) \leq F_{n(k)}(s) < F(x) + \epsilon$$

which is the desired conclusion. □

The last result raises a question: When can we conclude that no mass is lost in the limit in Theorem 3.2.12?

Theorem 3.2.13 *Every subsequential limit is the distribution function of a probability measure if and only if the sequence F_n is* **tight**, *i.e., for all $\epsilon > 0$ there is an M_ϵ so that*

$$\limsup_{n\to\infty} 1 - F_n(M_\epsilon) + F_n(-M_\epsilon) \leq \epsilon$$

Proof Suppose the sequence is tight and $F_{n(k)} \Rightarrow_v F$. Let $r < -M_\epsilon$ and $s > M_\epsilon$ be continuity points of F. Since $F_n(r) \to F(r)$ and $F_n(s) \to F(s)$, we have

$$\begin{aligned}1 - F(s) + F(r) &= \lim_{k\to\infty} 1 - F_{n(k)}(s) + F_{n(k)}(r) \\ &\leq \limsup_{n\to\infty} 1 - F_n(M_\epsilon) + F_n(-M_\epsilon) \leq \epsilon\end{aligned}$$

The last result implies $\limsup_{x\to\infty} 1 - F(x) + F(-x) \leq \epsilon$. Since ϵ is arbitrary, it follows that F is the distribution function of a probability measure.

To prove the converse now suppose F_n is not tight. In this case, there is an $\epsilon > 0$ and a subsequence $n(k) \to \infty$ so that

$$1 - F_{n(k)}(k) + F_{n(k)}(-k) \geq \epsilon$$

for all k. By passing to a further subsequence $F_{n(k_j)}$ we can suppose that $F_{n(k_j)} \Rightarrow_v F$. Let $r < 0 < s$ be continuity points of F.

$$\begin{aligned} 1 - F(s) + F(r) &= \lim_{j\to\infty} 1 - F_{n(k_j)}(s) + F_{n(k_j)}(r) \\ &\geq \liminf_{j\to\infty} 1 - F_{n(k_j)}(k_j) + F_{n(k_j)}(-k_j) \geq \epsilon \end{aligned}$$

Letting $s \to \infty$ and $r \to -\infty$, we see that F is not the distribution function of a probability measure. □

The following sufficient condition for tightness is often useful.

Theorem 3.2.14 *If there is a $\varphi \geq 0$ so that $\varphi(x) \to \infty$ as $|x| \to \infty$ and*

$$C = \sup_n \int \varphi(x) dF_n(x) < \infty$$

then F_n is tight.

Proof $1 - F_n(M) + F_n(-M) \leq C / \inf_{|x|\geq M} \varphi(x)$. □

Exercises 3.2.6 and 3.2.7 define metrics for convergence in distribution. The fact that convergence in distribution comes from a metric immediately implies:

Theorem 3.2.15 *If each subsequence of X_n has a further subsequence that converges to X, then $X_n \Rightarrow X$.*

We will prove this again at the end of the proof of Theorem 3.3.17.

Exercises

3.2.1 Give an example of random variables X_n with densities f_n so that $X_n \Rightarrow$ a uniform distribution on (0,1) but $f_n(x)$ does not converge to 1 for any $x \in [0, 1]$.

3.2.2 **Convergence of maxima.** Let $X_1, X_2, \ldots$ be independent with distribution F, and let $M_n = \max_{m\leq n} X_m$. Then $P(M_n \leq x) = F(x)^n$. Prove the following limit laws for M_n:

(i) If $F(x) = 1 - x^{-\alpha}$ for $x \geq 1$, where $\alpha > 0$, then for $y > 0$

$$P(M_n/n^{1/\alpha} \leq y) \to \exp(-y^{-\alpha})$$

(ii) If $F(x) = 1 - |x|^\beta$ for $-1 \leq x \leq 0$, where $\beta > 0$, then for $y < 0$

$$P(n^{1/\beta} M_n \leq y) \to \exp(-|y|^\beta)$$

(iii) If $F(x) = 1 - e^{-x}$ for $x \geq 0$, then for all $y \in (-\infty, \infty)$

$$P(M_n - \log n \leq y) \to \exp(-e^{-y})$$

The limits that appear here are called the **extreme value distributions**. The last one is called the **double exponential** or **Gumbel distribution**. Necessary and sufficient conditions for $(M_n - b_n)/a_n$ to converge to these limits were obtained by Gnedenko (1943). For a recent treatment, see Resnick (1987).

3.2.3 Let $X_1, X_2, \ldots$ be i.i.d. and have the standard normal distribution. (i) From Theorem 1.2.6, we know

$$P(X_i > x) \sim \frac{1}{\sqrt{2\pi}\, x} e^{-x^2/2} \quad \text{as } x \to \infty$$

Use this to conclude that for any real number θ

$$P(X_i > x + (\theta/x))/P(X_i > x) \to e^{-\theta}$$

(ii) Show that if we define b_n by $P(X_i > b_n) = 1/n$

$$P(b_n(M_n - b_n) \leq x) \to \exp(-e^{-x})$$

(iii) Show that $b_n \sim (2\log n)^{1/2}$ and conclude $M_n/(2\log n)^{1/2} \to 1$ in probability.

3.2.4 **Fatou's lemma** Let $g \geq 0$ be continuous. If $X_n \Rightarrow X_\infty$, then

$$\liminf_{n\to\infty} Eg(X_n) \geq Eg(X_\infty)$$

3.2.5 **Integration to the limit** Suppose g, h are continuous with $g(x) > 0$, and $|h(x)|/g(x) \to 0$ as $|x| \to \infty$. If $F_n \Rightarrow F$ and $\int g(x)\, dF_n(x) \leq C < \infty$, then

$$\int h(x)\, dF_n(x) \to \int h(x) dF(x)$$

3.2.6 **The Lévy Metric** Show that

$$\rho(F, G) = \inf\{\epsilon : F(x - \epsilon) - \epsilon \leq G(x) \leq F(x + \epsilon) + \epsilon \text{ for all } x\}$$

defines a metric on the space of distributions and $\rho(F_n, F) \to 0$ if and only if $F_n \Rightarrow F$.

3.2.7 **The Ky Fan metric** on random variables is defined by

$$\alpha(X, Y) = \inf\{\epsilon \geq 0 : P(|X - Y| > \epsilon) \leq \epsilon\}$$

Show that if $\alpha(X, Y) = \alpha$, then the corresponding distributions have Lévy distance $\rho(F, G) \leq \alpha$.

3.2.8 Let $\alpha(X, Y)$ be the metric in the previous exercise and let $\beta(X, Y) = E(|X - Y|/(1 + |X - Y|))$ be the metric of Exercise 2.3.6. If $\alpha(X, Y) = a$, then

$$a^2/(1 + a) \leq \beta(X, Y) \leq a + (1 - a)a/(1 + a)$$

3.2.9 If $F_n \Rightarrow F$ and F is continuous, then $\sup_x |F_n(x) - F(x)| \to 0$.

3.2.10 If F is any distribution function, there is a sequence of distribution functions of the form $\sum_{m=1}^{n} a_{n,m} 1_{(x_{n,m} \le x)}$ with $F_n \Rightarrow F$.

3.2.11 Let X_n, $1 \le n \le \infty$, be integer valued. Show that $X_n \Rightarrow X_\infty$ if and only if $P(X_n = m) \to P(X_\infty = m)$ for all m.

3.2.12 Show that if $X_n \to X$ in probability, then $X_n \Rightarrow X$ and that, conversely, if $X_n \Rightarrow c$, where c is a constant, then $X_n \to c$ in probability.

3.2.13 **Converging together lemma** If $X_n \Rightarrow X$ and $Y_n \Rightarrow c$, where c is a constant, then $X_n + Y_n \Rightarrow X + c$. A useful consequence of this result is that if $X_n \Rightarrow X$ and $Z_n - X_n \Rightarrow 0$, then $Z_n \Rightarrow X$.

3.2.14 Suppose $X_n \Rightarrow X$, $Y_n \ge 0$, and $Y_n \Rightarrow c$, where $c > 0$ is a constant, then $X_n Y_n \Rightarrow cX$. This result is true without the assumptions $Y_n \ge 0$ and $c > 0$. We have imposed these only to make the proof less tedious.

3.2.15 Show that if $X_n = (X_n^1, \dots, X_n^n)$ is uniformly distributed over the surface of the sphere of radius $\sqrt{n}$ in $\mathbf{R}^n$, then $X_n^1 \Rightarrow$ a standard normal. Hint: Let $Y_1, Y_2, \dots$ be i.i.d. standard normals and let $X_n^i = Y_i (n / \sum_{m=1}^{n} Y_m^2)^{1/2}$.

3.2.16 Suppose $Y_n \ge 0$, $EY_n^\alpha \to 1$ and $EY_n^\beta \to 1$ for some $0 < \alpha < \beta$. Show that $Y_n \to 1$ in probability.

3.2.17 For each $K < \infty$ and $y < 1$ there is a $c_{y,K} > 0$ so that $EX^2 = 1$ and $EX^4 \le K$ implies $P(|X| > y) \ge c_{y,K}$.

3.3 Characteristic Functions

This long section is divided into five parts. The first three are required reading, and the last two are optional. In the first part, we show that the characteristic function $\varphi(t) = E\exp(itX)$ determines $F(x) = P(X \le x)$, and we give recipes for computing F from φ. In the second part, we relate weak convergence of distributions to the behavior of the corresponding characteristic functions. In the third part, we relate the behavior of $\varphi(t)$ at 0 to the moments of X. In the fourth part, we prove Polya's criterion and use it to construct some famous and some strange examples of characteristic functions. Finally, in the fifth part, we consider the moment problem, i.e., when is a distribution characterized by its moments?

3.3.1 Definition, Inversion Formula

If X is a random variable, we define its **characteristic function (ch.f.)** by

$$\varphi(t) = Ee^{itX} = E\cos tX + iE\sin tX$$

The last formula requires taking the expected value of a complex valued random variable, but as the second equality may suggest no new theory is required. If Z is complex valued, we define $EZ = E(\operatorname{Re} Z) + iE(\operatorname{Im} Z)$, where $\operatorname{Re}(a + bi) = a$ is the **real part** and $\operatorname{Im}(a + bi) = b$ is the **imaginary part**. Some other definitions we will need are: the **modulus** of the complex number $z = a + bi$ is $|a + bi| = (a^2 + b^2)^{1/2}$, and the **complex conjugate** of $z = a + bi$, $\bar{z} = a - bi$.

Theorem 3.3.1 *All characteristic functions have the following properties:*

(a) $\varphi(0) = 1$,

(b) $\varphi(-t) = \overline{\varphi(t)}$,

(c) $|\varphi(t)| = |Ee^{itX}| \le E|e^{itX}| = 1$

(d) $|\varphi(t+h) - \varphi(t)| \le E|e^{ihX} - 1|$, *so* $\varphi(t)$ *is uniformly continuous on* $(-\infty, \infty)$.

(e) $Ee^{it(aX+b)} = e^{itb}\varphi(at)$

Proof (a) is obvious. For (b) we note that

$$\varphi(-t) = E(\cos(-tX) + i\sin(-tX)) = E(\cos(tX) - i\sin(tX))$$

(c) follows from Exercise 1.6.2, since $\varphi(x, y) = (x^2 + y^2)^{1/2}$ is convex.

$$\begin{aligned}|\varphi(t+h) - \varphi(t)| &= |E(e^{i(t+h)X} - e^{itX})| \\ &\le E|e^{i(t+h)X} - e^{itX}| = E|e^{ihX} - 1|\end{aligned}$$

so uniform convergence follows from the bounded convergence theorem. For (e) we note $Ee^{it(aX+b)} = e^{itb}Ee^{i(ta)X} = e^{itb}\varphi(at)$. □

The main reason for introducing characteristic functions is the following:

Theorem 3.3.2 *If* X_1 *and* X_2 *are independent and have ch.f.'s* φ_1 *and* φ_2*, then* $X_1 + X_2$ *has ch.f.* $\varphi_1(t)\varphi_2(t)$.

Proof

$$Ee^{it(X_1+X_2)} = E(e^{itX_1}e^{itX_2}) = Ee^{itX_1}Ee^{itX_2}$$

since e^{itX_1} and e^{itX_2} are independent. □

The next order of business is to give some examples.

Example 3.3.3 (Coin flips) If $P(X = 1) = P(X = -1) = 1/2$, then

$$Ee^{itX} = (e^{it} + e^{-it})/2 = \cos t$$

Example 3.3.4 (Poisson distribution) If $P(X = k) = e^{-\lambda}\lambda^k/k!$ for $k = 0, 1, 2, \ldots$ then

$$Ee^{itX} = \sum_{k=0}^{\infty} e^{-\lambda}\frac{\lambda^k e^{itk}}{k!} = \exp(\lambda(e^{it} - 1))$$

Example 3.3.5 (Normal distribution)

Density	$(2\pi)^{-1/2}\exp(-x^2/2)$
Ch.f.	$\exp(-t^2/2)$

Combining this result with (e) of Theorem 3.3.1, we see that a normal distribution with mean μ and variance σ^2 has ch.f. $\exp(i\mu t - \sigma^2 t^2/2)$. Similar scalings can be applied to other examples, so we will often just give the ch.f. for one member of the family.

Physics Proof

$$\int e^{itx}(2\pi)^{-1/2}e^{-x^2/2}\,dx = e^{-t^2/2}\int (2\pi)^{-1/2}e^{-(x-it)^2/2}\,dx$$

The integral is 1, since the integrand is the normal density with mean *it* and variance 1. □

Math Proof Now that we have cheated and figured out the answer, we can verify it by a formal calculation that gives very little insight into why it is true. Let

$$\varphi(t) = \int e^{itx}(2\pi)^{-1/2}e^{-x^2/2}dx = \int \cos tx\,(2\pi)^{-1/2}e^{-x^2/2}dx$$

since $i \sin tx$ is an odd function. Differentiating with respect to t (referring to Theorem A.5.1 for the justification) and then integrating by parts gives

$$\begin{aligned}\varphi'(t) &= \int -x \sin tx\,(2\pi)^{-1/2}e^{-x^2/2}dx \\ &= -\int t \cos tx\,(2\pi)^{-1/2}e^{-x^2/2}dx = -t\varphi(t)\end{aligned}$$

This implies $\frac{d}{dt}\{\varphi(t)\exp(t^2/2)\} = 0$, so $\varphi(t)\exp(t^2/2) = \varphi(0) = 1$. □

In the next three examples, the density is 0 outside the indicated range.

Example 3.3.6 (Uniform distribution on (a,b))

Density	$1/(b-a)$	$x \in (a,b)$
Ch.f.	$(e^{itb} - e^{ita})/\,it(b-a)$	

In the special case $a = -c$, $b = c$ the ch.f. is $(e^{itc} - e^{-itc})/2cit = (\sin ct)/ct$.

Proof Once you recall that $\int_a^b e^{\lambda x}\,dx = (e^{\lambda b} - e^{\lambda a})/\lambda$ holds for complex λ, this is immediate. □

Example 3.3.7 (Triangular distribution)

Density	$1 - \lvert x\rvert$	$x \in (-1,1)$
Ch.f.	$2(1-\cos t)/t^2$	

Proof To see this, notice that if X and Y are independent and uniform on $(-1/2, 1/2)$, then $X+Y$ has a triangular distribution. Using Example 3.3.6 now and Theorem 3.3.2, it follows that the desired ch.f. is

$$\{(e^{it/2} - e^{-it/2})/it\}^2 = \{2\sin(t/2)/t\}^2$$

Using the trig identity $\cos 2\theta = 1 - 2\sin^2\theta$ with $\theta = t/2$ converts the answer into the form given here. □

Example 3.3.8 (Exponential distribution)

Density	e^{-x}	$x \in (0,\infty)$
Ch.f.	$1/(1-it)$	

Proof Integrating gives

$$\int_0^\infty e^{itx} e^{-x}\,dx = \left.\frac{e^{(it-1)x}}{it-1}\right|_0^\infty = \frac{1}{1-it}$$

since $\exp((it-1)x) \to 0$ as $x \to \infty$. □

For the next result we need the following fact, which follows from the fact that $\int f\,d(\mu+\nu) = \int f\,d\mu + \int f\,d\nu$.

Lemma 3.3.9 *If $F_1, \ldots, F_n$ have ch.f. $\varphi_1, \ldots, \varphi_n$ and $\lambda_i \geq 0$ have $\lambda_1 + \cdots + \lambda_n = 1$, then $\sum_{i=1}^n \lambda_i F_i$ has ch.f. $\sum_{i=1}^n \lambda_i \varphi_i$.*

Example 3.3.10 (Bilateral exponential)

Density	$\frac{1}{2}e^{-\lvert x\rvert}$	$x \in (-\infty,\infty)$
Ch.f.	$1/(1+t^2)$	

Proof This follows from Lemma 3.3.9 with F_1 the distribution of an exponential random variable X, F_2 the distribution of $-X$, and $\lambda_1 = \lambda_2 = 1/2$, then using (b) of Theorem 3.3.1 we see the desired ch.f. is

$$\frac{1}{2(1-it)} + \frac{1}{2(1+it)} = \frac{(1+it)+(1-it)}{2(1+t^2)} = \frac{1}{(1+t^2)}$$ □

The first issue to be settled is that the characteristic function uniquely determines the distribution. This and more is provided by:

Theorem 3.3.11 (The inversion formula) *Let $\varphi(t) = \int e^{itx}\mu(dx)$ where μ is a probability measure. If $a < b$, then*

$$\lim_{T\to\infty} (2\pi)^{-1} \int_{-T}^{T} \frac{e^{-ita} - e^{-itb}}{it}\varphi(t)\,dt = \mu(a,b) + \frac{1}{2}\mu(\{a,b\})$$

Remark The existence of the limit is part of the conclusion. If $\mu = \delta_0$, a point mass at 0, $\varphi(t) \equiv 1$. In this case, if $a = -1$ and $b = 1$, the integrand is $(2\sin t)/t$ and the integral does not converge absolutely.

Proof Let

$$I_T = \int_{-T}^{T} \frac{e^{-ita} - e^{-itb}}{it}\varphi(t)\,dt = \int_{-T}^{T}\int \frac{e^{-ita} - e^{-itb}}{it} e^{itx}\mu(dx)\,dt$$

The integrand may look bad near $t = 0$ but if we observe that

$$\frac{e^{-ita} - e^{-itb}}{it} = \int_a^b e^{-ity}\,dy$$

we see that the modulus of the integrand is bounded by $b - a$. Since μ is a probability measure and $[-T,T]$ is a finite interval, it follows from Fubini's theorem, $\cos(-x) = \cos x$, and $\sin(-x) = -\sin x$ that

$$I_T = \int \int_{-T}^{T} \frac{e^{-ita} - e^{-itb}}{it} e^{itx}\, dt\, \mu(dx)$$
$$= \int \left\{ \int_{-T}^{T} \frac{\sin(t(x-a))}{t}\, dt - \int_{-T}^{T} \frac{\sin(t(x-b))}{t}\, dt \right\} \mu(dx)$$

Introducing $R(\theta,T) = \int_{-T}^{T} (\sin \theta t)/t\, dt$, we can write the last result as

$$(*) \qquad I_T = \int \{R(x-a,T) - R(x-b,T)\} \mu(dx)$$

If we let $S(T) = \int_0^T (\sin x)/x\, dx$, then for $\theta > 0$ changing variables $t = x/\theta$ shows that

$$R(\theta,T) = 2 \int_0^{T\theta} \frac{\sin x}{x}\, dx = 2S(T\theta)$$

while for $\theta < 0$, $R(\theta,T) = -R(|\theta|,T)$. Introducing the function $\operatorname{sgn} x$, which is 1 if $x > 0$, -1 if $x < 0$, and 0 if $x = 0$, we can write the last two formulas together as

$$R(\theta,T) = 2(\operatorname{sgn} \theta) S(T|\theta|)$$

As $T \to \infty$, $S(T) \to \pi/2$ (see Exercise 1.7.5), so we have $R(\theta,T) \to \pi \operatorname{sgn} \theta$ and

$$R(x-a,T) - R(x-b,T) \to \begin{cases} 2\pi & a < x < b \\ \pi & x = a \text{ or } x = b \\ 0 & x < a \text{ or } x > b \end{cases}$$

$|R(\theta,T)| \le 2 \sup_y S(y) < \infty$, so using the bounded convergence theorem with $(*)$ implies

$$(2\pi)^{-1} I_T \to \mu(a,b) + \frac{1}{2}\mu(\{a,b\})$$

proving the desired result. □

Two trivial consequences of the inversion formula are:

Corollary 3.3.12 *If φ is real, then X and $-X$ have the same distribution.*

Corollary 3.3.13 *If X_i, $i = 1,2$ are independent and have normal distributions with mean 0 and variance σ_i^2, then $X_1 + X_2$ has a normal distribution with mean 0 and variance $\sigma_1^2 + \sigma_2^2$.*

The inversion formula is simpler when φ is integrable, but as the next result shows this only happens when the underlying measure is nice.

Theorem 3.3.14 *If $\int |\varphi(t)|\, dt < \infty$, then μ has bounded continuous density*

$$f(y) = \frac{1}{2\pi} \int e^{-ity} \varphi(t)\, dt$$

Proof As we observed in the proof of Theorem 3.3.11

$$\left| \frac{e^{-ita} - e^{-itb}}{it} \right| = \left| \int_a^b e^{-ity}\, dy \right| \le |b-a|$$

so the integral in Theorem 3.3.11 converges absolutely in this case and

$$\mu(a,b) + \frac{1}{2}\mu(\{a,b\}) = \frac{1}{2\pi}\int_{-\infty}^{\infty} \frac{e^{-ita} - e^{-itb}}{it}\,\varphi(t)\,dt \leq \frac{(b-a)}{2\pi}\int_{-\infty}^{\infty} |\varphi(t)|dt$$

The last result implies μ has no point masses and

$$\begin{aligned}
\mu(x,x+h) &= \frac{1}{2\pi}\int \frac{e^{-itx} - e^{-it(x+h)}}{it}\,\varphi(t)\,dt \\
&= \frac{1}{2\pi}\int \left(\int_x^{x+h} e^{-ity}\,dy\right)\varphi(t)\,dt \\
&= \int_x^{x+h} \left(\frac{1}{2\pi}\int e^{-ity}\varphi(t)\,dt\right)dy
\end{aligned}$$

by Fubini's theorem, so the distribution μ has density function

$$f(y) = \frac{1}{2\pi}\int e^{-ity}\varphi(t)\,dt$$

The dominated convergence theorem implies f is continuous and the proof is complete. □

Theorem 3.3.14 and the next result show that the behavior of φ at infinity is related to the smoothness of the underlying measure.

Applying the inversion formula Theorem 3.3.14 to the ch.f. in Examples 3.3.7 and 3.3.10 gives us two more examples of ch.f. The first one does not have an official name, so we gave it one to honor its role in the proof of Polya's criterion, see Theorem 3.3.22.

Example 3.3.15 (Polya's distribution)

Density	$(1-\cos x)/\pi x^2$
Ch.f.	$(1-\lvert t\rvert)^+$

Proof Theorem 3.3.14 implies

$$\frac{1}{2\pi}\int \frac{2(1-\cos s)}{s^2}e^{-isy}\,ds = (1-|y|)^+$$

Now let $s = x$, $y = -t$. □

Example 3.3.16 (The Cauchy distribution)

Density	$1/\pi(1+x^2)$
Ch.f.	$\exp(-\lvert t\rvert)$

Proof Theorem 3.3.14 implies

$$\frac{1}{2\pi}\int \frac{1}{1+s^2}e^{-isy}\,ds = \frac{1}{2}e^{-|y|}$$

Now let $s = x$, $y = -t$ and multiply each side by 2. □

Exercises

3.3.1 Show that if φ is a ch.f., then $\mathrm{Re}\,\varphi$ and $|\varphi|^2$ are also.

3.3.2 (i) Imitate the proof of Theorem 3.3.11 to show that

$$\mu(\{a\}) = \lim_{T\to\infty} \frac{1}{2T}\int_{-T}^{T} e^{-ita}\varphi(t)\,dt$$

(ii) If $P(X \in h\mathbf{Z}) = 1$, where $h > 0$, then its ch.f. has $\varphi(2\pi/h + t) = \varphi(t)$ so

$$P(X = x) = \frac{h}{2\pi}\int_{-\pi/h}^{\pi/h} e^{-itx}\varphi(t)\,dt \quad \text{for } x \in h\mathbf{Z}$$

(iii) If $X = Y + b$, then $E\exp(itX) = e^{itb}E\exp(itY)$. So if $P(X \in b + h\mathbf{Z}) = 1$, the inversion formula in (ii) is valid for $x \in b + h\mathbf{Z}$.

3.3.3 Suppose X and Y are independent and have ch.f. φ and distribution μ. Apply Exercise 3.3.2 to $X - Y$ and use Exercise 2.1.5 to get

$$\lim_{T\to\infty} \frac{1}{2T}\int_{-T}^{T} |\varphi(t)|^2\,dt = P(X - Y = 0) = \sum_x \mu(\{x\})^2$$

Remark The last result implies that if $\varphi(t) \to 0$ as $t \to \infty$, μ has no point masses. Exercise 3.3.11 gives an example to show that the converse is false. The Riemann-Lebesgue Lemma (Exercise 1.4.4) shows that if μ has a density, $\varphi(t) \to 0$ as $t \to \infty$.

3.3.4 Give an example of a measure μ with a density but for which $\int |\varphi(t)|dt = \infty$. Hint: Two of the previous examples have this property.

3.3.5 Show that if $X_1, \ldots, X_n$ are independent and uniformly distributed on $(-1,1)$, then for $n \geq 2$, $X_1 + \cdots + X_n$ has density

$$f(x) = \frac{1}{\pi}\int_0^\infty (\sin t/t)^n \cos tx\,dt$$

Although it is not obvious from the formula, f is a polynomial in each interval $(k, k+1)$, $k \in \mathbf{Z}$ and vanishes on $[-n,n]^c$.

3.3.6 Use the result in Example 3.3.16 to conclude that if $X_1, X_2, \ldots$ are independent and have the Cauchy distribution, then $(X_1 + \cdots + X_n)/n$ has the same distribution as X_1.

3.3.2 Weak Convergence

Our next step toward the central limit theorem is to relate convergence of characteristic functions to weak convergence.

Theorem 3.3.17 (Continuity theorem) *Let μ_n, $1 \leq n \leq \infty$ be probability measures with ch.f. φ_n. (i) If $\mu_n \Rightarrow \mu_\infty$, then $\varphi_n(t) \to \varphi_\infty(t)$ for all t. (ii) If $\varphi_n(t)$ converges pointwise to a limit $\varphi(t)$ that is continuous at 0, then the associated sequence of distributions μ_n is tight and converges weakly to the measure μ with characteristic function φ.*

Remark To see why continuity of the limit at 0 is needed in (ii), let μ_n have a normal distribution with mean 0 and variance n. In this case, $\varphi_n(t) = \exp(-nt^2/2) \to 0$ for $t \neq 0$, and $\varphi_n(0) = 1$ for all n, but the measures do not converge weakly, since $\mu_n((-\infty, x]) \to 1/2$ for all x.

Proof (i) is easy. e^{itx} is bounded and continuous, so if $\mu_n \Rightarrow \mu_\infty$, then Theorem 3.2.9 implies $\varphi_n(t) \to \varphi_\infty(t)$. To prove (ii), our first goal is to prove tightness. We begin with some calculations that may look mysterious but will prove to be very useful.

$$\int_{-u}^{u} 1 - e^{itx}\,dt = 2u - \int_{-u}^{u} (\cos tx + i \sin tx)\,dt = 2u - \frac{2 \sin ux}{x}$$

Dividing both sides by u, integrating $\mu_n(dx)$, and using Fubini's theorem on the left-hand side gives

$$u^{-1} \int_{-u}^{u} (1 - \varphi_n(t))\,dt = 2 \int \left(1 - \frac{\sin ux}{ux}\right) \mu_n(dx)$$

To bound the right-hand side, we note that

$$|\sin x| = \left| \int_0^x \cos(y)\,dy \right| \leq |x| \quad \text{for all } x$$

so we have $1 - (\sin ux/ux) \geq 0$. Discarding the integral over $(-2/u, 2/u)$ and using $|\sin ux| \leq 1$ on the rest, the right-hand side is

$$\geq 2 \int_{|x| \geq 2/u} \left(1 - \frac{1}{|ux|}\right) \mu_n(dx) \geq \mu_n(\{x : |x| > 2/u\})$$

Since $\varphi(t) \to 1$ as $t \to 0$,

$$u^{-1} \int_{-u}^{u} (1 - \varphi(t))\,dt \to 0 \text{ as } u \to 0$$

Pick u so that the integral is $< \epsilon$. Since $\varphi_n(t) \to \varphi(t)$ for each t, it follows from the bounded convergence theorem that for $n \geq N$

$$2\epsilon \geq u^{-1} \int_{-u}^{u} (1 - \varphi_n(t))\,dt \geq \mu_n\{x : |x| > 2/u\}$$

Since ϵ is arbitrary, the sequence μ_n is tight.

To complete the proof now we observe that if $\mu_{n(k)} \Rightarrow \mu$, then it follows from the first sentence of the proof that μ has ch.f. φ. The last observation and tightness imply that every subsequence has a further subsequence that converges to μ. I claim that this implies the whole sequence converges to μ. To see this, observe that we have shown that if f is bounded and continuous, then every subsequence of $\int f\,d\mu_n$ has a further subsequence that converges to $\int f\,d\mu$, so Theorem 2.3.3 implies that the whole sequence converges to that limit. This shows $\int f\,d\mu_n \to \int f\,d\mu$ for all bounded continuous functions f, so the desired result follows from Theorem 3.2.9. □

Exercises

3.3.7 Suppose that $X_n \Rightarrow X$ and X_n has a normal distribution with mean 0 and variance σ_n^2. Prove that $\sigma_n^2 \to \sigma^2 \in [0, \infty)$.

3.3.8 Show that if X_n and Y_n are independent for $1 \le n \le \infty$, $X_n \Rightarrow X_\infty$, and $Y_n \Rightarrow Y_\infty$, then $X_n + Y_n \Rightarrow X_\infty + Y_\infty$.

3.3.9 Let $X_1, X_2, \ldots$ be independent and let $S_n = X_1 + \cdots + X_n$. Let φ_j be the ch.f. of X_j and suppose that $S_n \to S_\infty$ a.s. Then S_∞ has ch.f. $\prod_{j=1}^\infty \varphi_j(t)$.

3.3.10 Using the identity $\sin t = 2\sin(t/2)\cos(t/2)$ repeatedly leads to $(\sin t)/t = \prod_{m=1}^\infty \cos(t/2^m)$. Prove the last identity by interpreting each side as a characteristic function.

3.3.11 Let $X_1, X_2, \ldots$ be independent taking values 0 and 1 with probability 1/2 each. $X = 2\sum_{j\ge 1} X_j/3^j$ has the Cantor distribution. Compute the ch.f. φ of X and notice that φ has the same value at $t = 3^k\pi$ for $k = 0, 1, 2, \ldots$

3.3.3 Moments and Derivatives

In the proof of Theorem 3.3.17, we derived the inequality

$$\mu\{x : |x| > 2/u\} \le u^{-1} \int_{-u}^{u} (1 - \varphi(t))\, dt \tag{3.3.1}$$

which shows that the smoothness of the characteristic function at 0 is related to the decay of the measure at ∞. We leave the proof to the reader. (Use Theorem A.5.1.)

Theorem 3.3.18 *If $\int |x|^n \mu(dx) < \infty$, then its characteristic function φ has a continuous derivative of order n given by $\varphi^{(n)}(t) = \int (ix)^n e^{itx}\mu(dx)$.*

Proof This is proved by repeatedly differentiating under the integral and using Theorem A.5.1 to justify this. □

The result in Theorem 3.3.18 shows that if $E|X|^n < \infty$, then its characteristic function is n times differentiable at 0, and $\varphi^n(0) = E(iX)^n$. Expanding φ in a Taylor series about 0 leads to

$$\varphi(t) = \sum_{m=0}^{n} \frac{E(itX)^m}{m!} + o(t^n)$$

where $o(t^n)$ indicates a quantity $g(t)$ that has $g(t)/t^n \to 0$ as $t \to 0$. For our purposes, it will be important to have a good estimate on the error term, so we will now derive the last result. The starting point is a little calculus.

Lemma 3.3.19

$$\left| e^{ix} - \sum_{m=0}^{n} \frac{(ix)^m}{m!} \right| \le \min\left(\frac{|x|^{n+1}}{(n+1)!}, \frac{2|x|^n}{n!} \right) \tag{3.3.2}$$

The first term on the right is the usual order of magnitude we expect in the correction term. The second is better for large $|x|$ and will help us prove the central limit theorem without assuming finite third moments.

Proof Integrating by parts gives

$$\int_0^x (x-s)^n e^{is}\,ds = \frac{x^{n+1}}{n+1} + \frac{i}{n+1}\int_0^x (x-s)^{n+1} e^{is}\,ds$$

When $n = 0$, this says

$$\int_0^x e^{is}\,ds = x + i\int_0^x (x-s)e^{is}\,ds$$

The left-hand side is $(e^{ix}-1)/i$, so rearranging gives

$$e^{ix} = 1 + ix + i^2\int_0^x (x-s)e^{is}ds$$

Using the result for $n = 1$ now gives

$$e^{ix} = 1 + ix + \frac{i^2x^2}{2} + \frac{i^3}{2}\int_0^x (x-s)^2 e^{is}ds$$

and iterating we arrive at

$$\text{(a)} \qquad e^{ix} - \sum_{m=0}^{n}\frac{(ix)^m}{m!} = \frac{i^{n+1}}{n!}\int_0^x (x-s)^n e^{is}ds$$

To prove the result now it only remains to estimate the "error term" on the right-hand side. Since $|e^{is}| \le 1$ for all s,

$$\text{(b)} \qquad \left|\frac{i^{n+1}}{n!}\int_0^x (x-s)^n e^{is}ds\right| \le |x|^{n+1}/(n+1)!$$

The last estimate is good when x is small. The next is designed for large x. Integrating by parts

$$\frac{i}{n}\int_0^x (x-s)^n e^{is}ds = -\frac{x^n}{n} + \int_0^x (x-s)^{n-1}e^{is}ds$$

Noticing $x^n/n = \int_0^x (x-s)^{n-1}ds$ now gives

$$\frac{i^{n+1}}{n!}\int_0^x (x-s)^n e^{is}ds = \frac{i^n}{(n-1)!}\int_0^x (x-s)^{n-1}(e^{is}-1)ds$$

and since $|e^{ix}-1| \le 2$, it follows that

$$\text{(c)} \qquad \left|\frac{i^{n+1}}{n!}\int_0^x (x-s)^n e^{is}ds\right| \le \left|\frac{2}{(n-1)!}\int_0^x (x-s)^{n-1}\,ds\right| \le 2|x|^n/n!$$

Combining (a), (b), and (c) we have the desired result. □

Taking expected values, using Jensen's inequality, applying Lemma 3.3.2 to $x = tX$, gives

$$\begin{aligned}\left|Ee^{itX} - \sum_{m=0}^{n} E\frac{(itX)^m}{m!}\right| &\le E\left|e^{itX} - \sum_{m=0}^{n}\frac{(itX)^m}{m!}\right| \\ &\le E\min\left(|tX|^{n+1}, 2|tX|^n\right) \qquad (3.3.3)\end{aligned}$$

where in the second step we have dropped the denominators to make the bound simpler.

In the next section, the following special case will be useful.

Theorem 3.3.20 *If* $E|X|^2 < \infty$ *then*

$$\varphi(t) = 1 + itEX - t^2 E(X^2)/2 + o(t^2)$$

Proof The error term is $\le t^2 E(|t|\cdot|X|^3 \wedge 2|X|^2)$. The variable in parentheses is smaller than $2|X|^2$ and converges to 0 as $t \to 0$, so the desired conclusion follows from the dominated convergence theorem. □

Remark The point of the estimate in (3.3.3), which involves the minimum of two terms rather than just the first one that would result from a naive application of Taylor series, is that we get the conclusion in Theorem 3.3.20 under the assumption $E|X|^2 < \infty$, i.e., we do not have to assume $E|X|^3 < \infty$.

The next result shows that the existence of second derivatives implies the existence of second moments.

Theorem 3.3.21 *If* $\limsup_{h\downarrow 0}\{\varphi(h) - 2\varphi(0) + \varphi(-h)\}/h^2 > -\infty$, *then* $E|X|^2 < \infty$.

Proof $(e^{ihx} - 2 + e^{-ihx})/h^2 = -2(1 - \cos hx)/h^2 \le 0$ and $2(1 - \cos hx)/h^2 \to x^2$ as $h \to 0$ so Fatou's lemma and Fubini's theorem imply

$$\begin{aligned}\int x^2\, dF(x) &\le 2 \liminf_{h\to 0} \int \frac{1 - \cos hx}{h^2}\, dF(x)\\ &= -\limsup_{h\to 0} \frac{\varphi(h) - 2\varphi(0) + \varphi(-h)}{h^2} < \infty\end{aligned}$$

which proves the desired result. □

Exercises

3.3.12 Use Theorem 3.3.18 and the series expansion for $e^{-t^2/2}$ to show that the standard normal distribution has

$$EX^{2n} = (2n)!/2^n n! = (2n-1)(2n-3)\cdots 3\cdot 1 \equiv (2n-1)!!$$

3.3.13 (i) Suppose that the family of measures $\{\mu_i, i \in I\}$ is tight, i.e., $\sup_i \mu_i([-M, M]^c) \to 0$ as $M \to \infty$. Use (d) in Theorem 3.3.1 and (3.3.3) with $n = 0$ to show that their ch.f.'s φ_i are equicontinuous, i.e., if $\epsilon > 0$ we can pick $\delta > 0$ so that if $|h| < \delta$, then $|\varphi_i(t+h) - \varphi_i(t)| < \epsilon$. (ii) Suppose $\mu_n \Rightarrow \mu_\infty$. Use Theorem 3.3.17 and equicontinuity to conclude that the ch.f.'s $\varphi_n \to \varphi_\infty$ uniformly on compact sets. [Argue directly. You don't need to go to AA.] (iii) Give an example to show that the convergence need not be uniform on the whole real line.

3.3.14 Let $X_1, X_2, \ldots$ be i.i.d. with characteristic function φ. (i) If $\varphi'(0) = ia$ and $S_n = X_1 + \cdots + X_n$, then $S_n/n \to a$ in probability. (ii) If $S_n/n \to a$ in probability, then $\varphi(t/n)^n \to e^{iat}$ as $n \to \infty$ through the integers. (iii) Use (ii) and the uniform continuity established in (d) of Theorem 3.3.1 to show that $(\varphi(h) - 1)/h \to -ia$ as $h \to 0$ through the positive reals. Thus the weak law holds if and only if $\varphi'(0)$ exists. This result is due to E.J.G. Pitman (1956), with a little help from John Walsh who pointed out that we should prove (iii).

The last exercise in combination with Exercise 2.2.4 shows that $\varphi'(0)$ may exist when $E|X| = \infty$.

3.3.15 $2\int_0^\infty (1 - \text{Re}\,\varphi(t))/(\pi t^2)\,dt = \int |y| dF(y)$. Hint: Change variables $x = |y|t$ in the density function of Example 3.3.15, which integrates to 1.

3.3.16 Show that if $\lim_{t\downarrow 0}(\varphi(t) - 1)/t^2 = c > -\infty$, then $EX = 0$ and $E|X|^2 = -2c < \infty$. In particular, if $\varphi(t) = 1 + o(t^2)$, then $\varphi(t) \equiv 1$.

3.3.17 If Y_n are r.v.'s with ch.f.'s φ_n, then $Y_n \Rightarrow 0$ if and only if there is a $\delta > 0$ so that $\varphi_n(t) \to 1$ for $|t| \le \delta$.

3.3.18 Let $X_1, X_2, \ldots$ be independent. If $S_n = \sum_{m\le n} X_m$ converges in distribution then it converges in probability (and hence a.s. by Exercise 2.5.10). Hint: The last exercise implies that if $m, n \to \infty$, then $S_m - S_n \to 0$ in probability. Now use Exercise 2.5.11.

3.3.4 Polya's Criterion*

The next result is useful for constructing examples of ch.f.'s.

Theorem 3.3.22 (Polya's criterion) *Let $\varphi(t)$ be real nonnegative and have $\varphi(0) = 1$, $\varphi(t) = \varphi(-t)$, and φ is decreasing and convex on $(0,\infty)$ with*

$$\lim_{t\downarrow 0} \varphi(t) = 1, \qquad \lim_{t\uparrow\infty} \varphi(t) = 0$$

Then there is a probability measure ν on $(0,\infty)$, so that

$$(*) \qquad \varphi(t) = \int_0^\infty \left(1 - \left|\frac{t}{s}\right|\right)^+ \nu(ds)$$

and hence φ is a characteristic function.

Remark Before we get lost in the details of the proof, the reader should note that $(*)$ displays φ as a convex combination of ch.f.'s of the form given in Example 3.3.15, so an extension of Lemma 3.3.9 (to be proved later) implies that this is a ch.f.

The assumption that $\lim_{t\to 0}\varphi(t) = 1$ is necessary because the function $\varphi(t) = 1_{\{0\}}(t)$, which is 1 at 0 and 0 otherwise, satisfies all the other hypotheses. We could allow $\lim_{t\to\infty}\varphi(t) = c > 0$ by having a point mass of size c at 0, but we leave this extension to the reader.

Proof Let φ' be the right derivative of ϕ, i.e.,

$$\varphi'(t) = \lim_{h\downarrow 0} \frac{\varphi(t+h) - \varphi(t)}{h}$$

Since φ is convex, this exists and is right continuous and increasing. So we can let μ be the measure on $(0,\infty)$ with $\mu(a,b] = \varphi'(b) - \varphi'(a)$ for all $0 \le a < b < \infty$, and let ν be the measure on $(0,\infty)$ with $d\nu/d\mu = s$.

Now $\varphi'(t) \to 0$ as $t \to \infty$ (for if $\varphi'(t) \downarrow -\epsilon$, we would have $\varphi(t) \le 1 - \epsilon t$ for all t), so Exercise A.4.7 implies

$$-\varphi'(s) = \int_s^\infty r^{-1}\nu(dr)$$

Integrating again and using Fubini's theorem, we have for $t \geq 0$

$$\varphi(t) = \int_t^\infty \int_s^\infty r^{-1} \nu(dr)\, ds = \int_t^\infty r^{-1} \int_t^r ds\, \nu(dr)$$
$$= \int_t^\infty \left(1 - \frac{t}{r}\right) \nu(dr) = \int_0^\infty \left(1 - \frac{t}{r}\right)^+ \nu(dr)$$

Using $\varphi(-t) = \varphi(t)$ to extend the formula to $t \leq 0$, we have $(*)$. Setting $t = 0$ in $(*)$ shows ν has total mass 1.

If φ is piecewise linear, ν has a finite number of atoms and the result follows from Example 3.3.15 and Lemma 3.3.9. To prove the general result, let ν_n be a sequence of measures on $(0, \infty)$ with a finite number of atoms that converges weakly to ν (see Exercise 3.2.10) and let

$$\varphi_n(t) = \int_0^\infty \left(1 - \left|\frac{t}{s}\right|\right)^+ \nu_n(ds)$$

Since $s \to (1 - |t/s|)^+$ is bounded and continuous, $\varphi_n(t) \to \varphi(t)$, and the desired result follows from part (ii) of Theorem 3.3.17. □

A classic application of Polya's criterion is:

Exercise 3.3.19 Show that $\exp(-|t|^\alpha)$ is a characteristic function for $0 < \alpha \leq 1$.

(The case $\alpha = 1$ corresponds to the Cauchy distribution.) The next argument, which we learned from Frank Spitzer, proves that this is true for $0 < \alpha \leq 2$. The case $\alpha = 2$ corresponds to a normal distribution, so that case can be safely ignored in the proof.

Example 3.3.23 $\exp(-|t|^\alpha)$ is a characteristic function for $0 < \alpha < 2$.

Proof A little calculus shows that for any β and $|x| < 1$

$$(1 - x)^\beta = \sum_{n=0}^\infty \binom{\beta}{n} (-x)^n$$

where

$$\binom{\beta}{n} = \frac{\beta(\beta - 1) \cdots (\beta - n + 1)}{1 \cdot 2 \cdots n}$$

Let $\psi(t) = 1 - (1 - \cos t)^{\alpha/2} = \sum_{n=1}^\infty c_n (\cos t)^n$, where

$$c_n = \binom{\alpha/2}{n} (-1)^{n+1}$$

$c_n \geq 0$ (here we use $\alpha < 2$), and $\sum_{n=1}^\infty c_n = 1$ (take $t = 0$ in the definition of ψ). $\cos t$ is a characteristic function (see Example 3.3.3), so an easy extension of Lemma 3.3.9 shows that ψ is a ch.f. We have $1 - \cos t \sim t^2/2$ as $t \to 0$, so

$$1 - \cos(t \cdot 2^{1/2} \cdot n^{-1/\alpha}) \sim n^{-2/\alpha} t^2$$

Using Lemma 3.1.1 and (ii) of Theorem 3.3.17 now, it follows that

$$\exp(-|t|^\alpha) = \lim_{n \to \infty} \{\psi(t \cdot 2^{1/2} \cdot n^{-1/\alpha})\}^n$$

is a ch.f. □

Exercise 3.3.16 shows that $\exp(-|t|^\alpha)$ is not a ch.f. when $\alpha > 2$. A reason for interest in these characteristic functions is explained by the following generalization of Exercise 3.3.15.

Exercise 3.3.20 If $X_1, X_2, \ldots$ are independent and have characteristic function $\exp(-|t|^\alpha)$, then $(X_1 + \cdots + X_n)/n^{1/\alpha}$ has the same distribution as X_1.

We will return to this topic in Section 3.8.

Exercise 3.3.21 Let φ_1 and φ_2 be ch.f's. Show that $A = \{t : \varphi_1(t) = \varphi_2(t)\}$ is closed, contains 0, and is symmetric about 0. Show that if A is a set with these properties and $\varphi_1(t) = e^{-|t|}$, there is a φ_2 so that $\{t : \varphi_1(t) = \varphi_2(t)\} = A$.

Example 3.3.24 For some purposes, it is nice to have an explicit example of two ch.f.'s that agree on $[-1, 1]$. From Example 3.3.15, we know that $(1 - |t|)^+$ is the ch.f. of the density $(1 - \cos x)/\pi x^2$. Define $\psi(t)$ to be equal to φ on $[-1, 1]$ and periodic with period 2, i.e., $\psi(t) = \psi(t + 2)$. The Fourier series for ψ is

$$\psi(u) = \frac{1}{2} + \sum_{n=-\infty}^{\infty} \frac{2}{\pi^2(2n-1)^2} \exp(i(2n-1)\pi u)$$

The right-hand side is the ch.f. of a discrete distribution with

$$P(X = 0) = 1/2 \quad \text{and} \quad P(X = (2n-1)\pi) = 2\pi^{-2}(2n-1)^{-2} \quad n \in \mathbf{Z}.$$

Exercise 3.3.22 Find independent r.v.'s X, Y, and Z so that Y and Z do not have the same distribution but $X + Y$ and $X + Z$ do.

Exercise 3.3.23 Show that if X and Y are independent and $X + Y$ and X have the same distribution, then $Y = 0$ a.s.

For more curiosities, see Feller, Vol. II (1971), Section XV.2a.

*3.3.5 The Moment Problem**

Suppose $\int x^k dF_n(x)$ has a limit μ_k for each k. Then the sequence of distributions is tight by Theorem 3.2.14 and every subsequential limit has the moments μ_k by Exercise 3.2.5, so we can conclude the sequence converges weakly if there is only one distribution with these moments. It is easy to see that this is true if F is concentrated on a finite interval $[-M, M]$, since every continuous function can be approximated uniformly on $[-M, M]$ by polynomials. The result is false in general.

Counterexample 1 Heyde (1963) Consider the **lognormal density**

$$f_0(x) = (2\pi)^{-1/2} x^{-1} \exp(-(\log x)^2/2) \qquad x \geq 0$$

and for $-1 \leq a \leq 1$ let

$$f_a(x) = f_0(x)\{1 + a \sin(2\pi \log x)\}$$

To see that f_a is a density and has the same moments as f_0, it suffices to show that

$$\int_0^\infty x^r f_0(x) \sin(2\pi \log x)\, dx = 0 \text{ for } r = 0, 1, 2, \ldots$$

Changing variables $x = \exp(s + r)$, $s = \log x - r$, $ds = dx/x$ the integral becomes

$$(2\pi)^{-1/2} \int_{-\infty}^{\infty} \exp(rs + r^2) \exp(-(s + r)^2/2) \sin(2\pi(s + r))\, ds$$
$$= (2\pi)^{-1/2} \exp(r^2/2) \int_{-\infty}^{\infty} \exp(-s^2/2) \sin(2\pi s)\, ds = 0$$

The two equalities holding because r is an integer and the integrand is odd. From the proof, it should be clear that we could let

$$g(x) = f_0(x) \left\{ 1 + \sum_{k=1}^{\infty} a_k \sin(k\pi \log x) \right\} \quad \text{if} \sum_{k=1}^{\infty} |a_k| \le 1$$

to get a large family of densities having the same moments as the lognormal.

The moments of the lognormal are easy to compute. Recall that if χ has the standard normal distribution, then Exercise 1.2.6 implies $\exp(\chi)$ has the lognormal distribution.

$$EX^n = E \exp(n\chi) = \int e^{nx} (2\pi)^{-1/2} e^{-x^2/2}\, dx$$
$$= e^{n^2/2} \int (2\pi)^{-1/2} e^{-(x-n)^2/2}\, dx = \exp(n^2/2)$$

since the last integrand is the density of the normal with mean n and variance 1. Somewhat remarkably, there is a family of discrete random variables with these moments. Let $a > 0$ and

$$P(Y_a = ae^k) = a^{-k} \exp(-k^2/2)/c_a \quad \text{for } k \in \mathbf{Z}$$

where c_a is chosen to make the total mass 1.

$$\exp(-n^2/2) E Y_a^n = \exp(-n^2/2) \sum_k (ae^k)^n a^{-k} \exp(-k^2/2)/c_a$$
$$= \sum_k a^{-(k-n)} \exp(-(k - n)^2/2)/c_a = 1$$

by the definition of c_a.

The lognormal density decays like $\exp(-(\log x)^2/2)$ as $|x| \to \infty$. The next counterexample has more rapid decay. Since the exponential distribution, e^{-x} for $x \ge 0$, is determined by its moments (see Exercise 3.3.25), we cannot hope to do much better than this.

Counterexample 2 Let $\lambda \in (0, 1)$ and for $-1 \le a \le 1$ let

$$f_{a,\lambda}(x) = c_\lambda \exp(-|x|^\lambda)\{1 + a \sin(\beta |x|^\lambda \operatorname{sgn}(x))\}$$

where $\beta = \tan(\lambda\pi/2)$ and $1/c_\lambda = \int \exp(-|x|^\lambda)\, dx$. To prove that these are density functions and that for a fixed value of λ they have the same moments, it suffices to show

$$\int x^n \exp(-|x|^\lambda) \sin(\beta |x|^\lambda \operatorname{sgn}(x))\, dx = 0 \quad \text{for } n = 0, 1, 2, \ldots$$

This is clear for even n, since the integrand is odd. To prove the result for odd n, it suffices to integrate over $[0, \infty)$. Using the identity

$$\int_0^\infty t^{p-1} e^{-qt} dt = \Gamma(p)/q^p \quad \text{when } \mathrm{Re}\, q > 0$$

with $p = (n+1)/\lambda$, $q = 1 + \beta i$, and changing variables $t = x^\lambda$, we get

$$\begin{aligned}
&\Gamma((n+1)/\lambda)/(1+\beta\, i)^{(n+1)/\lambda} \\
&= \int_0^\infty x^{\lambda\{(n+1)/\lambda - 1\}} \exp(-(1+\beta i)x^\lambda)\lambda\, x^{\lambda-1}\, dx \\
&= \lambda \int_0^\infty x^n \exp(-x^\lambda)\cos(\beta x^\lambda) dx - i\lambda \int_0^\infty x^n \exp(-x^\lambda)\sin(\beta x^\lambda)\, dx
\end{aligned}$$

Since $\beta = \tan(\lambda\pi/2)$

$$(1+\beta i)^{(n+1)/\lambda} = (\cos \lambda\pi/2)^{-(n+1)/\lambda}(\exp(i\lambda\pi/2))^{(n+1)/\lambda}$$

The right-hand side is real, since $\lambda < 1$ and $(n+1)$ is even, so

$$\int_0^\infty x^n \exp(-x^\lambda)\sin(\beta x^\lambda)\, dx = 0$$

A useful sufficient condition for a distribution to be determined by its moments is:

Theorem 3.3.25 *If* $\limsup_{k\to\infty} \mu_{2k}^{1/2k}/2k = r < \infty$, *then there is at most one d.f.* F *with* $\mu_k = \int x^k dF(x)$ *for all positive integers* k.

Remark This is slightly stronger than **Carleman's condition**

$$\sum_{k=1}^\infty 1/\mu_{2k}^{1/2k} = \infty$$

which is also sufficient for the conclusion of Theorem 3.3.25.

Proof Let F be any d.f. with the moments μ_k and let $\nu_k = \int |x|^k dF(x)$. The Cauchy-Schwarz inequality implies $\nu_{2k+1}^2 \le \mu_{2k}\mu_{2k+2}$, so

$$\limsup_{k\to\infty} (\nu_k^{1/k})/k = r < \infty$$

Taking $x = tX$ in Lemma 3.3.2 and multiplying by $e^{i\theta X}$, we have

$$\left| e^{i\theta X}\left(e^{itX} - \sum_{m=0}^{n-1} \frac{(itX)^m}{m!} \right) \right| \le \frac{|tX|^n}{n!}$$

Taking expected values and using Exercise 3.3.18 gives

$$\left| \varphi(\theta+t) - \varphi(\theta) - t\varphi'(\theta)\ldots - \frac{t^{n-1}}{(n-1)!}\varphi^{(n-1)}(\theta) \right| \le \frac{|t|^n}{n!}\nu_n$$

Using the last result, the fact that $\nu_k \le (r+\epsilon)^k k^k$ for large k, and the trivial bound $e^k \ge k^k/k!$ (expand the left-hand side in its power series), we see that for any θ

$$(*) \qquad \varphi(\theta+t) = \varphi(\theta) + \sum_{m=1}^{\infty} \frac{t^m}{m!}\varphi^{(m)}(\theta) \quad \text{for } |t| < 1/er$$

Let G be another distribution with the given moments and ψ its ch.f. Since $\varphi(0) = \psi(0) = 1$, it follows from $(*)$ and induction that $\varphi(t) = \psi(t)$ for $|t| \le k/3r$ for all k, so the two ch.f.'s coincide and the distributions are equal. □

Combining Theorem 3.3.25 with the discussion that began our consideration of the moment problem.

Theorem 3.3.26 *Suppose $\int x^k dF_n(x)$ has a limit μ_k for each k and*

$$\limsup_{k\to\infty} \mu_{2k}^{1/2k}/2k < \infty$$

then F_n converges weakly to the unique distribution with these moments.

Exercise 3.3.24 Let $G(x) = P(|X| < x)$, $\lambda = \sup\{x : G(x) < 1\}$, and $\nu_k = E|X|^k$. Show that $\nu_k^{1/k} \to \lambda$, so the assumption of Theorem 3.3.26 holds if $\lambda < \infty$.

Exercise 3.3.25 Suppose $|X|$ has density $Cx^\alpha \exp(-x^\lambda)$ on $(0,\infty)$. Changing variables $y = x^\lambda$, $dx = (1/\lambda)x^{1/\lambda - 1}\, dx$

$$E|X|^n = \int_0^\infty C\lambda^{-1} y^{(n+\alpha)/\lambda} \exp(-y) y^{1/\lambda-1} dy = C\lambda^{-1}\Gamma((n+\alpha+1)/\lambda)$$

Use the identity $\Gamma(x+1) = x\Gamma(x)$ for $x \ge 0$ to conclude that the assumption of Theorem 3.3.26 is satisfied for $\lambda \ge 1$ but not for $\lambda < 1$. This shows the normal ($\lambda = 2$) and gamma ($\lambda = 1$) distributions are determined by their moments.

Our results so far have been for the so-called **Hamburger moment problem**. If we assume *a priori* that the distribution is concentrated on $[0,\infty)$, we have the **Stieltjes moment problem**. There is a 1-1 correspondence between $X \ge 0$ and symmetric distributions on **R** given by $X \to \xi\sqrt{X}$, where $\xi \in \{-1, 1\}$ is independent of X and takes its two values with equal probability. From this we see that

$$\limsup_{k\to\infty} \nu_k^{1/2k}/2k < \infty$$

is sufficient for there to be a unique distribution on $[0,\infty)$ with the given moments. The next example shows that for nonnegative random variables, the last result is close to the best possible.

Counterexample 3 Let $\lambda \in (0, 1/2)$, $\beta = \tan(\lambda\pi)$, $-1 \le a \le 1$ and

$$f_a(x) = c_\lambda \exp(-x^\lambda)(1 + a\sin(\beta x^\lambda)) \quad \text{for } x \ge 0$$

where $1/c_\lambda = \int_0^\infty \exp(-x^\lambda)\, dx$.

By imitating the calculations in Counterexample 2, it is easy to see that the f_a are probability densities that have the same moments. This example seems to be due to Stoyanov (1987) pp. 92–93. The special case $\lambda = 1/4$ is widely known.

3.4 Central Limit Theorems

We are now ready for the main business of the chapter. We will first prove the central limit theorem for:

3.4.1 i.i.d. Sequences

Theorem 3.4.1 *Let $X_1, X_2, \ldots$ be i.i.d. with $EX_i = \mu$, $var(X_i) = \sigma^2 \in (0, \infty)$. If $S_n = X_1 + \cdots + X_n$, then*

$$(S_n - n\mu)/\sigma n^{1/2} \Rightarrow \chi$$

where χ has the standard normal distribution.

This notation is nonstandard but convenient. To see the logic, note that the square of a normal has a chi-squared distribution.

Proof By considering $X_i' = X_i - \mu$, it suffices to prove the result when $\mu = 0$. From Theorem 3.3.20

$$\varphi(t) = E\exp(itX_1) = 1 - \frac{\sigma^2 t^2}{2} + o(t^2)$$

so

$$E\exp(itS_n/\sigma n^{1/2}) = \left(1 - \frac{t^2}{2n} + o(n^{-1})\right)^n$$

From Lemma 3.1.1 it should be clear that the last quantity $\to \exp(-t^2/2)$, as $n \to \infty$, which with Theorem 3.3.17 completes the proof. However, Lemma 3.1.1 is a fact about real numbers, so we need to extend it to the complex case to complete the proof.

Theorem 3.4.2 *If $c_n \to c \in \mathbf{C}$, then $(1 + c_n/n)^n \to e^c$.*

Proof The proof is based on two simple facts:

Lemma 3.4.3 *Let $z_1, \ldots, z_n$ and $w_1, \ldots, w_n$ be complex numbers of modulus $\le \theta$. Then*

$$\left|\prod_{m=1}^n z_m - \prod_{m=1}^n w_m\right| \le \theta^{n-1} \sum_{m=1}^n |z_m - w_m|$$

Proof The result is true for $n = 1$. To prove it for $n > 1$ observe that

$$\left|\prod_{m=1}^n z_m - \prod_{m=1}^n w_m\right| \le \left|z_1 \prod_{m=2}^n z_m - z_1 \prod_{m=2}^n w_m\right| + \left|z_1 \prod_{m=2}^n w_m - w_1 \prod_{m=2}^n w_m\right|$$
$$\le \theta \left|\prod_{m=2}^n z_m - \prod_{m=2}^n w_m\right| + \theta^{n-1}|z_1 - w_1|$$

and use induction. □

Lemma 3.4.4 *If b is a complex number with $|b| \le 1$, then $|e^b - (1 + b)| \le |b|^2$.*

Proof $e^b - (1+b) = b^2/2! + b^3/3! + b^4/4! + \cdots$ so if $|b| \le 1$, then

$$|e^b - (1+b)| \le \frac{|b|^2}{2}(1 + 1/2 + 1/2^2 + \cdots) = |b|^2$$ □

Proof of Theorem 3.4.2. Let $z_m = (1 + c_n/n)$, $w_m = \exp(c_n/n)$, and $\gamma > |c|$. For large n, $|c_n| < \gamma$. Since $1 + \gamma/n \le \exp(\gamma/n)$, it follows from Lemmas 3.4.3 and 3.4.4 that

$$|(1 + c_n/n)^n - e^{c_n}| \le \left(e^{\gamma/n}\right)^{n-1} n \left|\frac{c_n}{n}\right|^2 \le e^{\gamma} \frac{\gamma^2}{n} \to 0$$

as $n \to \infty$. □

To get a feel for what the central limit theorem says, we will look at some concrete cases.

Example 3.4.5 (Roulette) A roulette wheel has slots numbered 1–36 (18 red and 18 black) and two slots numbered 0 and 00 that are painted green. Players can bet \$1 that the ball will land in a red (or black) slot and win \$1 if it does. If we let X_i be the winnings on the ith play, then $X_1, X_2, \ldots$ are i.i.d. with $P(X_i = 1) = 18/38$ and $P(X_i = -1) = 20/38$.

$$EX_i = -1/19 \quad \text{and} \quad \text{var}(X) = EX^2 - (EX)^2 = 1 - (1/19)^2 = 0.9972$$

We are interested in

$$P(S_n \ge 0) = P\left(\frac{S_n - n\mu}{\sigma\sqrt{n}} \ge \frac{-n\mu}{\sigma\sqrt{n}}\right)$$

Taking $n = 361 = 19^2$ and replacing σ by 1 to keep computations simple,

$$\frac{-n\mu}{\sigma\sqrt{n}} = \frac{361 \cdot (1/19)}{\sqrt{361}} = 1$$

So the central limit theorem and our table of the normal distribution in the back of the book tells us that

$$P(S_n \ge 0) \approx P(\chi \ge 1) = 1 - 0.8413 = 0.1587$$

In words, after 361 spins of the roulette wheel the casino will have won \$19 of your money on the average, but there is a probability of about 0.16 that you will be ahead.

Example 3.4.6 (Coin flips) Let $X_1, X_2, \ldots$ be i.i.d. with $P(X_i = 0) = P(X_i = 1) = 1/2$. If $X_i = 1$ indicates that a heads occured on the ith toss, then $S_n = X_1 + \cdots + X_n$ is the total number of heads at time n.

$$EX_i = 1/2 \quad \text{and} \quad \text{var}(X) = EX^2 - (EX)^2 = 1/2 - 1/4 = 1/4$$

So the central limit theorem tells us $(S_n - n/2)/\sqrt{n/4} \Rightarrow \chi$. Our table of the normal distribution tells us that

$$P(\chi > 2) = 1 - 0.9773 = 0.0227$$

so $P(|\chi| \le 2) = 1 - 2(0.0227) = 0.9546$, or plugging into the central limit theorem

$$.95 \approx P((S_n - n/2)/\sqrt{n/4} \in [-2, 2]) = P(S_n - n/2 \in [-\sqrt{n}, \sqrt{n}])$$

Taking $n = 10{,}000$ this says that 95% of the time the number of heads will be between 4900 and 5100.

Example 3.4.7 (Normal approximation to the binomial) Let $X_1, X_2, \dots$ and S_n be as in the previous example. To estimate $P(S_{16} = 8)$ using the central limit theorem, we regard 8 as the interval $[7.5, 8.5]$. Since $\mu = 1/2$, and $\sigma\sqrt{n} = 2$ for $n = 16$

$$\begin{aligned} P(|S_{16} - 8| \le 0.5) &= P\left(\frac{|S_n - n\mu|}{\sigma\sqrt{n}} \le 0.25\right) \\ &\approx P(|\chi| \le 0.25) = 2(0.5987 - 0.5) = 0.1974 \end{aligned}$$

Even though n is small, this agrees well with the exact probability

$$\binom{16}{8}2^{-16} = \frac{13 \cdot 11 \cdot 10 \cdot 9}{65{,}536} = 0.1964.$$

These computations motivate the **histogram correction**, which is important in using the normal approximation for small n. For example, if we are going to approximate $P(S_{16} \le 11)$, then we regard this probability as $P(S_{16} \le 11.5)$. One obvious reason for doing this is to get the same answer if we regard $P(S_{16} \le 11) = 1 - P(S_{16} \ge 12)$.

Example 3.4.8 (Normal approximation to the Poisson) Let Z_λ have a Poisson distribution with mean λ. If $X_1, X_2, \dots$ are independent and have Poisson distributions with mean 1, then $S_n = X_1 + \cdots + X_n$ has a Poisson distribution with mean n. Since $\text{var}(X_i) = 1$, the central limit theorem implies:

$$(S_n - n)/n^{1/2} \Rightarrow \chi \quad \text{as } n \to \infty$$

To deal with values of λ that are not integers, let N_1, N_2, N_3 be independent Poisson with means $[\lambda]$, $\lambda - [\lambda]$, and $[\lambda] + 1 - \lambda$. If we let $S_{[\lambda]} = N_1$, $Z_\lambda = N_1 + N_2$ and $S_{[\lambda]+1} = N_1 + N_2 + N_3$, then $S_{[\lambda]} \le Z_\lambda \le S_{[\lambda]+1}$ and using the limit theorem for the S_n it follows that

$$(Z_\lambda - \lambda)/\lambda^{1/2} \Rightarrow \chi \quad \text{as } \lambda \to \infty$$

Example 3.4.9 **Pairwise independence** is good enough for the strong law of large numbers (see Theorem 2.4.1). It is not good enough for the central limit theorem. Let $\xi_1, \xi_2, \dots$ be i.i.d. with $P(\xi_i = 1) = P(\xi_i = -1) = 1/2$. We will arrange things so that for $n \ge 1$

$$S_{2^n} = \xi_1(1 + \xi_2) \cdots (1 + \xi_{n+1}) = \begin{cases} \pm 2^n & \text{with prob } 2^{-n-1} \\ 0 & \text{with prob } 1 - 2^{-n} \end{cases}$$

To do this we let $X_1 = \xi_1$, $X_2 = \xi_1\xi_2$, and for $m = 2^{n-1} + j$, $0 < j \le 2^{n-1}$, $n \ge 2$ let $X_m = X_j\xi_{n+1}$. Each X_m is a product of a different set of ξ_j's, so they are pairwise independent.

Exercises

3.4.1 Suppose you roll a die 180 times. Use the normal approximation (with the histogram correction) to estimate the probability you will get fewer than 25 sixes.

3.4.2 Let $X_1, X_2, \dots$ be i.i.d. with $EX_i = 0$, $0 < \text{var}(X_i) < \infty$, and let $S_n = X_1 + \cdots + X_n$. (a) Use the central limit theorem and Kolmogorov's zero-one law to conclude that $\limsup S_n/\sqrt{n} = \infty$ a.s. (b) Use an argument by contradiction to show that $S_n/\sqrt{n}$ does not converge in probability. Hint: Consider $n = m!$.

3.4.3 Let $X_1, X_2, \ldots$ be i.i.d. and let $S_n = X_1 + \cdots + X_n$. Assume that $S_n/\sqrt{n} \Rightarrow$ a limit and conclude that $EX_i^2 < \infty$. Sketch: Suppose $EX_i^2 = \infty$. Let $X_1', X_2', \ldots$ be an independent copy of the original sequence. Let $Y_i = X_i - X_i'$, $U_i = Y_i 1_{(|Y_i| \le A)}$, $V_i = Y_i 1_{(|Y_i| > A)}$, and observe that for any K

$$P\left(\sum_{m=1}^{n} Y_m \ge K\sqrt{n}\right) \ge P\left(\sum_{m=1}^{n} U_m \ge K\sqrt{n}, \sum_{m=1}^{n} V_m \ge 0\right)$$
$$\ge \frac{1}{2} P\left(\sum_{m=1}^{n} U_m \ge K\sqrt{n}\right) \ge \frac{1}{5}$$

for large n if A is large enough. Since K is arbitrary, this is a contradiction.

3.4.4 Let $X_1, X_2, \ldots$ be i.i.d. with $X_i \ge 0$, $EX_i = 1$, and $\text{var}(X_i) = \sigma^2 \in (0, \infty)$. Show that $2(\sqrt{S_n} - \sqrt{n}) \Rightarrow \sigma\chi$.

3.4.5 **Self-normalized sums.** Let $X_1, X_2, \ldots$ be i.i.d. with $EX_i = 0$ and $EX_i^2 = \sigma^2 \in (0, \infty)$. Then

$$\sum_{m=1}^{n} X_m \Big/ \left(\sum_{m=1}^{n} X_m^2\right)^{1/2} \Rightarrow \chi$$

3.4.6 **Random index central limit theorem.** Let $X_1, X_2, \ldots$ be i.i.d. with $EX_i = 0$ and $EX_i^2 = \sigma^2 \in (0, \infty)$, and let $S_n = X_1 + \cdots + X_n$. Let N_n be a sequence of nonnegative integer-valued random variables and a_n a sequence of integers with $a_n \to \infty$ and $N_n/a_n \to 1$ in probability. Show that

$$S_{N_n}/\sigma\sqrt{a_n} \Rightarrow \chi$$

Hint: Use Kolmogorov's inequality (Theorem 2.5.5) to conclude that if $Y_n = S_{N_n}/\sigma\sqrt{a_n}$ and $Z_n = S_{a_n}/\sigma\sqrt{a_n}$, then $Y_n - Z_n \to 0$ in probability.

3.4.7 **A central limit theorem in renewal theory.** Let $Y_1, Y_2, \ldots$ be i.i.d. positive random variables with $EY_i = \mu$ and $\text{var}(Y_i) = \sigma^2 \in (0, \infty)$. Let $S_n = Y_1 + \cdots + Y_n$ and $N_t = \sup\{m : S_m \le t\}$. Apply the previous exercise to $X_i = Y_i - \mu$ to prove that as $t \to \infty$

$$(\mu N_t - t)/(\sigma^2 t/\mu)^{1/2} \Rightarrow \chi$$

3.4.8 **A second proof of the renewal CLT.** Let $Y_1, Y_2, \ldots$, S_n, and N_t be as in the last exercise. Let $u = [t/\mu]$, $D_t = S_u - t$. Use Kolmogorov's inequality to show

$$P(|S_{u+m} - (S_u + m\mu)| > t^{2/5} \text{ for some } m \in [-t^{3/5}, t^{3/5}]) \to 0 \quad \text{as } t \to \infty$$

Conclude $|N_t - (t - D_t)/\mu| / t^{1/2} \to 0$ in probability and then obtain the result in the previous exercise.

3.4.2 Triangular Arrays

Our next step is to generalize the central limit theorem to:

Theorem 3.4.10 (The Lindeberg-Feller theorem) *For each n, let $X_{n,m}$, $1 \le m \le n$, be independent random variables with $EX_{n,m} = 0$. Suppose* *(i)* $\sum_{m=1}^{n} EX_{n,m}^2 \to \sigma^2 > 0$

(ii) For all $\epsilon > 0$, $\lim_{n\to\infty} \sum_{m=1}^{n} E(|X_{n,m}|^2; |X_{n,m}| > \epsilon) = 0$.

Then $S_n = X_{n,1} + \cdots + X_{n,n} \Rightarrow \sigma\chi$ *as* $n \to \infty$.

Remarks In words, the theorem says that a sum of a large number of small independent effects has approximately a normal distribution. To see that Theorem 3.4.10 contains our first central limit theorem, let $Y_1, Y_2 \ldots$ be i.i.d. with $EY_i = 0$ and $EY_i^2 = \sigma^2 \in (0,\infty)$, and let $X_{n,m} = Y_m/n^{1/2}$. Then $\sum_{m=1}^{n} EX_{n,m}^2 = \sigma^2$ and if $\epsilon > 0$

$$\sum_{m=1}^{n} E(|X_{n,m}|^2; |X_{n,m}| > \epsilon) = nE(|Y_1/n^{1/2}|^2; |Y_1/n^{1/2}| > \epsilon)$$

$$= E(|Y_1|^2; |Y_1| > \epsilon n^{1/2}) \to 0$$

by the dominated convergence theorem since $EY_1^2 < \infty$.

Proof Let $\varphi_{n,m}(t) = E\exp(itX_{n,m})$, $\sigma_{n,m}^2 = EX_{n,m}^2$. By Theorem 3.3.17, it suffices to show that

$$\prod_{m=1}^{n} \varphi_{n,m}(t) \to \exp(-t^2\sigma^2/2)$$

Let $z_{n,m} = \varphi_{n,m}(t)$ and $w_{n,m} = (1 - t^2\sigma_{n,m}^2/2)$. By (3.3.3)

$$\begin{aligned}|z_{n,m} - w_{n,m}| &\le E(|tX_{n,m}|^3 \wedge 2|tX_{n,m}|^2)\\ &\le E(|tX_{n,m}|^3; |X_{n,m}| \le \epsilon) + E(2|tX_{n,m}|^2; |X_{n,m}| > \epsilon)\\ &\le \epsilon t^3 E(|X_{n,m}|^2; |X_{n,m}| \le \epsilon) + 2t^2 E(|X_{n,m}|^2; |X_{n,m}| > \epsilon)\end{aligned}$$

Summing $m = 1$ to n, letting $n \to \infty$, and using (i) and (ii) gives

$$\limsup_{n\to\infty} \sum_{m=1}^{n} |z_{n,m} - w_{n,m}| \le \epsilon t^3 \sigma^2$$

Since $\epsilon > 0$ is arbitrary, it follows that the sequence converges to 0. Our next step is to use Lemma 3.4.3 with $\theta = 1$ to get

$$\left|\prod_{m=1}^{n} \varphi_{n,m}(t) - \prod_{m=1}^{n} (1 - t^2\sigma_{n,m}^2/2)\right| \to 0$$

To check the hypotheses of Lemma 3.4.3, note that since $\varphi_{n,m}$ is a ch.f., $|\varphi_{n,m}(t)| \le 1$ for all n, m. For the terms in the second product we note that

$$\sigma_{n,m}^2 \le \epsilon^2 + E(|X_{n,m}|^2 \; ; |X_{n,m}| > \epsilon)$$

and ϵ is arbitrary, so (ii) implies $\sup_m \sigma_{n,m}^2 \to 0$ and thus if n is large $1 \ge 1 - t^2\sigma_{n,m}^2/2 > -1$ for all m.

To complete the proof now, we apply Exercise 3.1.1 with $c_{m,n} = -t^2\sigma_{n,m}^2/2$. We have just shown $\sup_m \sigma_{n,m}^2 \to 0$. (i) implies

$$\sum_{m=1}^{n} c_{m,n} \to -\sigma^2 t^2/2$$

so $\prod_{m=1}^{n}(1 - t^2\sigma_{n,m}^2/2) \to \exp(-t^2\sigma^2/2)$ and the proof is complete. □

Example 3.4.11 (Cycles in a random permutation and record values) Continuing the analysis of Examples 2.2.8 and 2.3.10, let $Y_1, Y_2, \ldots$ be independent with $P(Y_m = 1) = 1/m$, and $P(Y_m = 0) = 1 - 1/m$. $EY_m = 1/m$ and $\text{var}\,(Y_m) = 1/m - 1/m^2$. So if $S_n = Y_1 + \cdots + Y_n$, then $ES_n \sim \log n$ and $\text{var}\,(S_n) \sim \log n$. Let

$$X_{n,m} = (Y_m - 1/m)/(\log\, n)^{1/2}$$

$EX_{n,m} = 0$, $\sum_{m=1}^n EX_{n,m}^2 \to 1$, and for any $\epsilon > 0$

$$\sum_{m=1}^n E(|X_{n,m}|^2; |X_{n,m}| > \epsilon) \to 0$$

since the sum is 0 as soon as $(\log n)^{-1/2} < \epsilon$. Applying Theorem 3.4.10 now gives

$$(\log n)^{-1/2}\left(S_n - \sum_{m=1}^n \frac{1}{m}\right) \Rightarrow \chi$$

Observing that

$$\sum_{m=1}^{n-1} \frac{1}{m} \geq \int_1^n x^{-1}\,dx = \log n \geq \sum_{m=2}^n \frac{1}{m}$$

shows $\left|\log n - \sum_{m=1}^n 1/m\right| \leq 1$ and the conclusion can be written as

$$(S_n - \log n)/(\log n)^{1/2} \Rightarrow \chi$$

Example 3.4.12 (The converse of the three series theorem) Recall the set up of Theorem 2.5.8. Let $X_1, X_2, \ldots$ be independent, let $A > 0$, and let $Y_m = X_m 1_{(|X_m| \leq A)}$. In order that $\sum_{n=1}^\infty X_n$ converges (i.e., $\lim_{N\to\infty} \sum_{n=1}^N X_n$ exists), it is necessary that:

$$\text{(i)} \sum_{n=1}^\infty P(|X_n| > A) < \infty, \text{ (ii)} \sum_{n=1}^\infty EY_n \text{ converges, and (iii)} \sum_{n=1}^\infty \text{var}\,(Y_n) < \infty$$

Proof The necessity of the first condition is clear. For if that sum is infinite, $P(|X_n| > A \text{ i.o.}) > 0$ and $\lim_{n\to\infty} \sum_{m=1}^n X_m$ cannot exist. Suppose next that the sum in (i) is finite but the sum in (iii) is infinite. Let

$$c_n = \sum_{m=1}^n \text{var}\,(Y_m) \quad \text{and} \quad X_{n,m} = (Y_m - EY_m)/c_n^{1/2}$$

$EX_{n,m} = 0$, $\sum_{m=1}^n EX_{n,m}^2 = 1$, and for any $\epsilon > 0$

$$\sum_{m=1}^n E(|X_{n,m}|^2; |X_{n,m}| > \epsilon) \to 0$$

since the sum is 0 as soon as $2A/c_n^{1/2} < \epsilon$. Applying Theorem 3.4.10 now gives that if $S_n = X_{n,1} + \cdots + X_{n,n}$, then $S_n \Rightarrow \chi$. Now

(i) if $\lim_{n\to\infty}\sum_{m=1}^n X_m$ exists, $\lim_{n\to\infty}\sum_{m=1}^n Y_m$ exists.

(ii) if we let $T_n = (\sum_{m\le n} Y_m)/c_n^{1/2}$, then $T_n \Rightarrow 0$.

The last two results and Exercise 3.2.13 imply $(S_n - T_n) \Rightarrow \chi$. Since

$$S_n - T_n = -\left(\sum_{m\le n} EY_m\right)/c_n^{1/2}$$

is not random, this is absurd.

Finally, assume the series in (i) and (iii) are finite. Theorem 2.5.6 implies that $\lim_{n\to\infty}\sum_{m=1}^n (Y_m - EY_m)$ exists, so if $\lim_{n\to\infty}\sum_{m=1}^n X_m$ and hence $\lim_{n\to\infty}\sum_{m=1}^n Y_m$ does, taking differences shows that (ii) holds. □

Example 3.4.13 (Infinite variance) Suppose $X_1, X_2, \dots$ are i.i.d. and have $P(X_1 > x) = P(X_1 < -x)$ and $P(|X_1| > x) = x^{-2}$ for $x \ge 1$.

$$E|X_1|^2 = \int_0^\infty 2xP(|X_1| > x)\,dx = \infty$$

but it turns out that when $S_n = X_1 + \cdots + X_n$ is suitably normalized it converges to a normal distribution. Let

$$Y_{n,m} = X_m 1_{(|X_m|\le n^{1/2}\log\log n)}$$

The truncation level $c_n = n^{1/2}\log\log n$ is chosen large enough to make

$$\sum_{m=1}^n P(Y_{n,m} \ne X_m) \le nP(|X_1| > c_n) \to 0$$

However, we want the variance of $Y_{n,m}$ to be as small as possible, so we keep the truncation close to the lowest possible level.

Our next step is to show $EY_{n,m}^2 \sim \log n$. For this we need upper and lower bounds. Since $P(|Y_{n,m}| > x) \le P(|X_1| > x)$ and is 0 for $x > c_n$, we have

$$\begin{aligned} EY_{n,m}^2 &\le \int_0^{c_n} 2yP(|X_1| > y)\,dy = 1 + \int_1^{c_n} 2/y\,dy \\ &= 1 + 2\log c_n = 1 + \log n + 2\log\log\log n \sim \log n \end{aligned}$$

In the other direction, we observe $P(|Y_{n,m}| > x) = P(|X_1| > x) - P(|X_1| > c_n)$ and the right-hand side is $\ge (1 - (\log\log n)^{-2})P(|X_1| > x)$ when $x \le \sqrt{n}$, so

$$EY_{n,m}^2 \ge (1 - (\log\log n)^{-2}) \int_1^{\sqrt{n}} 2/y\,dy \sim \log n$$

If $S_n' = Y_{n,1} + \cdots + Y_{n,n}$, then $\text{var}(S_n') \sim n\log n$, so we apply Theorem 3.4.10 to $X_{n,m} = Y_{n,m}/(n\log n)^{1/2}$. Things have been arranged so that (i) is satisfied. Since $|Y_{n,m}| \le n^{1/2}\log\log n$, the sum in (ii) is 0 for large n, and it follows that $S_n'/(n\log n)^{1/2} \Rightarrow \chi$. Since the choice of c_n guarantees $P(S_n \ne S_n') \to 0$, the same result holds for S_n.

Remark In Section 3.6, we will see that if we replace $P(|X_1| > x) = x^{-2}$ in Example 3.4.13 by $P(|X_1| > x) = x^{-\alpha}$, where $0 < \alpha < 2$, then $S_n/n^{1/\alpha} \Rightarrow$ to a limit which is not χ. The last word on convergence to the normal distribution is the next result due to Lévy.

Theorem 3.4.14 *Let $X_1, X_2, \dots$ be i.i.d. and $S_n = X_1 + \cdots + X_n$. In order that there exist constants a_n and $b_n > 0$ so that $(S_n - a_n)/b_n \Rightarrow \chi$, it is necessary and sufficient that*

$$y^2 P(|X_1| > y)/E(|X_1|^2; |X_1| \le y) \to 0.$$

A proof can be found in Gnedenko and Kolmogorov (1954), a reference that contains the last word on many results about sums of independent random variables.

Exercises

In the next five problems $X_1, X_2, \dots$ are independent and $S_n = X_1 + \cdots + X_n$.

3.4.9 Suppose $P(X_m = m) = P(X_m = -m) = m^{-2}/2$, and for $m \ge 2$

$$P(X_m = 1) = P(X_m = -1) = (1 - m^{-2})/2$$

Show that $\text{var}\,(S_n)/n \to 2$ but $S_n/\sqrt{n} \Rightarrow \chi$. The trouble here is that $X_{n,m} = X_m/\sqrt{n}$ does not satisfy (ii) of Theorem 3.4.10.

3.4.10 Show that if $|X_i| \le M$ and $\sum_n \text{var}\,(X_n) = \infty$, then

$$(S_n - ES_n)/\sqrt{\text{var}\,(S_n)} \Rightarrow \chi$$

3.4.11 Suppose $EX_i = 0$, $EX_i^2 = 1$ and $E|X_i|^{2+\delta} \le C$ for some $0 < \delta, C < \infty$. Show that $S_n/\sqrt{n} \Rightarrow \chi$.

3.4.12 Prove **Lyapunov's Theorem.** Let $\alpha_n = \{\text{var}\,(S_n)\}^{1/2}$. If there is a $\delta > 0$ so that

$$\lim_{n\to\infty} \alpha_n^{-(2+\delta)} \sum_{m=1}^{n} E(|X_m - EX_m|^{2+\delta}) = 0$$

then $(S_n - ES_n)/\alpha_n \Rightarrow \chi$. Note that the previous exercise is a special case of this result.

3.4.13 Suppose $P(X_j = j) = P(X_j = -j) = 1/2j^\beta$ and $P(X_j = 0) = 1 - j^{-\beta}$, where $\beta > 0$. Show that (i) If $\beta > 1$, then $S_n \to S_\infty$ a.s. (ii) if $\beta < 1$, then $S_n/n^{(3-\beta)/2} \Rightarrow c\chi$. (iii) if $\beta = 1$, then $S_n/n \Rightarrow \aleph$, where

$$E\exp(it\aleph) = \exp\left(-\int_0^1 x^{-1}(1 - \cos xt)\,dx\right)$$

*3.4.3 Prime Divisors (Erdös-Kac)**

Our aim here is to prove that an integer picked at random from $\{1, 2, \dots, n\}$ has about

$$\log\log n + \chi(\log\log n)^{1/2}$$

prime divisors. Since $\exp(e^4) = 5.15 \times 10^{23}$, this result does not apply to most numbers we encounter in "everyday life." The first step in deriving this result is to give a:

Second proof of Theorem 3.4.10. The first step is to let

$$h_n(\epsilon) = \sum_{m=1}^{n} E(X_{n,m}^2; |X_{n,m}| > \epsilon)$$

and observe

Lemma 3.4.15 $h_n(\epsilon) \to 0$ *for each fixed* $\epsilon > 0$*, so we can pick* $\epsilon_n \to 0$ *so that* $h_n(\epsilon_n) \to 0$.

Proof Let N_m be chosen so that $h_n(1/m) \le 1/m$ for $n \ge N_m$ and $m \to N_m$ is increasing. Let $\epsilon_n = 1/m$ for $N_m \le n < N_{m+1}$, and $= 1$ for $n < N_1$. When $N_m \le n < N_{m+1}$, $\epsilon_n = 1/m$, so $|h_n(\epsilon_n)| = |h_n(1/m)| \le 1/m$ and the desired result follows. □

Let $X'_{n,m} = X_{n,m} 1_{(|X_{n,m}|>\epsilon_n)}$, $Y_{n,m} = X_{n,m} 1_{(|X_{n,m}|\le\epsilon_n)}$, and $Z_{n,m} = Y_{n,m} - EY_{n,m}$. Clearly, $|Z_{n,m}| \le 2\epsilon_n$. Using $X_{n,m} = X'_{n,m} + Y_{n,m}$, $Z_{n,m} = Y_{n,m} - EY_{n,m}$, $EY_{n,m} = -EX'_{n,m}$, the variance of the sum is the sum of the variances, and $\text{var}(W) \le EW^2$, we have

$$E\left(\sum_{m=1}^{n} X_{n,m} - \sum_{m=1}^{n} Z_{n,m}\right)^2 = E\left(\sum_{m=1}^{n} X'_{n,m} - EX'_{n,m}\right)^2$$
$$= \sum_{m=1}^{n} E(X'_{n,m} - EX'_{n,m})^2 \le \sum_{m=1}^{n} E(X'_{n,m})^2 \to 0$$

as $n \to \infty$, by the choice of ϵ_n.

Let $S_n = \sum_{m=1}^{n} X_{n,m}$ and $T_n = \sum_{m=1}^{n} Z_{n,m}$. The last computation shows $S_n - T_n \to 0$ in L^2 and hence in probability by Lemma 2.2.2. Thus, by Exercise 3.2.13, it suffices to show $T_n \Rightarrow \sigma\chi$. (i) implies $ES_n^2 \to \sigma^2$. We have just shown that $E(S_n - T_n)^2 \to 0$, so the triangle inequality for the L^2 norm implies $ET_n^2 \to \sigma^2$. To compute higher moments, we observe

$$T_n^r = \sum_{k=1}^{r} \sum_{r_i} \frac{r!}{r_1! \cdots r_k!} \frac{1}{k!} \sum_{i_j} Z_{n,i_1}^{r_1} \cdots Z_{n,i_k}^{r_k}$$

where $\sum_{r_i}$ extends over all k-tuples of positive integers with $r_1 + \cdots + r_k = r$ and $\sum_{i_j}$ extends over all k-tuples of distinct integers with $1 \le i \le n$. If we let

$$A_n(r_1, \ldots, r_k) = \sum_{i_j} EZ_{n,i_1}^{r_1} \cdots EZ_{n,i_k}^{r_k}$$

then

$$ET_n^r = \sum_{k=1}^{r} \sum_{r_i} \frac{r!}{r_1! \cdots r_k!} \frac{1}{k!} A_n(r_1, \ldots r_k)$$

To evaluate the limit of ET_n^r, we observe:

(a) If some $r_j = 1$, then $A_n(r_1, \ldots r_k) = 0$ since $EZ_{n,i_j} = 0$.

(b) If all $r_j = 2$, then

$$\sum_{i_j} EZ_{n,i_1}^2 \cdots EZ_{n,i_k}^2 \le \left(\sum_{m=1}^{n} EZ_{n,m}^2\right)^k \to \sigma^{2k}$$

To argue the other inequality, we note that for any $1 \le a < b \le k$ we can estimate the sum over all the $i_1, \dots, i_k$ with $i_a = i_b$ by replacing EZ^2_{n,i_a} by $(2\epsilon_n)^2$ to get (the factor $\binom{k}{2}$ giving the number of ways to pick $1 \le a < b \le k$)

$$\left(\sum_{m=1}^{n} EZ^2_{n,m}\right)^k - \sum_{i_j} EZ^2_{n,i_1} \cdots EZ^2_{n,i_k} \le \binom{k}{2}(2\epsilon_n)^2 \left(\sum_{m=1}^{n} EZ^2_{n,m}\right)^{k-1} \to 0$$

(c) If all the $r_i \ge 2$ but some $r_j > 2$, then using

$$E|Z_{n,i_j}|^{r_j} \le (2\epsilon_n)^{r_j-2} EZ^2_{n,i_j}$$

we have

$$\begin{aligned} |A_n(r_1, \dots r_k)| &\le \sum_{i_j} E|Z_{n,i_1}|^{r_1} \cdots E|Z_{n,i_k}|^{r_k} \\ &\le (2\epsilon_n)^{r-2k} A_n(2, \dots 2) \to 0 \end{aligned}$$

When r is odd, some r_j must be $= 1$ or ≥ 3 so $ET^r_n \to 0$ by (a) and (c). If $r = 2k$ is even, (a)–(c) imply

$$ET^r_n \to \frac{\sigma^{2k}(2k)!}{2^k k!} = E(\sigma\chi)^r$$

and the result follows from Theorem 3.3.26. □

Turning to the result for prime divisors, let P_n denote the uniform distribution on $\{1, \dots, n\}$. If $P_\infty(A) \equiv \lim P_n(A)$ exists, the limit is called the density of $A \subset \mathbf{Z}$. Let A_p be the set of integers divisible by p. Clearly, if p is a prime $P_\infty(A_p) = 1/p$ and $q \ne p$ is another prime

$$P_\infty(A_p \cap A_q) = 1/pq = P_\infty(A_p)P_\infty(A_q)$$

Even though P_∞ is not a probability measure (since $P(\{i\}) = 0$ for all i), we can interpret this as saying that the events of being divisible by p and q are independent. Let $\delta_p(n) = 1$ if n is divisible by p, and $= 0$ otherwise, and

$$g(n) = \sum_{p \le n} \delta_p(n) \quad \text{be the number of prime divisors of } n$$

this and future sums on p being over the primes. Intuitively, the $\delta_p(n)$ behave like X_p that are i.i.d. with

$$P(X_p = 1) = 1/p \quad \text{and} \quad P(X_p = 0) = 1 - 1/p$$

The mean and variance of $\sum_{p \le n} X_p$ are

$$\sum_{p \le n} 1/p \quad \text{and} \quad \sum_{p \le n} 1/p(1 - 1/p)$$

respectively. It is known that

$$(*) \qquad \sum_{p \le n} 1/p = \log\log n + O(1)$$

(see Hardy and Wright (1959), chapter XXII), while anyone can see $\sum_p 1/p^2 < \infty$, so applying Theorem 3.4.10 to X_p and making a small leap of faith gives us:

Theorem 3.4.16 (Erdös-Kac central limit theorem) *As* $n \to \infty$

$$P_n\left(m \le n : g(m) - \log\log n \le x(\log\log n)^{1/2}\right) \to P(\chi \le x)$$

Proof We begin by showing that we can ignore the primes "near" n. Let

$$\begin{aligned} \alpha_n &= n^{1/\log\log n} \\ \log\alpha_n &= \log n/\log\log n \\ \log\log\alpha_n &= \log\log n - \log\log\log n \end{aligned}$$

The sequence α_n has two nice properties:

(a) $\left(\sum_{\alpha_n < p \le n} 1/p\right)/(\log\log n)^{1/2} \to 0$ by $(*)$

Proof of (a). By $(*)$

$$\begin{aligned} \sum_{\alpha_n < p \le n} 1/p &= \sum_{p \le n} 1/p - \sum_{p \le \alpha_n} 1/p \\ &= \log\log n - \log\log\alpha_n + O(1) \\ &= \log\log\log n + O(1) \end{aligned}$$

(b) If $\epsilon > 0$, then $\alpha_n \le n^\epsilon$ for large n and hence $\alpha_n^r/n \to 0$ for all $r < \infty$.

Proof of (b). $1/\log\log n \to 0$ as $n \to \infty$. □

Let $g_n(m) = \sum_{p \le \alpha_n} \delta_p(m)$ and let E_n denote expected value w.r.t. P_n.

$$E_n\left(\sum_{\alpha_n < p \le n} \delta_p\right) = \sum_{\alpha_n < p \le n} P_n(m : \delta_p(m) = 1) \le \sum_{\alpha_n < p \le n} 1/p$$

so by (a) it is enough to prove the result for g_n. Let

$$S_n = \sum_{p \le \alpha_n} X_p$$

where the X_p are the independent random variables introduced earlier. Let $b_n = ES_n$ and $a_n^2 = \text{var}(S_n)$. (a) tells us that b_n and a_n^2 are both

$$\log\log n + o((\log\log n)^{1/2})$$

so it suffices to show

$$P_n(m : g_n(m) - b_n \le xa_n) \to P(\chi \le x)$$

An application of Theorem 3.4.10 shows $(S_n - b_n)/a_n \Rightarrow \chi$, and since $|X_p| \le 1$, it follows from the second proof of Theorem 3.4.10 that

$$E\left((S_n - b_n)/a_n\right)^r \to E\chi^r \quad \text{for all } r$$

Using notation from that proof (and replacing i_j by p_j)

$$ES_n^r = \sum_{k=1}^{r} \sum_{r_i} \frac{r!}{r_1! \cdots r_k!} \frac{1}{k!} \sum_{p_j} E(X_{p_1}^{r_1} \cdots X_{p_k}^{r_k})$$

Since $X_p \in \{0, 1\}$, the summand is

$$E(X_{p_1} \cdots X_{p_k}) = 1/(p_1 \cdots p_k)$$

A little thought reveals that

$$E_n(\delta_{p_1} \cdots \delta_{p_k}) \leq \frac{1}{n} [n/(p_1 \cdots p_k)]$$

The two moments differ by $\leq 1/n$, so

$$|E(S_n^r) - E_n(g_n^r)| = \sum_{k=1}^{r} \sum_{r_i} \frac{r!}{r_1! \cdots r_k!} \frac{1}{k!} \sum_{p_j} \frac{1}{n}$$

$$\leq 13n\Big(\sum_{p \leq \alpha_n} 1\Big)^r \leq \frac{\alpha_n^r}{n} \to 0$$

by (b). Now

$$E(S_n - b_n)^r = \sum_{m=0}^{r} \binom{r}{m} ES_n^m (-b_n)^{r-m}$$

$$E(g_n - b_n)^r = \sum_{m=0}^{r} \binom{r}{m} Eg_n^m (-b_n)^{r-m}$$

so subtracting and using our bound on $|E(S_n^r) - E_n(g_n^r)|$ with $r = m$

$$|E(S_n - b_n)^r - E(g_n - b_n)^r| \leq \sum_{m=0}^{r} \binom{r}{m} \frac{1}{n} \alpha_n^m b_n^{r-m} = (\alpha_n + b_n)^r / n \to 0$$

since $b_n \leq \alpha_n$. This is more than enough to conclude that

$$E\left((g_n - b_n)/a_n\right)^r \to E\chi^r$$

and the desired result follows from Theorem 3.3.26. □

3.4.4 Rates of Convergence (Berry-Esseen)*

Theorem 3.4.17 *Let $X_1, X_2, \ldots$ be i.i.d. with $EX_i = 0$, $EX_i^2 = \sigma^2$, and $E|X_i|^3 = \rho < \infty$. If $F_n(x)$ is the distribution of $(X_1 + \cdots + X_n)/\sigma\sqrt{n}$ and $\mathcal{N}(x)$ is the standard normal distribution, then*

$$|F_n(x) - \mathcal{N}(x)| \leq 3\rho/\sigma^3\sqrt{n}$$

Remarks The reader should note that the inequality holds for all n and x, but since $\rho \geq \sigma^3$, it only has nontrivial content for $n \geq 10$. It is easy to see that the rate cannot be faster than $n^{-1/2}$. When $P(X_i = 1) = P(X_i = -1) = 1/2$, symmetry and (1.4) imply

$$F_{2n}(0) = \frac{1}{2}\{1 + P(S_{2n} = 0)\} = \frac{1}{2}(1 + (\pi n)^{-1/2}) + o(n^{-1/2})$$

The constant 3 is not the best known (van Beek (1972) gets 0.8), but as Feller brags, "our streamlined method yields a remarkably good bound even though it avoids the usual messy numerical calculations." The hypothesis $E|X|^3$ is needed to get the rate $n^{-1/2}$. Heyde (1967) has shown that for $0 < \delta < 1$

$$\sum_{n=1}^{\infty} n^{-1+\delta/2} \sup_x |F_n(x) - \mathcal{N}(x)| < \infty$$

if and only if $E|X|^{2+\delta} < \infty$. For this and more on rates of convergence, see Hall (1982).

Proof Since neither side of the inequality is affected by scaling, we can suppose without loss of generality that $\sigma^2 = 1$. The first phase of the argument is to derive an inequality, Lemma 3.4.19, that relates the difference between the two distributions to the distance between their ch.f.'s. Polya's density (see Example 3.3.15 and use (e) of Theorem 3.3.1)

$$h_L(x) = \frac{1 - \cos Lx}{\pi L x^2}$$

has ch.f. $\omega_L(\theta) = (1 - |\theta/L|)^+$ for $|\theta| \le L$. We will use H_L for its distribution function. We will convolve the distributions under consideration with H_L to get ch.f. that have compact support. The first step is to show that convolution with H_L does not reduce the difference between the distributions too much.

Lemma 3.4.18 *Let F and G be distribution functions with $G'(x) \le \lambda < \infty$. Let $\Delta(x) = F(x) - G(x)$, $\eta = \sup|\Delta(x)|$, $\Delta_L = \Delta * H_L$, and $\eta_L = \sup|\Delta_L(x)|$. Then*

$$\eta_L \ge \frac{\eta}{2} - \frac{12\lambda}{\pi L} \quad or \quad \eta \le 2\eta_L + \frac{24\lambda}{\pi L}$$

Proof Δ goes to 0 at $\pm\infty$, G is continuous, and F is a d.f., so there is an x_0 with $\Delta(x_0) = \eta$ or $\Delta(x_0-) = -\eta$. By looking at the d.f.'s of (-1) times the r.v.'s in the second case, we can suppose without loss of generality that $\Delta(x_0) = \eta$. Since $G'(x) \le \lambda$ and F is nondecreasing, $\Delta(x_0 + s) \ge \eta - \lambda s$. Letting $\delta = \eta/2\lambda$, and $t = x_0 + \delta$, we have

$$\Delta(t - x) \ge \begin{cases} (\eta/2) + \lambda x & \text{for } |x| \le \delta \\ -\eta & \text{otherwise} \end{cases}$$

To estimate the convolution Δ_L, we observe

$$2\int_\delta^\infty h_L(x)\,dx \le 2\int_\delta^\infty 2/(\pi L x^2)dx = 4/(\pi L\delta)$$

Looking at $(-\delta, \delta)$ and its complement separately and noticing symmetry implies $\int_{-\delta}^{\delta} x h_L(x)\,dx = 0$, we have

$$\eta_L \ge \Delta_L(t) \ge \frac{\eta}{2}\left(1 - \frac{4}{\pi L\delta}\right) - \eta\frac{4}{\pi L\delta} = \frac{\eta}{2} - \frac{6\eta}{\pi L\delta} = \frac{\eta}{2} - \frac{12\lambda}{\pi L}$$

which proves the lemma. □

Lemma 3.4.19 *Let K_1 and K_2 be d.f. with mean 0 whose ch.f. κ_i are integrable*

$$K_1(x) - K_2(x) = (2\pi)^{-1}\int -e^{-itx}\frac{\kappa_1(t) - \kappa_2(t)}{it}\,dt$$

Proof Since the κ_i are integrable, the inversion formula, Theorem 3.3.11, implies that the density $k_i(x)$ has

$$k_i(y) = (2\pi)^{-1} \int e^{-ity} \kappa_i(t)\, dt$$

Subtracting the last expression with $i = 2$ from the one with $i = 1$ then integrating from a to x and letting $\Delta K = K_1 - K_2$ gives

$$\begin{aligned} \Delta K(x) - \Delta K(a) &= (2\pi)^{-1} \int_a^x \int e^{-ity} \{\kappa_1(t) - \kappa_2(t)\}\, dt\, dy \\ &= (2\pi)^{-1} \int \{e^{-ita} - e^{-itx}\} \frac{\kappa_1(t) - \kappa_2(t)}{it}\, dt \end{aligned}$$

the application of Fubini's theorem being justified since the κ_i are integrable in t and we are considering a bounded interval in y.

The factor $1/it$ could cause problems near zero, but we have supposed that the K_i have mean 0, so $\{1 - \kappa_i(t)\}/t \to 0$ by Exercise 3.3.18, and hence $(\kappa_1(t) - \kappa_2(t))/it$ is bounded and continuous. The factor $1/it$ improves the integrability for large t, so $(\kappa_1(t) - \kappa_2(t))/it$ is integrable. Letting $a \to -\infty$ and using the Riemann-Lebesgue lemma (Exercise 1.4.4) proves the result. □

Let φ_F and φ_G be the ch.f.'s of F and G. Applying Lemma 3.4.19 to $F_L = F * H_L$ and $G_L = G * H_L$, gives

$$\begin{aligned} |F_L(x) - G_L(x)| &\le \frac{1}{2\pi} \int |\varphi_F(t)\omega_L(t) - \varphi_G(t)\omega_L(t)|\, \frac{dt}{|t|} \\ &\le \frac{1}{2\pi} \int_{-L}^{L} |\varphi_F(t) - \varphi_G(t)|\, \frac{dt}{|t|} \end{aligned}$$

since $|\omega_L(t)| \le 1$. Using Lemma 3.4.18 now, we have

$$|F(x) - G(x)| \le \frac{1}{\pi} \int_{-L}^{L} |\varphi_F(\theta) - \varphi_G(\theta)|\, \frac{d\theta}{|\theta|} + \frac{24\lambda}{\pi L}$$

where $\lambda = \sup_x G'(x)$. Plugging in $F = F_n$ and $G = \mathcal{N}$ gives

$$|F_n(x) - \mathcal{N}(x)| \le \frac{1}{\pi} \int_{-L}^{L} |\varphi^n(\theta/\sqrt{n}) - \psi(\theta)|\, \frac{d\theta}{|\theta|} + \frac{24\lambda}{\pi L} \tag{3.4.1}$$

and it remains to estimate the right-hand side. This phase of the argument is fairly routine, but there is a fair amount of algebra. To save the reader from trying to improve the inequalities along the way in hopes of getting a better bound, we would like to observe that we have used the fact that $C = 3$ to get rid of the cases $n \le 9$, and we use $n \ge 10$ in (e).

To estimate the second term in (3.4.1), we observe that

(a) $$\sup_x G'(x) = G'(0) = (2\pi)^{-1/2} = 0.39894 < 2/5$$

For the first, we observe that if $|\alpha|, |\beta| \le \gamma$

(b) $$|\alpha^n - \beta^n| \le \sum_{m=0}^{n-1} |\alpha^{n-m}\beta^m - \alpha^{n-m-1}\beta^{m+1}| \le n|\alpha - \beta|\gamma^{n-1}$$

Using (3.3.3) now gives (recall we are supposing $\sigma^2 = 1$)

(c) $$|\varphi(t) - 1 + t^2/2| \le \rho|t|^3/6$$

so if $t^2 \le 2$

(d) $$|\varphi(t)| \le 1 - t^2/2 + \rho|t|^3/6$$

Let $L = 4\sqrt{n}/3\rho$. If $|\theta| \le L$, then by (d) and the fact $\rho|\theta|/\sqrt{n} \le 4/3$

$$\begin{aligned}|\varphi(\theta/\sqrt{n})| &\le 1 - \theta^2/2n + \rho|\theta|^3/6n^{3/2}\\ &\le 1 - 5\theta^2/18n \le \exp(-5\theta^2/18n)\end{aligned}$$

since $1 - x \le e^{-x}$. We will now apply (b) with

$$\alpha = \varphi(\theta/\sqrt{n}) \qquad \beta = \exp(-\theta^2/2n) \qquad \gamma = \exp(-5\theta^2/18n)$$

Since we are supposing $n \ge 10$

(e) $$\gamma^{n-1} \le \exp(-\theta^2/4)$$

For the other part of (b), we write

$$n|\alpha - \beta| \le n|\varphi(\theta/\sqrt{n}) - 1 + \theta^2/2n| + n|1 - \theta^2/2n - \exp(-\theta^2/2n)|$$

To bound the first term on the right-hand side, observe (c) implies

$$n|\varphi(\theta/\sqrt{n}) - 1 + \theta^2/2n| \le \rho|\theta|^3/6n^{1/2}$$

For the second term, note that if $0 < x < 1$, then we have an alternating series with decreasing terms so

$$|e^{-x} - (1 - x)| = \left| -\frac{x^2}{2!} + \frac{x^3}{3!} - \dots \right| \le \frac{x^2}{2}$$

Taking $x = \theta^2/2n$, it follows that for $|\theta| \le L \le \sqrt{2n}$

$$n|1 - \theta^2/2n - \exp(-\theta^2/2n)| \le \theta^4/8n$$

Combining this with our estimate on the first term gives

(f) $$n|\alpha - \beta| \le \rho|\theta|^3/6n^{1/2} + \theta^4/8n$$

Using (f) and (e) in (b) gives

$$\begin{aligned}\frac{1}{|\theta|}|\varphi^n(\theta/\sqrt{n}) - \exp(-\theta^2/2)| &\le \exp(-\theta^2/4)\left\{\frac{\rho\theta^2}{6n^{1/2}} + \frac{|\theta|^3}{8n}\right\}\\ &\le \frac{1}{L}\exp(-\theta^2/4)\left\{\frac{2\theta^2}{9} + \frac{|\theta|^3}{18}\right\}\end{aligned}$$

since $\rho/\sqrt{n} = 4/3L$, and $1/n = 1/\sqrt{n} \cdot 1/\sqrt{n} \le 4/3L \cdot 1/3$, since $\rho \ge 1$ and $n \ge 10$. Using the last result and (a) in Lemma 3.4.19 gives

$$\pi L|F_n(x) - \mathcal{N}(x)| \le \int \exp(-\theta^2/4)\left\{\frac{2\theta^2}{9} + \frac{|\theta|^3}{18}\right\} d\theta + 9.6$$

Recalling $L = 4\sqrt{n}/3\rho$, we see that the last result is of the form $|F_n(x) - \mathcal{N}(x)| \le C\rho/\sqrt{n}$. To evaluate the constant, we observe

$$\int (2\pi a)^{-1/2} x^2 \exp(-x^2/2a) dx = a$$

and writing $x^3 = 2x^2 \cdot x/2$ and integrating by parts

$$2\int_0^\infty x^3 \exp(-x^2/4)\, dx = 2\int_0^\infty 4x \exp(-x^2/4)\, dx$$
$$= -16e^{-x^2/4}\Big|_0^\infty = 16$$

This gives us

$$|F_n(x) - \mathcal{N}(x)| \le \frac{1}{\pi} \cdot \frac{3}{4}\left(\frac{2}{9} \cdot 2 \cdot \sqrt{4\pi} + \frac{16}{18} + 9.6\right)\frac{\rho}{\sqrt{n}} < 3\frac{\rho}{\sqrt{n}}$$

For the last step, you have to get out your calculator or trust Feller. □

3.5 Local Limit Theorems*

In Section 3.1 we saw that if $X_1, X_2, \ldots$ are i.i.d. with $P(X_1 = 1) = P(X_1 = -1) = 1/2$ and k_n is a sequence of integers with $2k_n/(2n)^{1/2} \to x$, then

$$P(S_{2n} = 2k_n) \sim (\pi n)^{-1/2} \exp(-x^2/2)$$

In this section we will prove two theorems that generalize the last result. We begin with two definitions. A random variable X has a **lattice distribution** if there are constants b and $h > 0$ so that $P(X \in b + h\mathbf{Z}) = 1$, where $b + h\mathbf{Z} = \{b + hz : z \in \mathbf{Z}\}$. The largest h for which the last statement holds is called the **span** of the distribution.

Example 3.5.1 If $P(X = 1) = P(X = -1) = 1/2$, then X has a lattice distribution with span 2. When h is 2, one possible choice is $b = -1$.

The next result relates the last definition to the characteristic function. To check (ii) in its statement, note that in the last example $E(e^{itX}) = \cos t$ has $|\cos(t)| = 1$ when $t = n\pi$.

Theorem 3.5.2 *Let $\varphi(t) = Ee^{itX}$. There are only three possibilities.*

(i) $|\varphi(t)| < 1$ for all $t \neq 0$.

(ii) There is a $\lambda > 0$ so that $|\varphi(\lambda)| = 1$ and $|\varphi(t)| < 1$ for $0 < t < \lambda$. In this case, X has a lattice distribution with span $2\pi/\lambda$.

(iii) $|\varphi(t)| = 1$ for all t. In this case, $X = b$ a.s. for some b.

Proof We begin with (ii). It suffices to show that $|\varphi(t)| = 1$ if and only if $P(X \in b + (2\pi/t)\mathbf{Z}) = 1$ for some b. First, if $P(X \in b + (2\pi/t)\mathbf{Z}) = 1$, then

$$\varphi(t) = Ee^{itX} = e^{itb}\sum_{n\in\mathbf{Z}} e^{i2\pi n} P(X = b + (2\pi/t)n) = e^{itb}$$

Conversely, if $|\varphi(t)| = 1$, then there is equality in the inequality $|Ee^{itX}| \le E|e^{itX}|$, so by Exercise 1.6.1 the distribution of e^{itX} must be concentrated at some point e^{itb}, and $P(X \in b + (2\pi/t)\mathbf{Z}) = 1$.

To prove trichotomy now, we suppose that (i) and (ii) do not hold, i.e., there is a sequence $t_n \downarrow 0$ so that $|\varphi(t_n)| = 1$. The first paragraph shows that there is a b_n so that $P(X \in b_n + (2\pi/t_n)\mathbf{Z}) = 1$. Without loss of generality, we can pick $b_n \in (-\pi/t_n, \pi/t_n]$. As $n \to \infty$, $P(X \notin (-\pi/t_n, \pi/t_n]) \to 0$, so it follows that $P(X = b_n) \to 1$. This is only possible if $b_n = b$ for $n \ge N$, and $P(X = b) = 1$. □

We call the three cases in Theorem 3.5.2: (i) **nonlattice**, (ii) **lattice**, and (iii) **degenerate**. The reader should notice that this means that lattice random variables are by definition nondegenerate. Before we turn to the main business of this section, we would like to introduce one more special case. If X is a lattice distribution and we can take $b = 0$, i.e., $P(X \in h\mathbf{Z}) = 1$, then X is said to be **arithmetic**. In this case, if $\lambda = 2\pi/h$, then $\varphi(\lambda) = 1$ and φ is periodic: $\varphi(t + \lambda) = \varphi(t)$.

Our first local limit theorem is for the lattice case. Let $X_1, X_2, \ldots$ be i.i.d. with $EX_i = 0$, $EX_i^2 = \sigma^2 \in (0, \infty)$, and having a common lattice distribution with span h. If $S_n = X_1 + \cdots + X_n$ and $P(X_i \in b + h\mathbf{Z}) = 1$, then $P(S_n \in nb + h\mathbf{Z}) = 1$. We put

$$p_n(x) = P(S_n/\sqrt{n} = x) \quad \text{for } x \in \mathcal{L}_n = \{(nb + hz)/\sqrt{n} : z \in \mathbf{Z}\}$$

and

$$n(x) = (2\pi\sigma^2)^{-1/2} \exp(-x^2/2\sigma^2) \quad \text{for } x \in (-\infty, \infty)$$

Theorem 3.5.3 *Under these hypotheses, as $n \to \infty$*

$$\sup_{x \in \mathcal{L}_n} \left| \frac{n^{1/2}}{h} p_n(x) - n(x) \right| \to 0$$

Remark To explain the statement, note that if we followed the approach in Example 3.4.7, then we would conclude that for $x \in \mathcal{L}_n$

$$p_n(x) \approx \int_{x-h/2\sqrt{n}}^{x+h/2\sqrt{n}} n(y)\, dy \approx \frac{h}{\sqrt{n}} n(x)$$

Proof Let Y be a random variable with $P(Y \in a + \theta\mathbf{Z}) = 1$ and $\psi(t) = E\exp(itY)$. It follows from part (iii) of Exercise 3.3.2 that

$$P(Y = x) = \frac{1}{2\pi/\theta} \int_{-\pi/\theta}^{\pi/\theta} e^{-itx} \psi(t)\, dt$$

Using this formula with $\theta = h/\sqrt{n}$, $\psi(t) = E\exp(itS_n/\sqrt{n}) = \varphi^n(t/\sqrt{n})$, and then multiplying each side by $1/\theta$ gives

$$\frac{n^{1/2}}{h} p_n(x) = \frac{1}{2\pi} \int_{-\pi\sqrt{n}/h}^{\pi\sqrt{n}/h} e^{-itx} \varphi^n(t/\sqrt{n})\, dt$$

Using the inversion formula, Theorem 3.3.14, for $n(x)$, which has ch.f. $\exp(-\sigma^2 t^2/2)$, gives

$$n(x) = \frac{1}{2\pi} \int e^{-itx} \exp(-\sigma^2 t^2/2)\, dt$$

Subtracting the last two equations gives (recall $\pi > 1$, $|e^{-itx}| \le 1$)

$$\left|\frac{n^{1/2}}{h}p_n(x) - n(x)\right| \le \int_{-\pi\sqrt{n}/h}^{\pi\sqrt{n}/h} |\varphi^n(t/\sqrt{n}) - \exp(-\sigma^2 t^2/2)|\, dt$$
$$+ \int_{\pi\sqrt{n}/h}^{\infty} \exp(-\sigma^2 t^2/2)\, dt$$

The right-hand side is independent of x, so to prove Theorem 3.5.3 it suffices to show that it approaches 0. The second integral clearly $\to 0$. To estimate the first integral, we observe that $\varphi^n(t/\sqrt{n}) \to \exp(-\sigma^2 t^2/2)$, so the integrand goes to 0 and it is now just a question of "applying the dominated convergence theorem."

To do this, we will divide the integral into three pieces. The bounded convergence theorem implies that for any $A < \infty$ the integral over $(-A, A)$ approaches 0. To estimate the integral over $(-A, A)^c$, we observe that since $EX_i = 0$ and $EX_i^2 = \sigma^2$, formula (3.3.3) and the triangle inequality imply that

$$|\varphi(u)| \le |1 - \sigma^2 u^2/2| + \frac{u^2}{2} E(\min(|u| \cdot |X|^3, 6|X|^2))$$

The last expected value $\to 0$ as $u \to 0$. This means we can pick $\delta > 0$ so that if $|u| < \delta$, it is $\le \sigma^2/2$ and hence

$$|\varphi(u)| \le 1 - \sigma^2 u^2/2 + \sigma^2 u^2/4 = 1 - \sigma^2 u^2/4 \le \exp(-\sigma^2 u^2/4)$$

since $1 - x \le e^{-x}$. Applying the last result to $u = t/\sqrt{n}$, we see that for $t \le \delta\sqrt{n}$

$$(*) \qquad |\varphi(t/\sqrt{n})^n| \le \exp(-\sigma^2 t^2/4)$$

So the integral over $(-\delta\sqrt{n}, \delta\sqrt{n}) - (-A, A)$ is smaller than

$$2\int_A^{\delta\sqrt{n}} \exp(-\sigma^2 t^2/4)\, dt$$

which is small if A is large.

To estimate the rest of the integral, we observe that since X has span h, Theorem 3.5.2 implies $|\varphi(u)| \ne 1$ for $u \in [\delta, \pi/h]$. φ is continuous, so there is an $\eta < 1$ so that $|\varphi(u)| \le \eta < 1$ for $|u| \in [\delta, \pi/h]$. Letting $u = t/\sqrt{n}$ again, we see that the integral over $[-\pi\sqrt{n}/h, \pi\sqrt{n}/h] - (-\delta\sqrt{n}, \delta\sqrt{n})$ is smaller than

$$2\int_{\delta\sqrt{n}}^{\pi\sqrt{n}/h} \eta^n + \exp(-\sigma^2 t^2/2)\, dt$$

which $\to 0$ as $n \to \infty$. This completes the proof. □

We turn now to the nonlattice case. Let $X_1, X_2, \ldots$ be i.i.d. with $EX_i = 0$, $EX_i^2 = \sigma^2 \in (0, \infty)$, and having a common characteristic function $\varphi(t)$ that has $|\varphi(t)| < 1$ for all $t \ne 0$. Let $S_n = X_1 + \cdots + X_n$ and $n(x) = (2\pi\sigma^2)^{-1/2}\exp(-x^2/2\sigma^2)$.

Theorem 3.5.4 *Under the previous hypotheses, if $x_n/\sqrt{n} \to x$ and $a < b$*

$$\sqrt{n}P(S_n \in (x_n + a, x_n + b)) \to (b - a)n(x)$$

Remark The proof of this result has to be a little devious because the assumption does not give us much control over the behavior of φ. For a bad example, let $q_1, q_2, \dots$ be an enumeration of the positive rationals that has $q_n \le n$. Suppose

$$P(X = q_n) = P(X = -q_n) = 1/2^{n+1}$$

In this case, $EX = 0$, $EX^2 < \infty$, and the distribution is nonlattice. However, the characteristic function has $\limsup_{t\to\infty} |\varphi(t)| = 1$.

Proof To tame bad ch.f.'s we use a trick. Let $\delta > 0$

$$h_0(y) = \frac{1}{\pi} \cdot \frac{1 - \cos \delta y}{\delta y^2}$$

be the density of the Polya's distribution and let $h_\theta(x) = e^{i\theta x} h_0(x)$. If we introduce the Fourier transform

$$\hat{g}(u) = \int e^{iuy} g(y)\, dy$$

then it follows from Example 3.3.15 that

$$\hat{h}_0(u) = \begin{cases} 1 - |u/\delta| & \text{if } |u| \le \delta \\ 0 & \text{otherwise} \end{cases}$$

and it is easy to see that $\hat{h}_\theta(u) = \hat{h}_0(u + \theta)$. We will show that for any θ

(a) $$\sqrt{n}\, E h_\theta(S_n - x_n) \to n(x) \int h_\theta(y)\, dy$$

Before proving (a), we will show it implies Theorem 3.5.4. Let

$$\mu_n(A) = \sqrt{n} P(S_n - x_n \in A), \quad \text{and} \quad \mu(A) = n(x)|A|$$

where $|A| =$ the Lebesgue measure of A. Let

$$\alpha_n = \sqrt{n}\, E h_0(S_n - x_n) \quad \text{and} \quad \alpha = n(x) \int h_0(y)\, dy = n(x)$$

Finally, define probability measures by

$$\nu_n(B) = \frac{1}{\alpha_n} \int_B h_0(y) \mu_n(dy), \quad \text{and} \quad \nu(B) = \frac{1}{\alpha} \int_B h_0(y) \mu(dy)$$

Taking $\theta = 0$ in (a), we see $\alpha_n \to \alpha$ and so (a) implies

(b) $$\int e^{i\theta y} \nu_n(dy) \to \int e^{i\theta y} \nu(dy)$$

Since this holds for all θ, it follows from Theorem 3.3.17 that $\nu_n \Rightarrow \nu$. Now if $|a|, |b| < 2\pi/\delta$, then the function

$$k(y) = \frac{1}{h_0(y)} \cdot 1_{(a,b)}(y)$$

is bounded and continuous a.s. with respect to ν, so it follows from Theorem 3.2.10 that

$$\int k(y) \nu_n(dy) \to \int k(y) \nu(dy)$$

Since $\alpha_n \to \alpha$, this implies

$$\sqrt{n}P(S_n \in (x_n + a, x_n + b)) \to (b - a)n(x)$$

which is the conclusion of Theorem 3.5.4.

Turning now to the proof of (a), the inversion formula, Theorem 3.3.14, implies

$$h_0(x) = \frac{1}{2\pi} \int e^{-iux} \hat{h}_0(u)\, du$$

Recalling the definition of h_θ, using the last result, and changing variables $u = v + \theta$, we have

$$\begin{aligned} h_\theta(x) = e^{i\theta x} h_0(x) &= \frac{1}{2\pi} \int e^{-i(u-\theta)x} \hat{h}_0(u)\, du \\ &= \frac{1}{2\pi} \int e^{-ivx} \hat{h}_\theta(v)\, dv \end{aligned}$$

since $\hat{h}_\theta(v) = \hat{h}_0(v + \theta)$. Letting F_n be the distribution of $S_n - x_n$ and integrating gives

$$\begin{aligned} Eh_\theta(S_n - x_n) &= \frac{1}{2\pi} \int \int e^{-iux} \hat{h}_\theta(u)\, du\, dF_n(x) \\ &= \frac{1}{2\pi} \int \int e^{-iux}\, dF_n(x) \hat{h}_\theta(u)\, du \end{aligned}$$

by Fubini's theorem. (Recall $\hat{h}_\theta(u)$ has compact support and F_n is a distribution function.) Using (e) of Theorem 3.3.1, we see that the last expression

$$= \frac{1}{2\pi} \int \varphi(-u)^n e^{iux_n} \hat{h}_\theta(u)\, du$$

To take the limit as $n \to \infty$ of this integral, let $[-M, M]$ be an interval with $\hat{h}_\theta(u) = 0$ for $u \notin [-M, M]$. By (∗) above, we can pick δ so that for $|u| < \delta$

$$\text{(c)} \qquad |\varphi(u)| \le \exp(-\sigma^2 u^2/4)$$

Let $I = [-\delta, \delta]$ and $J = [-M, M] - I$. Since $|\varphi(u)| < 1$ for $u \neq 0$ and φ is continuous, there is a constant $\eta < 1$ so that $|\varphi(u)| \le \eta < 1$ for $u \in J$. Since $|\hat{h}_\theta(u)| \le 1$, this implies that

$$\left| \frac{\sqrt{n}}{2\pi} \int_J \varphi(-u)^n e^{iux_n} \hat{h}_\theta(u)\, du \right| \le \frac{\sqrt{n}}{2\pi} \cdot 2M\eta^n \to 0$$

as $n \to \infty$. For the integral over I, change variables $u = t/\sqrt{n}$ to get

$$\frac{1}{2\pi} \int_{-\delta\sqrt{n}}^{\delta\sqrt{n}} \varphi(-t/\sqrt{n})^n e^{itx_n/\sqrt{n}} \hat{h}_\theta(t/\sqrt{n})\, dt$$

The central limit theorem implies $\varphi(-t/\sqrt{n})^n \to \exp(-\sigma^2 t^2/2)$. Using (c) now and the dominated convergence theorem gives (recall $x_n/\sqrt{n} \to x$)

$$\begin{aligned} \frac{\sqrt{n}}{2\pi} \int_I \varphi(-u)^n e^{iux_n} \hat{h}_\theta(u)\, du &\to \frac{1}{2\pi} \int \exp(-\sigma^2 t^2/2) e^{itx} \hat{h}_\theta(0)\, dt \\ &= n(x) \hat{h}_\theta(0) = n(x) \int h_\theta(y)\, dy \end{aligned}$$

by the inversion formula, Theorem 3.3.14, and the definition of $\hat{h}_\theta(0)$. This proves (a) and completes the proof of Theorem 3.5.4. □

3.6 Poisson Convergence

3.6.1 The Basic Limit Theorem

Our first result is sometimes facetiously called the "weak law of small numbers" or the "law of rare events." These names derive from the fact that the Poisson appears as the limit of a sum of indicators of events that have small probabilities.

Theorem 3.6.1 *For each n let $X_{n,m}$, $1 \le m \le n$ be independent random variables with $P(X_{n,m} = 1) = p_{n,m}$, $P(X_{n,m} = 0) = 1 - p_{n,m}$. Suppose*

(i) $\sum_{m=1}^{n} p_{n,m} \to \lambda \in (0, \infty)$,

and (ii) $\max_{1 \le m \le n} p_{n,m} \to 0$.

If $S_n = X_{n,1} + \cdots + X_{n,n}$ then $S_n \Rightarrow Z$ where Z is Poisson(λ).

Here Poisson(λ) is shorthand for Poisson distribution with mean λ, that is,

$$P(Z = k) = e^{-\lambda}\lambda^k/k!$$

Note that in the spirit of the Lindeberg-Feller theorem, no single term contributes very much to the sum. In contrast to that theorem, the contributions, when positive, are not small.

First proof. Let $\varphi_{n,m}(t) = E(\exp(itX_{n,m})) = (1 - p_{n,m}) + p_{n,m}e^{it}$ and let $S_n = X_{n,1} + \cdots + X_{n,n}$. Then

$$E\exp(itS_n) = \prod_{m=1}^{n}(1 + p_{n,m}(e^{it} - 1))$$

Let $0 \le p \le 1$. $|\exp(p(e^{it} - 1))| = \exp(p\,\mathrm{Re}\,(e^{it} - 1)) \le 1$ and $|1 + p(e^{it} - 1)| \le 1$, since it is on the line segment connecting 1 to e^{it}. Using Lemma 3.4.3 with $\theta = 1$ and then Lemma 3.4.4, which is valid when $\max_m p_{n,m} \le 1/2$ since $|e^{it} - 1| \le 2$,

$$\begin{aligned}
&\left|\exp\left(\sum_{m=1}^{n} p_{n,m}(e^{it} - 1)\right) - \prod_{m=1}^{n}\{1 + p_{n,m}(e^{it} - 1)\}\right| \\
&\le \sum_{m=1}^{n}\left|\exp(p_{n,m}(e^{it} - 1)) - \{1 + p_{n,m}(e^{it} - 1)\}\right| \\
&\le \sum_{m=1}^{n} p_{n,m}^2|e^{it} - 1|^2
\end{aligned}$$

Using $|e^{it} - 1| \le 2$ again, it follows that the last expression

$$\le 4\left(\max_{1 \le m \le n} p_{n,m}\right)\sum_{m=1}^{n} p_{n,m} \to 0$$

by assumptions (i) and (ii). The last conclusion and $\sum_{m=1}^n p_{n,m} \to \lambda$ imply

$$E\exp(itS_n) \to \exp(\lambda(e^{it} - 1))$$

To complete the proof now, we consult Example 3.3.4 for the ch.f. of the Poisson distribution and apply Theorem 3.3.17. □

We will now consider some concrete situations in which Theorem 3.6.1 can be applied. In each case we are considering a situation in which $p_{n,m} = c/n$, so we approximate the distribution of the sum by a Poisson with mean c.

Example 3.6.2 In a calculus class with 400 students, the number of students who have their birthday on the day of the final exam has approximately a Poisson distribution with mean $400/365 = 1.096$. This means that the probability no one was born on that date is about $e^{-1.096} = 0.334$. Similar reasoning shows that the number of babies born on a given day or the number of people who arrive at a bank between 1:15 and 1:30 should have a Poisson distribution.

Example 3.6.3 Suppose we roll two dice 36 times. The probability of "double ones" (one on each die) is $1/36$ so the number of times this occurs should have approximately a Poisson distribution with mean 1. Comparing the Poisson approximation with exact probabilities shows that the agreement is good even though the number of trials is small.

k	0	1	2	3
Poisson	0.3678	0.3678	0.1839	0.0613
exact	0.3627	0.3730	0.1865	0.0604

After we give the second proof of Theorem 3.6.1, we will discuss rates of convergence. Those results will show that for large n the largest discrepancy occurs for $k = 1$ and is about $1/2en$ ($= 0.0051$ in this case).

Our second proof of Theorem 3.6.1 requires a little more work but provides information about the rate of convergence. We begin by defining the **total variation distance** between two measures on a countable set S.

$$\|\mu - \nu\| \equiv \frac{1}{2}\sum_z |\mu(z) - \nu(z)| = \sup_{A \subset S} |\mu(A) - \nu(A)|$$

The first equality is a definition. To prove the second, note that for any A

$$\sum_z |\mu(z) - \nu(z)| \geq |\mu(A) - \nu(A)| + |\mu(A^c) - \nu(A^c)| = 2|\mu(A) - \nu(A)|$$

and there is equality when $A = \{z : \mu(z) \geq \nu(z)\}$.

Lemma 3.6.4 *(i) $d(\mu, \nu) = \|\mu - \nu\|$ defines a metric on probability measures on* **Z** *and (ii) $\|\mu_n - \mu\| \to 0$ if and only if $\mu_n(x) \to \mu(x)$ for each $x \in$* **Z***, which by Exercise 3.2.11 is equivalent to $\mu_n \Rightarrow \mu$.*

Proof (i) Clearly, $d(\mu, \nu) = d(\nu, \mu)$ and $d(\mu, \nu) = 0$ if and only if $\mu = \nu$. To check the triangle inequality we note that the triangle inequality for real numbers implies

$$|\mu(x) - \nu(x)| + |\nu(x) - \pi(x)| \geq |\mu(x) - \pi(x)|$$

then sum over x.

(ii) One direction is trivial. We cannot have $\|\mu_n - \mu\| \to 0$ unless $\mu_n(x) \to \mu(x)$ for each x. To prove the converse, note that if $\mu_n(x) \to \mu(x)$

$$\sum_x |\mu_n(x) - \mu(x)| = 2\sum_x (\mu(x) - \mu_n(x))^+ \to 0$$

by the dominated convergence theorem. □

Exercise 3.6.1 Show that $\|\mu - \nu\| \le 2\delta$ if and only if there are random variables X and Y with distributions μ and ν so that $P(X \ne Y) \le \delta$.

The next three lemmas are the keys to our second proof.

Lemma 3.6.5 *If $\mu_1 \times \mu_2$ denotes the product measure on $\mathbf{Z} \times \mathbf{Z}$ that has $(\mu_1 \times \mu_2)(x, y) = \mu_1(x)\mu_2(y)$, then*

$$\|\mu_1 \times \mu_2 - \nu_1 \times \nu_2\| \le \|\mu_1 - \nu_1\| + \|\mu_2 - \nu_2\|$$

Proof $2\|\mu_1 \times \mu_2 - \nu_1 \times \nu_2\| = \sum_{x,y} |\mu_1(x)\mu_2(y) - \nu_1(x)\nu_2(y)|$

$$\le \sum_{x,y} |\mu_1(x)\mu_2(y) - \nu_1(x)\mu_2(y)| + \sum_{x,y} |\nu_1(x)\mu_2(y) - \nu_1(x)\nu_2(y)|$$

$$= \sum_y \mu_2(y) \sum_x |\mu_1(x) - \nu_1(x)| + \sum_x \nu_1(x) \sum_y |\mu_2(y) - \nu_2(y)|$$

$$= 2\|\mu_1 - \nu_1\| + 2\|\mu_2 - \nu_2\|$$

which gives the desired result. □

Lemma 3.6.6 *If $\mu_1 * \mu_2$ denotes the convolution of μ_1 and μ_2, that is,*

$$\mu_1 * \mu_2(x) = \sum_y \mu_1(x - y)\mu_2(y)$$

then $\|\mu_1 * \mu_2 - \nu_1 * \nu_2\| \le \|\mu_1 \times \mu_2 - \nu_1 \times \nu_2\|$

Proof $2\|\mu_1 * \mu_2 - \nu_1 * \nu_2\| = \sum_x \left|\sum_y \mu_1(x - y)\mu_2(y) - \sum_y \nu_1(x - y)\nu_2(y)\right|$

$$\le \sum_x \sum_y |\mu_1(x - y)\mu_2(y) - \nu_1(x - y)\nu_2(y)|$$

$$= 2\|\mu_1 \times \mu_2 - \nu_1 \times \nu_2\|$$

which gives the desired result. □

Lemma 3.6.7 *Let μ be the measure with $\mu(1) = p$ and $\mu(0) = 1 - p$. Let ν be a Poisson distribution with mean p. Then $\|\mu - \nu\| \le p^2$.*

Proof $2\|\mu - \nu\| = |\mu(0) - \nu(0)| + |\mu(1) - \nu(1)| + \sum_{n \ge 2} \nu(n)$

$$= |1 - p - e^{-p}| + |p - p\,e^{-p}| + 1 - e^{-p}(1 + p)$$

Since $1 - x \le e^{-x} \le 1$ for $x \ge 0$, the previous expression

$$= e^{-p} - 1 + p + p(1 - e^{-p}) + 1 - e^{-p} - pe^{-p}$$

$$= 2p(1 - e^{-p}) \le 2p^2$$

which gives the desired result. □

Second proof of Theorem 3.6.1 Let $\mu_{n,m}$ be the distribution of $X_{n,m}$. Let μ_n be the distribution of S_n. Let $\nu_{n,m}$, ν_n, and ν be Poisson distributions with means $p_{n,m}$, $\lambda_n = \sum_{m\le n} p_{n,m}$, and λ, respectively. Since $\mu_n = \mu_{n,1} * \cdots * \mu_{n,n}$ and $\nu_n = \nu_{n,1} * \cdots * \nu_{n,n}$, Lemmas 3.6.6, 3.6.5, and 3.6.7 imply

$$\|\mu_n - \nu_n\| \le \sum_{m=1}^{n} \|\mu_{n,m} - \nu_{n,m}\| \le 2\sum_{m=1}^{n} p_{n,m}^2 \tag{3.6.1}$$

Using the definition of total variation distance now gives

$$\sup_A |\mu_n(A) - \nu_n(A)| \le \sum_{m=1}^{n} p_{n,m}^2$$

Assumptions (i) and (ii) imply that the right-hand side $\to 0$. Since $\nu_n \Rightarrow \nu$ as $n \to \infty$, the result follows. □

Remark This proof is due to Hodges and Le Cam (1960). By different methods, C. Stein (1987) (see (43) on p. 89) has proved

$$\sup_A |\mu_n(A) - \nu_n(A)| \le (\lambda \vee 1)^{-1} \sum_{m=1}^{n} p_{n,m}^2$$

Rates of convergence When $p_{n,m} = 1/n$, (3.6.1) becomes

$$\sup_A |\mu_n(A) - \nu_n(A)| \le 1/n$$

To assess the quality of this bound, we will compare the Poisson and binomial probabilities for k successes.

k	Poisson	Binomial
0	e^{-1}	$\left(1-\frac{1}{n}\right)^n$
1	e^{-1}	$n\cdot n^{-1}\left(1-\frac{1}{n}\right)^{n-1} = \left(1-\frac{1}{n}\right)^{n-1}$
2	$e^{-1}/2!$	$\binom{n}{2}n^{-2}\left(1-\frac{1}{n}\right)^{n-2} = \left(1-\frac{1}{n}\right)^{n-1}/2!$
3	$e^{-1}/3!$	$\binom{n}{3}n^{-3}\left(1-\frac{1}{n}\right)^{n-3} = \left(1-\frac{2}{n}\right)\left(1-\frac{1}{n}\right)^{n-2}\Big/ 3!$

Since $(1-x) \le e^{-x}$, we have $\mu_n(0) - \nu_n(0) \le 0$. Expanding

$$\log(1+x) = x - \frac{x^2}{2} + \frac{x^3}{3} - \ldots$$

gives

$$(n-1)\log\left(1-\frac{1}{n}\right) = -\frac{n-1}{n} - \frac{n-1}{2n^2} - \ldots = -1 + \frac{1}{2n} + O(n^{-2})$$

So

$$n\left(\left(1-\frac{1}{n}\right)^{n-1} - e^{-1}\right) = ne^{-1}\left(\exp\{1/2n + O(n^{-2})\} - 1\right) \to e^{-1}/2$$

and it follows that

$$n(\mu_n(1) - \nu_n(1)) \to e^{-1}/2$$
$$n(\mu_n(2) - \nu_n(2)) \to e^{-1}/4$$

For $k \ge 3$, using $(1 - 2/n) \le (1 - 1/n)^2$ and $(1 - x) \le e^{-x}$ shows $\mu_n(k) - \nu_n(k) \le 0$, so

$$\sup_{A \subset \mathbf{Z}} |\mu_n(A) - \nu_n(A)| \approx 3/4en$$

There is a large literature on Poisson approximations for dependent events. Here we consider:

3.6.2 Two Examples with Dependence

Example 3.6.8 (Matching) Let π be a random permutation of $\{1, 2, \dots, n\}$, let $X_{n,m} = 1$ if m is a fixed point (0 otherwise), and let $S_n = X_{n,1} + \cdots + X_{n,n}$ be the number of fixed points. We want to compute $P(S_n = 0)$. (For a more exciting story consider men checking hats or wives swapping husbands.) Let $A_{n,m} = \{X_{n,m} = 1\}$. The inclusion-exclusion formula implies

$$\begin{aligned} P\left(\cup_{m=1}^n A_m\right) &= \sum_m P(A_m) - \sum_{\ell<m} P(A_\ell \cap A_m) \\ &\quad + \sum_{k<\ell<m} P(A_k \cap A_\ell \cap A_m) - \dots \\ &= n \cdot \frac{1}{n} - \binom{n}{2}\frac{(n-2)!}{n!} + \binom{n}{3}\frac{(n-3)!}{n!} - \dots \end{aligned}$$

since the number of permutations with k specified fixed points is $(n-k)!$ Canceling some factorials gives

$$P(S_n > 0) = \sum_{m=1}^n \frac{(-1)^{m-1}}{m!} \quad \text{so} \quad P(S_n = 0) = \sum_{m=0}^n \frac{(-1)^m}{m!}$$

Recognizing the second sum as the first $n+1$ terms in the expansion of e^{-1} gives

$$\begin{aligned} |P(S_n = 0) - e^{-1}| &= \left| \sum_{m=n+1}^\infty \frac{(-1)^m}{m!} \right| \\ &\le \frac{1}{(n+1)!} \left| \sum_{k=0}^\infty (n+2)^{-k} \right| = \frac{1}{(n+1)!} \cdot \left(1 - \frac{1}{n+2}\right)^{-1} \end{aligned}$$

a much better rate of convergence than $1/n$. To compute the other probabilities, we observe that by considering the locations of the fixed points

$$\begin{aligned} P(S_n = k) &= \binom{n}{k} \frac{1}{n(n-1)\cdots(n-k+1)} P(S_{n-k} = 0) \\ &= \frac{1}{k!} P(S_{n-k} = 0) \to e^{-1}/k! \end{aligned}$$

Example 3.6.9 (Occupancy problem) Suppose that r balls are placed at random into n boxes. It follows from the Poisson approximation to the Binomial that if $n \to \infty$ and $r/n \to c$, then the number of balls in a given box will approach a Poisson distribution with mean c. The last observation should explain why the fraction of empty boxes approached e^{-c} in Example 2.2.10. Here we will show:

Theorem 3.6.10 *If $ne^{-r/n} \to \lambda \in [0,\infty)$, the number of empty boxes approaches a Poisson distribution with mean λ.*

Proof To see where the answer comes from, notice that in the Poisson approximation the probability that a given box is empty is $e^{-r/n} \approx \lambda/n$, so if the occupancy of the various boxes were independent, the result would follow from Theorem 3.6.1. To prove the result, we begin by observing

$$P(\text{ boxes } i_1, i_2, \ldots, i_k \text{ are empty }) = \left(1 - \frac{k}{n}\right)^r$$

If we let $p_m(r,n)$ = the probability exactly m boxes are empty when r balls are put in n boxes, then $P(\text{ no empty box }) = 1 - P$ (at least one empty box). So by inclusion-exclusion

$$\text{(a)} \qquad p_0(r,n) = \sum_{k=0}^{n} (-1)^k \binom{n}{k} \left(1 - \frac{k}{n}\right)^r$$

By considering the locations of the empty boxes

$$\text{(b)} \qquad p_m(r,n) = \binom{n}{m} \left(1 - \frac{m}{n}\right)^r p_0(r, n-m)$$

To evaluate the limit of $p_m(r,n)$, we begin by showing that if $ne^{-r/n} \to \lambda$, then

$$\text{(c)} \qquad \binom{n}{m} \left(1 - \frac{m}{n}\right)^r \to \lambda^m/m!$$

One half of this is easy. Since $(1-x) \le e^{-x}$ and $ne^{-r/n} \to \lambda$

$$\text{(d)} \qquad \binom{n}{m} \left(1 - \frac{m}{n}\right)^r \le \frac{n^m}{m!} e^{-mr/n} \to \lambda^m/m!$$

For the other direction, observe $\binom{n}{m} \ge (n-m)^m/m!$ so

$$\binom{n}{m} \left(1 - \frac{m}{n}\right)^r \ge \left(1 - \frac{m}{n}\right)^{m+r} n^m/m!$$

Now $(1-m/n)^m \to 1$ as $n \to \infty$ and $1/m!$ is a constant. To deal with the rest, we note that if $0 \le t \le 1/2$, then

$$\begin{aligned} \log(1-t) &= -t - t^2/2 - t^3/3 \ldots \\ &\ge -t - \frac{t^2}{2}\left(1 + 2^{-1} + 2^{-2} + \cdots\right) = -t - t^2 \end{aligned}$$

so we have

$$\log\left(n^m \left(1 - \frac{m}{n}\right)^r\right) \ge m \log n - rm/n - r(m/n)^2$$

Our assumption $ne^{-r/n} \to \lambda$ means

$$r = n \log n - n \log \lambda + o(n)$$

so $r(m/n)^2 \to 0$. Multiplying the last display by m/n and rearranging gives $m \log n - rm/n \to m \log \lambda$. Combining the last two results shows

$$\liminf_{n\to\infty} n^m \left(1 - \frac{m}{n}\right)^r \geq \lambda^m$$

and (c) follows. From (a), (c), and the dominated convergence theorem (using (d) to get the domination) we get

(e) if $ne^{-r/n} \to \lambda$ then $p_0(r,n) \to \sum_{k=0}^{\infty}(-1)^k \frac{\lambda^k}{k!} = e^{-\lambda}$

For fixed m, $(n - m)e^{-r/(n-m)} \to \lambda$, so it follows from (e) that $p_0(r,n - m) \to e^{-\lambda}$. Combining this with (b) and (c) completes the proof. □

Example 3.6.11 (Coupon collector's problem) Let $X_1, X_2, \ldots$ be i.i.d. uniform on $\{1,2,\ldots,n\}$ and $T_n = \inf\{m : \{X_1, \ldots X_m\} = \{1,2,\ldots,n\}\}$. Since $T_n \leq m$ if and only if m balls fill up all n boxes, it follows from Theorem 3.6.10 that

$$P(T_n - n \log n \leq nx) \to \exp(-e^{-x})$$

Proof If $r = n \log n + nx$ then $ne^{-r/n} \to e^{-x}$. □

Note that T_n is the sum of n independent random variables (see Example 2.2.7), but T_n does not converge to the normal distribution. The problem is that the last few terms in the sum are of order n, so the hypotheses of the Lindeberg-Feller theorem are not satisfied.

For a concrete instance of the previous result consider: What is the probability that in a village of 2190 ($= 6 \cdot 365$) people all birthdays are represented? Do you think the answer is much different for 1825 ($= 5 \cdot 365$) people?

Solution. Here $n = 365$, so $365 \log 365 = 2153$ and

$$\begin{aligned} P(T_{365} \leq 2190) &= P((T_{365} - 2153)/365 \leq 37/365) \\ &\approx \exp(-e^{-0.1014}) = \exp(-0.9036) = 0.4051 \\ P(T_{365} \leq 1825) &= P((T_{365} - 2153)/365 \leq -328/365) \\ &\approx \exp(-e^{0.8986}) = \exp(-2.4562) = 0.085 \end{aligned}$$

3.7 Poisson Processes

Theorem 3.6.1 generalizes trivially to give the following result.

Theorem 3.7.1 *Let $X_{n,m}$, $1 \leq m \leq n$ be independent nonnegative integer valued random variables with $P(X_{n,m} = 1) = p_{n,m}$, $P(X_{n,m} \geq 2) = \epsilon_{n,m}$.*

(i) $\sum_{m=1}^{n} p_{n,m} \to \lambda \in (0,\infty)$,

(ii) $\max_{1\leq m\leq n} p_{n,m} \to 0$,

and (iii) $\sum_{m=1}^{n} \epsilon_{n,m} \to 0$.

If $S_n = X_{n,1} + \cdots + X_{n,n}$, then $S_n \Rightarrow Z$ where Z is Poisson(λ).

Proof Let $X'_{n,m} = 1$ if $X_{n,m} = 1$, and 0 otherwise. Let $S'_n = X'_{n,1} + \cdots + X'_{n,n}$. (i)-(ii) and Theorem 3.6.1 imply $S'_n \Rightarrow Z$, (iii) tells us $P(S_n \neq S'_n) \to 0$ and the result follows from the converging together lemma, Exercise 3.2.13. □

The next result, which uses Theorem 3.7.1, explains why the Poisson distribution comes up so frequently in applications. Let $N(s,t)$ be the number of arrivals at a bank or an ice cream parlor in the time interval $(s,t]$. Suppose

(i) the numbers of arrivals in disjoint intervals are independent,

(ii) the distribution of $N(s,t)$ only depends on $t - s$,

(iii) $P(N(0,h) = 1) = \lambda h + o(h)$,

and (iv) $P(N(0,h) \geq 2) = o(h)$.

Here, the two $o(h)$ stand for functions $g_1(h)$ and $g_2(h)$ with $g_i(h)/h \to 0$ as $h \to 0$.

Theorem 3.7.2 *If (i)–(iv) hold, then $N(0,t)$ has a Poisson distribution with mean λt.*

Proof Let $X_{n,m} = N((m-1)t/n, mt/n)$ for $1 \leq m \leq n$ and apply Theorem 3.7.1. □

A family of random variables N_t, $t \geq 0$ satisfying:

(i) if $0 = t_0 < t_1 < \ldots < t_n$, $N(t_k) - N(t_{k-1})$, $1 \leq k \leq n$ are independent,

(ii) $N(t) - N(s)$ is Poisson($\lambda(t-s)$),

is called a **Poisson process with rate** λ. To understand how N_t behaves, it is useful to have another method to construct it. Let $\xi_1, \xi_2, \ldots$ be independent random variables with $P(\xi_i > t) = e^{-\lambda t}$ for $t \geq 0$. Let $T_n = \xi_1 + \cdots + \xi_n$ and $N_t = \sup\{n : T_n \leq t\}$, where $T_0 = 0$. In the language of renewal theory (see Theorem 2.4.7), T_n is the time of the nth arrival and N_t is the number of arrivals by time t. To check that N_t is a Poisson process, we begin by recalling (see Theorem 2.1.18):

$$f_{T_n}(s) = \frac{\lambda^n s^{n-1}}{(n-1)!} e^{-\lambda s} \text{ for } s \geq 0$$

i.e., the distribution of T_n has a density given by the right-hand side. Now

$$P(N_t = 0) = P(T_1 > t) = e^{-\lambda t}$$

and for $n \geq 1$

$$P(N_t = n) = P(T_n \leq t < T_{n+1}) = \int_0^t P(T_n = s) P(\xi_{n+1} > t - s)\, ds$$

$$= \int_0^t \frac{\lambda^n s^{n-1}}{(n-1)!} e^{-\lambda s} e^{-\lambda(t-s)}\, ds = e^{-\lambda t} \frac{(\lambda t)^n}{n!}$$

The last two formulas show that N_t has a Poisson distribution with mean λt. To check that the number of arrivals in disjoint intervals is independent, we observe

$$P(T_{n+1} \geq u | N_t = n) = P(T_{n+1} \geq u, T_n \leq t)/P(N_t = n)$$

To compute the numerator, we observe

$$P(T_{n+1} \geq u, T_n \leq t) = \int_0^t f_{T_n}(s) P(\xi_{n+1} \geq u - s)\, ds$$
$$= \int_0^t \frac{\lambda^n s^{n-1}}{(n-1)!} e^{-\lambda s} e^{-\lambda(u-s)}\, ds = e^{-\lambda u} \frac{(\lambda t)^n}{n!}$$

The denominator is $P(N_t = n) = e^{-\lambda t}(\lambda t)^n / n!$, so

$$P(T_{n+1} \geq u | N_t = n) = e^{-\lambda u} / e^{-\lambda t} = e^{-\lambda(u-t)}$$

or rewriting things $P(T_{n+1} - t \geq s | N_t = n) = e^{-\lambda s}$. Let $T_1' = T_{N(t)+1} - t$, and $T_k' = T_{N(t)+k} - T_{N(t)+k-1}$ for $k \geq 2$. The last computation shows that T_1' is independent of N_t. If we observe that

$$P(T_n \leq t,\, T_{n+1} \geq u, T_{n+k} - T_{n+k-1} \geq v_k, k = 2, \ldots, K)$$
$$= P(T_n \leq t, T_{n+1} \geq u) \prod_{k=2}^{K} P(\xi_{n+k} \geq v_k)$$

then it follows that

(a) $T_1', T_2', \ldots$ are i.i.d. and independent of N_t.

The last observation shows that the arrivals after time t are independent of N_t and have the same distribution as the original sequence. From this it follows easily that:

(b) If $0 = t_0 < t_1 \ldots < t_n$, then $N(t_i) - N(t_{i-1}), i = 1, \ldots, n$ are independent.

To see this, observe that the vector $(N(t_2) - N(t_1), \ldots, N(t_n) - N(t_{n-1}))$ is $\sigma(T_k', k \geq 1)$ measurable and hence is independent of $N(t_1)$. Then use induction to conclude

$$P(N(t_i) - N(t_{i-1}) = k_i, i = 1, \ldots, n) = \prod_{i=1}^{n} \exp(-\lambda(t_i - t_{i-1})) \frac{\lambda(t_i - t_{i-1}))^{k_i}}{k_i!}$$

Remark The key to the proof of (a) is the lack of memory property of the exponential distribution:

$$P(T > t + s | T > t) = P(T > s) \tag{3.7.1}$$

which implies that the location of the first arrival after t is independent of what occurred before time t and has an exponential distribution.

Exercise 3.7.1 Show that if $P(T > 0) = 1$ and (3.7.1) holds, then there is a $\lambda > 0$ so that $P(T > t) = e^{-\lambda t}$ for $t \geq 0$. Hint: First show that this holds for $t = m2^{-n}$.

Exercise 3.7.2 If $S =$ exponential(λ) and $T =$ exponentail(μ) are independent, then $V = \min\{S, T\}$ is exponential($\lambda + \mu$) and $P(U = S) = \lambda/(\lambda + \mu)$.

Exercise 3.7.3 Suppose $T_i =$ exponential(λ_i). Let $V = \min(T_1, \ldots, T_n)$ and I be the the (random) index of the T_i that is smallest.

$$P(V > t) = \exp(-(\lambda_1 + \cdots + \lambda_n)t)$$

$$P(I = i) = \frac{\lambda_i}{\lambda_1 + \cdots + \lambda_n}$$

I and $V = \min\{T_1, \ldots T_n\}$ are independent.

3.7.1 Compound Poisson Processes

Suppose that between 12:00 and 1:00 cars arrive at a fast food restuarant according to a Poisson process $N(t)$ with rate λ. Let Y_i be the number of people in the ith vehicle, which we assume to be i.i.d. and independent o $N(t)$. Having introduced the Y_i's, it is natural to consider the sum of the Y_i's we have seen up to time t:

$$S(t) = Y_1 + \cdots + Y_{N(t)}$$

where we set $S(t) = 0$ if $N(t) = 0$. In the motivating example, $S(t)$ gives the number of customers that have arrived up to time t.

Theorem 3.7.3 *Let $Y_1, Y_2, \ldots$ be independent and identically distributed, let N be an independent nonnegative integer valued random variable, and let $S = Y_1 + \cdots + Y_N$ with $S = 0$ when $N = 0$.*

(i) If $E|Y_i|$, $EN < \infty$, then $ES = EN \cdot EY_i$.

(ii) If EY_i^2, $EN^2 < \infty$, then $var(S) = EN\, var(Y_i) + var(N)(EY_i)^2$.

(iii) If N is Poisson(λ), then $var(S) = \lambda EY_i^2$.

Why is this reasonable? The first of these is natural since if $N = n$ is nonrandom, $ES = nEY_i$. (i) then results by setting $n = EN$. This fact is known as Wald's equation. The formula in (ii) is more complicated but it clearly has two of the necessary properties:

If $N = n$ is nonrandom, $\text{var}(S) = n\,\text{var}(Y_i)$.

If $Y_i = c$ is nonrandom, $\text{var}(S) = c^2\,\text{var}(N)$.

Combining these two observations, we see that $EN\,\text{var}(Y_i)$ is the contribution to the variance from the variability of the Y_i, while $\text{var}(N)(EY_i)^2$ is the contribution from the variability of N.

Proof When $N = n$, $S = X_1 + \cdots + X_n$ has $ES = nEY_i$. Breaking things down according to the value of N,

$$ES = \sum_{n=0}^{\infty} E(S|N = n) \cdot P(N = n)$$

$$= \sum_{n=0}^{\infty} nEY_i \cdot P(N = n) = EN \cdot EY_i$$

For the second formula we note that when $N = n$, $S = X_1 + \cdots + X_n$ has $\text{var}(S) = n\,\text{var}(Y_i)$ and hence,

$$E(S^2|N = n) = n\,\text{var}(Y_i) + (nEY_i)^2$$

Computing as before we get

$$\begin{aligned} ES^2 &= \sum_{n=0}^{\infty} E(S^2|N = n) \cdot P(N = n) \\ &= \sum_{n=0}^{\infty} \{n \cdot \text{var}(Y_i) + n^2(EY_i)^2\} \cdot P(N = n) \\ &= (EN) \cdot \text{var}(Y_i) + EN^2 \cdot (EY_i)^2 \end{aligned}$$

To compute the variance now, we observe that

$$\begin{aligned} \text{var}(S) &= ES^2 - (ES)^2 \\ &= (EN) \cdot \text{var}(Y_i) + EN^2 \cdot (EY_i)^2 - (EN \cdot EY_i)^2 \\ &= (EN) \cdot \text{var}(Y_i) + \text{var}(N) \cdot (EY_i)^2 \end{aligned}$$

where in the last step we have used $\text{var}(N) = EN^2 - (EN)^2$ to combine the second and third terms.

For part (iii), we note that in the special case of the Poisson, we have $EN = \lambda$ and $\text{var}(N) = \lambda$, so the result follows from $\text{var}(Y_i) + (EY_i)^2 = EY_i^2$. □

3.7.2 Thinning

In the previous subsection, we added up the Y_i's associated with the arrivals in our Poisson process to see how many customers, etc. we had accumulated by time t. In this one we will use the Y_i to split the Poisson process into several, i.e., we let $N_j(t)$ be the number of $i \le N(t)$ with $Y_i = j$. The somewhat surprising fact is:

Theorem 3.7.4 *$N_j(t)$ are independent rate $\lambda P(Y_i = j)$ Poisson processes.*

Why is this surprising? There are two reasons: (i) the resulting processes are Poisson and (ii) they are independent. To drive the point home, consider a Poisson process with rate 10 per hour, and then flip coins to determine whether the arriving customers are male or female. One might think that seeing 40 men arrive in one hour would be indicative of a large volume of business and hence a larger than normal number of women, but Theorem 3.7.4 tells us that the number of men and the number of women that arrive per hour are independent.

Proof To begin, we suppose that $P(Y_i = 1) = p$ and $P(Y_i = 2) = 1 - p$, so there are only two Poisson processes to consider: $N_1(t)$ and $N_2(t)$. It should be clear that the independent increments property of the Poisson process implies that the pairs of increments

$$(N_1(t_i) - N_1(t_{i-1}), N_2(t_i) - N_2(t_{i-1})), \quad 1 \le i \le n$$

are independent of each other. Since $N_1(0) = N_2(0) = 0$ by definition, it only remains to check that the components $X_i = N_i(t + s) - N_i(s)$ are independent and have the right Poisson distributions. To do this, we note that if $X_1 = j$ and $X_2 = k$, then there must have been $j + k$ arrivals between s and $s + t$, j of which were assigned 1's and k of which were assigned 2's, so

$$\begin{aligned} P(X_1 = j, X_2 = k) &= e^{-\lambda t}\frac{(\lambda t)^{j+k}}{(j+k)!} \cdot \frac{(j+k)!}{j!\,k!} p^j (1-p)^k \\ &= e^{-\lambda pt}\frac{(\lambda pt)^j}{j!} e^{-\lambda(1-p)t}\frac{(\lambda(1-p)t)^k}{k!} \end{aligned} \tag{3.7.2}$$

so $X_1 = \text{Poisson}(\lambda pt)$ and $X_2 = \text{Poisson}(\lambda(1-p)t)$. For the general case, we use the multinomial to conclude that if $p_j = P(Y_i = j)$ for $1 \le j \le m$, then

$$\begin{aligned} &P(X_1 = k_1, \dots X_m = k_m) \\ &= e^{-\lambda t}\frac{(\lambda t)^{k_1+\cdots k_m}}{(k_1 + \cdots k_m)!} \frac{(k_1 + \cdots k_m)!}{k_1! \cdots k_m!} p_1^{k_1} \cdots p_m^{k_m} \\ &= \prod_{j=1}^{m} e^{-\lambda p_j t}\frac{(\lambda p_j)^{k_j}}{k_j!} \end{aligned}$$

which proves the desired result. □

The thinning results generalize easily to the nonhomogeneous case:

Theorem 3.7.5 *Suppose that in a Poisson process with rate λ, we keep a point that lands at s with probability $p(s)$. Then the result is a nonhomogeneous Poisson process with rate $\lambda p(s)$.*

For an application of this consider:

Exercise 3.7.4 (M/G/∞ queue) As one walks around the Duke campus it seems that every student is talking on their smartphone. The argument for arrivals at the ATM implies that the beginnings of calls follow a Poisson process. As for the calls themselves, while many people on the telephone show a lack of memory, there is no reason to suppose that the duration of a call has an exponential distribution, so we use a general distribution function G with $G(0) = 0$ and mean μ. Show that in the long run the number of calls in the system will be Poisson with mean

$$\lambda \int_{r=0}^{\infty} (1 - G(r))\, dr = \lambda\mu \tag{3.7.3}$$

Example 3.7.6 (Poissonization and the occupancy problem) If we put a Poisson number of balls with mean r in n boxes and let N_i be the number of balls in box i, then the last exercise implies $N_1, \dots, N_n$ are independent and have a Poisson distribution with mean r/n. Use this observation to prove Theorem 3.6.10.

To obtain a result for a fixed number of balls n, let $r = n \log n - (\log \lambda)n + o(n)$ and $s_i = n \log n - (\log \mu_i)n$ with $\mu_2 < \lambda < \mu_1$. The normal approximation to the Poisson tells us $P(\text{Poisson}(s_1) < r < \text{Poisson}(s_2)) \to 1$ as $n \to \infty$. If the number of balls has a Poisson distribution with mean $s = n \log n - n(\log \mu)$, then the number of balls in box i, N_i, are independent with mean $s/n = \log(n/\mu)$ and hence they are vacant with probability $\exp(-s/n) = \mu/n$. Letting $X_{n,i} = 1$ if the ith box is vacant, 0 otherwise and using Theorem 3.6.1 it follows that the number of vacant sites converges to a Poisson with mean μ.

To prove the result for a fixed number of balls, we note that the central limit theorem implies

$$P(\text{Poisson}(s_1) < r < \text{Poisson}(s_2)) \to 1$$

Since the number of vacant boxes is decreased when the number of balls increases, the desired result follows.

Exercise 3.7.5 Suppose that a Poisson number of Duke students with mean 2190 will show up to watch the next women's basketball game. What is the probability that for all of the 365 days there is at least one person in the crowd who has that birthday. (Pretend all birthdays have equal probability and February 29th does not exist.)

At times, e.g., in Section 3.8 the following generalization is useful:

Example 3.7.7 **A Poisson process on a measure space** $(S, \mathcal{S}, \mu)$ is a random map $m : \mathcal{S} \to \{0, 1, \ldots\}$ that for each ω is a measure on $\mathcal{S}$ and has the following property: if $A_1, \ldots, A_n$ are disjoint sets with $\mu(A_i) < \infty$, then $m(A_1), \ldots, m(A_n)$ are independent and have Poisson distributions with means $\mu(A_i)$. μ is called the **mean measure** of the process.

If $\mu(S) < \infty$, then it follows from Theorem 3.7.4 that we can construct m by the following recipe: let $X_1, X_2, \ldots$ be i.i.d. elements of S with distribution $\nu(\cdot) = \mu(\cdot)/\mu(S)$, let N be an independent Poisson random variable with mean $\mu(S)$, and let $m(A) = |\{j \le N : X_j \in A\}|$. To extend the construction to infinite measure spaces, e.g., $S = \mathbf{R}^d$, $\mathcal{S} =$ Borel sets, $\mu =$ Lebesgue measure, divide the space up into disjoint sets of finite measure and put independent Poisson processes on each set.

Our next example is a Poisson process on a square:

Example 3.7.8 **Flying bomb hits** in the South of London during World War II fit a Poisson distribution. As Feller, Vol. I (1968), pp.160–161 reports, the area was divided into 576 areas of 1/4 square kilometers each. The total number of hits was 537 for an average of 0.9323 per cell. The following table compares N_k the number of cells with k hits with the predictions of the Poisson approximation.

k	0	1	2	3	4	≥ 5
N_k	229	211	93	35	7	1
Poisson	226.74	211.39	98.54	30.62	7.14	1.57

3.7.3 Conditioning

Theorem 3.7.9 *Let T_n be the time of the nth arrival in a rate λ Poisson process. Let $U_1, U_2, \ldots, U_n$ be independent uniform on $(0,t)$ and let V_k^n be the kth smallest number in $\{U_1, \ldots, U_n\}$. If we condition on $N(t) = n$, the vectors $V = (V_1^n, \ldots, V_n^n)$ and $T = (T_1, \ldots, T_n)$ have the same distribution.*

Proof On $\{N9t) = n$ the joint density

$$f_T(t_1, \ldots, t_n) = \left(\prod_{m=1}^{n} \lambda e^{-\lambda(t_m - t_{m-1})} \right) \lambda e^{-\lambda(t-t_n)} = \lambda e^{-\lambda t}$$

If we divide by $P(N(t) = n) = e^{-\lambda t}(\lambda t^n/n!$, the result is $n!/t^n$, which is the joint distribution of v. □

Corollary 3.7.10 *If* $0 < s < t$, *then*

$$P(N(s) = m|N(t) = n) = \binom{n}{m}\left(\frac{s}{t}\right)^m \left(1 - \frac{s}{t}\right)^{n-m}$$

Proof Once we observe that the set of values $\{U_1, \dots U_n\}$ and $\{V_1, \dots V_n\}$ have the same distribution, This is an immediate conseqeunce of Theorem 3.7.9. □

The following variant of Theorem 3.7.9:

Theorem 3.7.11 *Let T_n be the time of the nth arrival in a rate λ Poisson process. Let $U_1, U_2, \dots, U_n$ be independent uniform on (0,1) and let V_k^n be the kth smallest number in $\{U_1, \dots, U_n\}$. The vectors $(V_1^n, \dots, V_n^n)$ and $(T_1/T_{n+1}, \dots, T_n/T_{n+1})$ have the same distribution.*

Proof We change variables $v = r(t)$, where $v_i = t_i/t_{n+1}$ for $i \le n$, $v_{n+1} = t_{n+1}$. The inverse function is

$$s(v) = (v_1 v_{n+1}, \dots, v_n v_{n+1}, v_{n+1})$$

which has matrix of partial derivatives $\partial s_i/\partial v_j$ given by

$$\begin{pmatrix} v_{n+1} & 0 & \dots & 0 & v_1 \\ 0 & v_{n+1} & \dots & 0 & v_2 \\ \vdots & \vdots & \ddots & \vdots & \vdots \\ 0 & 0 & \dots & v_{n+1} & v_n \\ 0 & 0 & \dots & 0 & 1 \end{pmatrix}$$

The determinant of this matrix is v_{n+1}^n, so if we let $W = (V_1, \dots, V_{n+1}) = r(T_1, \dots, T_{n+1})$, the change of variables formula implies W has joint density

$$f_W(v_1, \dots, v_n, v_{n+1}) = \left(\prod_{m=1}^{n} \lambda e^{-\lambda v_{n+1}(v_m - v_{m-1})}\right) \lambda e^{-\lambda v_{n+1}(1-v_n)} v_{n+1}^n$$

To find the joint density of $V = (V_1, \dots, V_n)$, we simplify the preceding formula and integrate out the last coordinate to get

$$f_V(v_1, \dots, v_n) = \int_0^\infty \lambda^{n+1} v_{n+1}^n e^{-\lambda v_{n+1}}\, dv_{n+1} = n!$$

for $0 < v_1 < v_2 \dots < v_n < 1$, which is the desired joint density. □

Spacings The last result can be used to study the spacings between the order statistics of i.i.d. uniforms. We use notation of Theorem 3.7.11 in the next four exercises, taking $\lambda = 1$ and letting $V_0^n = 0$, and $V_{n+1}^n = 1$.

Exercise 3.7.6 Smirnov (1949) $n V_k^n \Rightarrow T_k$.

Exercise 3.7.7 Weiss (1955) $n^{-1} \sum_{m=1}^n 1_{(n(V_i^n - V_{i-1}^n) > x)} \to e^{-x}$ in probability.

Exercise 3.7.8 $(n/\log n) \max_{1 \le m \le n+1} V_m^n - V_{m-1}^n \to 1$ in probability.

Exercise 3.7.9 $P(n^2 \min_{1 \le m \le n} V_m^n - V_{m-1}^n > x) \to e^{-x}$.

3.8 Stable Laws*

Let $X_1, X_2, \ldots$ be i.i.d. and $S_n = X_1 + \cdots + X_n$. Theorem 3.4.1 showed that if $EX_i = \mu$ and $\text{var}(X_i) = \sigma^2 \in (0, \infty)$, then

$$(S_n - n\mu)/\sigma n^{1/2} \Rightarrow \chi$$

In this section, we will investigate the case $EX_1^2 = \infty$ and give necessary and sufficient conditions for the existence of constants a_n and b_n so that

$$(S_n - b_n)/a_n \Rightarrow Y \quad \text{where } Y \text{ is nondegenerate}$$

We begin with an example. Suppose the distribution of X_i has

$$P(X_1 > x) = P(X_1 < -x) = x^{-\alpha}/2 \quad \text{for } x \geq 1 \tag{3.8.1}$$

where $0 < \alpha < 2$. If $\varphi(t) = E\exp(itX_1)$, then

$$\begin{aligned}
1 - \varphi(t) &= \int_1^\infty (1 - e^{itx}) \frac{\alpha}{2|x|^{\alpha+1}}\, dx + \int_{-\infty}^{-1} (1 - e^{itx}) \frac{\alpha}{2|x|^{\alpha+1}}\, dx \\
&= \alpha \int_1^\infty \frac{1 - \cos(tx)}{x^{\alpha+1}}\, dx
\end{aligned}$$

Changing variables $tx = u$, $dx = du/t$, the last integral becomes

$$= \alpha \int_t^\infty \frac{1 - \cos u}{(u/t)^{\alpha+1}} \frac{du}{t} = t^\alpha \alpha \int_t^\infty \frac{1 - \cos u}{u^{\alpha+1}}\, du$$

As $u \to 0$, $1 - \cos u \sim u^2/2$. So $(1 - \cos u)/u^{\alpha+1} \sim u^{-\alpha+1}/2$, which is integrable, since $\alpha < 2$ implies $-\alpha + 1 > -1$. If we let

$$C = \alpha \int_0^\infty \frac{1 - \cos u}{u^{\alpha+1}} du < \infty$$

and observe (3.8.1) implies $\varphi(t) = \varphi(-t)$, then the results show

$$1 - \varphi(t) \sim C|t|^\alpha \text{ as } t \to 0 \tag{3.8.2}$$

Let $X_1, X_2, \ldots$ be i.i.d. with the distribution given in (3.8.1) and let $S_n = X_1 + \cdots + X_n$.

$$E\exp(itS_n/n^{1/\alpha}) = \varphi(t/n^{1/\alpha})^n = (1 - \{1 - \varphi(t/n^{1/\alpha})\})^n$$

As $n \to \infty$, $n(1 - \varphi(t/n^{1/\alpha})) \to C|t|^\alpha$, so it follows from Theorem 3.4.2 that

$$E\exp(itS_n/n^{1/\alpha}) \to \exp(-C|t|^\alpha)$$

From part (ii) of Theorem 3.3.17, it follows that the expression on the right is the characteristic function of some Y and

$$S_n/n^{1/\alpha} \Rightarrow Y \tag{3.8.3}$$

To prepare for our general result, we will now give another proof of (3.8.3). If $0 < a < b$ and $an^{1/\alpha} > 1$, then

$$P(an^{1/\alpha} < X_1 < bn^{1/\alpha}) = \frac{1}{2}(a^{-\alpha} - b^{-\alpha})n^{-1}$$

so it follows from Theorem 3.6.1 that

$$N_n(a,b) \equiv |\{m \le n : X_m/n^{1/\alpha} \in (a,b)\}| \Rightarrow N(a,b)$$

where $N(a,b)$ has a Poisson distribution with mean $(a^{-\alpha} - b^{-\alpha})/2$. An easy extension of the last result shows that if $A \subset \mathbf{R} - (-\delta,\delta)$ and $\delta n^{1/\alpha} > 1$, then

$$P(X_1/n^{1/\alpha} \in A) = n^{-1}\int_A \frac{\alpha}{2|x|^{\alpha+1}}\,dx$$

so $N_n(A) \equiv |\{m \le n : X_m/n^{1/\alpha} \in A\}| \Rightarrow N(A)$, where $N(A)$ has a Poisson distribution with mean

$$\mu(A) = \int_A \frac{\alpha}{2|x|^{\alpha+1}}\,dx < \infty$$

The limiting family of random variables $N(A)$ is called a **Poisson process on** $(-\infty,\infty)$ **with mean measure** μ. (See Example 3.7.7 for more on this process.) Notice that for any $\epsilon > 0$, $\mu(\epsilon,\infty) = \epsilon^{-\alpha}/2 < \infty$, so $N(\epsilon,\infty) < \infty$.

The last paragraph describes the limiting behavior of the random set

$$\mathcal{X}_n = \{X_m/n^{1/\alpha} : 1 \le m \le n\}$$

To describe the limit of $S_n/n^{1/\alpha}$, we will "sum up the points." Let $\epsilon > 0$ and

$$I_n(\epsilon) = \{m \le n : |X_m| > \epsilon n^{1/\alpha}\}$$

$$\hat{S}_n(\epsilon) = \sum_{m\in I_n(\epsilon)} X_m \qquad \bar{S}_n(\epsilon) = S_n - \hat{S}_n(\epsilon)$$

$I_n(\epsilon)$ = the indices of the "big terms," i.e., those $> \epsilon n^{1/\alpha}$ in magnitude. $\hat{S}_n(\epsilon)$ is the sum of the big terms, and $\bar{S}_n(\epsilon)$ is the rest of the sum. The first thing we will do is show that the contribution of $\bar{S}_n(\epsilon)$ is small if ϵ is. Let

$$\bar{X}_m(\epsilon) = X_m 1_{(|X_m| \le \epsilon n^{1/\alpha})}$$

Symmetry implies $E\bar{X}_m(\epsilon) = 0$, so $E(\bar{S}_n(\epsilon)^2) = nE\bar{X}_1(\epsilon)^2$.

$$\begin{aligned} E\bar{X}_1(\epsilon)^2 &= \int_0^\infty 2yP(|\bar{X}_1(\epsilon)| > y)\,dy \le \int_0^1 2y\,dy + \int_1^{\epsilon n^{1/\alpha}} 2y\,y^{-\alpha}\,dy \\ &= 1 + \frac{2}{2-\alpha}\epsilon^{2-\alpha}n^{2/\alpha-1} - \frac{2}{2-\alpha} \le \frac{2\epsilon^{2-\alpha}}{2-\alpha}n^{2/\alpha-1} \end{aligned}$$

where we have used $\alpha < 2$ in computing the integral and $\alpha > 0$ in the final inequality. From this it follows that

$$E(\bar{S}_n(\epsilon)/n^{1/\alpha})^2 \le \frac{2\epsilon^{2-\alpha}}{2-\alpha} \tag{3.8.4}$$

To compute the limit of $\hat{S}_n(\epsilon)/n^{1/\alpha}$, we observe that $|I_n(\epsilon)|$ has a binomial distribution with success probability $p = \epsilon^{-\alpha}/n$. Given $|I_n(\epsilon)| = m$, $\hat{S}_n(\epsilon)/n^{1/\alpha}$ is the sum of m independent random variables with a distribution F_n^ϵ that is symmetric about 0 and has

$$1 - F_n^\epsilon(x) = P(X_1/n^{1/\alpha} > x \mid |X_1|/n^{1/\alpha} > \epsilon) = x^{-\alpha}/2\epsilon^{-\alpha} \quad \text{for } x \ge \epsilon$$

The last distribution is the same as that of ϵX_1, so if $\varphi(t) = E\exp(itX_1)$, the distribution F_n^ϵ has characteristic function $\varphi(\epsilon t)$. Combining the observations in this paragraph gives

$$E\exp(it\hat{S}_n(\epsilon)/n^{1/\alpha}) = \sum_{m=0}^{n} \binom{n}{m}(\epsilon^{-\alpha}/n)^m(1-\epsilon^{-\alpha}/n)^{n-m}\varphi(\epsilon t)^m$$

Writing

$$\binom{n}{m}\frac{1}{n^m} = \frac{1}{m!}\frac{n(n-1)\cdots(n-m+1)}{n^m} \le \frac{1}{m!}$$

noting $(1-\epsilon^{-\alpha}/n)^n \le \exp(-\epsilon^{-\alpha})$ and using the dominated convergence theorem

$$\begin{aligned} E\exp(it\hat{S}_n(\epsilon)/n^{1/\alpha}) &\to \sum_{m=0}^{\infty}\exp(-\epsilon^{-\alpha})(\epsilon^{-\alpha})^m\varphi(\epsilon t)^m/m! \\ &= \exp(-\epsilon^{-\alpha}\{1-\varphi(\epsilon t)\}) \end{aligned} \tag{3.8.5}$$

To get (3.8.3) now, we use the following generalization of Lemma 3.4.15.

Lemma 3.8.1 *If $h_n(\epsilon) \to g(\epsilon)$ for each $\epsilon > 0$ and $g(\epsilon) \to g(0)$ as $\epsilon \to 0$, then we can pick $\epsilon_n \to 0$ so that $h_n(\epsilon_n) \to g(0)$.*

Proof Let N_m be chosen so that $|h_n(1/m) - g(1/m)| \le 1/m$ for $n \ge N_m$ and $m \to N_m$ is increasing. Let $\epsilon_n = 1/m$ for $N_m \le n < N_{m+1}$ and $= 1$ for $n < N_1$. When $N_m \le n < N_{m+1}$, $\epsilon_n = 1/m$, so it follows from the triangle inequality and the definition of ϵ_n that

$$\begin{aligned} |h_n(\epsilon_n) - g(0)| &\le |h_n(1/m) - g(1/m)| + |g(1/m) - g(0)| \\ &\le 1/m + |g(1/m) - g(0)| \end{aligned}$$

When $n \to \infty$, we have $m \to \infty$ and the result follows. □

Let $h_n(\epsilon) = E\exp(it\hat{S}_n(\epsilon)/n^{1/\alpha})$ and $g(\epsilon) = \exp(-\epsilon^{-\alpha}\{1-\varphi(\epsilon t)\})$. (3.8.2) implies $1-\varphi(t) \sim C|t|^\alpha$ as $t \to 0$, so

$$g(\epsilon) \to \exp(-C|t|^\alpha) \quad \text{as } \epsilon \to 0$$

and Lemma 3.8.1 implies we can pick $\epsilon_n \to 0$ with $h_n(\epsilon_n) \to \exp(-C|t|^\alpha)$. Introducing Y with $E\exp(itY) = \exp(-C|t|^\alpha)$, it follows that $\hat{S}_n(\epsilon_n)/n^{1/\alpha} \Rightarrow Y$. If $\epsilon_n \to 0$, then (3.8.4) implies

$$\bar{S}_n(\epsilon_n)/n^{1/\alpha} \Rightarrow 0$$

and (3.8.3) follows from the converging together lemma, Exercise 3.2.13. □

Once we give one final definition, we will state and prove the general result alluded to earlier. L is said to be **slowly varying**, if

$$\lim_{x\to\infty} L(tx)/L(x) = 1 \quad \text{for all } t > 0$$

Note that $L(t) = \log t$ is slowly varying but t^ϵ is not if $\epsilon \ne 0$.

Theorem 3.8.2 *Suppose $X_1, X_2, \ldots$ are i.i.d. with a distribution that satisfies*

(i) $\lim_{x\to\infty} P(X_1 > x)/P(|X_1| > x) = \theta \in [0,1]$

(ii) $P(|X_1| > x) = x^{-\alpha} L(x)$

where $\alpha < 2$ and L is slowly varying. Let $S_n = X_1 + \cdots + X_n$

$$a_n = \inf\{x : P(|X_1| > x) \le n^{-1}\} \quad \textit{and} \quad b_n = nE(X_1 1_{(|X_1| \le a_n)})$$

As $n \to \infty$, $(S_n - b_n)/a_n \Rightarrow Y$, where Y has a nondegenerate distribution.

Remark This is not much of a generalization of the example, but the conditions are necessary for the existence of constants a_n and b_n so that $(S_n - b_n)/a_n \Rightarrow Y$, where Y is nondegenerate. Proofs of necessity can be found in chapter 9 of Breiman (1968) or in Gnedenko and Kolmogorov (1954). (3.8.11) gives the ch.f. of Y. The reader has seen the main ideas in the second proof of (3.8.3) and so can skip to that point without much loss.

Proof It is not hard to see that (ii) implies

$$nP(|X_1| > a_n) \to 1 \tag{3.8.6}$$

To prove this, note that $nP(|X_1| > a_n) \le 1$ and let $\epsilon > 0$. Taking $x = a_n/(1+\epsilon)$ and $t = 1 + 2\epsilon$, (ii) implies

$$(1+2\epsilon)^{-\alpha} = \lim_{n\to\infty} \frac{P(|X_1| > (1+2\epsilon)a_n/(1+\epsilon))}{P(|X_1| > a_n/(1+\epsilon))} \le \liminf_{n\to\infty} \frac{P(|X_1| > a_n)}{1/n}$$

proving (3.8.6), since ϵ is arbitrary. Combining (3.8.6) with (i) and (ii) gives

$$nP(X_1 > xa_n) \to \theta x^{-\alpha} \quad \text{for } x > 0 \tag{3.8.7}$$

so $|\{m \le n : X_m > xa_n\}| \Rightarrow \text{Poisson}(\theta x^{-\alpha})$. The last result leads, as before, to the conclusion that $\mathcal{X}_n = \{X_m/a_n : 1 \le m \le n\}$ converges to a Poisson process on $(-\infty, \infty)$ with mean measure

$$\mu(A) = \int_{A\cap(0,\infty)} \theta\alpha|x|^{-(\alpha+1)}\,dx + \int_{A\cap(-\infty,0)} (1-\theta)\alpha|x|^{-(\alpha+1)}\,dx$$

To sum up the points, let $I_n(\epsilon) = \{m \le n : |X_m| > \epsilon a_n\}$

$$\hat{\mu}(\epsilon) = EX_m 1_{(\epsilon a_n < |X_m| \le a_n)} \quad \hat{S}_n(\epsilon) = \sum_{m\in I_n(\epsilon)} X_m$$

$$\bar{\mu}(\epsilon) = EX_m 1_{(|X_m| \le \epsilon a_n)}$$

$$\bar{S}_n(\epsilon) = (S_n - b_n) - (\hat{S}_n(\epsilon) - n\hat{\mu}(\epsilon)) = \sum_{m=1}^{n} \{X_m 1_{(|X_m| \le \epsilon a_n)} - \bar{\mu}(\epsilon)\}$$

If we let $\bar{X}_m(\epsilon) = X_m 1_{(|X_m| \le \epsilon a_n)}$, then

$$E(\bar{S}_n(\epsilon)/a_n)^2 = n \operatorname{var}(\bar{X}_1(\epsilon)/a_n) \le nE(\bar{X}_1(\epsilon)/a_n)^2$$

$$E(\bar{X}_1(\epsilon)/a_n)^2 \le \int_0^\epsilon 2yP(|X_1| > ya_n)\,dy$$

$$= P(|X_1| > a_n) \int_0^\epsilon 2y \frac{P(|X_1| > ya_n)}{P(|X_1| > a_n)}\,dy$$

We would like to use (3.8.7) and (ii) to conclude

$$nE(\bar{X}_1(\epsilon)/a_n)^2 \to \int_0^\epsilon 2y\, y^{-\alpha}\, dy = \frac{2}{2-\alpha}\epsilon^{2-\alpha}$$

and hence

$$\limsup_{n\to\infty} E(\bar{S}_n(\epsilon)/a_n)^2 \le \frac{2\epsilon^{2-\alpha}}{2-\alpha} \tag{3.8.8}$$

To justify interchanging the limit and the integral and complete the proof of (3.8.8), we show the following (take $\delta < 2-\alpha$):

Lemma 3.8.3 *For any $\delta > 0$ there is C so that for all $t \ge t_0$ and $y \le 1$*

$$P(|X_1| > yt)/P(|X_1| > t) \le Cy^{-\alpha-\delta}$$

Proof (ii) implies that as $t \to \infty$

$$P(|X_1| > t/2)/P(|X_1| > t) \to 2^\alpha$$

so for $t \ge t_0$ we have

$$P(|X_1| > t/2)/P(|X_1| > t) \le 2^{\alpha+\delta}$$

Iterating and stopping the first time $t/2^m < t_0$, we have for all $n \ge 1$

$$P(|X_1| > t/2^n)/P(|X_1| > t) \le C2^{(\alpha+\delta)n}$$

where $C = 1/P(|X_1| > t_0)$. Applying the last result to the first n with $1/2^n < y$ and noticing $y \le 1/2^{n-1}$, we have

$$P(|X_1| > yt)/P(|X_1| > t) \le C2^{\alpha+\delta}y^{-\alpha-\delta}$$

which proves the lemma. □

To compute the limit of $\hat{S}_n(\epsilon)$, we observe that $|I_n(\epsilon)| \Rightarrow \text{Poisson}(\epsilon^{-\alpha})$. Given $|I_n(\epsilon)| = m$, $\hat{S}_n(\epsilon)/a_n$ is the sum of m independent random variables with distribution F_n^ϵ that has

$$1 - F_n^\epsilon(x) = P(X_1/a_n > x \mid |X_1|/a_n > \epsilon) \to \theta x^{-\alpha}/\epsilon^{-\alpha}$$
$$F_n^\epsilon(-x) = P(X_1/a_n < -x \mid |X_1|/a_n > \epsilon) \to (1-\theta)|x|^{-\alpha}/\epsilon^{-\alpha}$$

for $x \ge \epsilon$. If we let $\psi_n^\epsilon(t)$ denote the characteristic function of F_n^ϵ, then Theorem 3.3.17 implies

$$\psi_n^\epsilon(t) \to \psi^\epsilon(t) = \int_\epsilon^\infty e^{itx}\theta\epsilon^\alpha\alpha x^{-(\alpha+1)}dx + \int_{-\infty}^{-\epsilon} e^{itx}(1-\theta)\epsilon^\alpha\alpha|x|^{-(\alpha+1)}\,dx$$

as $n \to \infty$. So repeating the proof of (3.8.5) gives

$$\begin{aligned} E\exp(it\hat{S}_n(\epsilon)/a_n) &\to \exp(-\epsilon^{-\alpha}\{1 - \psi^\epsilon(t)\}) \\ &= \exp\left(\int_\epsilon^\infty (e^{itx} - 1)\theta\alpha x^{-(\alpha+1)}\,dx \right. \\ &\quad \left. + \int_{-\infty}^{-\epsilon} (e^{itx} - 1)(1-\theta)\alpha|x|^{-(\alpha+1)}dx\right) \end{aligned}$$

where we have used $\epsilon^{-\alpha} = \int_\epsilon^\infty \alpha x^{-(\alpha+1)}\,dx$. To bring in

$$\hat{\mu}(\epsilon) = EX_m 1_{(\epsilon a_n < |X_m| \le a_n)}$$

we observe that (3.8.7) implies $nP(xa_n < X_m \le ya_n) \to \theta(x^{-\alpha} - y^{-\alpha})$. So

$$n\hat{\mu}(\epsilon)/a_n \to \int_\epsilon^1 x\theta\alpha x^{-(\alpha+1)}\,dx + \int_{-1}^{-\epsilon} x(1-\theta)\alpha|x|^{-(\alpha+1)}\,dx$$

From this it follows that $E\exp(it\{\hat{S}_n(\epsilon) - n\hat{\mu}(\epsilon)\}/a_n) \to$

$$\begin{aligned}\exp\Bigg(&\int_1^\infty (e^{itx}-1)\theta\alpha x^{-(\alpha+1)}\,dx \\ &+ \int_\epsilon^1 (e^{itx}-1-itx)\theta\alpha x^{-(\alpha+1)}\,dx \\ &+ \int_{-1}^{-\epsilon} (e^{itx}-1-itx)(1-\theta)\alpha|x|^{-(\alpha+1)}\,dx \\ &+ \int_{-\infty}^{-1} (e^{itx}-1)(1-\theta)\alpha|x|^{-(\alpha+1)}\,dx\Bigg)\end{aligned} \tag{3.8.9}$$

The last expression is messy, but $e^{itx} - 1 - itx \sim -t^2x^2/2$ as $t \to 0$, so we need to subtract the itx to make

$$\int_0^1 (e^{itx} - 1 - itx)x^{-(\alpha+1)}dx \quad \text{converge when } \alpha \ge 1$$

To reduce the number of integrals from four to two, we can write the limit as $\epsilon \to 0$ of the right-hand side of (3.8.9) as

$$\begin{aligned}\exp\Bigg(itc &+ \int_0^\infty \left(e^{itx} - 1 - \frac{itx}{1+x^2}\right)\theta\alpha x^{-(\alpha+1)}\,dx \\ &+ \int_{-\infty}^0 \left(e^{itx} - 1 - \frac{itx}{1+x^2}\right)(1-\theta)\alpha|x|^{-(\alpha+1)}\,dx\Bigg)\end{aligned} \tag{3.8.10}$$

where c is a constant. Combining (3.8.6) and (3.8.9) using Lemma 3.8.1, it follows easily that $(S_n - b_n)/a_n \Rightarrow Y$, where Ee^{itY} is given in (3.8.10). □

By doing some calculus (see Breiman (1968), pp. 204–206) one can rewrite (3.8.10) as

$$\exp(itc - b|t|^\alpha\{1 + i\kappa\ \text{sgn}(t)w_\alpha(t)\}) \tag{3.8.11}$$

where $-1 \le \kappa \le 1$, ($\kappa = 2\theta - 1$) and

$$w_\alpha(t) = \begin{cases} \tan(\pi\alpha/2) & \text{if } \alpha \ne 1 \\ (2/\pi)\log|t| & \text{if } \alpha = 1 \end{cases}$$

The reader should note that while we have assumed $0 < \alpha < 2$ throughout these developments, if we set $\alpha = 2$, then the term with κ vanishes and (3.8.11) reduces to the characteristic function of the normal distribution with mean c and variance $2b$.

The distributions whose characteristic functions are given in (3.8.11) are called **stable laws**. α is commonly called the **index**. When $\alpha = 1$ and $\kappa = 0$, we have the Cauchy distribution. Apart from the Cauchy and the normal, there is only one other case in which the density is known: When $\alpha = 1/2$, $\kappa = 1$, $c = 0$, and $b = 1$, the density is

$$(2\pi y^3)^{-1/2} \exp(-1/2y) \quad \text{for } y \geq 0 \tag{3.8.12}$$

One can calculate the ch.f. and verify our claim. However, later (see Section 7.4) we will be able to check the claim without effort, so we leave the somewhat tedious calculation to the reader.

We are now finally ready to treat some examples

Example 3.8.4 Let $X_1, X_2, \ldots$ be i.i.d. with a density that is symmetric about 0, and continuous and positive at 0. We claim that

$$\frac{1}{n}\left(\frac{1}{X_1} + \cdots + \frac{1}{X_n}\right) \Rightarrow \text{a Cauchy distribution } (\alpha = 1, \kappa = 0)$$

To verify this, note that

$$P(1/X_i > x) = P(0 < X_i < x^{-1}) = \int_0^{x^{-1}} f(y)\,dy \sim f(0)/x$$

as $x \to \infty$. A similar calculation shows $P(1/X_i < -x) \sim f(0)/x$, so in (i) in Theorem 3.8.2 holds with $\theta = 1/2$, and (ii) holds with $\alpha = 1$. The scaling constant $a_n \sim 2f(0)n$, while the centering constant vanishes, since we have supposed the distribution of X is symmetric about 0.

Remark Readers who want a challenge should try to drop the symmetry assumption, assuming for simplicity that f is differentiable at 0.

Example 3.8.5 Let $X_1, X_2, \ldots$ be i.i.d. with $P(X_i = 1) = P(X_i = -1) = 1/2$, let $S_n = X_1 + \cdots + X_n$, and let $\tau = \inf\{n \geq 1 : S_n = 0\}$. In Chapter 4 (see the discussion after (4.9.2)) we will show

$$P(\tau > 2n) \sim \pi^{-1/2} n^{-1/2} \quad \text{as } n \to \infty$$

Let $\tau_1, \tau_2, \ldots$ be independent with the same distribution as τ, and let $T_n = \tau_1 + \cdots + \tau_n$. Results in Section 4.1 imply that T_n has the same distribution as the nth time S_m hits 0. We claim that T_n/n^2 converges to the stable law with $\alpha = 1/2$, $\kappa = 1$ and note that this is the key to the derivation of (3.8.12). To prove the claim, note that in (i) in Theorem 3.8.2 holds with $\theta = 1$ and (ii) holds with $\alpha = 1/2$. The scaling constant $a_n \sim Cn^2$. Since $\alpha < 1$, Exercise 3.8.1 implies the centering constant is unnecessary.

Example 3.8.6 Assume n objects $X_{n,1}, \ldots, X_{n,n}$ are placed independently and at random in $[-n, n]$. Let

$$F_n = \sum_{m=1}^{n} \operatorname{sgn}(X_{n,m})/|X_{n,m}|^p \tag{3.8.13}$$

be the net force exerted on 0. We will now show that if $p > 1/2$, then

$$\lim_{n\to\infty} E\exp(itF_n) = \exp(-c|t|^{1/p})$$

To do this, it is convenient to let $X_{n,m} = nY_m$, where the Y_i are i.i.d. on $[-1, 1]$. Then

$$F_n = n^{-p} \sum_{m=1}^{n} \text{sgn}(Y_m)/|Y_m|^p$$

Letting $Z_m = \text{sgn}(Y_m)/|Y_m|^p$, Z_m is symmetric about 0 with $P(|Z_m| > x) = P(|Y_m| < x^{-1/p})$, so in (i) in Theorem 3.8.2 holds with $\theta = 1/2$ and (ii) holds with $\alpha = 1/p$. The scaling constant $a_n \sim Cn^p$ and the centering constant is 0 by symmetry.

Example 3.8.7 In these examples, we have had $b_n = 0$. To get a feel for the centering constants consider $X_1, X_2, \ldots$ i.i.d. with

$$P(X_i > x) = \theta x^{-\alpha} \qquad P(X_i < -x) = (1-\theta)x^{-\alpha}$$

where $0 < \alpha < 2$. In this case $a_n = n^{1/\alpha}$ and

$$b_n = n \int_1^{n^{1/\alpha}} (2\theta - 1)\alpha x^{-\alpha}\, dx \sim \begin{cases} cn & \alpha > 1 \\ cn \log n & \alpha = 1 \\ cn^{1/\alpha} & \alpha < 1 \end{cases}$$

When $\alpha < 1$, the centering is the same size as the scaling and can be ignored. When $\alpha > 1$, $b_n \sim n\mu$ where $\mu = EX_i$.

Our next result explains the name **stable laws**. A random variable Y is said to have a **stable law** if for every integer $k > 0$ there are constants a_k and b_k so that if $Y_1, \ldots, Y_k$ are i.i.d. and have the same distribution as Y, then $(Y_1 + \cdots + Y_k - b_k)/a_k =_d Y$. The last definition makes half of the next result obvious.

Theorem 3.8.8 *Y is the limit of $(X_1 + \cdots + X_k - b_k)/a_k$ for some i.i.d. sequence X_i if and only if Y has a stable law.*

Proof If Y has a stable law, we can take $X_1, X_2, \ldots$ i.i.d. with distribution Y. To go the other way, let

$$Z_n = (X_1 + \cdots + X_n - b_n)/a_n$$

and $S_n^j = X_{(j-1)n+1} + \cdots + X_{jn}$. A little arithmetic shows

$$\begin{aligned} Z_{nk} &= (S_n^1 + \cdots + S_n^k - b_{nk})/a_{nk} \\ a_{nk}Z_{nk} &= (S_n^1 - b_n) + \cdots + (S_n^k - b_n) + (kb_n - b_{nk}) \\ a_{nk}Z_{nk}/a_n &= (S_n^1 - b_n)/a_n + \cdots + (S_n^k - b_n)/a_n + (kb_n - b_{nk})/a_n \end{aligned}$$

The first k terms on the right-hand side $\Rightarrow Y_1 + \cdots + Y_k$ as $n \to \infty$, where $Y_1, \ldots, Y_k$ are independent and have the same distribution as Y, and $Z_{nk} \Rightarrow Y$. Taking $W_n = Z_{nk}$ and

$$W_n' = \frac{a_{kn}}{a_n} Z_{nk} - \frac{kb_n - b_{nk}}{a_n}$$

gives the desired result. □

Theorem 3.8.9 (Convergence of types theorem) *If $W_n \Rightarrow W$ and there are constants $\alpha_n > 0$, β_n so that $W_n' = \alpha_n W_n + \beta_n \Rightarrow W'$, where W and W' are nondegenerate, then there are constants α and β so that $\alpha_n \to \alpha$ and $\beta_n \to \beta$.*

Proof Let $\varphi_n(t) = E\exp(itW_n)$.

$$\psi_n(t) = E\exp(it(\alpha_n W_n + \beta_n)) = \exp(it\beta_n)\varphi_n(\alpha_n t)$$

If φ and ψ are the characteristic functions of W and W', then

(a) $$\varphi_n(t) \to \varphi(t) \qquad \psi_n(t) = \exp(it\beta_n)\varphi_n(\alpha_n t) \to \psi(t)$$

Take a subsequence $\alpha_{n(m)}$ that converges to a limit $\alpha \in [0,\infty]$. Our first step is to observe $\alpha = 0$ is impossible. If this happens, then using the uniform convergence proved in Exercise 3.3.13

(b) $$|\psi_n(t)| = |\varphi_n(\alpha_n t)| \to 1$$

$|\psi(t)| \equiv 1$, and the limit is degenerate by Theorem 3.5.2. Letting $t = u/\alpha_n$ and interchanging the roles of φ and ψ shows $\alpha = \infty$ is impossible. If α is a subsequential limit, then arguing as in (b) gives $|\psi(t)| = |\varphi(\alpha t)|$. If there are two subsequential limits $\alpha' < \alpha$, using the last equation for both limits implies $|\varphi(u)| = |\varphi(u\alpha'/\alpha)|$. Iterating gives $|\varphi(u)| = |\varphi(u(\alpha'/\alpha)^k)| \to 1$ as $k \to \infty$, contradicting our assumption that W' is nondegenerate, so $\alpha_n \to \alpha \in [0,\infty)$.

To conclude that $\beta_n \to \beta$ now, we observe that (ii) of Exercise 3.3.13 implies $\varphi_n \to \varphi$ uniformly on compact sets, so $\varphi_n(\alpha_n t) \to \varphi(\alpha t)$. If δ is small enough so that $|\varphi(\alpha t)| > 0$ for $|t| \le \delta$, it follows from (a) and another use of Exercise 3.3.13 that

$$\exp(it\beta_n) = \frac{\psi_n(t)}{\varphi_n(\alpha t)} \to \frac{\psi(t)}{\varphi(\alpha t)}$$

uniformly on $[-\delta,\delta]$. $\exp(it\beta_n)$ is the ch.f. of a point mass at β_n. Using (3.3.1) now as in the proof of Theorem 3.3.17, it follows that the sequence of distributions that are point masses at β_n is tight, i.e., β_n is bounded. If $\beta_{n_m} \to \beta$, then $\exp(it\beta) = \psi(t)/\varphi(\alpha t)$ for $|t| \le \delta$, so there can only be one subsequential limit. □

Theorem 3.8.8 justifies calling the distributions with characteristic functions given by (3.8.11) or (3.8.10) stable laws. To complete the story, we should mention that these are the only stable laws. Again, see chapter 9 of Breiman (1968) or Gnedenko and Kolmogorov (1954). The next example shows that it is sometimes useful to know what all the possible limits are.

Example 3.8.10 (The Holtsmark distribution) ($\alpha = 3/2$, $\kappa = 0$). Suppose stars are distributed in space according to a Poisson process with density t and their masses are i.i.d. Let X_t be the x-component of the gravitational force at 0 when the density is t. A change of density $1 \to t$ corresponds to a change of length $1 \to t^{-1/3}$, and gravitational attraction follows an inverse square law, so

$$X_t \stackrel{d}{=} t^{3/2} X_1 \tag{3.8.14}$$

If we imagine thinning the Poisson process by rolling an n-sided die, then Theorem 3.7.4 implies

$$X_t \stackrel{d}{=} X^1_{t/n} + \cdots + X^n_{t/n}$$

where the random variables on the right-hand side are independent and have the same distribution as $X_{t/n}$. It follows from Theorem 3.8.8 that X_t has a stable law. The scaling property (3.8.14) implies $\alpha = 3/2$. Since $X_t =_d -X_t$, $\kappa = 0$.

Exercises

3.8.1 Show that when $\alpha < 1$, centering in Theorem 3.8.2 is unnecessary, i.e., we can let $b_n = 0$.

3.8.2 Consider F_n defined in (3.8.13). Show that (i) If $p < 1/2$, then $F_n/n^{1/2-p} \Rightarrow c\chi$, where χ is a standard normal. (ii) If $p = 1/2$, then $F_n/(\log n)^{1/2} \Rightarrow c\chi$.

3.8.3 Let Y be a stable law with $\kappa = 1$. Use the limit theorem, Theorem 3.8.2, to conclude that $Y \geq 0$ if $\alpha < 1$.

3.8.4 Let X be symmetric stable with index α. (i) Use (3.3.1) to show that $E|X|^p < \infty$ for $p < \alpha$. (ii) Use the second proof of (3.8.3) to show that $P(|X| \geq x) \geq Cx^{-\alpha}$ so $E|X|^\alpha = \infty$.

3.8.5 Let $Y, Y_1, Y_2, \ldots$ be independent and have a stable law with index α. Theorem 3.8.8 implies there are constants α_k and β_k so that $Y_1 + \cdots + Y_k$ and $\alpha_k Y + \beta_k$ have the same distribution. Use the proof of Theorem 3.8.8, Theorem 3.8.2, and Exercise 3.8.1 to conclude that (i) $\alpha_k = k^{1/\alpha}$, (ii) if $\alpha < 1$, then $\beta_k = 0$.

3.8.6 Let Y be a stable law with index $\alpha < 1$ and $\kappa = 1$. Exercise 3.8.3 implies that $Y \geq 0$, so we can define its Laplace transform $\psi(\lambda) = E\exp(-\lambda Y)$. The previous exercise implies that for any integer $n \geq 1$ we have $\psi(\lambda)^n = \psi(n^{1/\alpha}\lambda)$. Use this to conclude $E\exp(-\lambda Y) = \exp(-c\lambda^\alpha)$.

3.8.7 (i) Show that if X is symmetric stable with index α and $Y \geq 0$ is an independent stable with index $\beta < 1$, then $XY^{1/\alpha}$ is symmetric stable with index $\alpha\beta$. (ii) Let W_1 and W_2 be independent standard normals. Check that $1/W_2^2$ has the density given in (3.8.12) and use this to conclude that W_1/W_2 has a Cauchy distribution.

3.9 Infinitely Divisible Distributions*

In the last section we identified the distributions that can appear as the limit of normalized sums of i.i.d.r.v.'s. In this section we will describe those that are limits of sums

$$(*) \qquad S_n = X_{n,1} + \cdots + X_{n,n}$$

where the $X_{n,m}$ are i.i.d. Note the verb "describe." We will prove almost nothing in this section, just state some of the most important facts to bring the reader up to cocktail party literacy.

A sufficient condition for Z to be a limit of sums of the form $(*)$ is that Z has an **infinitely divisible distribution**, i.e., for each n there is an i.i.d. sequence $Y_{n,1}, \ldots, Y_{n,n}$ so that

$$Z \stackrel{d}{=} Y_{n,1} + \cdots + Y_{n,n}$$

Our first result shows that this condition is also necessary:

Theorem 3.9.1 *Z is a limit of sums of type $(*)$ if and only if Z has an infinitely divisible distribution.*

Proof As remarked previously, we only have to prove necessity. Write

$$S_{2n} = (X_{2n,1} + \cdots + X_{2n,n}) + (X_{2n,n+1} + \cdots + X_{2n,2n}) \equiv Y_n + Y_n'$$

The random variables Y_n and Y_n' are independent and have the same distribution. If $S_n \Rightarrow Z$, then the distributions of Y_n are a tight sequence, since

$$P(Y_n > y)^2 = P(Y_n > y)P(Y_n' > y) \le P(S_{2n} > 2y)$$

and similarly, $P(Y_n < -y)^2 \le P(S_{2n} < -2y)$. If we take a subsequence n_k so that $Y_{n_k} \Rightarrow Y$ (and hence $Y_{n_k}' \Rightarrow Y'$), then $Z =_d Y + Y'$. A similar argument shows that Z can be divided into $n > 2$ pieces and the proof is complete. □

With Theorem 3.9.1 established, we turn now to examples. In the first three cases, the distribution is infinitely divisible because it is a limit of sums of the form $(*)$. The number gives the relevant limit theorem.

Example 3.9.2 Normal distribution. Theorem 3.4.1.

Example 3.9.3 Stable Laws. Theorem 3.8.2.

Example 3.9.4 Poisson distribution. Theorem 3.6.1.

Example 3.9.5 Compound Poisson distribution. Let $\xi_1, \xi_2, \ldots$ be i.i.d. and $N(\lambda)$ be an independent Poisson r.v. with mean λ. Then $Z = \xi_1 + \cdots + \xi_{N(\lambda)}$ has an infinitely divisible distribution. (Let $X_{n,j} =_d \xi_1 + \cdots + \xi_{N(\lambda/n)}$.) For the following developments, we would like to observe that if $\varphi(t) = E\exp(it\xi_i)$, then

$$E\exp(itZ) = \sum_{n=0}^{\infty} e^{-\lambda}\frac{\lambda^n}{n!}\varphi(t)^n = \exp(-\lambda(1-\varphi(t))) \tag{3.9.1}$$

Example 3.9.5 is a son of 3.9.4 but a father of 3.9.2 and 3.9.3. To explain this remark, we observe that if $\xi = \epsilon$ and $-\epsilon$ with probability 1/2 each, then $\varphi(t) = (e^{i\epsilon t} + e^{-i\epsilon t})/2 = \cos(\epsilon t)$. So if $\lambda = \epsilon^{-2}$, then (3.9.1) implies

$$E\exp(itZ) = \exp(-\epsilon^{-2}(1-\cos(\epsilon t))) \to \exp(-t^2/2)$$

as $\epsilon \to 0$. In words, the normal distribution is a limit of compound Poisson distributions. To see that stable laws are also a special case (using the notation from the proof of Theorem 3.8.2), let

$$I_n(\epsilon) = \{m \le n : |X_m| > \epsilon a_n\}$$
$$\hat{S}_n(\epsilon) = \sum_{m \in I_n(\epsilon)} X_m$$
$$\bar{S}_n(\epsilon) = S_n - \hat{S}_n(\epsilon)$$

If $\epsilon_n \to 0$, then $\bar{S}_n(\epsilon_n)/a_n \Rightarrow 0$. If ϵ is fixed, then as $n \to \infty$ we have $|I_n(\epsilon)| \Rightarrow \text{Poisson}(\epsilon^{-\alpha})$ and $\hat{S}_n(\epsilon)/a_n \Rightarrow$ a compound Poisson distribution:

$$E\exp(it\hat{S}_n(\epsilon)/a_n) \to \exp(-\epsilon^{-\alpha}\{1 - \psi^{\epsilon}(t)\})$$

Combining the last two observations and using the proof of Theorem 3.8.2 shows that stable laws are limits of compound Poisson distributions. The formula (3.8.10) for the limiting ch.f.

$$\exp\left(itc + \int_0^\infty \left(e^{itx} - 1 - \frac{itx}{1+x^2}\right)\theta\alpha x^{-(\alpha+1)}\,dx + \int_{-\infty}^0 \left(e^{itx} - 1 - \frac{itx}{1+x^2}\right)(1-\theta)\alpha|x|^{-(\alpha+1)}\,dx\right) \tag{3.9.2}$$

helps explain:

Theorem 3.9.6 (Lévy-Khinchin Theorem) *Z has an infinitely divisible distribution if and only if its characteristic function has*

$$\log\varphi(t) = ict - \frac{\sigma^2 t^2}{2} + \int \left(e^{itx} - 1 - \frac{itx}{1+x^2}\right)\mu(dx)$$

where μ is a measure with $\mu(\{0\}) = 0$ and $\int \frac{x^2}{1+x^2}\mu(dx) < \infty$.

For a proof, see Breiman (1968), section 9.5., or Feller II (1971), section XVII.2. μ is called the **Lévy measure** of the distribution. Comparing with (3.9.2) and recalling the proof of Theorem 3.8.2 suggests the following interpretation of μ: If $\sigma^2 = 0$, then Z can be built up by making a Poisson process on $\mathbf{R}$ with mean measure μ and then summing up the points. As in the case of stable laws, we have to sum the points in $[-\epsilon, \epsilon]^c$, subtract an appropriate constant, and let $\epsilon \to 0$.

The theory of infinitely divisible distributions is simpler in the case of finite variance. In this case, we have:

Theorem 3.9.7 (Kolmogorov's Theorem) *Z has an infinitely divisible distribution with mean 0 and finite variance if and only if its ch.f. has*

$$\log\varphi(t) = \int (e^{itx} - 1 - itx)x^{-2}\,\nu(dx)$$

Here the integrand is $-t^2/2$ at 0, ν is called the **canonical measure**, *and* $var(Z) = \nu(\mathbf{R})$.

To explain the formula, note that if Z_λ has a Poisson distribution with mean λ

$$E\exp(itx(Z_\lambda - \lambda)) = \exp(\lambda(e^{itx} - 1 - itx))$$

so the measure for $Z = x(Z_\lambda - \lambda)$ has $\nu(\{x\}) = \lambda x^2$.

Exercises

3.9.1 Show that the gamma distribution is infinitely divisible.

3.9.2 Show that the distribution of a bounded r.v. Z is infinitely divisible if and only if Z is constant. Hint: Show $var(Z) = 0$.

3.9.3 Show that if μ is infinitely divisible, its ch.f. φ never vanishes. Hint: Look at $\psi = |\varphi|^2$, which is also infinitely divisible, to avoid taking nth roots of complex numbers then use Exercise 3.3.17.

3.9.4 What is the Lévy measure for the limit $\aleph$ in part (iii) of Exercise 3.4.13?

3.10 Limit Theorems in $\mathbf{R}^d$

We begin by considering weak convergence in a general metric space (S, ρ). In this setting we say $X_n \Rightarrow X_\infty$ if and only if $Ef(X_n) \to Ef(X_\infty)$ for all bounded continuous f. As in Section 3.2, it will be useful to have several equivalent definitions of weak convergence. f is said to be **Lipschitz continuous** if there is a constant C so that $|f(x) - f(y)| \le C\rho(x, y)$.

Theorem 3.10.1 *The following statements are equivalent:*

(i) $Ef(X_n) \to Ef(X_\infty)$ *for all bounded continuous* f.

(ii) $Ef(X_n) \to Ef(X_\infty)$ *for all bounded Lipschitz continuous* f.

(iii) For all closed sets K, $\limsup_{n\to\infty} P(X_n \in K) \le P(X_\infty \in K)$.

(iv) For all open sets G, $\liminf_{n\to\infty} P(X_n \in G) \ge P(X_\infty \in G)$.

(v) For all sets A *with* $P(X_\infty \in \partial A) = 0$, $\lim_{n\to\infty} P(X_n \in A) = P(X_\infty \in A)$.

(vi) Let D_f *= the set of discontinuities of* f. *For all bounded functions* f *with* $P(X_\infty \in D_f) = 0$, *we have* $Ef(X_n) \to Ef(X_\infty)$.

Proof *(i) implies (ii):* Trivial.

(ii) implies (iii): Let $\rho(x, K) = \inf\{\rho(x, y) : y \in K\}$, $\varphi_j(r) = (1 - jr)^+$, and $f_j(x) = \varphi_j(\rho(x, K))$. f_j is Lipschitz continuous, has values in [0,1], and $\downarrow 1_K(x)$ as $j \uparrow \infty$. So

$$\limsup_{n\to\infty} P(X_n \in K) \le \lim_{n\to\infty} Ef_j(X_n) = Ef_j(X_\infty) \downarrow P(X_\infty \in K) \text{ as } j \uparrow \infty$$

(iii) is equivalent to (iv): As in the proof of Theorem 3.2.11, this follows easily from two facts: A is open if and only if A^c is closed; $P(A) + P(A^c) = 1$.

(iii) and (iv) imply (v): Let $K = \bar{A}$, $G = A^o$, and reason as in the proof of Theorem 3.2.11.

(v) implies (vi): Suppose $|f(x)| \le K$ and pick $\alpha_0 < \alpha_1 < \ldots < \alpha_\ell$ so that $P(f(X_\infty) = \alpha_i) = 0$ for $0 \le i \le \ell$, $\alpha_0 < -K < K < \alpha_\ell$, and $\alpha_i - \alpha_{i-1} < \epsilon$. This is always possible, since $\{\alpha : P(f(X_\infty) = \alpha) > 0\}$ is a countable set. Let $A_i = \{x : \alpha_{i-1} < f(x) \le \alpha_i\}$. $\partial A_i \subset \{x : f(x) \in \{\alpha_{i-1}, \alpha_i\}\} \cup D_f$, so $P(X_\infty \in \partial A_i) = 0$, and it follows from (v) that

$$\sum_{i=1}^{\ell} \alpha_i P(X_n \in A_i) \to \sum_{i=1}^{\ell} \alpha_i P(X_\infty \in A_i)$$

The definition of the α_i implies

$$0 \le \sum_{i=1}^{\ell} \alpha_i P(X_n \in A_i) - Ef(X_n) \le \epsilon \quad \text{for } 1 \le n \le \infty$$

Since ϵ is arbitrary, it follows that $Ef(X_n) \to Ef(X_\infty)$.

(vi) implies (i): Trivial. □

Specializing now to $\mathbf{R}^d$, let $X = (X_1, \dots, X_d)$ be a random vector. We define its **distribution function** by $F(x) = P(X \le x)$. Here $x \in \mathbf{R}^d$, and $X \le x$ means $X_i \le x_i$ for $i = 1, \dots, d$. As in one dimension, F has three obvious properties:

(i) It is nondecreasing, i.e., if $x \le y$, then $F(x) \le F(y)$.

(ii) $\lim_{x\to\infty} F(x) = 1, \quad \lim_{x_i\to-\infty} F(x) = 0$.

(iii) F is right continuous, i.e., $\lim_{y\downarrow x} F(y) = F(x)$.

Here $x \to \infty$ means each coordinate x_i goes to ∞, $x_i \to -\infty$ means we let $x_i \to -\infty$ keeping the other coordinates fixed, and $y \downarrow x$ means each coordinate $y_i \downarrow x_i$.

As discussed in Section 1.1, an additional condition is needed to guarantee that F is the distribution function of a probability measure, let

$$\begin{aligned} A &= (a_1, b_1] \times \cdots \times (a_d, b_d] \\ V &= \{a_1, b_1\} \times \cdots \times \{a_d, b_d\} \end{aligned}$$

$V =$ the vertices of the rectangle A. If $v \in V$, let

$$\text{sgn}\,(v) = (-1)^{\#\text{ of } a\text{'s in } v}$$

The inclusion-exclusion formula implies

$$P(X \in A) = \sum_{v \in V} \text{sgn}\,(v) F(v)$$

So if we use $\Delta_A F$ to denote the right-hand side, we need

(iv) $\Delta_A F \ge 0$ for all rectangles A.

The last condition guarantees that the measure assigned to each rectangle is ≥ 0. At this point we have defined the measure on the semialgebra $\mathcal{S}_d$ defined in Example 1.1.5. Theorem 1.1.11 now implies that there is a unique probability measure with distribution F.

If F_n and F are distribution functions on $\mathbf{R}^d$, we say that F_n **converges weakly** to F, and write $F_n \Rightarrow F$, if $F_n(x) \to F(x)$ at all continuity points of F. Our first task is to show that there are enough continuity points for this to be a sensible definition. For a concrete example, consider

$$F(x,y) = \begin{cases} 1 & \text{if } x \ge 0,\ y \ge 1 \\ y & \text{if } x \ge 0,\ 0 \le y < 1 \\ 0 & \text{otherwise} \end{cases}$$

F is the distribution function of $(0, Y)$, where Y is uniform on (0,1). Notice that this distribution has no atoms, but F is discontinuous at $(0, y)$ when $y > 0$.

Keeping the last example in mind, observe that if $x_n < x$, i.e., $x_{n,i} < x_i$ for all coordinates i, and $x_n \uparrow x$ as $n \to \infty$, then

$$F(x) - F(x_n) = P(X \le x) - P(X \le x_n) \downarrow P(X \le x) - P(X < x)$$

In $d = 2$, the last expression is the probability X lies in

$$\{(a, x_2) : a \le x_1\} \cup \{(x_1, b) : b \le x_2\}$$

Let $H_c^i = \{x : x_i = c\}$ be the hyperplane where the ith coordinate is c. For each i, the H_c^i are disjoint so $D^i = \{c : P(X \in H_c^i) > 0\}$ is at most countable. It is easy to see that if x has $x_i \notin D^i$ for all i, then F is continuous at x. This gives us more than enough points to reconstruct F.

Theorem 3.10.2 *On* $\mathbf{R}^d$ *weak convergence defined in terms of convergence of distribution* $F_n \Rightarrow F$ *is equivalent to notion of weak convergence defined for a general metric spce.*

Proof *(v) implies* $F_n \Rightarrow$: If F is continuous at x, then $A = (-\infty, x_1] \times \ldots \times (-\infty, x_d]$ has $\mu(\partial A) = 0$, so $F_n(x) = P(X_n \in A) \to P(X_\infty \in A) = F(x)$.

$F_N \Rightarrow F$ *implies (iv):* Let $D^i = \{c : P(X_\infty \in H_c^i) > 0\}$, where $H_c^i = \{x : x^i = c\}$. We say a rectangle $A = (a_1, b_1] \times \ldots \times (a_d, b_d]$ is good if a_i, $b_i \notin D^i$ for all i. ($\Rightarrow$) implies that for all good rectangles $P(X_n \in A) \to P(X_\infty \in A)$. This is also true for B that are a finite disjoint union of good rectangles. Now any open set G is an increasing limit of B_k's that are a finite disjoint union of good rectangles, so

$$\liminf_{n\to\infty} P(X_n \in G) \geq \liminf_{n\to\infty} P(X_n \in B_k) = P(X_\infty \in B_k) \uparrow P(X_\infty \in G)$$

as $k \to \infty$. The proof is complete. □

Remark In Section 3.2, we proved that (i)–(v) are consequences of weak convergence by constructing r.v's with the given distributions so that $X_n \to X_\infty$ a.s. This can be done in $\mathbf{R}^d$ (or any complete separable metric space), but the construction is rather messy. See Billingsley (1979), pp. 337–340 for a proof in $\mathbf{R}^d$.

A sequence of probability measures μ_n is said to be **tight** if for any $\epsilon > 0$, there is an M so that $\liminf_{n\to\infty} \mu_n([-M, M]^d) \geq 1 - \epsilon$.

Theorem 3.10.3 *If* μ_n *is tight, then there is a weakly convergent subsequence.*

Proof Let F_n be the associated distribution functions, and let $q_1, q_2, \ldots$ be an enumeration of $\mathbf{Q}^d$ = the points in $\mathbf{R}^d$ with rational coordinates. By a diagonal argument like the one in the proof of Theorem 3.2.12, we can pick a subsequence so that $F_{n(k)}(q) \to G(q)$ for all $q \in \mathbf{Q}^d$. Let

$$F(x) = \inf\{G(q) : q \in \mathbf{Q}^d, q > x\}$$

where $q > x$ means $q_i > x_i$ for all i. It is easy to see that F is right continuous. To check that it is a distribution function, we observe that if A is a rectangle with vertices in $\mathbf{Q}^d$, then $\Delta_A F_n \geq 0$ for all n, so $\Delta_A G \geq 0$, and taking limits we see that the last conclusion holds for F for all rectangles A. Tightness implies that F has properties (i) and (ii) of a distribution F. We leave it to the reader to check that $F_n \Rightarrow F$. The proof of Theorem 3.2.12 works if you read inequalities such as $r_1 < r_2 < x < s$ as the corresponding relations between vectors. □

The **characteristic function** of $(X_1, \ldots, X_d)$ is $\varphi(t) = E\exp(it \cdot X)$, where $t \cdot X = t_1 X_1 + \cdots + t_d X_d$ is the usual dot product of two vectors.

Theorem 3.10.4 (Inversion formula) *If $A = [a_1, b_1] \times \ldots \times [a_d, b_d]$ with $\mu(\partial A) = 0$, then*

$$\mu(A) = \lim_{T \to \infty} (2\pi)^{-d} \int_{[-T,T]^d} \prod_{j=1}^{d} \psi_j(t_j)\varphi(t)\, dt$$

where $\psi_j(s) = (\exp(-isa_j) - \exp(-isb_j))/is$.

Proof Fubini's theorem implies

$$\int_{[-T,T]^d} \int \prod_{j=1}^{d} \psi_j(t_j) \exp(it_j x_j)\, \mu(dx)\, dt$$
$$= \int \prod_{j=1}^{d} \int_{-T}^{T} \psi_j(t_j) \exp(it_j x_j)\, dt_j\, \mu(dx)$$

It follows from the proof of Theorem 3.3.11 that

$$\int_{-T}^{T} \psi_j(t_j) \exp(it_j x_j)\, dt_j \to \pi \left(1_{(a_j,b_j)}(x) + 1_{[a_j,b_j]}(x)\right)$$

so the desired conclusion follows from the bounded convergence theorem. □

Theorem 3.10.5 (Convergence theorem) *Let X_n, $1 \le n \le \infty$ be random vectors with ch.f. φ_n. A necessary and sufficient condition for $X_n \Rightarrow X_\infty$ is that $\varphi_n(t) \to \varphi_\infty(t)$.*

Proof $\exp(it \cdot x)$ is bounded and continuous, so if $X_n \Rightarrow X_\infty$, then $\varphi_n(t) \to \varphi_\infty(t)$. To prove the other direction it suffices, as in the proof of Theorem 3.3.17, to prove that the sequence is tight. To do this, we observe that if we fix $\theta \in \mathbf{R}^d$, then for all $s \in \mathbf{R}$, $\varphi_n(s\theta) \to \varphi_\infty(s\theta)$, so it follows from Theorem 3.3.17 that the distributions of $\theta \cdot X_n$ are tight. Applying the last observation to the d unit vectors $e_1, \ldots, e_d$ shows that the distributions of X_n are tight and completes the proof. □

Remark As before, if $\varphi_n(t) \to \varphi_\infty(t)$ with $\varphi_\infty(t)$ continuous at 0, then $\varphi_\infty(t)$ is the ch.f. of some X_∞ and $X_n \Rightarrow X_\infty$.

Theorem 3.10.5 has an important corollary.

Theorem 3.10.6 (Cramér-Wold device) *A sufficient condition for $X_n \Rightarrow X_\infty$ is that $\theta \cdot X_n \Rightarrow \theta \cdot X_\infty$ for all $\theta \in \mathbf{R}^d$.*

Proof The indicated condition implies $E \exp(i\theta \cdot X_n) \to E \exp(i\theta \cdot X_\infty)$ for all $\theta \in \mathbf{R}^d$. □

Theorem 3.10.6 leads immediately to:

Theorem 3.10.7 (The central limit theorem in $\mathbf{R}^d$) *Let $X_1, X_2, \ldots$ be i.i.d. random vectors with $EX_n = \mu$, and finite covariances*

$$\Gamma_{ij} = E((X_{n,i} - \mu_i)(X_{n,j} - \mu_j))$$

If $S_n = X_1 + \cdots + X_n$*, then* $(S_n - n\mu)/n^{1/2} \Rightarrow \chi$*, where* χ *has a multivariate normal distribution with mean 0 and covariance* Γ*, i.e.,*

$$E \exp(i\theta \cdot \chi) = \exp\left(-\sum_i \sum_j \theta_i \theta_j \Gamma_{ij}/2 \right)$$

Proof By considering $X'_n = X_n - \mu$, we can suppose without loss of generality that $\mu = 0$. Let $\theta \in \mathbf{R}^d$. $\theta \cdot X_n$ is a random variable with mean 0 and variance

$$E\Big(\sum_i \theta_i X_{n,i}\Big)^2 = \sum_i \sum_j E\left(\theta_i \theta_j X_{n,i} X_{n,j}\right) = \sum_i \sum_j \theta_i \theta_j \Gamma_{ij}$$

so it follows from the one-dimensional central limit theorem and Theorem 3.10.6 that $S_n/n^{1/2} \Rightarrow \chi$, where

$$E \exp(i\theta \cdot \chi) = \exp\left(-\sum_i \sum_j \theta_i \theta_j \Gamma_{ij}/2 \right) \qquad \square$$

which proves the desired result. $\square$

To illustrate the use of Theorem 3.10.7, we consider two examples. In each $e_1, \ldots, e_d$ are the d unit vectors.

Example 3.10.8 (Simple random walk on $\mathbf{Z}^d$) Let $X_1, X_2, \ldots$ be i.i.d. with

$$P(X_n = +e_i) = P(X_n = -e_i) = 1/2d \quad \text{for } i = 1, \ldots, d$$

$EX^i_n = 0$ and if $i \neq j$, then $EX^i_n X^j_n = 0$, since both components cannot be nonzero simultaneously. So the covariance matrix is $\Gamma_{ij} = (1/2d)I$.

Example 3.10.9 Let $X_1, X_2, \ldots$ be i.i.d. with $P(X_n = e_i) = 1/6$ for $i = 1, 2, \ldots, 6$. In words, we are rolling a die and keeping track of the numbers that come up. $EX_{n,i} = 1/6$ and $EX_{n,i} X_{n,j} = 0$ for $i \neq j$, so $\Gamma_{ij} = (1/6)(5/6)$ when $i = j$ and $= -(1/6)^2$ when $i \neq j$. In this case, the limiting distribution is concentrated on $\{x : \sum_i x_i = 0\}$.

Our treatment of the central limit theorem would not be complete without some discussion of the multivariate normal distribution. We begin by observing that $\Gamma_{ij} = \Gamma_{ji}$ and if $EX_i = 0$ and $EX_i X_j = \Gamma_{i,j}$

$$\sum_i \sum_j \theta_i \theta_j \Gamma_{ij} = E\left(\sum_i \theta_i X_i\right)^2 \geq 0$$

so Γ is symmetric and nonnegative definite. A well-known result implies that there is an orthogonal matrix U (i.e., one with $U^t U = I$, the identity matrix) so that $\Gamma = U^t V U$, where $V \geq 0$ is a diagonal matrix. Let W be the nonnegative diagonal matrix with $W^2 = V$. If we let $A = WU$, then $\Gamma = A^t A$. Let Y be a d-dimensional vector whose components are independent and have normal distributions with mean 0 and variance 1. If we view vectors as $1 \times d$ matrices and let $\chi = YA$, then χ has the desired normal distribution. To check this, observe that

$$\theta \cdot YA = \sum_i \theta_i \sum_j Y_j A_{ji}$$

has a normal distribution with mean 0 and variance

$$\sum_j \left(\sum_i A_{ji}\theta_i \right)^2 = \sum_j \left(\sum_i \theta_i A^t_{ij} \right) \left(\sum_k A_{jk}\theta_k \right) = \theta A^t A \theta^t = \theta \Gamma \theta^t$$

so $E(\exp(i\theta \cdot \chi)) = \exp(-(\theta\Gamma\theta^t)/2)$.

If the covariance matrix has rank d, we say that the normal distribution is **nondegenerate.** In this case, its density function is given by

$$(2\pi)^{-d/2}(\det \Gamma)^{-1/2} \exp\left(-\sum_{i,j} y_i \Gamma^{-1}_{ij} y_j/2 \right)$$

The joint distribution in degenerate cases can be computed by using a linear transformation to reduce to the nondegenerate case. For instance, in Example 3.10.9 we can look at the distribution of $(X_1, \ldots, X_5)$.

Exercises

3.10.1 If F is the distribution of $(X_1, \ldots, X_d)$ then $F_i(x) = P(X_i \le x)$ are its **marginal distributions**. How can they be obtained from F?

3.10.2 Let $F_1, \ldots, F_d$ be distributions on $\mathbf{R}$. Show that for any $\alpha \in [-1, 1]$

$$F(x_1, \ldots, x_d) = \left\{ 1 + \alpha \prod_{i=1}^d (1 - F_i(x_i)) \right\} \prod_{j=1}^d F_j(x_j)$$

is a d.f. with the given marginals. The case $\alpha = 0$ corresponds to independent r.v.'s.

3.10.3 A distribution F is said to have a **density** f if

$$F(x_1, \ldots, x_k) = \int_{-\infty}^{x_1} \cdots \int_{-\infty}^{x_k} f(y)\, dy_k \ldots dy_1$$

Show that if f is continuous, $\partial^k F/\partial x_1 \ldots \partial x_k = f$.

3.10.4 Let X_n be random vectors. Show that if $X_n \Rightarrow X$, then the coordinates $X_{n,i} \Rightarrow X_i$.

3.10.5 Let φ be the ch.f. of a distribution F on $\mathbf{R}$. What is the distribution on $\mathbf{R}^d$ that corresponds to the ch.f. $\psi(t_1, \ldots, t_d) = \varphi(t_1 + \cdots + t_d)$?

3.10.6 Show that random variables $X_1, \ldots, X_k$ are independent if and only if

$$\varphi_{X_1, \ldots X_k}(t) = \prod_{j=1}^k \varphi_{X_j}(t_j)$$

3.10.7 Suppose $(X_1, \dots, X_d)$ has a multivariate normal distribution with mean vector θ and covariance Γ. Show $X_1, \dots, X_d$ are independent if and only if $\Gamma_{ij} = 0$ for $i \neq j$. In words, uncorrelated random variables with a joint normal distribution are independent.

3.10.8 Show that $(X_1, \dots, X_d)$ has a multivariate normal distribution with mean vector θ and covariance Γ if and only if every linear combination $c_1 X_1 + \cdots + c_d X_d$ has a normal distribution with mean $c\theta^t$ and variance $c\Gamma c^t$.

4

Martingales

A martingale X_n can be thought of as the fortune at time n of a player who is betting on a fair game; submartingales (supermartingales) as the outcome of betting on a favorable (unfavorable) game. There are two basic facts about martingales. The first is that you cannot make money betting on them (see Theorem 4.2.8), and in particular if you choose to stop playing at some bounded time N, then your expected winnings EX_N are equal to your initial fortune X_0. (We are supposing for the moment that X_0 is not random.) Our second fact, Theorem 4.2.11, concerns submartingales. To use a heuristic we learned from Mike Brennan, "They are the stochastic analogues of nondecreasing sequences and so if they are bounded above (to be precise, $\sup_n EX_n^+ < \infty$) they converge almost surely." As the material in Section 4.3 shows, this result has diverse applications. Later sections give sufficient conditions for martingales to converge in L^p, $p > 1$ (Section 4.4) and in L^1 (Section 4.6); study the special case of square integrable martingales (Section 4.5); and consider martingales indexed by $n \le 0$ (Section 4.7). We give sufficient conditions for $EX_N = EX_0$ to hold for unbounded stopping times (Section 4.8). These results are quite useful for studying the behavior of random walks. Section 4.9 complements the random walk results derived from martingale arguments in Section 4.8.1 by giving combinatorial proofs.

4.1 Conditional Expectation

We begin with a definition that is important for this chapter and the next one. After giving the definition, we will consider several examples to explain it. Given are a probability space $(\Omega, \mathcal{F}_o, P)$, a σ-field $\mathcal{F} \subset \mathcal{F}_o$, and a random variable $X \in \mathcal{F}_o$ with $E|X| < \infty$. We define the **conditional expectation of** X **given** $\mathcal{F}$, $E(X|\mathcal{F})$, to be any random variable Y that has

(i) $Y \in \mathcal{F}$, i.e., is $\mathcal{F}$ measurable

(ii) for all $A \in \mathcal{F}$, $\int_A X\,dP = \int_A Y\,dP$

Any Y satisfying (i) and (ii) is said to be a **version of** $E(X|\mathcal{F})$. The first thing to be settled is that the conditional expectation exists and is unique. We tackle the second claim first but start with a technical point.

Lemma 4.1.1 *If Y satisfies (i) and (ii), then it is integrable.*

Proof Letting $A = \{Y > 0\} \in \mathcal{F}$, using (ii) twice, and then adding

$$\int_A Y\,dP = \int_A X\,dP \le \int_A |X|\,dP$$
$$\int_{A^c} -Y\,dP = \int_{A^c} -X\,dP \le \int_{A^c} |X|\,dP$$

So we have $E|Y| \le E|X|$. □

Uniqueness If Y' also satisfies (i) and (ii), then

$$\int_A Y\,dP = \int_A Y'\,dP \quad \text{for all } A \in \mathcal{F}$$

Taking $A = \{Y - Y' \ge \epsilon > 0\}$, we see

$$0 = \int_A X - X\,dP = \int_A Y - Y'\,dP \ge \epsilon P(A)$$

so $P(A) = 0$. Since this holds for all ϵ, we have $Y \le Y'$ a.s., and interchanging the roles of Y and Y', we have $Y = Y'$ a.s. Technically, all equalities such as $Y = E(X|\mathcal{F})$ should be written as $Y = E(X|\mathcal{F})$ a.s., but we have ignored this point in previous chapters and will continue to do so.

Repeating the last argument gives:

Theorem 4.1.2 *If $X_1 = X_2$ on $B \in \mathcal{F}$, then $E(X_1|\mathcal{F}) = E(X_2|\mathcal{F})$ a.s. on B.*

Proof Let $Y_1 = E(X_1|\mathcal{F})$ and $Y_2 = E(X_2|\mathcal{F})$. Taking $A = \{Y_1 - Y_2 \ge \epsilon > 0\}$, we see

$$0 = \int_{A\cap B} X_1 - X_2\,dP = \int_{A\cap B} Y_1 - Y_2\,dP \ge \epsilon P(A)$$

so $P(A) = 0$, and the conclusion follows as before. □

Existence To start, we recall ν is said to be **absolutely continuous with respect to** μ (abbreviated $\nu \ll \mu$) if $\mu(A) = 0$ implies $\nu(A) = 0$, and we use Theorem A.4.8:

Radon-Nikodym Theorem Let μ and ν be σ-finite measures on $(\Omega, \mathcal{F})$. If $\nu \ll \mu$, there is a function $f \in \mathcal{F}$ so that for all $A \in \mathcal{F}$

$$\int_A f\,d\mu = \nu(A)$$

f is usually denoted $d\nu/d\mu$ and called the **Radon-Nikodym derivative**.

The last theorem easily gives the existence of conditional expectation. Suppose first that $X \ge 0$. Let $\mu = P$ and

$$\nu(A) = \int_A X\,dP \quad \text{for } A \in \mathcal{F}$$

The dominated convergence theorem implies ν is a measure (see Exercise 1.5.4) and the definition of the integral implies $\nu \ll \mu$. The Radon- Nikodym derivative $d\nu/d\mu \in \mathcal{F}$ and for any $A \in \mathcal{F}$ has

$$\int_A X\,dP = \nu(A) = \int_A \frac{d\nu}{d\mu}\,dP$$

Taking $A = \Omega$, we see that $d\nu/d\mu \geq 0$ is integrable, and we have shown that $d\nu/d\mu$ is a version of $E(X|\mathcal{F})$.

To treat the general case now, write $X = X^+ - X^-$, let $Y_1 = E(X^+|\mathcal{F})$ and $Y_2 = E(X^-|\mathcal{F})$. Now $Y_1 - Y_2 \in \mathcal{F}$ is integrable, and for all $A \in \mathcal{F}$ we have

$$\begin{aligned}\int_A X\,dP &= \int_A X^+\,dP - \int_A X^-\,dP \\ &= \int_A Y_1\,dP - \int_A Y_2\,dP = \int_A (Y_1 - Y_2)\,dP\end{aligned}$$

This shows $Y_1 - Y_2$ is a version of $E(X|\mathcal{F})$ and completes the proof. □

4.1.1 Examples

Intuitively, we think of $\mathcal{F}$ as describing the information we have at our disposal - for each $A \in \mathcal{F}$, we know whether or not A has occurred. $E(X|\mathcal{F})$ is then our "best guess" of the value of X given the information we have. Some examples should help to clarify this and connect $E(X|\mathcal{F})$ with other definitions of conditional expectation.

Example 4.1.3 If $X \in \mathcal{F}$, then $E(X|\mathcal{F}) = X$; i.e., if we know X, then our "best guess" is X itself. Since X always satisfies (ii), the only thing that can keep X from being $E(X|\mathcal{F})$ is condition (i). A special case of this example is $X = c$, where c is a constant.

Example 4.1.4 At the other extreme from perfect information is no information. Suppose X is independent of $\mathcal{F}$, i.e., for all $B \in \mathcal{R}$ and $A \in \mathcal{F}$

$$P(\{X \in B\} \cap A) = P(X \in B)P(A)$$

We claim that, in this case, $E(X|\mathcal{F}) = EX$; i.e., if you don't know anything about X, then the best guess is the mean EX. To check the definition, note that $EX \in \mathcal{F}$ so (i) holds. To verify (ii), we observe that if $A \in \mathcal{F}$, then since X and $1_A \in \mathcal{F}$ are independent, Theorem 2.1.13 implies

$$\int_A X\,dP = E(X1_A) = EX\,E1_A = \int_A EX\,dP$$

The reader should note that here and in what follows the game is "guess and verify." We come up with a formula for the conditional expectation and then check that it satisfies (i) and (ii).

Example 4.1.5 In this example, we relate the new definition of conditional expectation to the first one taught in an undergraduate probability course. Suppose $\Omega_1, \Omega_2, \ldots$ is a finite or infinite partition of Ω into disjoint sets, each of which has positive probability, and let $\mathcal{F} = \sigma(\Omega_1, \Omega_2, \ldots)$ be the σ-field generated by these sets. Then

$$E(X|\mathcal{F}) = \frac{E(X;\Omega_i)}{P(\Omega_i)} \quad \text{on } \Omega_i$$

In words, the information in Ω_i tells us which element of the partition our outcome lies in and, given this information, the best guess for X is the average value of X over Ω_i. To prove our guess is correct, observe that the proposed formula is constant on each Ω_i, so it is measurable with respect to $\mathcal{F}$. To verify (ii), it is enough to check the equality for $A = \Omega_i$, but this is trivial:

$$\int_{\Omega_i} \frac{E(X;\Omega_i)}{P(\Omega_i)}\,dP = E(X;\Omega_i) = \int_{\Omega_i} X\,dP$$

A degenerate but important special case is $\mathcal{F} = \{\emptyset, \Omega\}$, the trivial σ-field. In this case, $E(X|\mathcal{F}) = EX$.

To continue the connection with undergraduate notions, let

$$P(A|\mathcal{G}) = E(1_A|\mathcal{G})$$
$$P(A|B) = P(A \cap B)/P(B)$$

and observe that in the last example $P(A|\mathcal{F}) = P(A|\Omega_i)$ on Ω_i.

The definition of conditional expectation given a σ-field contains conditioning on a random variable as a special case. We define

$$E(X|Y) = E(X|\sigma(Y))$$

where $\sigma(Y)$ is the σ-field generated by Y.

Example 4.1.6 To continue making connection with definitions of conditional expectation from undergraduate probability, suppose X and Y have joint density $f(x, y)$, i.e.,

$$P((X,Y) \in B) = \int_B f(x,y)\,dx\,dy \quad \text{for } B \in \mathcal{R}^2$$

and suppose for simplicity that $\int f(x,y)\,dx > 0$ for all y. We claim that in this case, if $E|g(X)| < \infty$, then $E(g(X)|Y) = h(Y)$, where

$$h(y) = \int g(x) f(x,y)\,dx \Big/ \int f(x,y)\,dx$$

To "guess" this formula, note that treating the probability densities $P(Y = y)$ as if they were real probabilities

$$P(X = x|Y = y) = \frac{P(X = x, Y = y)}{P(Y = y)} = \frac{f(x,y)}{\int f(x,y)\,dx}$$

so, integrating against the conditional probability density, we have

$$E(g(X)|Y = y) = \int g(x) P(X = x|Y = y)\,dx$$

To "verify" the proposed formula now, observe $h(Y) \in \sigma(Y)$, so (i) holds. To check (ii), observe that if $A \in \sigma(Y)$, then $A = \{\omega : Y(\omega) \in B\}$ for some $B \in \mathcal{R}$, so

$$\begin{aligned} E(h(Y); A) &= \int_B \int h(y) f(x,y)\,dx\,dy = \int_B \int g(x) f(x,y)\,dx\,dy \\ &= E(g(X) 1_B(Y)) = E(g(X); A) \end{aligned}$$

Remark To drop the assumption that $\int f(x,y)\,dx > 0$, define h by

$$h(y)\int f(x,y)\,dx = \int g(x)f(x,y)\,dx$$

(i.e., h can be anything where $\int f(x,y)\,dx = 0$), and observe this is enough for the proof.

Example 4.1.7 Suppose X and Y are independent. Let φ be a function with $E|\varphi(X,Y)| < \infty$ and let $g(x) = E(\varphi(x,Y))$. We will now show that

$$E(\varphi(X,Y)|X) = g(X)$$

Proof It is clear that $g(X) \in \sigma(X)$. To check (ii), note that if $A \in \sigma(X)$, then $A = \{X \in C\}$, so using the change of variables formula (Theorem 1.6.9) and the fact that the distribution of (X,Y) is product measure (Theorem 2.1.11), then the definition of g, and change of variables again,

$$\begin{aligned}\int_A \varphi(X,Y)\,dP &= E\{\varphi(X,Y)1_C(X)\} \\ &= \int\int \phi(x,y)1_C(x)\,\nu(dy)\,\mu(dx) \\ &= \int 1_C(x)g(x)\,\mu(dx) = \int_A g(X)\,dP\end{aligned}$$

which proves the desired result. □

Example 4.1.8 (Borel's paradox) Let X be a randomly chosen point on the earth, let θ be its longitude, and φ be its latitude. It is customary to take $\theta \in [0,2\pi)$ and $\varphi \in (-\pi/2, \pi/2]$ but we can equally well take $\theta \in [0,\pi)$ and $\varphi \in (-\pi,\pi]$. In words, the new longitude specifies the great circle on which the point lies and then φ gives the angle.

At first glance it might seem that if X is uniform on the globe, then θ and the angle φ on the great circle should both be uniform over their possible values. θ is uniform but φ is not. The paradox completely evaporates once we realize that in the new or in the traditional formulation φ is independent of θ, so the conditional distribution is the unconditional one, which is not uniform since there is more land near the equator than near the North Pole.

4.1.2 Properties

Conditional expectation has many of the same properties that ordinary expectation does.

Theorem 4.1.9 *In the first two parts we assume $E|X|, E|Y| < \infty$.*
(a) Conditional expectation is linear:

$$E(aX + Y|\mathcal{F}) = aE(X|\mathcal{F}) + E(Y|\mathcal{F}) \tag{4.1.1}$$

(b) If $X \le Y$, then

$$E(X|\mathcal{F}) \le E(Y|\mathcal{F}). \tag{4.1.2}$$

(c) If $X_n \ge 0$ and $X_n \uparrow X$ with $EX < \infty$, then

$$E(X_n|\mathcal{F}) \uparrow E(X|\mathcal{F}) \tag{4.1.3}$$

Remark By applying the last result to $Y_1 - Y_n$, we see that if $Y_n \downarrow Y$ and we have $E|Y_1|, E|Y| < \infty$, then $E(Y_n|\mathcal{F}) \downarrow E(Y|\mathcal{F})$.

Proof To prove (a), we need to check that the right-hand side is a version of the left. It clearly is $\mathcal{F}$-measurable. To check (ii), we observe that if $A \in \mathcal{F}$, then by linearity of the integral and the defining properties of $E(X|\mathcal{F})$ and $E(Y|\mathcal{F})$,

$$\begin{aligned}\int_A \{aE(X|\mathcal{F}) + E(Y|\mathcal{F})\}\, dP &= a\int_A E(X|\mathcal{F})\, dP + \int_A E(Y|\mathcal{F})\, dP \\ &= a\int_A X\, dP + \int_A Y\, dP = \int_A aX + Y\, dP\end{aligned}$$

which proves (4.1.1).

Using the definition

$$\int_A E(X|\mathcal{F})\, dP = \int_A X\, dP \le \int_A Y\, dP = \int_A E(Y|\mathcal{F})\, dP$$

Letting $A = \{E(X|\mathcal{F}) - E(Y|\mathcal{F}) \ge \epsilon > 0\}$, we see that the indicated set has probability 0 for all $\epsilon > 0$, and we have proved (4.1.2).

Let $Y_n = X - X_n$. It suffices to show that $E(Y_n|\mathcal{F}) \downarrow 0$. Since $Y_n \downarrow$, (4.1.2) implies $Z_n \equiv E(Y_n|\mathcal{F}) \downarrow$ a limit Z_∞. If $A \in \mathcal{F}$, then

$$\int_A Z_n\, dP = \int_A Y_n\, dP$$

Letting $n \to \infty$, noting $Y_n \downarrow 0$, and using the dominated convergence theorem gives that $\int_A Z_\infty\, dP = 0$ for all $A \in \mathcal{F}$, so $Z_\infty \equiv 0$. □

Theorem 4.1.10 *If φ is convex and $E|X|$, $E|\varphi(X)| < \infty$, then*

$$\varphi(E(X|\mathcal{F})) \le E(\varphi(X)|\mathcal{F}) \tag{4.1.4}$$

Proof If φ is linear, the result is trivial, so we will suppose φ is not linear. We do this so that if we let $S = \{(a,b) : a,b \in \mathbf{Q},\ ax + b \le \varphi(x) \text{ for all } x\}$, then $\varphi(x) = \sup\{ax + b : (a,b) \in S\}$. See the proof of Theorem 1.6.2 for more details. If $\varphi(x) \ge ax + b$, then (4.1.2) and (4.1.1) imply

$$E(\varphi(X)|\mathcal{F}) \ge a\, E(X|\mathcal{F}) + b \quad \text{a.s.}$$

Taking the sup over $(a,b) \in S$ gives

$$E(\varphi(X)|\mathcal{F}) \ge \varphi(E(X|\mathcal{F})) \quad \text{a.s.}$$

which proves the desired result. □

Remark Here we have written a.s. by the inequalities to stress that there is an exceptional set for each a, b, so we have to take the sup over a countable set.

Theorem 4.1.11 *Conditional expectation is a contraction in L^p, $p \ge 1$.*

Proof (4.1.4) implies $|E(X|\mathcal{F})|^p \le E(|X|^p|\mathcal{F})$. Taking expected values gives

$$E(|E(X|\mathcal{F})|^p) \le E(E(|X|^p|\mathcal{F})) = E|X|^p$$

□

In the last equality, we have used an identity that is an immediate consequence of the definition (use property (ii) in the definition with $A = \Omega$).

$$E(E(Y|\mathcal{F})) = E(Y) \tag{4.1.5}$$

Conditional expectation also has properties, like (4.1.5), that have no analogue for "ordinary" expectation.

Theorem 4.1.12 *If $\mathcal{F} \subset \mathcal{G}$ and $E(X|\mathcal{G}) \in \mathcal{F}$, then $E(X|\mathcal{F}) = E(X|\mathcal{G})$.*

Proof By assumption $E(X|\mathcal{G}) \in \mathcal{F}$. To check the other part of the definition we note that if $A \in \mathcal{F} \subset \mathcal{G}$, then

$$\int_A X\,dP = \int_A E(X|\mathcal{G})\,dP \qquad \square$$

Theorem 4.1.13 *If $\mathcal{F}_1 \subset \mathcal{F}_2$, then (i) $E(E(X|\mathcal{F}_1)|\mathcal{F}_2) = E(X|\mathcal{F}_1)$ (ii) $E(E(X|\mathcal{F}_2)|\mathcal{F}_1) = E(X|\mathcal{F}_1)$.*

In words, the smaller σ-field always wins. As the proof will show, the first equality is trivial. The second is easy to prove, but in combination with Theorem 4.1.14 is a powerful tool for computing conditional expectations. I have seen it used several times to prove results that are false.

Proof Once we notice that $E(X|\mathcal{F}_1) \in \mathcal{F}_2$, (i) follows from Example 4.1.3. To prove (ii), notice that $E(X|\mathcal{F}_1) \in \mathcal{F}_1$, and if $A \in \mathcal{F}_1 \subset \mathcal{F}_2$, then

$$\int_A E(X|\mathcal{F}_1)\,dP = \int_A X\,dP = \int_A E(X|\mathcal{F}_2)\,dP \qquad \square$$

The next result shows that for conditional expectation with respect to $\mathcal{F}$, random variables $X \in \mathcal{F}$ are like constants. They can be brought outside the "integral."

Theorem 4.1.14 *If $X \in \mathcal{F}$ and $E|Y|$, $E|XY| < \infty$, then*

$$E(XY|\mathcal{F}) = XE(Y|\mathcal{F}).$$

Proof The right-hand side $\in \mathcal{F}$, so we have to check (ii). To do this, we use the usual four-step procedure. First, suppose $X = 1_B$ with $B \in \mathcal{F}$. In this case, if $A \in \mathcal{F}$

$$\int_A 1_B E(Y|\mathcal{F})\,dP = \int_{A\cap B} E(Y|\mathcal{F})\,dP = \int_{A\cap B} Y\,dP = \int_A 1_B Y\,dP$$

so (ii) holds. The last result extends to simple X by linearity. If $X, Y \geq 0$, let X_n be simple random variables that $\uparrow X$, and use the monotone convergence theorem to conclude that

$$\int_A XE(Y|\mathcal{F})\,dP = \int_A XY\,dP$$

To prove the result in general, split X and Y into their positive and negative parts. $\square$

Theorem 4.1.15 *Suppose $EX^2 < \infty$. $E(X|\mathcal{F})$ is the variable $Y \in \mathcal{F}$ that minimizes the "mean square error" $E(X - Y)^2$.*

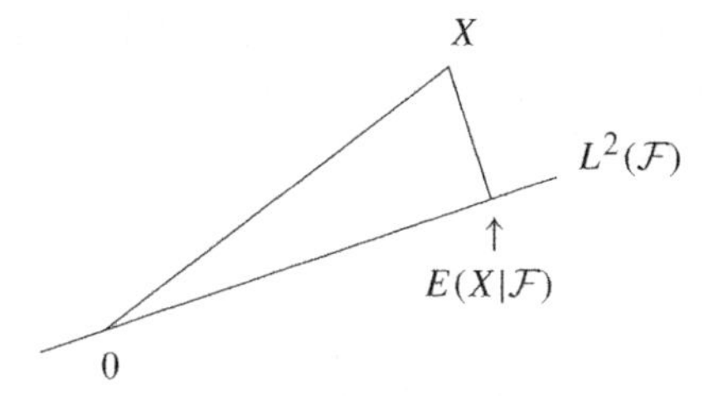

Figure 4.1 Conditional expectation as projection in L^2.

Remark This result gives a "geometric interpretation" of $E(X|\mathcal{F})$. $L^2(\mathcal{F}_o) = \{Y \in \mathcal{F}_o : EY^2 < \infty\}$ is a Hilbert space, and $L^2(\mathcal{F})$ is a closed subspace. In this case, $E(X|\mathcal{F})$ is the projection of X onto $L^2(\mathcal{F})$. That is, the point in the subspace closest to X.

Proof We begin by observing that if $Z \in L^2(\mathcal{F})$, then Theorem 4.1.14 implies

$$ZE(X|\mathcal{F}) = E(ZX|\mathcal{F})$$

($E|XZ| < \infty$ by the Cauchy-Schwarz inequality.) Taking expected values gives

$$E(ZE(X|\mathcal{F})) = E(E(ZX|\mathcal{F})) = E(ZX)$$

or, rearranging,

$$E[Z(X - E(X|\mathcal{F}))] = 0 \quad \text{for } Z \in L^2(\mathcal{F})$$

If $Y \in L^2(\mathcal{F})$ and $Z = E(X|\mathcal{F}) - Y$, then

$$E(X-Y)^2 = E\{X - E(X|\mathcal{F}) + Z\}^2 = E\{X - E(X|\mathcal{F})\}^2 + EZ^2$$

since the cross-product term vanishes. From the last formula, it is easy to see $E(X-Y)^2$ is minimized when $Z = 0$. □

4.1.3 Regular Conditional Probabilities*

Let $(\Omega, \mathcal{F}, P)$ be a probability space, $X : (\Omega, \mathcal{F}) \to (S, \mathcal{S})$ a measurable map, and $\mathcal{G}$ a σ-field $\subset \mathcal{F}$. $\mu : \Omega \times \mathcal{S} \to [0,1]$ is said to be a **regular conditional distribution** for X given $\mathcal{G}$ if

(i) For each A, $\omega \to \mu(\omega, A)$ is a version of $P(X \in A|\mathcal{G})$.

(ii) For a.e. ω, $A \to \mu(\omega, A)$ is a probability measure on $(S, \mathcal{S})$.

When $S = \Omega$ and X is the identity map, μ is called a **regular conditional probability**.

Continuation of Example 4.1.6 Suppose X and Y have a joint density $f(x,y) > 0$. If

$$\mu(y, A) = \int_A f(x,y)\,dx \Big/ \int f(x,y)\,dx$$

then $\mu(Y(\omega), A)$ is a r.c.d. for X given $\sigma(Y)$.

(i) in the definition follows by taking $h = 1_A$ in Example 4.1.1. To check (ii) note that the dominated convergence theorem implies that $A \to \mu(y, A)$ is a probability measure.

Regular conditional distributions are useful because they allow us to simultaneously compute the conditional expectation of all functions of X and to generalize properties of ordinary expectation in a more straightforward way.

Theorem 4.1.16 *Let $\mu(\omega, A)$ be a r.c.d. for X given $\mathcal{F}$. If $f : (S, \mathcal{S}) \to (\mathbf{R}, \mathcal{R})$ has $E|f(X)| < \infty$, then*

$$E(f(X)|\mathcal{F}) = \int \mu(\omega, dx) f(x) \quad a.s.$$

Proof If $f = 1_A$, this follows from the definition. Linearity extends the result to simple f and monotone convergence to nonnegative f. Finally, we get the result in general by writing $f = f^+ - f^-$. □

Unfortunately, r.c.d.'s do not always exist. The first example was due to Dieudonné (1948). See Doob (1953), p. 624, or Faden (1985) for more recent developments. Without going into the details of the example, it is easy to see the source of the problem. If $A_1, A_2, \ldots$ are disjoint, then (4.1.1) and (4.1.3) imply

$$P(X \in \cup_n A_n|\mathcal{G}) = \sum_n P(X \in A_n|\mathcal{G}) \quad \text{a.s.}$$

but if $\mathcal{S}$ contains enough countable collections of disjoint sets, the exceptional sets may pile up. Fortunately,

Theorem 4.1.17 *r.c.d.'s exist if $(S, \mathcal{S})$ is nice.*

Proof By definition, there is a 1-1 map $\varphi : S \to \mathbf{R}$ so that φ and φ^{-1} are measurable. Using monotonicity (4.1.2) and throwing away a countable collection of null sets, we find there is a set Ω_o with $P(\Omega_o) = 1$ and a family of random variables $G(q, \omega)$, $q \in \mathbf{Q}$ so that $q \to G(q, \omega)$ is nondecreasing and $\omega \to G(q, \omega)$ is a version of $P(\varphi(X) \le q|\mathcal{G})$. Let $F(x, \omega) = \inf\{G(q, \omega) : q > x\}$. The notation may remind the reader of the proof of Theorem 3.2.12. The argument given there shows F is a distribution function. Since $G(q_n, \omega) \downarrow F(x, \omega)$, the remark after Theorem 4.1.9 implies that $F(x, \omega)$ is a version of $P(\varphi(X) \le x|\mathcal{G})$.

Now, for each $\omega \in \Omega_o$, there is a unique measure $\nu(\omega, \cdot)$ on $(\mathbf{R}, \mathcal{R})$ so that $\nu(\omega, (-\infty, x]) = F(x, \omega)$. To check that for each $B \in \mathcal{R}$, $\nu(\omega, B)$ is a version of $P(\varphi(X) \in B|\mathcal{G})$, we observe that the class of B for which this statement is true (this includes the measurability of $\omega \to \nu(\omega, B)$) is a λ-system that contains all sets of the form $(a_1, b_1] \cup \cdots (a_k, b_k]$, where $-\infty \le a_i < b_i \le \infty$, so the desired result follows from the $\pi - \lambda$ theorem. To extract the desired r.c.d., notice that if $A \in \mathcal{S}$ and $B = \varphi(A)$, then $B = (\varphi^{-1})^{-1}(A) \in \mathcal{R}$, and set $\mu(\omega, A) = \nu(\omega, B)$. □

The following generalization of Theorem 4.1.17 will be needed in Section 6.1.

Theorem 4.1.18 *Suppose X and Y take values in a nice space $(S, \mathcal{S})$ and $\mathcal{G} = \sigma(Y)$. There is a function $\mu : S \times \mathcal{S} \to [0, 1]$ so that*

(i) for each A, $\mu(Y(\omega), A)$ is a version of $P(X \in A|\mathcal{G})$

(ii) for a.e. ω, $A \to \mu(Y(\omega), A)$ is a probability measure on $(S, \mathcal{S})$.

Proof As in the proof of Theorem 4.1.17, we find there is a set Ω_o with $P(\Omega_o) = 1$ and a family of random variables $G(q,\omega)$, $q \in \mathbf{Q}$ so that $q \to G(q,\omega)$ is nondecreasing and $\omega \to G(q,\omega)$ is a version of $P(\varphi(X) \le q|\mathcal{G})$. Since $G(q,\omega) \in \sigma(Y)$, we can write $G(q,\omega) = H(q,Y(\omega))$. Let $F(x,y) = \inf\{G(q,y) : q > x\}$. The argument given in the proof of Theorem 4.1.17 shows that there is a set A_0 with $P(Y \in A_0) = 1$ so that when $y \in A_0$, F is a distribution function and that $F(x,Y(\omega))$ is a version of $P(\varphi(X) \le x|Y)$.

For each $y \in A_o$, there is a unique measure $\nu(y,\cdot)$ on $(\mathbf{R},\mathcal{R})$ so that $\nu(y,(-\infty,x]) = F(x,y))$. To check that for each $B \in \mathcal{R}$, $\nu(Y(\omega),B)$ is a version of $P(\varphi(X) \in B|Y)$, we observe that the class of B for which this statement is true (this includes the measurability of $\omega \to \nu(Y(\omega),B))$ is a λ-system that contains all sets of the form $(a_1,b_1] \cup \cdots (a_k,b_k]$, where $-\infty \le a_i < b_i \le \infty$, so the desired result follows from the $\pi - \lambda$ theorem. To extract the desired r.c.d. notice that if $A \in \mathcal{S}$, and $B = \varphi(A)$, then $B = (\varphi^{-1})^{-1}(A) \in \mathcal{R}$, and set $\mu(y,A) = \nu(y,B)$. □

Exercises

4.1.1 **Bayes' formula.** Let $G \in \mathcal{G}$ and show that

$$P(G|A) = \int_G P(A|\mathcal{G})\,dP \Big/ \int_\Omega P(A|\mathcal{G})\,dP$$

When $\mathcal{G}$ is the σ-field generated by a partition, this reduces to the usual Bayes' formula

$$P(G_i|A) = P(A|G_i)P(G_i) \Big/ \sum_j P(A|G_j)P(G_j)$$

4.1.2 Prove **Chebyshev's inequality.** If $a > 0$, then

$$P(|X| \ge a|\mathcal{F}) \le a^{-2}E(X^2|\mathcal{F})$$

4.1.3 Imitate the proof in the remark after Theorem 1.5.2 to prove the conditional Cauchy-Schwarz inequality.

$$E(XY|\mathcal{G})^2 \le E(X^2|\mathcal{G})E(Y^2|\mathcal{G})$$

4.1.4 Use regular conditional probability to get the conditional Hölder inequality from the unconditional one, i.e., show that if $p,q \in (1,\infty)$ with $1/p + 1/q = 1$, then

$$E(|XY||\mathcal{G}) \le E(|X|^p|\mathcal{G})^{1/p}E(|Y|^q|\mathcal{G})^{1/q}$$

4.1.5 Give an example on $\Omega = \{a,b,c\}$ in which

$$E(E(X|\mathcal{F}_1)|\mathcal{F}_2) \ne E(E(X|\mathcal{F}_2)|\mathcal{F}_1)$$

4.1.6 Show that if $\mathcal{G} \subset \mathcal{F}$ and $EX^2 < \infty$, then

$$E(\{X - E(X|\mathcal{F})\}^2) + E(\{E(X|\mathcal{F}) - E(X|\mathcal{G})\}^2) = E(\{X - E(X|\mathcal{G})\}^2)$$

Dropping the second term on the left, we get an inequality that says geometrically, the larger the subspace the closer the projection is, or statistically, more information means a smaller mean square error.

4.1.7 An important special case of the previous result occurs when $\mathcal{G} = \{\emptyset, \Omega\}$. Let $\text{var}(X|\mathcal{F}) = E(X^2|\mathcal{F}) - E(X|\mathcal{F})^2$. Show that

$$\text{var}(X) = E(\text{var}(X|\mathcal{F})) + \text{var}(E(X|\mathcal{F}))$$

4.1.8 Let $Y_1, Y_2, \ldots$ be i.i.d. with mean μ and variance σ^2, N an independent positive integer valued r.v. with $EN^2 < \infty$ and $X = Y_1 + \cdots + Y_N$. Show that $\text{var}(X) = \sigma^2 EN + \mu^2 \text{var}(N)$. To understand and help remember the formula, think about the two special cases in which N or Y is constant.

4.1.9 Show that if X and Y are random variables with $E(Y|\mathcal{G}) = X$ and $EY^2 = EX^2 < \infty$, then $X = Y$ a.s.

4.1.10 The result in the last exercise implies that if $EY^2 < \infty$ and $E(Y|\mathcal{G})$ has the same distribution as Y, then $E(Y|\mathcal{G}) = Y$ a.s. Prove this under the assumption $E|Y| < \infty$. Hint: The trick is to prove that $\text{sgn}(X) = \text{sgn}(E(X|\mathcal{G}))$ a.s., and then take $X = Y - c$ to get the desired result.

4.2 Martingales, Almost Sure Convergence

In this section we will define martingales and their cousins supermartingales and submartingales, and take the first steps in developing their theory. Let $\mathcal{F}_n$ be a **filtration**, i.e., an increasing sequence of σ-fields. A sequence X_n is said to be **adapted** to $\mathcal{F}_n$ if $X_n \in \mathcal{F}_n$ for all n. If X_n is sequence with

(i) $E|X_n| < \infty$,

(ii) X_n is adapted to $\mathcal{F}_n$,

(iii) $E(X_{n+1}|\mathcal{F}_n) = X_n$ for all n,

then X is said to be a **martingale** (with respect to $\mathcal{F}_n$). If in the last definition $=$ is replaced by $\leq$ or $\geq$, then X is said to be a **supermartingale** or **submartingale**, respectively.

We begin by describing three examples related to random walk. Let $\xi_1, \xi_2, \ldots$ be independent and identically distributed. Let $S_n = S_0 + \xi_1 + \cdots + \xi_n$, where S_0 is a constant. Let $\mathcal{F}_n = \sigma(\xi_1, \ldots, \xi_n)$ for $n \geq 1$ and take $\mathcal{F}_0 = \{\emptyset, \Omega\}$.

Example 4.2.1 (Linear martingale) If $\mu = E\xi_i = 0$, then $S_n, n \geq 0$, is a martingale with respect to $\mathcal{F}_n$.

To prove this, we observe that $S_n \in \mathcal{F}_n$, $E|S_n| < \infty$, and ξ_{n+1} is independent of $\mathcal{F}_n$, so using the linearity of conditional expectation, (4.1.1), and Example 4.1.4,

$$E(S_{n+1}|\mathcal{F}_n) = E(S_n|\mathcal{F}_n) + E(\xi_{n+1}|\mathcal{F}_n) = S_n + E\xi_{n+1} = S_n$$

If $\mu \leq 0$, then the computation just completed shows $E(X_{n+1}|\mathcal{F}_n) \leq X_n$, i.e., X_n is a supermartingale. In this case, X_n corresponds to betting on an unfavorable game, so there is nothing "super" about a supermartingale. The name comes from the fact that if f is superharmonic (i.e., f has continuous derivatives of order ≤ 2 and $\partial^2 f/\partial x_1^2 + \cdots + \partial^2 f/\partial x_d^2 \leq 0$), then

$$f(x) \geq \frac{1}{|B(x,r)|} \int_{B(x,r)} f(y)\, dy \tag{4.2.1}$$

where $B(x,r) = \{y : |x - y| \leq r\}$ is the ball of radius r, and $|B(x,r)|$ is the volume of the ball.

If $\mu \geq 0$, then S_n is a submartingale. Applying the first result to $\xi_i' = \xi_i - \mu$, we see that $S_n - n\mu$ is a martingale.

Example 4.2.2 (Quadratic martingale) Suppose now that $\mu = E\xi_i = 0$ and $\sigma^2 = \text{var}(\xi_i) < \infty$. In this case, $S_n^2 - n\sigma^2$ is a martingale.

Since $(S_n + \xi_{n+1})^2 = S_n^2 + 2S_n\xi_{n+1} + \xi_{n+1}^2$ and ξ_{n+1} is independent of $\mathcal{F}_n$, we have

$$\begin{aligned} E(S_{n+1}^2 - (n+1)\sigma^2|\mathcal{F}_n) &= S_n^2 + 2S_n E(\xi_{n+1}|\mathcal{F}_n) + E(\xi_{n+1}^2|\mathcal{F}_n) - (n+1)\sigma^2 \\ &= S_n^2 + 0 + \sigma^2 - (n+1)\sigma^2 = S_n^2 - n\sigma^2 \end{aligned}$$

Example 4.2.3 (Exponential martingale) Let $Y_1, Y_2, \ldots$ be nonnegative i.i.d. random variables with $EY_m = 1$. If $\mathcal{F}_n = \sigma(Y_1, \ldots, Y_n)$, then $M_n = \prod_{m \leq n} Y_m$ defines a martingale. To prove this note that

$$E(M_{n+1}|\mathcal{F}_n) = M_n E(X_{n+1}|\mathcal{F}_n) = Y_n$$

Suppose now that $Y_i = e^{\theta\xi_i}$ and $\phi(\theta) = Ee^{\theta\xi_i} < \infty$. $Y_i = \exp(\theta\xi)/\phi(\theta)$ has mean 1 so $EY_i = 1$ and

$$M_n = \prod_{i=1}^{n} Y_i = \exp(\theta S_n)/\phi(\theta)^n \quad \text{is a martingale.}$$

We will see many other examples below, so we turn now to deriving properties of martingales. Our first result is an immediate consequence of the definition of a supermartingale. We could take the conclusion of the result as the definition of supermartingale, but then the definition would be harder to check.

Theorem 4.2.4 *If X_n is a supermartingale, then for $n > m$, $E(X_n|\mathcal{F}_m) \leq X_m$.*

Proof The definition gives the result for $n = m + 1$. Suppose $n = m + k$ with $k \geq 2$. By Theorem 4.1.2,

$$E(X_{m+k}|\mathcal{F}_m) = E(E(X_{m+k}|\mathcal{F}_{m+k-1})|\mathcal{F}_m) \leq E(X_{m+k-1}|\mathcal{F}_m)$$

by the definition and (4.1.2). The desired result now follows by induction. □

Theorem 4.2.5 *(i) If X_n is a submartingale, then for $n > m$, $E(X_n|\mathcal{F}_m) \geq X_m$.*
(ii) If X_n is a martingale, then for $n > m$, $E(X_n|\mathcal{F}_m) = X_m$.

Proof To prove (i), note that $-X_n$ is a supermartingale and use (4.1.1). For (ii), observe that X_n is a supermartingale and a submartingale. □

Remark The idea in the proof of Theorem 4.2.5 will be used many times in what follows. To keep from repeating ourselves, we will just state the result for either supermartingales or submartingales and leave it to the reader to translate the result for the other two.

Theorem 4.2.6 *If X_n is a martingale w.r.t. $\mathcal{F}_n$ and φ is a convex function with $E|\varphi(X_n)| < \infty$ for all n, then $\varphi(X_n)$ is a submartingale w.r.t. $\mathcal{F}_n$. Consequently, if $p \geq 1$ and $E|X_n|^p < \infty$ for all n, then $|X_n|^p$ is a submartingale w.r.t. $\mathcal{F}_n$.*

Proof By Jensen's inequality and the definition

$$E(\varphi(X_{n+1})|\mathcal{F}_n) \geq \varphi(E(X_{n+1}|\mathcal{F}_n)) = \varphi(X_n) \qquad \square$$

Theorem 4.2.7 *If X_n is a submartingale w.r.t. $\mathcal{F}_n$ and φ is an increasing convex function with $E|\varphi(X_n)| < \infty$ for all n, then $\varphi(X_n)$ is a submartingale w.r.t. $\mathcal{F}_n$. Consequently, (i) If X_n is a submartingale, then $(X_n - a)^+$ is a submartingale. (ii) If X_n is a supermartingale, then $X_n \wedge a$ is a supermartingale.*

Proof By Jensen's inequality and the assumptions

$$E(\varphi(X_{n+1})|\mathcal{F}_n) \geq \varphi(E(X_{n+1}|\mathcal{F}_n)) \geq \varphi(X_n) \qquad \square$$

Let $\mathcal{F}_n, n \geq 0$ be a filtration. $H_n, n \geq 1$ is said to be a **predictable sequence** if $H_n \in \mathcal{F}_{n-1}$ for all $n \geq 1$. In words, the value of H_n may be predicted (with certainty) from the information available at time $n-1$. In this section, we will be thinking of H_n as the amount of money a gambler will bet at time n. This can be based on the outcomes at times $1, \ldots, n-1$ but not on the outcome at time n!

Once we start thinking of H_n as a gambling system, it is natural to ask how much money we would make if we used it. Let X_n be the net amount of money you would have won at time n if you had bet one dollar each time. If you bet according to a gambling system H, then your winnings at time n would be

$$(H \cdot X)_n = \sum_{m=1}^{n} H_m(X_m - X_{m-1})$$

since if at time m you have wagered \$3, the change in your fortune would be 3 times that of a person who wagered \$1. Alternatively, you can think of X_m as the value of a stock and H_m the number of shares you hold from time $m-1$ to time m.

Suppose now that $\xi_m = X_m - X_{m-1}$ have $P(\xi_m = 1) = p$ and $P(\xi_m = -1) = 1 - p$. A famous gambling system called the "martingale" is defined by $H_1 = 1$ and for $n \geq 2$,

$$H_n = \begin{cases} 2H_{n-1} & \text{if } \xi_{n-1} = -1 \\ 1 & \text{if } \xi_{n-1} = 1 \end{cases}$$

In words, we double our bet when we lose, so that if we lose k times and then win, our net winnings will be 1. To see this, consider the following concrete situation

H_n	1	2	4	8	16
ξ_n	−1	−1	−1	−1	1
$(H \cdot X)_n$	−1	−3	−7	−15	1

This system seems to provide us with a "sure thing" as long as $P(\xi_m = 1) > 0$. However, the next result says there is no system for beating an unfavorable game.

Theorem 4.2.8 *Let X_n, $n \ge 0$, be a supermartingale. If $H_n \ge 0$ is predictable and each H_n is bounded, then $(H \cdot X)_n$ is a supermartingale.*

Proof Using the fact that conditional expectation is linear, $(H \cdot X)_n \in \mathcal{F}_n$, $H_n \in \mathcal{F}_{n-1}$, and (4.1.14), we have

$$\begin{aligned} E((H \cdot X)_{n+1}|\mathcal{F}_n) &= (H \cdot X)_n + E(H_{n+1}(X_{n+1} - X_n)|\mathcal{F}_n) \\ &= (H \cdot X)_n + H_{n+1}E((X_{n+1} - X_n)|\mathcal{F}_n) \le (H \cdot X)_n \end{aligned}$$

since $E((X_{n+1} - X_n)|\mathcal{F}_n) \le 0$ and $H_{n+1} \ge 0$. □

Remark The same result is obviously true for submartingales and for martingales (in the last case, without the restriction $H_n \ge 0$).

We will now consider a very special gambling system: bet \$1 each time $n \le N$ then stop playing. A random variable N is said to be a **stopping time** if $\{N = n\} \in \mathcal{F}_n$ for all $n < \infty$, i.e., the decision to stop at time n must be measurable with respect to the information known at that time. If we let $H_n = 1_{\{N \ge n\}}$, then $\{N \ge n\} = \{N \le n-1\}^c \in \mathcal{F}_{n-1}$, so H_n is predictable, and it follows from Theorem 4.2.8 that $(H \cdot X)_n = X_{N \wedge n} - X_0$ is a supermartingale. Since the constant sequence $Y_n = X_0$ is a supermartingale and the sum of two supermartingales is also, we have:

Theorem 4.2.9 *If N is a stopping time and X_n is a supermartingale, then $X_{N \wedge n}$ is a supermartingale.*

Although Theorem 4.2.8 implies that you cannot make money with gambling systems, you can prove theorems with them. Suppose X_n, $n \ge 0$, is a submartingale. Let $a < b$, let $N_0 = -1$, and for $k \ge 1$ let

$$\begin{aligned} N_{2k-1} &= \inf\{m > N_{2k-2} : X_m \le a\} \\ N_{2k} &= \inf\{m > N_{2k-1} : X_m \ge b\} \end{aligned}$$

The N_j are stopping times and $\{N_{2k-1} < m \le N_{2k}\} = \{N_{2k-1} \le m-1\} \cap \{N_{2k} \le m-1\}^c \in \mathcal{F}_{m-1}$, so

$$H_m = \begin{cases} 1 & \text{if } N_{2k-1} < m \le N_{2k} \text{ for some } k \\ 0 & \text{otherwise} \end{cases}$$

defines a predictable sequence. $X(N_{2k-1}) \le a$ and $X(N_{2k}) \ge b$, so between times N_{2k-1} and N_{2k}, X_m crosses from below a to above b. H_m is a gambling system that tries to take advantage of these "upcrossings." In stock market terms, we buy when $X_m \le a$ and sell when $X_m \ge b$, so every time an upcrossing is completed, we make a profit of $\ge (b-a)$. Finally, $U_n = \sup\{k : N_{2k} \le n\}$ is the number of upcrossings completed by time n.

Theorem 4.2.10 (Upcrossing inequality) *If X_m, $m \ge 0$, is a submartingale, then*

$$(b-a)EU_n \le E(X_n - a)^+ - E(X_0 - a)^+$$

Proof Let $Y_m = a + (X_m - a)^+$. By Theorem 4.2.7, Y_m is a submartingale. Clearly, it upcrosses $[a,b]$ the same number of times that X_m does, and we have $(b-a)U_n \le (H \cdot Y)_n$, since each upcrossing results in a profit $\ge (b-a)$ and a final incomplete upcrossing (if there

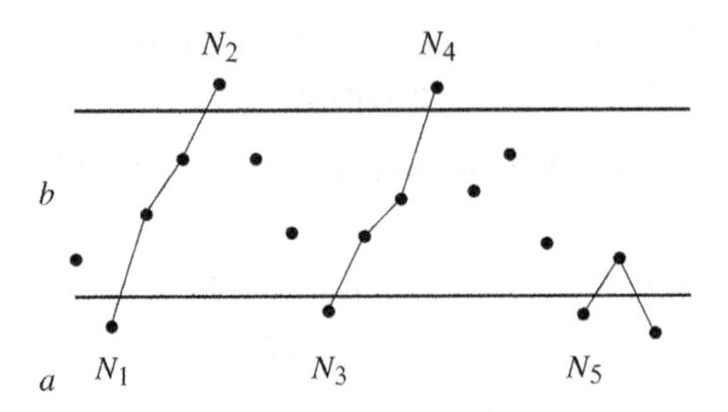

Figure 4.2 Upcrossings of (a,b). Lines indicate increments that are included in $(H \cdot X)_n$. In Y_n the points $< a$ are moved up to a.

is one) makes a nonnegative contribution to the right-hand side. It is for this reason we had to replace X_m by Y_m.

Let $K_m = 1 - H_m$. Clearly, $Y_n - Y_0 = (H \cdot Y)_n + (K \cdot Y)_n$, and it follows from Theorem 4.2.8 that $E(K \cdot Y)_n \geq E(K \cdot Y)_0 = 0$, so $E(H \cdot Y)_n \leq E(Y_n - Y_0)$, proving the desired inequality. □

We have proved the result in its classical form, even though this is a little misleading. The key fact is that $E(K \cdot Y)_n \geq 0$, i.e., no matter how hard you try you can't lose money betting on a submartingale. From the upcrossing inequality, we easily get:

Theorem 4.2.11 (Martingale convergence theorem) *If X_n is a submartingale with $\sup EX_n^+ < \infty$, then as $n \to \infty$, X_n converges a.s. to a limit X with $E|X| < \infty$.*

Proof Since $(X - a)^+ \leq X^+ + |a|$, Theorem 4.2.10 implies that

$$EU_n \leq (|a| + EX_n^+)/(b - a)$$

As $n \uparrow \infty$, $U_n \uparrow U$ the number of upcrossings of $[a,b]$ by the whole sequence, so if $\sup EX_n^+ < \infty$, then $EU < \infty$ and hence $U < \infty$ a.s. Since the last conclusion holds for all rational a and b,

$$\cup_{a,b \in \mathbf{Q}} \{\liminf X_n < a < b < \limsup X_n\} \quad \text{has probability 0}$$

and hence $\limsup X_n = \liminf X_n$ a.s., i.e., $\lim X_n$ exists a.s. Fatou's lemma guarantees $EX^+ \leq \liminf EX_n^+ < \infty$, so $X < \infty$ a.s. To see $X > -\infty$, we observe that

$$EX_n^- = EX_n^+ - EX_n \leq EX_n^+ - EX_0$$

(since X_n is a submartingale), so another application of Fatou's lemma shows

$$EX^- \leq \liminf_{n \to \infty} EX_n^- \leq \sup_n EX_n^+ - EX_0 < \infty$$

and completes the proof. □

Remark To prepare for the proof of Theorem 4.7.1, the reader should note that we have shown that if the number of upcrossings of (a,b) by X_n is finite for all $a, b \in \mathbf{Q}$, then the limit of X_n exists.

An important special case of Theorem 4.2.11 is:

Theorem 4.2.12 *If $X_n \geq 0$ is a supermartingale, then as $n \to \infty$, $X_n \to X$ a.s. and $EX \leq EX_0$.*

Proof $Y_n = -X_n \leq 0$ is a submartingale with $EY_n^+ = 0$. Since $EX_0 \geq EX_n$, the inequality follows from Fatou's lemma. □

In the next section we will give several applications of the last two results. We close this one by giving two "counterexamples."

Example 4.2.13 The first shows that the assumptions of Theorem 4.2.12 (or 4.2.11) do not guarantee convergence in L^1. Let S_n be a symmetric simple random walk with $S_0 = 1$, i.e., $S_n = S_{n-1} + \xi_n$, where $\xi_1, \xi_2, \dots$ are i.i.d. with $P(\xi_i = 1) = P(\xi_i = -1) = 1/2$. Let $N = \inf\{n : S_n = 0\}$ and let $X_n = S_{N \wedge n}$. Theorem 4.2.9 implies that X_n is a nonnegative martingale. Theorem 4.2.12 implies X_n converges to a limit $X_\infty < \infty$ that must be $\equiv 0$, since convergence to $k > 0$ is impossible. (If $X_n = k > 0$, then $X_{n+1} = k \pm 1$.) Since $EX_n = EX_0 = 1$ for all n and $X_\infty = 0$, convergence cannot occur in L^1.

Example 4.2.13 is an important counterexample to keep in mind as you read the rest of this chapter. The next one is not as important.

Example 4.2.14 We will now give an example of a martingale with $X_k \to 0$ in probability but not a.s. Let $X_0 = 0$. When $X_{k-1} = 0$, let $X_k = 1$ or -1 with probability $1/2k$ and $= 0$ with probability $1 - 1/k$. When $X_{k-1} \neq 0$, let $X_k = kX_{k-1}$ with probability $1/k$ and $= 0$ with probability $1 - 1/k$. From the construction, $P(X_k = 0) = 1 - 1/k$, so $X_k \to 0$ in probability. On the other hand, the second Borel-Cantelli lemma implies $P(X_k = 0$ for $k \geq K) = 0$, and values in $(-1, 1) - \{0\}$ are impossible, so X_k does not converge to 0 a.s.

Exercises

4.2.1 Suppose X_n is a martingale w.r.t. $\mathcal{G}_n$ and let $\mathcal{F}_n = \sigma(X_1, \dots, X_n)$. Then $\mathcal{G}_n \supset \mathcal{F}_n$ and X_n is a martingale w.r.t. $\mathcal{F}_n$.

4.2.2 Give an example of a submartingale X_n so that X_n^2 is a supermartingale. Hint: X_n does not have to be random.

4.2.3 Generalize (i) of Theorem 4.2.7 by showing that if X_n and Y_n are submartingales w.r.t. $\mathcal{F}_n$, then $X_n \vee Y_n$ is also.

4.2.4 Let X_n, $n \geq 0$, be a submartingale with $\sup X_n < \infty$. Let $\xi_n = X_n - X_{n-1}$ and suppose $E(\sup \xi_n^+) < \infty$. Show that X_n converges a.s.

4.2.5 Give an example of a martingale X_n with $X_n \to -\infty$ a.s. Hint: Let $X_n = \xi_1 + \cdots + \xi_n$, where the ξ_i are independent (but not identically distributed) with $E\xi_i = 0$.

4.2.6 Let $Y_1, Y_2, \dots$ be nonnegative i.i.d. random variables with $EY_m = 1$ and $P(Y_m = 1) < 1$. By example 4.2.3 that $X_n = \prod_{m \leq n} Y_m$ defines a martingale. (i) Use Theorem 4.2.12 and an argument by contradiction to show $X_n \to 0$ a.s. (ii) Use the strong law of large numbers to conclude $(1/n) \log X_n \to c < 0$.

4.2.7 Suppose $y_n > -1$ for all n and $\sum |y_n| < \infty$. Show that $\prod_{m=1}^{\infty}(1 + y_m)$ exists.

4.2.8 Let X_n and Y_n be positive integrable and adapted to $\mathcal{F}_n$. Suppose

$$E(X_{n+1}|\mathcal{F}_n) \leq (1 + Y_n)X_n$$

with $\sum Y_n < \infty$ a.s. Prove that X_n converges a.s. to a finite limit by finding a closely related supermartingale to which Theorem 4.2.12 can be applied.

4.2.9 **The switching principle.** Suppose X_n^1 and X_n^2 are supermartingales with respect to $\mathcal{F}_n$, and N is a stopping time so that $X_N^1 \geq X_N^2$. Then

$$Y_n = X_n^1 1_{(N>n)} + X_n^2 1_{(N\leq n)} \text{ is a supermartingale.}$$
$$Z_n = X_n^1 1_{(N\geq n)} + X_n^2 1_{(N<n)} \text{ is a supermartingale.}$$

4.2.10 **Dubins' inequality.** For every positive supermartingale X_n, $n \geq 0$, the number of upcrossings U of $[a,b]$ satisfies

$$P(U \geq k) \leq \left(\frac{a}{b}\right)^k E\min(X_0/a, 1)$$

To prove this, we let $N_0 = -1$ and for $j \geq 1$ let

$$N_{2j-1} = \inf\{m > N_{2j-2} : X_m \leq a\}$$
$$N_{2j} = \inf\{m > N_{2j-1} : X_m \geq b\}$$

Let $Y_n = 1$ for $0 \leq n < N_1$ and for $j \geq 1$

$$Y_n = \begin{cases} (b/a)^{j-1}(X_n/a) & \text{for } N_{2j-1} \leq n < N_{2j} \\ (b/a)^j & \text{for } N_{2j} \leq n < N_{2j+1} \end{cases}$$

(i) Use the switching principle in the previous exercise and induction to show that $Z_n^j = Y_{n\wedge N_j}$ is a supermartingale. (ii) Use $EY_{n\wedge N_{2k}} \leq EY_0$ and let $n \to \infty$ to get Dubins' inequality.

4.3 Examples

In this section, we will apply the martingale convergence theorem to generalize the second Borel-Cantelli lemma and to study Polya's urn scheme, Radon-Nikodym derivatives, and branching processes. The four topics are independent of each other and are taken up in the order indicated.

4.3.1 Bounded Increments

Our first result shows that martingales with bounded increments either converge or oscillate between $+\infty$ and $-\infty$.

Theorem 4.3.1 *Let $X_1, X_2, \ldots$ be a martingale with $|X_{n+1} - X_n| \leq M < \infty$. Let*

$$C = \{\lim X_n \textit{ exists and is finite}\}$$
$$D = \{\limsup X_n = +\infty \textit{ and } \liminf X_n = -\infty\}$$

Then $P(C \cup D) = 1$.

Proof Since $X_n - X_0$ is a martingale, we can without loss of generality suppose that $X_0 = 0$. Let $0 < K < \infty$ and let $N = \inf\{n : X_n \le -K\}$. $X_{n\wedge N}$ is a martingale with $X_{n\wedge N} \ge -K - M$ a.s., so applying Theorem 4.2.12 to $X_{n\wedge N} + K + M$ shows $\lim X_n$ exists on $\{N = \infty\}$. Letting $K \to \infty$, we see that the limit exists on $\{\liminf X_n > -\infty\}$. Applying the last conclusion to $-X_n$, we see that $\lim X_n$ exists on $\{\limsup X_n < \infty\}$ and the proof is complete. □

To prepare for an application of this result we need:

Theorem 4.3.2 (Doob's decomposition) *Any submartingale X_n, $n \ge 0$, can be written in a unique way as $X_n = M_n + A_n$, where M_n is a martingale and A_n is a predictable increasing sequence with $A_0 = 0$.*

Proof We want $X_n = M_n + A_n$, $E(M_n|\mathcal{F}_{n-1}) = M_{n-1}$, and $A_n \in \mathcal{F}_{n-1}$. So we must have

$$\begin{aligned} E(X_n|\mathcal{F}_{n-1}) &= E(M_n|\mathcal{F}_{n-1}) + E(A_n|\mathcal{F}_{n-1}) \\ &= M_{n-1} + A_n = X_{n-1} - A_{n-1} + A_n \end{aligned}$$

and it follows that

$$A_n - A_{n-1} = E(X_n|\mathcal{F}_{n-1}) - X_{n-1} \tag{4.3.1}$$

Since $A_0 = 0$, we have

$$A_n = \sum_{m=1}^{n} E(X_n - X_{n-1}|\mathcal{F}_{n-1}) \tag{4.3.2}$$

To check that our recipe works, we observe that $A_n - A_{n-1} \ge 0$ since X_n is a submartingale and $A_n \in \mathcal{F}_{n-1}$. To prove that $M_n = X_n - A_n$ is a martingale, we note that using $A_n \in \mathcal{F}_{n-1}$ and (4.3.1)

$$\begin{aligned} E(M_n|\mathcal{F}_{n-1}) &= E(X_n - A_n|\mathcal{F}_{n-1}) \\ &= E(X_n|\mathcal{F}_{n-1}) - A_n = X_{n-1} - A_{n-1} = M_{n-1} \end{aligned}$$

which completes the proof. □

To illustrate the use of this result, we do the following important example.

Example 4.3.3 Let us suppose $B_n \in \mathcal{F}_n$. Using (4.3.2)

$$M_n = \sum_{m=1}^{n} 1_{B_m} - E(1_{B_m}|\mathcal{F}_{m-1})$$

Theorem 4.3.4 (Second Borel-Cantelli lemma, II) *Let $\mathcal{F}_n$, $n \ge 0$ be a filtration with $\mathcal{F}_0 = \{\emptyset, \Omega\}$ and A_n, $n \ge 1$ a sequence of events with $B_n \in \mathcal{F}_n$. Then*

$$\{B_n \ i.o.\} = \left\{ \sum_{n=1}^{\infty} P(B_n|\mathcal{F}_{n-1}) = \infty \right\}$$

Proof If we let $X_0 = 0$ and $X_n = \sum_{m\le n} 1_{B_m}$, then X_n is a submartingale. (4.3.2) implies $B_n = \sum_{m=1}^{n} E(1_{A_m}|\mathcal{F}_{m-1})$, so if $M_0 = 0$ and

$$M_n = \sum_{m=1}^{n} 1_{A_m} - P(A_m|\mathcal{F}_{m-1})$$

for $n \geq 1$ then M_n is a martingale with $|M_n - M_{n-1}| \leq 1$. Using the notation of Theorem 4.3.1 we have:

$$\text{on } C, \quad \sum_{n=1}^{\infty} 1_{B_n} = \infty \quad \text{if and only if} \quad \sum_{n=1}^{\infty} P(B_n|\mathcal{F}_{n-1}) = \infty$$

$$\text{on } D, \quad \sum_{n=1}^{\infty} 1_{B_n} = \infty \quad \text{and} \quad \sum_{n=1}^{\infty} P(B_n|\mathcal{F}_{n-1}) = \infty$$

Since $P(C \cup D) = 1$, the result follows. □

4.3.2 Polya's Urn Scheme

An urn contains r red and g green balls. At each time we draw a ball out, then replace it, and add c more balls of the color drawn. Let X_n be the fraction of green balls after the nth draw. To check that X_n is a martingale, note that if there are i red balls and j green balls at time n, then

$$X_{n+1} = \begin{cases} (j+c)/(i+j+c) & \text{with probability } j/(i+j) \\ j/(i+j+c) & \text{with probability } i/(i+j) \end{cases}$$

and we have

$$\frac{j+c}{i+j+c} \cdot \frac{j}{i+j} + \frac{j}{i+j+c} \cdot \frac{i}{i+j} = \frac{(j+c+i)j}{(i+j+c)(i+j)} = \frac{j}{i+j}$$

Since $X_n \geq 0$, Theorem 4.2.12 implies that $X_n \to X_\infty$ a.s. To compute the distribution of the limit, we observe (a) the probability of getting green on the first m draws then red on the next $\ell = n - m$ draws is

$$\frac{g}{g+r} \cdot \frac{g+c}{g+r+c} \cdots \frac{g+(m-1)c}{g+r+(m-1)c} \cdot \frac{r}{g+r+mc} \cdots \frac{r+(\ell-1)c}{g+r+(n-1)c}$$

and (b) any other outcome of the first n draws with m green balls drawn and ℓ red balls drawn has the same probability, since the denominator remains the same and the numerator is permuted. Consider the special case $c = 1, g = 1, r = 1$. Let G_n be the number of green balls after the nth draw has been completed and the new ball has been added. It follows from (a) and (b) that

$$P(G_n = m+1) = \binom{n}{m} \frac{m!\,(n-m)!}{(n+1)!} = \frac{1}{n+1}$$

so X_∞ has a uniform distribution on (0,1).

If we suppose that $c = 1$, $g = 2$, and $r = 1$, then

$$P(G_n = m+2) = \frac{n!}{m!\,(n-m)!} \frac{(m+1)!\,(n-m)!}{(n+2)!\,/2} \to 2x$$

if $n \to \infty$ and $m/n \to x$. In general, the distribution of X_∞ has density

$$\frac{\Gamma((g+r)/c)}{\Gamma(g/c)\Gamma(r/c)} x^{(g/c)-1}(1-x)^{(r/c)-1}$$

This is the **beta distribution** with parameters g/c and r/c. In Example 4.5.6 we will see that the limit behavior changes drastically if, in addition to the c balls of the color chosen, we always add one ball of the opposite color.

4.3.3 Radon-Nikodym Derivatives

Let μ be a finite measure and ν a probability measure on $(\Omega, \mathcal{F})$. Let $\mathcal{F}_n \uparrow \mathcal{F}$ be σ-fields (i.e., $\sigma(\cup \mathcal{F}_n) = \mathcal{F}$). Let μ_n and ν_n be the restrictions of μ and ν to $\mathcal{F}_n$.

Theorem 4.3.5 *Suppose $\mu_n \ll \nu_n$ for all n. Let $X_n = d\mu_n/d\nu_n$ and let $X = \limsup X_n$. Then*

$$\mu(A) = \int_A X d\nu + \mu(A \cap \{X = \infty\})$$

Remark $\mu_r(A) \equiv \int_A X\, d\nu$ is a measure $\ll \nu$. Since Theorem 4.2.12 implies $\nu(X = \infty) = 0$, $\mu_s(A) \equiv \mu(A \cap \{X = \infty\})$ is singular w.r.t. ν. Thus $\mu = \mu_r + \mu_s$ gives the Lebesgue decomposition of μ (see Theorem A.4.7), and $X_\infty = d\mu_r/d\nu$, ν-a.s. Here and in the proof we need to keep track of the measure to which the a.s. refers.

Proof As the reader can probably anticipate:

Lemma 4.3.6 *X_n (defined on $(\Omega, \mathcal{F}, \nu)$) is a martingale w.r.t. $\mathcal{F}_n$.*

Proof We observe that, by definition, $X_n \in \mathcal{F}_n$. Let $A \in \mathcal{F}_n$. Since $X_n \in \mathcal{F}_n$ and ν_n is the restriction of ν to $\mathcal{F}_n$

$$\int_A X_n\, d\nu = \int_A X_n\, d\nu_n$$

Using the definition of X_n and Exercise A.4.7

$$\int_A X_n\, d\nu_n = \mu_n(A) = \mu(A)$$

the last equality holding, since $A \in \mathcal{F}_n$ and μ_n is the restriction of μ to $\mathcal{F}_n$. If $A \in \mathcal{F}_{m-1} \subset \mathcal{F}_m$, using the last result for $n = m$ and $n = m - 1$ gives

$$\int_A X_m d\nu = \mu(A) = \int_A X_{m-1} d\nu$$

so $E(X_m | \mathcal{F}_{m-1}) = X_{m-1}$. □

Since X_n is a nonnegative martingale, Theorem 4.2.12 implies that $X_n \to X$ ν-a.s. We want to check that the equality in the theorem holds. Dividing $\mu(A)$ by $\mu(\Omega)$, we can without loss of generality suppose μ is a probability measure. Let $\rho = (\mu + \nu)/2$, $\rho_n = (\mu_n + \nu_n)/2 =$ the restriction of ρ to $\mathcal{F}_n$. Let $Y_n = d\mu_n/d\rho_n$, $Z_n = d\nu_n/d\rho_n$. $Y_n, Z_n \geq 0$ and $Y_n + Z_n = 2$ (by Exercise A.4.6), so Y_n and Z_n are bounded martingales with limits Y and Z. As the reader can probably guess,

$$(*) \qquad Y = d\mu/d\rho \qquad Z = d\nu/d\rho$$

It suffices to prove the first equality. From the proof of Lemma 4.3.6, if $A \in \mathcal{F}_m \subset \mathcal{F}_n$

$$\mu(A) = \int_A Y_n \, d\rho \to \int_A Y \, d\rho$$

by the bounded convergence theorem. The last computation shows that

$$\mu(A) = \int_A Y \, d\rho \quad \text{for all } A \in \mathcal{G} = \cup_m \mathcal{F}_m$$

$\mathcal{G}$ is a π-system, so the $\pi - \lambda$ theorem implies the equality is valid for all $A \in \mathcal{F} = \sigma(\mathcal{G})$ and $(*)$ is proved.

It follows from Exercises A.4.8 and A.4.9 that $X_n = Y_n/Z_n$. At this point, the reader can probably leap to the conclusion that $X = Y/Z$. To get there carefully, note $Y + Z = 2$ ρ-a.s., so $\rho(Y = 0, Z = 0) = 0$. Having ruled out $0/0$, we have $X = Y/Z$ ρ-a.s. (Recall $X \equiv \limsup X_n$.) Let $W = (1/Z) \cdot 1_{(Z>0)}$. Using $(*)$, then $1 = ZW + 1_{(Z=0)}$, we have

(a) $$\mu(A) = \int_A Y \, d\rho = \int_A YWZ \, d\rho + \int_A 1_{(Z=0)} Y \, d\rho$$

Now $(*)$ implies $d\nu = Z \, d\rho$, and it follows from the definitions that

$$YW = X 1_{(Z>0)} = X \quad \nu\text{-a.s.}$$

the second equality holding, since $\nu(\{Z = 0\}) = 0$. Combining things, we have

(b) $$\int_A YWZ \, d\rho = \int_A X \, d\nu$$

To handle the other term, we note that $(*)$ implies $d\mu = Y \, d\rho$, and it follows from the definitions that $\{X = \infty\} = \{Z = 0\}$ μ-a.s. so

(c) $$\int_A 1_{(Z=0)} Y \, d\rho = \int_A 1_{(X=\infty)} \, d\mu$$

Combining (a), (b), and (c) gives the desired result. □

Example 4.3.7 Suppose $\mathcal{F}_n = \sigma(I_{k,n} : 0 \le k < K_n)$, where for each n, $I_{k,n}$ is a partition of Ω, and the $(n+1)$th partition is a refinement of the nth. In this case, the condition $\mu_n \ll \nu_n$ is $\nu(I_{k,n}) = 0$ implies $\mu(I_{k,n}) = 0$, and the martingale $X_n = \mu(I_{k,n})/\nu(I_{k,n})$ on $I_{k,n}$ is an approximation to the Radon-Nikodym derivative. For a concrete example, consider $\Omega = [0, 1)$, $I_{k,n} = [k2^{-n}, (k+1)2^{-n})$ for $0 \le k < 2^n$, and $\nu =$ Lebesgue measure.

Kakutani dichotomy for infinite product measures. Let μ and ν be measures on sequence space $(\mathbf{R}^{\mathbf{N}}, \mathcal{R}^{\mathbf{N}})$ that make the coordinates $\xi_n(\omega) = \omega_n$ independent. Let $F_n(x) = \mu(\xi_n \le x)$, $G_n(x) = \nu(\xi_n \le x)$. Suppose $F_n \ll G_n$ and let $q_n = dF_n/dG_n$. To avoid a problem, we will suppose $q_n > 0$, G_n-a.s.

Let $\mathcal{F}_n = \sigma(\xi_m : m \le n)$, let μ_n and ν_n be the restrictions of μ and ν to $\mathcal{F}_n$, and let

$$X_n = \frac{d\mu_n}{d\nu_n} = \prod_{m=1}^{n} q_m.$$

Theorem 4.3.5 implies that $X_n \to X$ ν-a.s. Thanks to our assumption $q_n > 0$, G_n-a.s. $\sum_{m=1}^{\infty} \log(q_m) > -\infty$is a tail event, so the Kolmogorov 0-1 law implies

$$\nu(X = 0) \in \{0, 1\} \tag{4.3.3}$$

and it follows from Theorem 4.3.5 that either $\mu \ll \nu$ or $\mu \perp \nu$. The next result gives a concrete criterion for which of the two alternatives occurs.

Theorem 4.3.8 *$\mu \ll \nu$ or $\mu \perp \nu$, according as $\prod_{m=1}^{\infty} \int \sqrt{q_m}\, dG_m > 0$ or $= 0$.*

Proof Jensen's inequality and Exercise A.4.7 imply

$$\left(\int \sqrt{q_m}\, dG_m\right)^2 \le \int q_m\, dG_m = \int dF_m = 1$$

so the infinite product of the integrals is well defined and ≤ 1. Let

$$X_n = \prod_{m \le n} q_m(\omega_m)$$

as above, and recall that $X_n \to X$ ν-a.s. If the infinite product is 0, then

$$\int X_n^{1/2}\, d\nu = \prod_{m=1}^{n} \int \sqrt{q_m}\, dG_m \to 0$$

Fatou's lemma implies

$$\int X^{1/2}\, d\nu \le \liminf_{n\to\infty} \int X_n^{1/2}\, d\nu = 0$$

so $X = 0$ ν-a.s., and Theorem 4.3.5 implies $\mu \perp \nu$. To prove the other direction, let $Y_n = X_n^{1/2}$. Now $\int q_m\, dG_m = 1$, so if we use E to denote expected value with respect to ν, then $EY_m^2 = EX_m = 1$, so

$$E(Y_{n+k} - Y_n)^2 = E(X_{n+k} + X_n - 2X_n^{1/2}X_{n+k}^{1/2}) = 2\left(1 - \prod_{m=n+1}^{n+k} \int \sqrt{q_m}\, dG_m\right)$$

Now $|a-b| = |a^{1/2} - b^{1/2}| \cdot (a^{1/2} + b^{1/2})$, so using Cauchy-Schwarz and the fact $(a+b)^2 \le 2a^2 + 2b^2$ gives

$$\begin{aligned} E|X_{n+k} - X_n| &= E(|Y_{n+k} - Y_n|(Y_{n+k} + Y_n)) \\ &\le \left(E(Y_{n+k} - Y_n)^2 E(Y_{n+k} + Y_n)^2\right)^{1/2} \\ &\le \left(4E(Y_{n+k} - Y_n)^2\right)^{1/2} \end{aligned}$$

From the last two equations, it follows that if the infinite product is > 0, then X_n converges to X in $L^1(\nu)$, so $\nu(X = 0) < 1$, (4.3.3) implies the probability is 0, and the desired result follows from Theorem 4.3.5. □

4.3.4 Branching Processes

Let ξ_i^n, $i,n \geq 1$, be i.i.d. nonnegative integer-valued random variables. Define a sequence $Z_n, n \geq 0$ by $Z_0 = 1$ and

$$Z_{n+1} = \begin{cases} \xi_1^{n+1} + \cdots + \xi_{Z_n}^{n+1} & \text{if } Z_n > 0 \\ 0 & \text{if } Z_n = 0 \end{cases} \tag{4.3.4}$$

Z_n is called a **Galton-Watson process**. The idea behind the definitions is that Z_n is the number of individuals in the nth generation, and each member of the nth generation gives birth independently to an identically distributed number of children. $p_k = P(\xi_i^n = k)$ is called the **offspring distribution**:

Lemma 4.3.9 *Let $\mathcal{F}_n = \sigma(\xi_i^m : i \geq 1, 1 \leq m \leq n)$ and $\mu = E\xi_i^m \in (0, \infty)$. Then Z_n/μ^n is a martingale w.r.t. $\mathcal{F}_n$.*

Proof Clearly, $Z_n \in \mathcal{F}_n$. Using Theorem 4.1.2 to conclude that on $\{Z_n = k\}$

$$E(Z_{n+1}|\mathcal{F}_n) = E(\xi_1^{n+1} + \cdots + \xi_k^{n+1}|\mathcal{F}_n) = k\mu = \mu Z_n$$

where in the second equality we used the fact that the ξ_k^{n+1} are independent of $\mathcal{F}_n$. □

Z_n/μ^n is a nonnegative martingale, so Theorem 4.2.12 implies $Z_n/\mu^n \to$ a limit a.s. We begin by identifying cases when the limit is trivial.

Theorem 4.3.10 *If $\mu < 1$, then $Z_n = 0$ for all n sufficiently large, so $Z_n/\mu^n \to 0$.*

Proof $E(Z_n/\mu^n) = E(Z_0) = 1$, so $E(Z_n) = \mu^n$. Now $Z_n \geq 1$ on $\{Z_n > 0\}$ so

$$P(Z_n > 0) \leq E(Z_n; Z_n > 0) = E(Z_n) = \mu^n \to 0$$

exponentially fast if $\mu < 1$. □

The last answer should be intuitive. If each individual on the average gives birth to less than one child, the species will die out. The next result shows that after we exclude the trivial case in which each individual has exactly one child, the same result holds when $\mu = 1$.

Theorem 4.3.11 *If $\mu = 1$ and $P(\xi_i^m = 1) < 1$, then $Z_n = 0$ for all n sufficiently large.*

Proof When $\mu = 1$, Z_n is itself a nonnegative martingale. Since Z_n is integer valued and by Theorem 4.2.12 converges to an a.s. finite limit Z_∞, we must have $Z_n = Z_\infty$ for large n. If $P(\xi_i^m = 1) < 1$ and $k > 0$, then $P(Z_n = k \text{ for all } n \geq N) = 0$ for any N, so we must have $Z_\infty \equiv 0$. □

When $\mu \leq 1$, the limit of Z_n/μ^n is 0 because the branching process dies out. Our next step is to show that if $\mu > 1$, then $P(Z_n > 0 \text{ for all } n) > 0$. For $s \in [0,1]$, let $\varphi(s) = \sum_{k \geq 0} p_k s^k$, where $p_k = P(\xi_i^m = k)$. φ is the **generating function** for the offspring distribution p_k.

Theorem 4.3.12 *Suppose $\mu > 1$. If $Z_0 = 1$, then $P(Z_n = 0 \text{ for some } n) = \rho$ the only solution of $\varphi(\rho) = \rho$ in $[0, 1)$.*

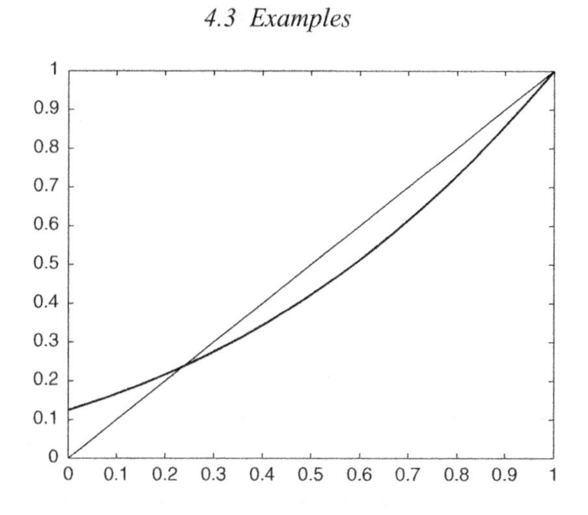

Figure 4.3 Generating function for Binomial(3,1/2).

Proof $\phi(1) = 1$ Differentiating and referring to Theorem A.5.3 for the justification gives for $s < 1$

$$\varphi'(s) = \sum_{k=1}^{\infty} k\, p_k s^{k-1} \geq 0,$$

so ϕ is increasing. We may have $\phi(s) = \infty$ when $s > 1$, so we have to work carefully.

$$\lim_{s \uparrow 1} \varphi'(s) = \sum_{k=1}^{\infty} k p_k = \mu$$

Integrating, we have

$$\phi(1) - \phi(1-h) = \int_{1-h}^{1} \phi'(s)\, ds \sim \mu h$$

as $h \to 0$, so if h is small $\phi(1-h) < 1 - h$. $\phi(0) \geq 0$, so there must be a solution of $\phi(x) = x$ in $[0, 1)$.

To prove uniqueness we note that for $s < 1$

$$\varphi''(s) = \sum_{k=2}^{\infty} k(k-1) p_k s^{k-2} > 0$$

since $\mu > 1$ implies that $p_k > 0$ for some $k \geq 2$. Let ρ be the smallest solution of $\phi(\rho) = \rho$ in $[0, 1)$. Since $\phi(1) = 1$ and ϕ is strictly convex, we have $\phi(x) < x$ for $x \in (\rho, 1)$, so there is only one solution of $\phi(\rho) = \rho$ in $[0, 1)$.

Combining the next two results will complete the proof.

(a) If $\theta_m = P(Z_m = 0)$, then $\theta_m = \sum_{k=0}^{\infty} p_k (\theta_{m-1})^k = \phi(\theta_{m-1})$

Proof of (a). If $Z_1 = k$, an event with probability p_k, then $Z_m = 0$ if and only if all k families die out in the remaining $m - 1$ units of time, an independent event with probability θ_{m-1}^k. Summing over the disjoint possibilities for each k gives the desired result.

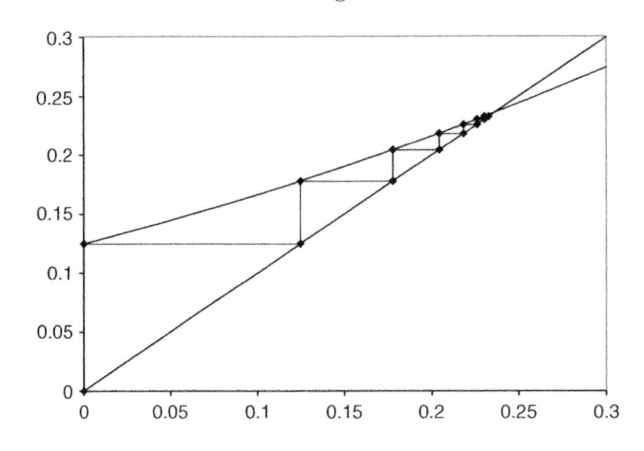

Figure 4.4 Iteration as in part (c) for the Binomial(3,1/2) generating function.

(b) As $m \uparrow \infty, \theta_m \uparrow \rho$.

Proof of (b). Clearly $\theta_m = P(Z_m = 0)$ is increasing. To show by induction that $\theta_m \le \rho$, we note that $\theta_0 = 0 \le \rho$, and if the result is true for $m - 1$

$$\theta_m = \varphi(\theta_{m-1}) \le \varphi(\rho) = \rho.$$

Taking limits in $\theta_m = \varphi(\theta_{m-1})$, we see $\theta_\infty = \varphi(\theta_\infty)$. Since $\theta_\infty \le \rho$, it follows that $\theta_\infty = \rho$. □

The last result shows that when $\mu > 1$, the limit of Z_n/μ^n has a chance of being nonzero. The best result on this question is due to Kesten and Stigum:

Theorem 4.3.13 $W = \lim Z_n/\mu^n$ *is not* $\equiv 0$ *if and only if* $\sum p_k k \log k < \infty$.

For a proof, see Athreya and Ney (1972), pp. 24–29. In the next section, we will show that $\sum k^2 p_k < \infty$ is sufficient for a nontrivial limit.

Exercises

4.3.1 Give an example of a martingale X_n with $\sup_n |X_n| < \infty$ and $P(X_n = a \text{ i.o.}) = 1$ for $a = -1, 0, 1$. This example shows that it is not enough to have $\sup |X_{n+1} - X_n| < \infty$ in Theorem 4.3.1.

4.3.2 (Assumes familiarity with finite state Markov chains.) Fine tune the example for the previous problem so that $P(X_n = 0) \to 1 - 2p$ and $P(X_n = -1), P(X_n = 1) \to p$, where p is your favorite number in $(0, 1/2)$, i.e., you are asked to do this for one value of p that you may choose. This example shows that a martingale can converge in distribution without converging a.s. (or in probability).

4.3.3 Let X_n and Y_n be positive integrable and adapted to $\mathcal{F}_n$. Suppose $E(X_{n+1}|\mathcal{F}_n) \le X_n + Y_n$, with $\sum Y_n < \infty$ a.s. Prove that X_n converges a.s. to a finite limit. Hint: Let $N = \inf_k \sum_{m=1}^k Y_m > M$, and stop your supermartingale at time N.

4.3.4 Let $p_m \in [0, 1)$. Use the Borel-Cantelli lemmas to show that

$$\prod_{m=1}^{\infty} (1 - p_m) = 0 \quad \text{if and only if} \quad \sum_{m=1}^{\infty} p_m = \infty.$$

4.3.5 Show $\sum_{n=2}^{\infty} P(A_n | \cap_{m=1}^{n-1} A_m^c) = \infty$ implies $P(\cap_{m=1}^{\infty} A_m^c) = 0$.

4.3.6 Check by direct computation that the X_n in Example 4.3.7 is a martingale. Show that if we drop the condition $\mu_n \ll \nu_n$ and set $X_n = 0$ when $\nu(I_{k,n}) = 0$, then $E(X_{n+1}|\mathcal{F}_n) \le X_n$.

4.3.7 Apply Theorem 4.3.5 to Example 4.3.7 to get a "probabilistic" proof of the Radon-Nikodym theorem. To be precise, suppose $\mathcal{F}$ is **countably generated** (i.e., there is a sequence of sets A_n so that $\mathcal{F} = \sigma(A_n : n \ge 1)$) and show that if μ and ν are σ-finite measures and $\mu \ll \nu$, then there is a function g so that $\mu(A) = \int_A g \, d\nu$. Before you object to this as circular reasoning (the Radon-Nikodym theorem was used to define conditional expectation!), observe that the conditional expectations that are needed for Example 4.3.7 have elementary definitions.

Bernoulli product measures For the next three exercises, suppose F_n, G_n are concentrated on $\{0, 1\}$ and have $F_n(0) = 1 - \alpha_n$, $G_n(0) = 1 - \beta_n$.

4.3.8 (i) Use Theorem 4.3.8 to find a necessary and sufficient condition for $\mu \ll \nu$. (ii) Suppose that $0 < \epsilon \le \alpha_n, \beta_n \le 1 - \epsilon < 1$. Show that in this case the condition is simply $\sum (\alpha_n - \beta_n)^2 < \infty$.

4.3.9 Show that if $\sum \alpha_n < \infty$ and $\sum \beta_n = \infty$ in the previous exercise, then $\mu \perp \nu$. This shows that the condition $\sum (\alpha_n - \beta_n)^2 < \infty$ is not sufficient for $\mu \ll \nu$ in general.

4.3.10 Suppose $0 < \alpha_n, \beta_n < 1$. Show that $\sum |\alpha_n - \beta_n| < \infty$ is sufficient for $\mu \ll \nu$ in general.

4.3.11 Show that if $P(\lim Z_n/\mu^n = 0) < 1$, then it is $= \rho$ and hence

$$\{\lim Z_n/\mu^n > 0\} = \{Z_n > 0 \text{ for all } n\} \quad \text{a.s.}$$

4.3.12 Let Z_n be a branching process with offspring distribution p_k, defined in part d of Section 4.3, and let $\varphi(\theta) = \sum p_k \theta^k$. Suppose $\rho < 1$ has $\varphi(\rho) = \rho$. Show that ρ^{Z_n} is a martingale and use this to conclude $P(Z_n = 0 \text{ for some } n \ge 1 | Z_0 = x) = \rho^x$.

4.3.13 Galton and Watson, who invented the process that bears their names, were interested in the survival of family names. Suppose each family has exactly three children but coin flips determine their sex. In the 1800s, only male children kept the family name, so following the male offspring leads to a branching process with $p_0 = 1/8$, $p_1 = 3/8$, $p_2 = 3/8$, $p_3 = 1/8$. Compute the probability ρ that the family name will die out when $Z_0 = 1$.

4.4 Doob's Inequality, Convergence in L^p, $p > 1$

We begin by proving a consequence of Theorem 4.2.9.

Theorem 4.4.1 *If X_n is a submartingale and N is a stopping time with $P(N \le k) = 1$, then*

$$EX_0 \le EX_N \le EX_k$$

Remark Let S_n be a simple random walk with $S_0 = 1$ and let $N = \inf\{n : S_n = 0\}$. (See Example 4.2.13 for more details.) $ES_0 = 1 > 0 = ES_N$, so the first inequality need not hold for unbounded stopping times. In Section 5.7 we will give conditions that guarantee $EX_0 \le EX_N$ for unbounded N.

Proof Theorem 4.2.9 implies $X_{N \wedge n}$ is a submartingale, so it follows that

$$EX_0 = EX_{N \wedge 0} \le EX_{N \wedge k} = EX_N$$

To prove the other inequality, let $K_n = 1_{\{N<n\}} = 1_{\{N \le n-1\}}$. K_n is predictable, so Theorem 4.2.8 implies $(K \cdot X)_n = X_n - X_{N \wedge n}$ is a submartingale and it follows that

$$EX_k - EX_N = E(K \cdot X)_k \ge E(K \cdot X)_0 = 0$$

□

We will see later that Theorem 4.4.1 is very useful. The first indication of this is:

Theorem 4.4.2 (Doob's inequality) *Let X_m be a submartingale,*

$$\bar{X}_n = \max_{0 \le m \le n} X_m^+$$

$\lambda > 0$, and $A = \{\bar{X}_n \ge \lambda\}$. Then

$$\lambda P(A) \le EX_n 1_A \le EX_n^+$$

Proof Let $N = \inf\{m : X_m \ge \lambda \text{ or } m = n\}$. Since $X_N \ge \lambda$ on A,

$$\lambda P(A) \le EX_N 1_A \le EX_n 1_A$$

The second inequality follows from the fact that Theorem 4.4.1 implies $EX_N \le EX_n$ and we have $X_N = X_n$ on A^c. The second inequality is trivial, so the proof is complete. □

Example 4.4.3 (Random walks) If we let $S_n = \xi_1 + \cdots + \xi_n$, where the ξ_m are independent and have $E\xi_m = 0, \sigma_m^2 = E\xi_m^2 < \infty$. S_n is a martingale, so Theorem 4.2.6 implies $X_n = S_n^2$ is a submartingale. If we let $\lambda = x^2$ and apply Theorem 4.4.2 to X_n, we get Kolmogorov's maximal inequality, Theorem 2.5.5:

$$P\left(\max_{1 \le m \le n} |S_m| \ge x\right) \le x^{-2} \operatorname{var}(S_n)$$

Integrating the inequality in Theorem 4.4.2 gives:

Theorem 4.4.4 (L^p maximum inequality) *If X_n is a submartingale, then for $1 < p < \infty$,*

$$E(\bar{X}_n^p) \le \left(\frac{p}{p-1}\right)^p E(X_n^+)^p$$

Consequently, if Y_n is a martingale and $Y_n^* = \max_{0 \le m \le n} |Y_m|$,

$$E|Y_n^*|^p \le \left(\frac{p}{p-1}\right)^p E(|Y_n|^p)$$

Proof The second inequality follows by applying the first to $X_n = |Y_n|$. To prove the first we will, for reasons that will become clear in a moment, work with $\bar{X}_n \wedge M$ rather than $\bar{X}_n$. Since $\{\bar{X}_n \wedge M \geq \lambda\}$ is always $\{\bar{X}_n \geq \lambda\}$ or $\emptyset$, this does not change the application of Theorem 4.4.2. Using Lemma 2.2.13, Theorem 4.4.2, Fubini's theorem, and a little calculus gives

$$\begin{aligned}
E((\bar{X}_n \wedge M)^p) &= \int_0^\infty p\lambda^{p-1} P(\bar{X}_n \wedge M \geq \lambda)\, d\lambda \\
&\leq \int_0^\infty p\lambda^{p-1} \left(\lambda^{-1} \int X_n^+ 1_{(\bar{X}_n \wedge M \geq \lambda)}\, dP \right) d\lambda \\
&= \int X_n^+ \int_0^{\bar{X}_n \wedge M} p\lambda^{p-2}\, d\lambda\, dP \\
&= \frac{p}{p-1} \int X_n^+ (\bar{X}_n \wedge M)^{p-1}\, dP
\end{aligned}$$

If we let $q = p/(p-1)$ be the exponent conjugate to p and apply Hölder's inequality, Theorem 1.6.3, we see that

$$\leq \left(\frac{p}{1-p} \right) (E|X_n^+|^p)^{1/p} (E|\bar{X}_n \wedge M|^p)^{1/q}$$

If we divide both sides of the last inequality by $(E|\bar{X}_n \wedge M|^p)^{1/q}$, which is finite thanks to the $\wedge M$, then take the pth power of each side, we get

$$E(|\bar{X}_n \wedge M|^p) \leq \left(\frac{p}{p-1} \right)^p E(X_n^+)^p$$

Letting $M \to \infty$ and using the monotone convergence theorem gives the desired result. □

Example 4.4.5 (There is no L^1 maximal inequality) Again, the counterexample is provided by Example 4.2.13. Let S_n be a simple random walk starting from $S_0 = 1$, $N = \inf\{n : S_n = 0\}$, and $X_n = S_{N \wedge n}$. Theorem 4.4.1 implies $EX_n = ES_{N \wedge n} = ES_0 = 1$ for all n. Using hitting probabilities for simple random walk from Theorem 4.8.7, we have

$$P\left(\max_m X_m \geq M \right) = \frac{1}{M} \tag{4.4.1}$$

so $E(\max_m X_m) = \sum_{M=1}^\infty P(\max_m X_m \geq M) = \sum_{M=1}^\infty 1/M = \infty$. The monotone convergence theorem implies that $E \max_{m \leq n} X_m \uparrow \infty$ as $n \uparrow \infty$.

From Theorem 4.4.4, we get the following:

Theorem 4.4.6 (L^p convergence theorem) *If X_n is a martingale with* $\sup E|X_n|^p < \infty$ *where $p > 1$, then $X_n \to X$ a.s. and in L^p.*

Proof $(EX_n^+)^p \leq (E|X_n|)^p \leq E|X_n|^p$, so it follows from the martingale convergence theorem (4.2.11) that $X_n \to X$ a.s. The second conclusion in Theorem 4.4.4 implies

$$E\left(\sup_{0 \leq m \leq n} |X_m| \right)^p \leq \left(\frac{p}{p-1} \right)^p E|X_n|^p$$

Letting $n \to \infty$ and using the monotone convergence theorem implies $\sup |X_n| \in L^p$. Since $|X_n - X|^p \le (2 \sup |X_n|)^p$, it follows from the dominated convergence theorem that $E|X_n - X|^p \to 0$. □

The most important special case of the results in this section occurs when $p = 2$. To treat this case, the next two results are useful.

Theorem 4.4.7 (Orthogonality of martingale increments) *Let X_n be a martingale with $EX_n^2 < \infty$ for all n. If $m \le n$ and $Y \in \mathcal{F}_m$ has $EY^2 < \infty$, then*

$$E((X_n - X_m)Y) = 0$$

and hence if $\ell < m < n$

$$E((X_n - X_m)(X_m - X_\ell) = 0$$

Proof The Cauchy-Schwarz inequality implies $E|(X_n - X_m)Y| < \infty$. Using (4.1.5), Theorem 4.1.14, and the definition of a martingale,

$$E((X_n - X_m)Y) = E[E((X_n - X_m)Y|\mathcal{F}_m)] = E[YE((X_n - X_m)|\mathcal{F}_m)] = 0$$ □

Theorem 4.4.8 (Conditional variance formula) *If X_n is a martingale with $EX_n^2 < \infty$ for all n,*

$$E((X_n - X_m)^2|\mathcal{F}_m) = E(X_n^2|\mathcal{F}_m) - X_m^2.$$

Remark This is the conditional analogue of $E(X - EX)^2 = EX^2 - (EX)^2$ and is proved in exactly the same way.

Proof Using the linearity of conditional expectation and then Theorem 4.1.14, we have

$$\begin{aligned} E(X_n^2 - 2X_nX_m + X_m^2|\mathcal{F}_m) &= E(X_n^2|\mathcal{F}_m) - 2X_m E(X_n|\mathcal{F}_m) + X_m^2 \\ &= E(X_n^2|\mathcal{F}_m) - 2X_m^2 + X_m^2 \end{aligned}$$

which gives the desired result. □

Example 4.4.9 (Branching processes) We continue the study begun at the end of the last section. Using the notation introduced there, we suppose $\mu = E(\xi_i^m) > 1$ and $\text{var}(\xi_i^m) = \sigma^2 < \infty$. Let $X_n = Z_n/\mu^n$. Taking $m = n - 1$ in Theorem 4.4.8 and rearranging, we have

$$E(X_n^2|\mathcal{F}_{n-1}) = X_{n-1}^2 + E((X_n - X_{n-1})^2|\mathcal{F}_{n-1})$$

To compute the second term, we observe

$$\begin{aligned} E((X_n - X_{n-1})^2|\mathcal{F}_{n-1}) &= E((Z_n/\mu^n - Z_{n-1}/\mu^{n-1})^2|\mathcal{F}_{n-1}) \\ &= \mu^{-2n} E((Z_n - \mu Z_{n-1})^2|\mathcal{F}_{n-1}) \end{aligned}$$

It follows from Exercise 4.1.2 that on $\{Z_{n-1} = k\}$,

$$E((Z_n - \mu Z_{n-1})^2|\mathcal{F}_{n-1}) = E\left(\left(\sum_{i=1}^k \xi_i^n - \mu k\right)^2 \middle| \mathcal{F}_{n-1}\right) = k\sigma^2 = Z_{n-1}\sigma^2$$

Combining the last three equations gives

$$EX_n^2 = EX_{n-1}^2 + E(Z_{n-1}\sigma^2/\mu^{2n}) = EX_{n-1}^2 + \sigma^2/\mu^{n+1}$$

since $E(Z_{n-1}/\mu^{n-1}) = EZ_0 = 1$. Now $EX_0^2 = 1$, so $EX_1^2 = 1 + \sigma^2/\mu^2$, and induction gives

$$EX_n^2 = 1 + \sigma^2 \sum_{k=2}^{n+1} \mu^{-k}$$

This shows $\sup EX_n^2 < \infty$, so $X_n \to X$ in L^2, and hence $EX_n \to EX$. $EX_n = 1$ for all n, so $EX = 1$ and X is not $\equiv 0$. It follows from Exercise 4.3.11 that $\{X > 0\} = \{Z_n > 0 \text{ for all n }\}$.

Exercises

4.4.1 Show that if $j \le k$, then $E(X_j; N = j) \le E(X_k; N = j)$ and sum over j to get a second proof of $EX_N \le EX_k$.

4.4.2 Generalize the proof of Theorem 4.4.1 to show that if X_n is a submartingale and $M \le N$ are stopping times with $P(N \le k) = 1$, then $EX_M \le EX_N$.

4.4.3 Suppose $M \le N$ are stopping times. If $A \in \mathcal{F}_M$, then

$$L = \begin{cases} M & \text{on } A \\ N & \text{on } A^c \end{cases} \quad \text{is a stopping time.}$$

4.4.4 Use the stopping times from the previous exercise to strengthen the conclusion of Exercise 4.4.2 to $X_M \le E(X_N|\mathcal{F}_M)$.

4.4.5 Prove the following variant of the conditional variance formula. If $\mathcal{F} \subset \mathcal{G}$, then

$$E\big(E[Y|\mathcal{G}] - E[Y|\mathcal{F}]\big)^2 = E(E[Y|\mathcal{G}])^2 - E(E[Y|\mathcal{F}])^2$$

4.4.6 Suppose in addition to the conditions introduced earlier that $|\xi_m| \le K$ and let $s_n^2 = \sum_{m \le n} \sigma_m^2$. Exercise 4.2.2 implies that $S_n^2 - s_n^2$ is a martingale. Use this and Theorem 4.4.1 to conclude

$$P\left(\max_{1 \le m \le n} |S_m| \le x\right) \le (x + K)^2 / \operatorname{var}(S_n)$$

4.4.7 The next result gives an extension of Theorem 4.4.2 to $p = 1$. Let X_n be a martingale with $X_0 = 0$ and $EX_n^2 < \infty$. Show that

$$P\left(\max_{1 \le m \le n} X_m \ge \lambda\right) \le EX_n^2/(EX_n^2 + \lambda^2)$$

Hint: Use the fact that $(X_n + c)^2$ is a submartingale and optimize over c.

4.4.8 Let X_n be a submartingale and $\log^+ x = \max(\log x, 0)$.

$$E\bar{X}_n \le (1 - e^{-1})^{-1}\{1 + E(X_n^+ \log^+(X_n^+))\}$$

Prove this by carrying out the following steps: (i) Imitate the proof of 4.4.2 but use the trivial bound $P(A) \le 1$ for $\lambda \le 1$ to show

$$E(\bar{X}_n \wedge M) \le 1 + \int X_n^+ \log(\bar{X}_n \wedge M)\, dP$$

(ii) Use calculus to show $a \log b \le a \log a + b/e \le a \log^+ a + b/e$.

4.4.9 Let X_n and Y_n be martingales with $EX_n^2 < \infty$ and $EY_n^2 < \infty$.

$$EX_nY_n - EX_0Y_0 = \sum_{m=1}^{n} E(X_m - X_{m-1})(Y_m - Y_{m-1})$$

4.4.10 Let X_n, $n \geq 0$, be a martingale and let $\xi_n = X_n - X_{n-1}$ for $n \geq 1$. If EX_0^2, $\sum_{m=1}^{\infty} E\xi_m^2 < \infty$, then $X_n \to X_\infty$ a.s. and in L^2.

4.4.11 Continuing with the notation from the previous problem. If $b_m \uparrow \infty$ and $\sum_{m=1}^{\infty} E\xi_m^2/b_m^2 < \infty$, then $X_n/b_n \to 0$ a.s. In particular, if $E\xi_n^2 \leq K < \infty$ and $\sum_{m=1}^{\infty} b_m^{-2} < \infty$, then $X_n/b_n \to 0$ a.s.

4.5 Square Integrable Martingales*

In this section, we will suppose

$$X_n \text{ is a martingale with } X_0 = 0 \text{ and } EX_n^2 < \infty \text{ for all } n$$

Theorem 4.2.6 implies X_n^2 is a submartingale. It follows from Doob's decomposition Theorem 4.3.2 that we can write $X_n^2 = M_n + A_n$, where M_n is a martingale, and from formulas in Theorems 4.3.2 and 4.4.8 that

$$A_n = \sum_{m=1}^{n} E(X_m^2|\mathcal{F}_{m-1}) - X_{m-1}^2 = \sum_{m=1}^{n} E((X_m - X_{m-1})^2|\mathcal{F}_{m-1})$$

A_n is called the **increasing process** associated with X_n. A_n can be thought of as a path by path measurement of the variance at time n, and $A_\infty = \lim A_n$ as the total variance in the path. Theorems 4.5.2 and 4.5.3 describe the behavior of the martingale on $\{A_n < \infty\}$ and $\{A_n = \infty\}$, respectively. The key to the proof of the first result is the following:

Theorem 4.5.1 $E\left(\sup_m |X_m|^2\right) \leq 4EA_\infty$.

Proof Applying the L^2 maximum inequality (Theorem 4.4.4) to X_n gives

$$E\left(\sup_{0\leq m\leq n} |X_m|^2\right) \leq 4EX_n^2 = 4EA_n$$

since $EX_n^2 = EM_n + EA_n$ and $EM_n = EM_0 = EX_0^2 = 0$. Using the monotone convergence theorem now gives the desired result. □

Theorem 4.5.2 $\lim_{n\to\infty} X_n$ *exists and is finite a.s. on* $\{A_\infty < \infty\}$.

Proof Let $a > 0$. Since $A_{n+1} \in \mathcal{F}_n$, $N = \inf\{n : A_{n+1} > a^2\}$ is a stopping time. Applying Theorem 4.5.1 to $X_{N\wedge n}$ and noticing $A_{N\wedge n} \leq a^2$ gives

$$E\left(\sup_n |X_{N\wedge n}|^2\right) \leq 4a^2$$

so the L^2 convergence theorem, 4.4.6, implies that $\lim X_{N\wedge n}$ exists and is finite a.s. Since a is arbitrary, the desired result follows. □

The next result is a variation on the theme of Exercise 4.4.11.

Theorem 4.5.3 *Let $f \geq 1$ be increasing with $\int_0^\infty f(t)^{-2}\,dt < \infty$. Then $X_n/f(A_n) \to 0$ a.s. on $\{A_\infty = \infty\}$.*

Proof $H_m = f(A_m)^{-1}$ is bounded and predictable, so Theorem 4.2.8 implies

$$Y_n \equiv (H \cdot X)_n = \sum_{m=1}^{n} \frac{X_m - X_{m-1}}{f(A_m)} \quad \text{is a martingale}$$

If B_n is the increasing process associated with Y_n, then

$$\begin{aligned} B_{n+1} - B_n &= E((Y_{n+1} - Y_n)^2 | \mathcal{F}_n) \\ &= E\left(\frac{(X_{n+1} - X_n)^2}{f(A_{n+1})^2} \middle| \mathcal{F}_n \right) = \frac{A_{n+1} - A_n}{f(A_{n+1})^2} \end{aligned}$$

since $f(A_{n+1}) \in \mathcal{F}_n$. Our hypotheses on f imply that

$$\sum_{n=0}^{\infty} \frac{A_{n+1} - A_n}{f(A_{n+1})^2} \leq \sum_{n=0}^{\infty} \int_{[A_n, A_{n+1})} f(t)^{-2}\,dt < \infty$$

so it follows from Theorem 4.5.2 that $Y_n \to Y_\infty$, and the desired conclusion follows from Kronecker's lemma, Theorem 2.5.9. □

Example 4.5.4 Let $\epsilon > 0$ and $f(t) = (t \log^{1+\epsilon} t)^{1/2} \vee 1$. Then f satisfies the hypotheses of Theorem 4.5.3. Let $\xi_1, \xi_2, \ldots$ be independent with $E\xi_m = 0$ and $E\xi_m^2 = \sigma_m^2$. In this case, $X_n = \xi_1 + \cdots + \xi_n$ is a square integrable martingale with $A_n = \sigma_1^2 + \cdots + \sigma_n^2$, so if $\sum_{i=1}^\infty \sigma_i^2 = \infty$, Theorem 4.5.3 implies $X_n/f(A_n) \to 0$ generalizing Theorem 2.5.11.

From Theorem 4.5.3 we get a result due to Dubins and Freedman (1965) that extends our two previous versions in Theorems 2.3.7 and 4.3.4.

Theorem 4.5.5 (Second Borel-Cantelli Lemma, III) *Suppose B_n is adapted to $\mathcal{F}_n$ and let $p_n = P(B_n|\mathcal{F}_{n-1})$. Then*

$$\sum_{m=1}^{n} 1_{B(m)} \Big/ \sum_{m=1}^{n} p_m \to 1 \quad \text{a.s. on} \quad \left\{ \sum_{m=1}^{\infty} p_m = \infty \right\}$$

Proof Define a martingale by $X_0 = 0$ and $X_n - X_{n-1} = 1_{B_n} - P(B_n|\mathcal{F}_{n-1})$ for $n \geq 1$ so that we have

$$\left(\sum_{m=1}^{n} 1_{B(m)} \Big/ \sum_{m=1}^{n} p_m \right) - 1 = X_n \Big/ \sum_{m=1}^{n} p_m$$

The increasing process associated with X_n has

$$\begin{aligned} A_n - A_{n-1} &= E((X_n - X_{n-1})^2 | \mathcal{F}_{n-1}) \\ &= E\left((1_{B_n} - p_n)^2 \middle| \mathcal{F}_{n-1} \right) = p_n - p_n^2 \leq p_n \end{aligned}$$

On $\{A_\infty < \infty\}$, $X_n \to$ a finite limit by Theorem 4.5.2, so on $\{A_\infty < \infty\} \cap \{\sum_m p_m = \infty\}$

$$X_n \Big/ \sum_{m=1}^{n} p_m \to 0$$

$\{A_\infty = \infty\} = \{\sum_m p_m(1 - p_m) = \infty\} \subset \{\sum_m p_m = \infty\}$, so on $\{A_\infty = \infty\}$ the desired conclusion follows from Theorem 4.5.3 with $f(t) = t \vee 1$. □

Remark The trivial example $B_n = \Omega$ for all n shows we may have $A_\infty < \infty$ and $\sum p_m = \infty$ a.s.

Example 4.5.6 (Bernard Friedman's urn) Consider a variant of Polya's urn (see Section 5.3) in which we add a balls of the color drawn and b balls of the opposite color where $a \geq 0$ and $b > 0$. We will show that if we start with g green balls and r red balls, where $g, r > 0$, then the fraction of green balls $g_n \to 1/2$. Let G_n and R_n be the number of green and red balls after the nth draw is completed. Let B_n be the event that the nth ball drawn is green, and let D_n be the number of green balls drawn in the first n draws. It follows from Theorem 4.5.5 that

$$(\star) \qquad D_n \Big/ \sum_{m=1}^{n} g_{m-1} \to 1 \quad \text{a.s. on} \quad \sum_{m=1}^{\infty} g_{m-1} = \infty$$

which always holds since $g_m \geq g/(g + r + (a+b)m)$. At this point, the argument breaks into three cases.

Case 1. $a = b = c$. In this case, the result is trivial since we always add c balls of each color.

Case 2. $a > b$. We begin with the observation

$$(*) \qquad g_{n+1} = \frac{G_{n+1}}{G_{n+1} + R_{n+1}} = \frac{g + aD_n + b(n - D_n)}{g + r + n(a+b)}$$

If $\limsup_{n\to\infty} g_n \leq x$, then $(\star)$ implies $\limsup_{n\to\infty} D_n/n \leq x$ and (since $a > b$)

$$\limsup_{n\to\infty} g_{n+1} \leq \frac{ax + b(1-x)}{a+b} = \frac{b + (a-b)x}{a+b}$$

The right-hand side is a linear function with slope < 1 and fixed point at $1/2$, so starting with the trivial upper bound $x = 1$ and iterating we conclude that $\limsup g_n \leq 1/2$. Interchanging the roles of red and green shows $\liminf_{n\to\infty} g_n \geq 1/2$, and the result follows.

Case 3. $a < b$. The result is easier to believe in this case, since we are adding more balls of the type not drawn but is a little harder to prove. The trouble is that when $b > a$ and $D_n \leq xn$, the right-hand side of $(*)$ is maximized by taking $D_n = 0$, so we need to also use the fact that if r_n is fraction of red balls, then

$$r_{n+1} = \frac{R_{n+1}}{G_{n+1} + R_{n+1}} = \frac{r + bD_n + a(n - D_n)}{g + r + n(a+b)}$$

Combining this with the formula for g_{n+1}, it follows that if $\limsup_{n\to\infty} g_n \leq x$ and $\limsup_{n\to\infty} r_n \leq y$, then

$$\limsup_{n\to\infty} g_n \leq \frac{a(1-y) + by}{a+b} = \frac{a + (b-a)y}{a+b}$$

$$\limsup_{n\to\infty} r_n \leq \frac{bx + a(1-x)}{a+b} = \frac{a + (b-a)x}{a+b}$$

Starting with the trivial bounds $x = 1$, $y = 1$ and iterating (observe the two upper bounds are always the same), we conclude as in Case 2 that both limsups are $\leq 1/2$. □

Remark B. Friedman (1949) considered a number of different urn models. The previous result is due to Freedman (1965), who proved the result by different methods. The previous proof is due to Ornstein and comes from a remark in Freedman's paper.

Theorem 4.5.1 came from using Theorem 4.4.4. If we use Theorem 4.4.2 instead, we get a slightly better result.

Theorem 4.5.7 $E(\sup_n |X_n|) \le 3EA_\infty^{1/2}$.

Proof As in the proof of Theorem 4.5.2, we let $a > 0$ and let $N = \inf\{n : A_{n+1} > a^2\}$. This time, however, our starting point is

$$P\left(\sup_m |X_m| > a\right) \le P(N < \infty) + P\left(\sup_m |X_{N\wedge m}| > a\right)$$

$P(N < \infty) = P(A_\infty > a^2)$. To bound the second term, we apply Theorem 4.4.2 to $X^2_{N\wedge m}$ with $\lambda = a^2$ to get

$$P\left(\sup_{m\le n} |X_{N\wedge m}| > a\right) \le a^{-2}EX^2_{N\wedge n} = a^{-2}EA_{N\wedge n} \le a^{-2}E(A_\infty \wedge a^2)$$

Letting $n \to \infty$ in the last inequality, substituting the result in the first one, and integrating gives

$$\int_0^\infty P\left(\sup_m |X_m| > a\right) da \le \int_0^\infty P(A_\infty > a^2)\, da + \int_0^\infty a^{-2}E(A_\infty \wedge a^2)\, da$$

Since $P(A_\infty > a^2) = P(A_\infty^{1/2} > a)$, the first integral is $EA_\infty^{1/2}$. For the second, we use Lemma 2.2.13 (in the first and fourth steps), Fubini's theorem, and calculus to get

$$\int_0^\infty a^{-2}E(A_\infty \wedge a^2)\, da = \int_0^\infty a^{-2}\int_0^{a^2} P(A_\infty > b)\, db\, da$$
$$= \int_0^\infty P(A_\infty > b)\int_{\sqrt{b}}^\infty a^{-2}\, da\, db = \int_0^\infty b^{-1/2}P(A_\infty > b)\, db = 2EA_\infty^{1/2}$$

which completes the proof. □

Example 4.5.8 Let $\xi_1, \xi_2, \ldots$ be i.i.d. with $P(\xi_i = 1) = P(\xi_i = -1)$. Let $S_n = \xi_1 + \cdots + \xi_n$. Theorem 4.4.1 implies that for any stopping time N, $ES_{N\wedge n} = 0$. Using Theorem 4.5.7, we can conclude that if $EN^{1/2} < \infty$, then $ES_N = 0$. Let $T = \inf\{n : S_n = -1\}$. Since $S_T = -1$ does not have mean 0, it follows that $ET^{1/2} = \infty$.

4.6 Uniform Integrability, Convergence in L^1

In this section, we will give necessary and sufficient conditions for a martingale to converge in L^1. The key to this is the following definition. A collection of random variables X_i, $i \in I$, is said to be **uniformly integrable** if

$$\lim_{M\to\infty}\left(\sup_{i\in I} E(|X_i|; |X_i| > M)\right) = 0$$

If we pick M large enough so that the sup < 1, it follows that

$$\sup_{i \in I} E|X_i| \le M + 1 < \infty$$

This remark will be useful several times later.

A trivial example of a uniformly integrable family is a collection of random variables that are dominated by an integrable random variable, i.e., $|X_i| \le Y$, where $EY < \infty$. Our first result gives an interesting example that shows that uniformly integrable families can be very large.

Theorem 4.6.1 *Given a probability space $(\Omega, \mathcal{F}_o, P)$ and an $X \in L^1$, then $\{E(X|\mathcal{F}) : \mathcal{F}$ is a σ-field $\subset \mathcal{F}_o\}$ is uniformly integrable.*

Proof If A_n is a sequence of sets with $P(A_n) \to 0$, then the dominated convergence theorem implies $E(|X|; A_n) \to 0$. From the last result, it follows that if $\epsilon > 0$, we can pick $\delta > 0$ so that if $P(A) \le \delta$, then $E(|X|; A) \le \epsilon$. (If not, there are sets A_n with $P(A_n) \le 1/n$ and $E(|X|; A_n) > \epsilon$, a contradiction.)

Pick M large enough so that $E|X|/M \le \delta$. Jensen's inequality and the definition of conditional expectation imply

$$\begin{aligned} E(\,|E(X|\mathcal{F})|\,;\,|E(X|\mathcal{F})| > M) &\le E(\,E(|X||\mathcal{F})\,;\,E(|X||\mathcal{F}) > M) \\ &= E(\,|X|\,;\,E(|X||\mathcal{F}) > M) \end{aligned}$$

since $\{E(|X||\mathcal{F}) > M\} \in \mathcal{F}$. Using Chebyshev's inequality and recalling the definition of M, we have

$$P\{E(|X||\mathcal{F}) > M\} \le E\{E(|X||\mathcal{F})\}/M = E|X|/M \le \delta$$

So, by the choice of δ, we have

$$E(|E(X|\mathcal{F})|; |E(X|\mathcal{F})| > M) \le \epsilon \qquad \text{for all } \mathcal{F}$$

Since ϵ was arbitrary, the collection is uniformly integrable. □

A common way to check uniform integrability is to use:

Theorem 4.6.2 *Let $\varphi \ge 0$ be any function with $\varphi(x)/x \to \infty$ as $x \to \infty$, e.g., $\varphi(x) = x^p$ with $p > 1$ or $\varphi(x) = x \log^+ x$. If $E\varphi(|X_i|) \le C$ for all $i \in I$, then $\{X_i : i \in I\}$ is uniformly integrable.*

Proof Let $\epsilon_M = \sup\{x/\phi(x) : x \ge M\}$. For $i \in I$

$$E(|X_i|; |X_i| > M) \le \epsilon_M E(\phi(|X_i|); |X_i| > M) \le C\epsilon_M$$

and $\epsilon_M \to 0$ as $M \to \infty$. □

The relevance of uniform integrability to convergence in L^1 is explained by:

Theorem 4.6.3 *Suppose that $E|X_n| < \infty$ for all n. If $X_n \to X$ in probability, then the following are equivalent:*
(i) $\{X_n : n \ge 0\}$ is uniformly integrable.
(ii) $X_n \to X$ in L^1.
(iii) $E|X_n| \to E|X| < \infty$.

Proof *(i) implies (ii).* Let

$$\varphi_M(x) = \begin{cases} M & \text{if } x \geq M \\ x & \text{if } |x| \leq M \\ -M & \text{if } x \leq -M \end{cases}$$

The triangle inequality implies

$$|X_n - X| \leq |X_n - \varphi_M(X_n)| + |\varphi_M(X_n) - \varphi_M(X)| + |\varphi_M(X) - X|$$

Since $|\varphi_M(Y) - Y)| = (|Y| - M)^+ \leq |Y|1_{(|Y|>M)}$, taking expected value gives

$$E|X_n - X| \leq E|\varphi_M(X_n) - \varphi_M(X)| + E(|X_n|; |X_n| > M) + E(|X|; |X| > M)$$

Theorem 2.3.4 implies that $\varphi_M(X_n) \to \varphi_M(X)$ in probability, so the first term $\to 0$ by the bounded convergence theorem. (See Exercise 2.3.5.) If $\epsilon > 0$ and M is large, uniform integrability implies that the second term $\leq \epsilon$. To bound the third term, we observe that uniform integrability implies $\sup E|X_n| < \infty$, so Fatou's lemma (in the form given in Exercise 2.3.4) implies $E|X| < \infty$, and by making M larger we can make the third term $\leq \epsilon$. Combining the last three facts shows $\limsup E|X_n - X| \leq 2\epsilon$. Since ϵ is arbitrary, this proves (ii).

(ii) implies (iii). Jensen's inequality implies

$$|E|X_n| - E|X|| \leq E||X_n| - |X|| \leq E|X_n - X| \to 0$$

(iii) implies (i). Let

$$\psi_M(x) = \begin{cases} x & \text{on } [0, M-1], \\ 0 & \text{on } [M, \infty) \\ \text{linear} & \text{on } [M-1, M] \end{cases}.$$

The dominated convergence theorem implies that if M is large, $E|X| - E\psi_M(|X|) \leq \epsilon/2$. As in the first part of the proof, the bounded convergence theorem implies $E\psi_M(|X_n|) \to E\psi_M(|X|)$, so using (iii), we get that if $n \geq n_0$

$$\begin{aligned} E(|X_n|; |X_n| > M) &\leq E|X_n| - E\psi_M(|X_n|) \\ &\leq E|X| - E\psi_M(|X|) + \epsilon/2 < \epsilon \end{aligned}$$

By choosing M larger, we can make $E(|X_n|; |X_n| > M) \leq \epsilon$ for $0 \leq n < n_0$, so X_n is uniformly integrable. □

We are now ready to state the main theorems of this section. We have already done all the work, so the proofs are short.

Theorem 4.6.4 *For a submartingale, the following are equivalent:*
(i) It is uniformly integrable.
(ii) It converges a.s. and in L^1.
(iii) It converges in L^1.

Proof *(i) implies (ii).* Uniform integrability implies $\sup E|X_n| < \infty$, so the martingale convergence theorem implies $X_n \to X$ a.s., and Theorem 4.6.3 implies $X_n \to X$ in L^1.

(ii) implies (iii). Trivial. *(iii) implies (i).* $X_n \to X$ in L^1 implies $X_n \to X$ in probability, (see Lemma 2.2.2), so this follows from Theorem 4.6.3. □

Before proving the analogue of Theorem 4.6.4 for martingales, we will isolate two parts of the argument that will be useful later.

Lemma 4.6.5 *If integrable random variables $X_n \to X$ in L^1, then*

$$E(X_n; A) \to E(X; A)$$

Proof $|EX_m 1_A - EX1_A| \le E|X_m 1_A - X1_A| \le E|X_m - X| \to 0$ □

Lemma 4.6.6 *If a martingale $X_n \to X$ in L^1, then $X_n = E(X|\mathcal{F}_n)$.*

Proof The martingale property implies that if $m > n$, $E(X_m|\mathcal{F}_n) = X_n$, so if $A \in \mathcal{F}_n$, $E(X_n; A) = E(X_m; A)$. Lemma 4.6.5 implies $E(X_m; A) \to E(X; A)$, so we have $E(X_n; A) = E(X; A)$ for all $A \in \mathcal{F}_n$. Recalling the definition of conditional expectation, it follows that $X_n = E(X|\mathcal{F}_n)$. □

Theorem 4.6.7 *For a martingale, the following are equivalent:*
(i) It is uniformly integrable.
(ii) It converges a.s. and in L^1.
(iii) It converges in L^1.
(iv) There is an integrable random variable X so that $X_n = E(X|\mathcal{F}_n)$.

Proof *(i) implies (ii).* Since martingales are also submartingales, this follows from Theorem 4.6.4. *(ii) implies (iii).* Trivial. *(iii) implies (iv).* This follows from Lemma 4.6.6. *(iv) implies (i).* This follows from Theorem 4.6.1. □

The next result is related to Lemma 4.6.6 but goes in the other direction.

Theorem 4.6.8 *Suppose $\mathcal{F}_n \uparrow \mathcal{F}_\infty$, i.e., $\mathcal{F}_n$ is an increasing sequence of σ-fields and $\mathcal{F}_\infty = \sigma(\cup_n \mathcal{F}_n)$. As $n \to \infty$,*

$$E(X|\mathcal{F}_n) \to E(X|\mathcal{F}_\infty) \quad \text{a.s. and in } L^1$$

Proof The first step is to note that if $m > n$, then Theorem 4.1.13 implies

$$E(E(X|\mathcal{F}_m)|\mathcal{F}_n) = E(X|\mathcal{F}_n)$$

so $Y_n = E(X|\mathcal{F}_n)$ is a martingale. Theorem 4.6.1 implies that Y_n is uniformly integrable, so Theorem 4.6.7 implies that Y_n converges a.s. and in L^1 to a limit Y_∞. The definition of Y_n and Lemma 4.6.6 imply $E(X|\mathcal{F}_n) = Y_n = E(Y_\infty|\mathcal{F}_n)$, and hence

$$\int_A X\,dP = \int_A Y_\infty\,dP \quad \text{for all } A \in \mathcal{F}_n$$

Since X and Y_∞ are integrable, and $\cup_n \mathcal{F}_n$ is a π-system, the $\pi - \lambda$ theorem implies that the last result holds for all $A \in \mathcal{F}_\infty$. Since $Y_\infty \in \mathcal{F}_\infty$, it follows that $Y_\infty = E(X|\mathcal{F}_\infty)$. □

An immediate consequence of Theorem 4.6.8 is:

Theorem 4.6.9 (Lévy's 0-1 law) *If $\mathcal{F}_n \uparrow \mathcal{F}_\infty$ and $A \in \mathcal{F}_\infty$, then $E(1_A|\mathcal{F}_n) \to 1_A$ a.s.*

To steal a line from Chung: *"The reader is urged to ponder over the meaning of this result and judge for himself whether it is obvious or incredible."* We will now argue for the two points of view.

"It is obvious." $1_A \in \mathcal{F}_\infty$, and $\mathcal{F}_n \uparrow \mathcal{F}_\infty$, so our best guess of 1_A given the information in $\mathcal{F}_n$ should approach 1_A (the best guess given $\mathcal{F}_\infty$).

"It is incredible." Let $X_1, X_2, \ldots$ be independent and suppose $A \in \mathcal{T}$, the tail σ-field. For each n, A is independent of $\mathcal{F}_n$, so $E(1_A|\mathcal{F}_n) = P(A)$. As $n \to \infty$, the left-hand side converges to 1_A a.s., so $P(A) = 1_A$ a.s., and it follows that $P(A) \in \{0, 1\}$, i.e., we have proved Kolmogorov's 0-1 law.

The last argument may not show that Theorem 4.6.9 is "too unusual or improbable to be possible," but this and other applications of Theorem 4.6.9 show that it is a very useful result.

A more technical consequence of Theorem 4.6.8 is:

Theorem 4.6.10 (Dominated convergence theorem for conditional expectations) *Suppose $Y_n \to Y$ a.s. and $|Y_n| \le Z$ for all n, where $EZ < \infty$. If $\mathcal{F}_n \uparrow \mathcal{F}_\infty$, then*

$$E(Y_n|\mathcal{F}_n) \to E(Y|\mathcal{F}_\infty) \quad a.s.$$

Proof Let $W_N = \sup\{|Y_n - Y_m| : n, m \ge N\}$. $W_N \le 2Z$, so $EW_N < \infty$. Using monotonicity (4.1.2) and applying Theorem 4.6.8 to W_N gives

$$\limsup_{n\to\infty} E(|Y_n - Y||\mathcal{F}_n) \le \lim_{n\to\infty} E(W_N|\mathcal{F}_n) = E(W_N|\mathcal{F}_\infty)$$

The last result is true for all N and $W_N \downarrow 0$ as $N \uparrow \infty$, so (4.1.3) implies $E(W_N|\mathcal{F}_\infty) \downarrow 0$, and Jensen's inequality gives us

$$|E(Y_n|\mathcal{F}_n) - E(Y|\mathcal{F}_n)| \le E(|Y_n - Y||\mathcal{F}_n) \to 0 \quad \text{a.s. as } n \to \infty$$

Theorem 4.6.8 implies $E(Y|\mathcal{F}_n) \to E(Y|\mathcal{F}_\infty)$ a.s. The desired result follows from the last two conclusions and the triangle inequality. □

Example 4.6.11 Suppose $X_1, X_2, \ldots$ are uniformly integrable and $\to X$ a.s. Theorem 4.6.3 implies $X_n \to X$ in L^1 and combining this with Exercise 4.6.7 shows $E(X_n|\mathcal{F}) \to E(X|\mathcal{F})$ in L^1. We will now show that $E(X_n|\mathcal{F})$ need not converge a.s. Let $Y_1, Y_2, \ldots$ and $Z_1, Z_2, \ldots$ be independent r.v.'s with

$$P(Y_n = 1) = 1/n \qquad P(Y_n = 0) = 1 - 1/n$$
$$P(Z_n = n) = 1/n \qquad P(Z_n = 0) = 1 - 1/n$$

Let $X_n = Y_n Z_n$. $P(X_n > 0) = 1/n^2$, so the Borel-Cantelli lemma implies $X_n \to 0$ a.s. $E(X_n; |X_n| \ge 1) = n/n^2$, so X_n is uniformly integrable. Let $\mathcal{F} = \sigma(Y_1, Y_2, \ldots)$.

$$E(X_n|\mathcal{F}) = Y_n E(Z_n|\mathcal{F}) = Y_n E Z_n = Y_n$$

Since $Y_n \to 0$ in L^1 but not a.s., the same is true for $E(X_n|\mathcal{F})$.

Exercises

4.6.1 Let $Z_1, Z_2, \dots$ be i.i.d. with $E|Z_i| < \infty$, let θ be an independent r.v. with finite mean, and let $Y_i = Z_i + \theta$. If Z_i is normal(0,1), then in statistical terms we have a sample from a normal population with variance 1 and unknown mean. The distribution of θ is called the **prior distribution**, and $P(\theta \in \cdot|Y_1, \dots, Y_n)$ is called the **posterior distribution** after n observations. Show that $E(\theta|Y_1, \dots, Y_n) \to \theta$ a.s.

In the next two exercises, $\Omega = [0,1)$, $I_{k,n} = [k2^{-n}, (k+1)2^{-n})$, and $\mathcal{F}_n = \sigma(I_{k,n} : 0 \le k < 2^n)$.

4.6.2 f is said to be **Lipschitz continuous** if $|f(t) - f(s)| \le K|t-s|$ for $0 \le s,t < 1$. Show that $X_n = (f((k+1)2^{-n}) - f(k2^{-n}))/2^{-n}$ on $I_{k,n}$ defines a martingale, $X_n \to X_\infty$ a.s. and in L^1, and

$$f(b) - f(a) = \int_a^b X_\infty(\omega)\, d\omega$$

4.6.3 Suppose f is integrable on [0,1). $E(f|\mathcal{F}_n)$ is a step function and $\to f$ in L^1. From this it follows immediately that if $\epsilon > 0$, there is a step function g on [0,1] with $\int |f - g|\, dx < \epsilon$. This approximation is much simpler than the bare-hands approach we used in Exercise 1.4.3, but of course we are using a lot of machinery.

4.6.4 Let X_n be r.v.'s taking values in $[0, \infty)$. Let $D = \{X_n = 0 \text{ for some } n \ge 1\}$ and assume

$$P(D|X_1, \dots, X_n) \ge \delta(x) > 0 \quad \text{a.s. on } \{X_n \le x\}$$

Use Theorem 4.6.9 to conclude that $P(D \cup \{\lim_n X_n = \infty\}) = 1$.

4.6.5 Let Z_n be a branching process with offspring distribution p_k (see the end of Section 5.3 for definitions). Use the last result to show that if $p_0 > 0$, then $P(\lim_n Z_n = 0 \text{ or } \infty) = 1$.

4.6.6 Let $X_n \in [0,1]$ be adapted to $\mathcal{F}_n$. Let $\alpha, \beta > 0$ with $\alpha + \beta = 1$ and suppose

$$P(X_{n+1} = \alpha + \beta X_n|\mathcal{F}_n) = X_n \qquad P(X_{n+1} = \beta X_n|\mathcal{F}_n) = 1 - X_n$$

Show $P(\lim_n X_n = 0 \text{ or } 1) = 1$ and if $X_0 = \theta$, then $P(\lim_n X_n = 1) = \theta$.

4.6.7 Show that if $\mathcal{F}_n \uparrow \mathcal{F}_\infty$ and $Y_n \to Y$ in L^1, then $E(Y_n|\mathcal{F}_n) \to E(Y|\mathcal{F}_\infty)$ in L^1.

4.7 Backwards Martingales

A **backwards martingale** (some authors call them reversed) is a martingale indexed by the negative integers, i.e., X_n, $n \le 0$, adapted to an increasing sequence of σ-fields $\mathcal{F}_n$ with

$$E(X_{n+1}|\mathcal{F}_n) = X_n \qquad \text{for } n \le -1$$

Because the σ-fields decrease as $n \downarrow -\infty$, the convergence theory for backwards martingales is particularly simple.

Theorem 4.7.1 *$X_{-\infty} = \lim_{n\to-\infty} X_n$ exists a.s. and in L^1.*

Proof Let U_n be the number of upcrossings of $[a,b]$ by $X_{-n},\dots,X_0$. The upcrossing inequality, Theorem 4.2.10, implies $(b-a)EU_n \le E(X_0-a)^+$. Letting $n \to \infty$ and using the monotone convergence theorem, we have $EU_\infty < \infty$, so by the remark after the proof of Theorem 4.2.11, the limit exists a.s. The martingale property implies $X_n = E(X_0|\mathcal{F}_n)$, so Theorem 4.6.1 implies X_n is uniformly integrable and Theorem 4.6.3 tells us that the convergence occurs in L^1. □

The next result identifies the limit in Theorem 4.7.1.

Theorem 4.7.2 *If* $X_{-\infty} = \lim_{n\to-\infty} X_n$ *and* $\mathcal{F}_{-\infty} = \cap_n \mathcal{F}_n$, *then* $X_{-\infty} = E(X_0|\mathcal{F}_{-\infty})$.

Proof Clearly, $X_{-\infty} \in \mathcal{F}_{-\infty}$. $X_n = E(X_0|\mathcal{F}_n)$, so if $A \in \mathcal{F}_{-\infty} \subset \mathcal{F}_n$, then

$$\int_A X_n\, dP = \int_A X_0\, dP$$

Theorem 4.7.1 and Lemma 4.6.5 imply $E(X_n;A) \to E(X_{-\infty};A)$, so

$$\int_A X_{-\infty}\, dP = \int_A X_0\, dP$$

for all $A \in \mathcal{F}_{-\infty}$, proving the desired conclusion. □

The next result is Theorem 4.6.8 backwards.

Theorem 4.7.3 *If* $\mathcal{F}_n \downarrow \mathcal{F}_{-\infty}$ *as* $n \downarrow -\infty$ *(i.e.,* $\mathcal{F}_{-\infty} = \cap_n \mathcal{F}_n$*), then*

$$E(Y|\mathcal{F}_n) \to E(Y|\mathcal{F}_{-\infty}) \quad \text{a.s. and in } L^1$$

Proof $X_n = E(Y|\mathcal{F}_n)$ is a backwards martingale, so Theorem 4.7.1 and 4.7.2 imply that as $n \downarrow -\infty$, $X_n \to X_{-\infty}$ a.s. and in L^1, where

$$X_{-\infty} = E(X_0|\mathcal{F}_{-\infty}) = E(E(Y|\mathcal{F}_0)|\mathcal{F}_{-\infty}) = E(Y|\mathcal{F}_{-\infty})$$

□

Even though the convergence theory for backwards martingales is easy, there are some nice applications. For the rest of the section, we return to the special space utilized in Section 4.1, so we can utilize definitions given there. That is, we suppose

$$\Omega = \{(\omega_1,\omega_2,\dots) : \omega_i \in S\}$$
$$\mathcal{F} = \mathcal{S} \times \mathcal{S} \times \dots$$
$$X_n(\omega) = \omega_n$$

Let $\mathcal{E}_n$ be the σ-field generated by events that are invariant under permutations that leave $n+1, n+2, \dots$ fixed and let $\mathcal{E} = \cap_n \mathcal{E}_n$ be the exchangeable σ-field.

Example 4.7.4 (Strong law of large numbers) Let $\xi_1,\xi_2,\dots$ be i.i.d. with $E|\xi_i| < \infty$. Let $S_n = \xi_1 + \dots + \xi_n$, let $X_{-n} = S_n/n$, and let

$$\mathcal{F}_{-n} = \sigma(S_n, S_{n+1}, S_{n+2}, \dots) = \sigma(S_n, \xi_{n+1}, \xi_{n+2}, \dots)$$

To compute $E(X_{-n}|\mathcal{F}_{-n-1})$, we observe that if $j,k \le n+1$, symmetry implies $E(\xi_j|\mathcal{F}_{-n-1}) = E(\xi_k|\mathcal{F}_{-n-1})$, so

$$E(\xi_{n+1}|\mathcal{F}_{-n-1}) = \frac{1}{n+1}\sum_{k=1}^{n+1} E(\xi_k|\mathcal{F}_{-n-1})$$
$$= \frac{1}{n+1}E(S_{n+1}|\mathcal{F}_{-n-1}) = \frac{S_{n+1}}{n+1}$$

Since $X_{-n} = (S_{n+1} - \xi_{n+1})/n$, it follows that

$$E(X_{-n}|\mathcal{F}_{-n-1}) = E(S_{n+1}/n|\mathcal{F}_{-n-1}) - E(\xi_{n+1}/n|\mathcal{F}_{-n-1})$$
$$= \frac{S_{n+1}}{n} - \frac{S_{n+1}}{n(n+1)} = \frac{S_{n+1}}{n+1} = X_{-n-1}$$

The last computation shows X_{-n} is a backwards martingale, so it follows from Theorems 4.7.1 and 4.7.2 that $\lim_{n\to\infty} S_n/n = E(X_{-1}|\mathcal{F}_{-\infty})$. Since $\mathcal{F}_{-n} \subset \mathcal{E}_n$, $\mathcal{F}_{-\infty} \subset \mathcal{E}$. The Hewitt-Savage 0-1 law (Theorem 2.5.4) says $\mathcal{E}$ is trivial, so we have

$$\lim_{n\to\infty} S_n/n = E(X_{-1}) \quad \text{a.s.}$$

Example 4.7.5 (Ballot theorem) Let $\{\xi_j, 1 \le j \le n\}$ be i.i.d. nonnegative integer-valued r.v.'s, let $S_k = \xi_1 + \cdots + \xi_k$, and let $G = \{S_j < j \text{ for } 1 \le j \le n\}$. Then

$$P(G|S_n) = (1 - S_n/n)^+ \tag{4.7.1}$$

Remark To explain the name, consider an election in which candidate B gets β votes and A gets $\alpha > \beta$ votes. Let $\xi_1, \xi_2, \ldots, \xi_n$ be i.i.d. and take values 0 or 2 with probability 1/2 each. Interpreting 0's and 2's as votes for candidates A and B, we see that $G = \{A \text{ leads } B \text{ throughout the counting}\}$, so if $n = \alpha + \beta$

$$P(G|B \text{ gets } \beta \text{ votes }) = \left(1 - \frac{2\beta}{n}\right)^+ = \frac{\alpha - \beta}{\alpha + \beta}$$

the result in Theorem 4.9.2.

Proof The result is trivial when $S_n \ge n$, so suppose $S_n < n$. Computations in Example 4.7.4 show that $X_{-j} = S_j/j$ is a martingale w.r.t. $\mathcal{F}_{-j} = \sigma(S_j, \ldots, S_n)$. Let $T = \inf\{k \ge -n : X_k \ge 1\}$ and set $T = -1$ if the set is $\emptyset$. We claim that $X_T = 1$ on G^c. To check this, note that if $S_{j+1} < j+1$, then the fact that the ξ_i are nonnegative integer values implies $S_j \le S_{j+1} \le j$. Since $G \subset \{T = -1\}$ and $S_1 < 1$ implies $S_1 = 0$, we have $X_T = 0$ on G. Noting $\mathcal{F}_{-n} = \sigma(S_n)$ and using Exercise 4.4.4, we see that on $\{S_n < n\}$

$$P(G^c|S_n) = E(X_T|\mathcal{F}_{-n}) = X_{-n} = S_n/n$$

Subtracting from 1 and recalling that this computation has been done under the assumption $S_n < n$ gives the desired result. □

Example 4.7.6 (Hewitt-Savage 0-1 law) If $X_1, X_2, \ldots$ are i.i.d. and $A \in \mathcal{E}$, then $P(A) \in \{0, 1\}$.

The key to the new proof is:

Lemma 4.7.7 *Suppose $X_1, X_2, \ldots$ are i.i.d. and let*

$$A_n(\varphi) = \frac{1}{(n)_k} \sum_i \varphi(X_{i_1}, \ldots, X_{i_k})$$

where the sum is over all sequences of distinct integers $1 \le i_1, \ldots, i_k \le n$ and

$$(n)_k = n(n-1)\cdots(n-k+1)$$

is the number of such sequences. If φ is bounded, $A_n(\varphi) \to E\varphi(X_1, \ldots, X_k)$ a.s.

Proof $A_n(\varphi) \in \mathcal{E}_n$, so

$$A_n(\varphi) = E(A_n(\varphi)|\mathcal{E}_n) = \frac{1}{(n)_k} \sum_i E(\varphi(X_{i_1}, \ldots, X_{i_k})|\mathcal{E}_n)$$
$$= E(\varphi(X_1, \ldots, X_k)|\mathcal{E}_n)$$

since all the terms in the sum are the same. Theorem 4.7.3 with $\mathcal{F}_{-m} = \mathcal{E}_m$ for $m \ge 1$ implies that

$$E(\varphi(X_1, \ldots, X_k)|\mathcal{E}_n) \to E(\varphi(X_1, \ldots, X_k)|\mathcal{E})$$

We want to show that the limit is $E(\varphi(X_1, \ldots, X_k))$. The first step is to observe that there are $k(n-1)_{k-1}$ terms in $A_n(\varphi)$ involving X_1 and φ is bounded, so if we let $1 \in i$ denote the sum over sequences that contain 1

$$\frac{1}{(n)_k} \sum_{1 \in i} \varphi(X_{i_1}, \ldots, X_{i_k}) \le \frac{k(n-1)_{k-1}}{(n)_k} \sup \phi \to 0$$

This shows that

$$E(\varphi(X_1, \ldots, X_k)|\mathcal{E}) \in \sigma(X_2, X_3, \ldots)$$

Repeating the argument for $2, 3, \ldots, k$ shows

$$E(\varphi(X_1, \ldots, X_k)|\mathcal{E}) \in \sigma(X_{k+1}, X_{k+2}, \ldots)$$

Intuitively, if the conditional expectation of a r.v. is independent of the r.v., then

(a) $$E(\varphi(X_1, \ldots, X_k)|\mathcal{E}) = E(\varphi(X_1, \ldots, X_k))$$

To show this, we prove:

(b) If $EX^2 < \infty$ and $E(X|\mathcal{G}) \in \mathcal{F}$ with X independent of $\mathcal{F}$, then $E(X|\mathcal{G}) = EX$.

Proof Let $Y = E(X|\mathcal{G})$ and note that Theorem 4.1.11 implies $EY^2 \le EX^2 < \infty$. By independence, $EXY = EX\,EY = (EY)^2$ since $EY = EX$. From the geometric interpretation of conditional expectation, Theorem 4.1.15, $E((X-Y)Y) = 0$, so $EY^2 = EXY = (EY)^2$ and $\text{var}\,(Y) = EY^2 - (EY)^2 = 0$. □

(a) holds for all bounded φ, so $\mathcal{E}$ is independent of $\mathcal{G}_k = \sigma(X_1, \ldots, X_k)$. Since this holds for all k, and $\cup_k \mathcal{G}_k$ is a π-system that contains Ω, Theorem 2.1.6 implies $\mathcal{E}$ is independent of $\sigma(\cup_k \mathcal{G}_k) \supset \mathcal{E}$, and we get the usual 0-1 law punch line. If $A \in \mathcal{E}$, it is independent of itself, and hence $P(A) = P(A \cap A) = P(A)P(A)$, i.e., $P(A) \in \{0, 1\}$. □

Example 4.7.8 (de Finetti's Theorem) A sequence $X_1, X_2, \ldots$ is said to be **exchangeable** if for each n and permutation π of $\{1, \ldots, n\}$, $(X_1, \ldots, X_n)$ and $(X_{\pi(1)}, \ldots, X_{\pi(n)})$ have the same distribution.

Theorem 4.7.9 (de Finetti's Theorem) *If $X_1, X_2, \ldots$ are exchangeable then conditional on $\mathcal{E}$, $X_1, X_2, \ldots$ are independent and identically distributed.*

Proof Repeating the first calculation in the proof of Lemma 4.7.7 and using the notation introduced there shows that for any exchangeable sequence:

$$
\begin{aligned}
A_n(\varphi) = E(A_n(\varphi)|\mathcal{E}_n) &= \frac{1}{(n)_k} \sum_i E(\varphi(X_{i_1}, \ldots, X_{i_k})|\mathcal{E}_n) \\
&= E(\varphi(X_1, \ldots, X_k)|\mathcal{E}_n)
\end{aligned}
$$

since all the terms in the sum are the same. Again, Theorem 4.7.3 implies that

$$
A_n(\varphi) \to E(\varphi(X_1, \ldots, X_k)|\mathcal{E}) \tag{4.7.2}
$$

This time, however, $\mathcal{E}$ may be nontrivial, so we cannot hope to show that the limit is $E(\varphi(X_1, \ldots, X_k))$.

Let f and g be bounded functions on $\mathbf{R}^{k-1}$ and $\mathbf{R}$, respectively. If we let $I_{n,k}$ be the set of all sequences of distinct integers $1 \le i_1, \ldots, i_k \le n$, then

$$
\begin{aligned}
(n)_{k-1} A_n(f)\, n A_n(g) &= \sum_{i \in I_{n,k-1}} f(X_{i_1}, \ldots, X_{i_{k-1}}) \sum_m g(X_m) \\
&= \sum_{i \in I_{n,k}} f(X_{i_1}, \ldots, X_{i_{k-1}}) g(X_{i_k}) \\
&\quad + \sum_{i \in I_{n,k-1}} \sum_{j=1}^{k-1} f(X_{i_1}, \ldots, X_{i_{k-1}}) g(X_{i_j})
\end{aligned}
$$

If we let $\varphi(x_1, \ldots, x_k) = f(x_1, \ldots, x_{k-1}) g(x_k)$, note that

$$
\frac{(n)_{k-1} n}{(n)_k} = \frac{n}{(n-k+1)} \quad \text{and} \quad \frac{(n)_{k-1}}{(n)_k} = \frac{1}{(n-k+1)}
$$

then rearrange, we have

$$
A_n(\varphi) = \frac{n}{n-k+1} A_n(f) A_n(g) - \frac{1}{n-k+1} \sum_{j=1}^{k-1} A_n(\varphi_j)
$$

where $\varphi_j(x_1, \ldots, x_{k-1}) = f(x_1, \ldots, x_{k-1}) g(x_j)$. Applying (4.7.2) to φ, f, g, and all the φ_j gives

$$
E(f(X_1, \ldots, X_{k-1}) g(X_k)|\mathcal{E}) = E(f(X_1, \ldots, X_{k-1})|\mathcal{E}) E(g(X_k)|\mathcal{E})
$$

It follows by induction that

$$
E\left(\prod_{j=1}^k f_j(X_j) \middle| \mathcal{E} \right) = \prod_{j=1}^k E(f_j(X_j)|\mathcal{E})
$$

□

When the X_i take values in a nice space, there is a regular conditional distribution for $(X_1, X_2, \ldots)$ given $\mathcal{E}$, and the sequence can be represented as a mixture of i.i.d. sequences. Hewitt and Savage (1956) call the sequence **presentable** in this case. For the usual measure theoretic problems, the last result is not valid when the X_i take values in an arbitrary measure space. See Dubins and Freedman (1979) and Freedman (1980) for counterexamples.

The simplest special case of Theorem 4.7.9 occurs when the $X_i \in \{0, 1\}$. In this case:

Theorem 4.7.10 *If $X_1, X_2, \ldots$ are exchangeable and take values in $\{0, 1\}$, then there is a probability distribution on $[0, 1]$ so that*

$$P(X_1 = 1, \ldots, X_k = 1, X_{k+1} = 0, \ldots, X_n = 0) = \int_0^1 \theta^k (1 - \theta)^{n-k} \, dF(\theta)$$

This result is useful for people concerned about the foundations of statistics (see Section 3.7 of Savage (1972)), since from the palatable assumption of symmetry one gets the powerful conclusion that the sequence is a mixture of i.i.d. sequences. Theorem 4.7.10 has been proved in a variety of different ways. See Feller, Vol. II (1971), pp. 228–229 for a proof that is related to the moment problem. Diaconis and Freedman (1980) have a nice proof that starts with the trivial observation that the distribution of a finite exchangeable sequence X_m, $1 \le m \le n$ has the form $p_0 H_{0,n} + \cdots + p_n H_{n,n}$, where $H_{m,n}$ is "drawing without replacement from an urn with m ones and $n - m$ zeros." If $m \to \infty$ and $m/n \to p$, then $H_{m,n}$ approaches product measure with density p. Theorem 4.7.10 follows easily from this, and one can get bounds on the rate of convergence.

Exercises

4.7.1 Show that if a backwards martingale has $X_0 \in L^p$, the convergence occurs in L^p.

4.7.2 Prove the backwards analogue of Theorem 4.6.10. Suppose $Y_n \to Y_{-\infty}$ a.s. as $n \to -\infty$ and $|Y_n| \le Z$ a.s., where $EZ < \infty$. If $\mathcal{F}_n \downarrow \mathcal{F}_{-\infty}$, then $E(Y_n|\mathcal{F}_n) \to E(Y_{-\infty}|\mathcal{F}_{-\infty})$ a.s.

4.7.3 Prove directly from the definition that if $X_1, X_2, \ldots \in \{0, 1\}$ are exchangeable

$$P(X_1 = 1, \ldots, X_k = 1 | S_n = m) = \binom{n-k}{n-m} \Big/ \binom{n}{m}$$

4.7.4 If $X_1, X_2, \ldots \in \mathbf{R}$ are exchangeable with $EX_i^2 < \infty$, then $E(X_1 X_2) \ge 0$.

4.7.5 Use the first few lines of the proof of Lemma 4.7.7 to conclude that if $X_1, X_2, \ldots$ are i.i.d. with $EX_i = \mu$ and $\text{var}(X_i) = \sigma^2 < \infty$, then

$$\binom{n}{2}^{-1} \sum_{1 \le i < j \le n} (X_i - X_j)^2 \to 2\sigma^2$$

4.8 Optional Stopping Theorems

In this section, we will prove a number of results that allow us to conclude that if X_n is a submartingale and $M \le N$ are stopping times, then $EX_M \le EX_N$. Example 4.2.13 shows that this is not always true, but Exercise 4.4.2 shows this is true if N is bounded, so our attention will be focused on the case of unbounded N.

Theorem 4.8.1 *If X_n is a uniformly integrable submartingale, then for any stopping time N, $X_{N\wedge n}$ is uniformly integrable.*

As in Theorem 4.2.5, the last result implies one for supermartingales with $\geq$ and one for martingales with $=$. This is true for the next two theorems as well.

Proof X_n^+ is a submartingale, so Theorem 4.4.1 implies $EX_{N\wedge n}^+ \leq EX_n^+$. Since X_n^+ is uniformly integrable, it follows from the remark after the definition that

$$\sup_n EX_{N\wedge n}^+ \leq \sup_n EX_n^+ < \infty$$

Using the martingale convergence theorem (4.2.11) now gives $X_{N\wedge n} \to X_N$ a.s. (here $X_\infty = \lim_n X_n$) and $E|X_N| < \infty$. With this established, the rest is easy. We write

$$\begin{aligned} E(|X_{N\wedge n}|; |X_{N\wedge n}| > K) &= E(|X_N|; |X_N| > K, N \leq n) \\ &\quad + E(|X_n|; |X_n| > K, N > n) \end{aligned}$$

Since $E|X_N| < \infty$ and X_n is uniformly integrable, if K is large, then each term is $< \epsilon/2$. □

From the last computation in the proof of Theorem 4.8.1, we get:

Theorem 4.8.2 *If $E|X_N| < \infty$ and $X_n 1_{(N>n)}$ is uniformly integrable, then $X_{N\wedge n}$ is uniformly integrable and hence $EX_0 \leq EX_N$.*

Theorem 4.8.3 *If X_n is a uniformly integrable submartingale, then for any stopping time $N \leq \infty$, we have $EX_0 \leq EX_N \leq EX_\infty$, where $X_\infty = \lim X_n$.*

Proof Theorem 4.4.1 implies $EX_0 \leq EX_{N\wedge n} \leq EX_n$. Letting $n \to \infty$ and observing that Theorem 4.8.1 and 4.6.4 imply $X_{N\wedge n} \to X_N$ and $X_n \to X_\infty$ in L^1 gives the desired result. □

The next result does not require uniform integrability.

Theorem 4.8.4 *If X_n is a nonnegative supermartingale and $N \leq \infty$ is a stopping time, then $EX_0 \geq EX_N$, where $X_\infty = \lim X_n$, which exists by Theorem 4.2.12.*

Proof Using Theorem 4.4.1 and Fatou's Lemma,

$$EX_0 \geq \liminf_{n\to\infty} EX_{N\wedge n} \geq EX_N$$ □

The next result is useful in some situations.

Theorem 4.8.5 *Suppose X_n is a submartingale and $E(|X_{n+1} - X_n| | \mathcal{F}_n) \leq B$ a.s. If N is a stopping time with $EN < \infty$, then $X_{N\wedge n}$ is uniformly integrable and hence $EX_N \geq EX_0$.*

Proof We begin by observing that

$$|X_{N\wedge n}| \leq |X_0| + \sum_{m=0}^{\infty} |X_{m+1} - X_m| 1_{(N>m)}$$

To prove uniform integrability, it suffices to show that the right-hand side has finite expectation for then $|X_{N\wedge n}|$ is dominated by an integrable r.v. Now, $\{N > m\} \in \mathcal{F}_m$, so

$$E(|X_{m+1} - X_m|; N > m) = E(E(|X_{m+1} - X_m||\mathcal{F}_m); N > m) \le BP(N > m)$$

and $E\sum_{m=0}^{\infty} |X_{m+1} - X_m| 1_{(N>m)} \le B\sum_{m=0}^{\infty} P(N > m) = BEN < \infty$. □

4.8.1 Applications to Random Walks

Let $\xi_1, \xi_2, \ldots$ be i.i.d., $S_n = S_0 + \xi_1 + \cdots + \xi_n$, where S_0 is a constant, and let $\mathcal{F}_n = \sigma(\xi_1, \ldots, \xi_n)$. We will now derive some result by using the three martingales from Section 4.2.

Linear martingale If we let $\mu = E\xi_i$, then $X_n = S_n - n\mu$ is a martingale. (See Example 4.2.1.)

Using the linear martingale with Theorem 4.8.5 gives:

Theorem 4.8.6 (Wald's equation) *If $\xi_1, \xi_2, \ldots$ are i.i.d. with $E\xi_i = \mu$, $S_n = \xi_1 + \cdots + \xi_n$ and N is a stopping time with $EN < \infty$, then $ES_N = \mu EN$.*

Proof Let $X_n = S_n - n\mu$ and note that $E(|X_{n+1} - X_n||\mathcal{F}_n) = E|\xi_i - \mu|$. □

Quadratic martingale Suppose $E\xi_i = 0$ and $E\xi_i^2 = \sigma^2 \in (0,\infty)$. Then $X_n = S_n - n\sigma^2$ is a martingale. (See Example 4.2.2.)

Exponential martingale Suppose that $\phi(\theta) = E\exp(\theta\xi_i) < \infty$. Then $X_n = \exp(\theta S_n)/\phi(\theta)^n$ is martingale. (See Example 4.2.3.)

Theorem 4.8.7 (Symmetric simple random walk) *refers to the special case in which $P(\xi_i = 1) = P(\xi_i = -1) = 1/2$. Suppose $S_0 = x$ and let $N = \min\{n : S_n \notin (a,b)\}$. Writing a subscript x to remind us of the starting point*

$$(a) \qquad P_x(S_N = a) = \frac{b - x}{b - a} \qquad P_x(S_N = b) = \frac{x - a}{b - a}$$

(b) $E_0 N = -ab$ and hence $E_x N = (b - x)(x - a)$.

Let $T_x = \min\{n : S_n = x\}$. Taking $a = 0$, $x = 1$ and $b = M$ we have

$$P_1(T_M < T_0) = \frac{1}{M} \qquad P_1(T_M < T_0) = \frac{M - 1}{M}$$

The first result proves (4.4.1). Letting $M \to \infty$ in the second we have $P_1(T_0 < \infty) = 1$.

Proof (a) To see that $P(N < \infty) = 1$ note that if we have $(b - a)$ consecutive steps of size $+1$, we will exit the interval. From this it follows that

$$P(N > m(b - a)) \le (1 - 2^{-(b-a)})^m$$

so $EN < \infty$.

Clearly, $E|S_N| < \infty$ and $S_n 1_{\{N>n\}}$ are uniformly integrable, so using Theorem 4.8.2, we have

$$x = ES_N = aP_x(S_N = a) + b[1 - P_x(S_N = a)]$$

Rearranging, we have $P_x(S_N = a) = (b - x)/(b - a)$ subtracting this from 1, $P_x(S_N = b) = (x - a)/(b - a)$.

(b) The second result is an immediate consequence of the first.

Using the stopping theorem for the bounded stopping time $N \wedge n$, we have

$$0 = E_0 S_{N \wedge n}^2 - E_0(N \wedge n)$$

The monotone convergence theorem implies that $E_0(N \wedge N) \uparrow E_0 N$. Using the bounded convergence theorem and the result of (a) with $x = 0$ implies

$$\begin{aligned} E_0 S_{N \wedge n}^2 &\to a^2 \frac{b}{b-a} + b^2 \frac{-a}{b-a} \\ &= -ab \left[\frac{-a}{b-a} + \frac{b}{b-a} \right] = -ab \end{aligned}$$

which completes the proof. □

Remark The reader should study the technique in this proof of (b) because it is useful in a number of situations. We apply Theorem 4.4.1 to the bounded stopping time $T_b \wedge n$, then let $n \to \infty$, and use an appropriate convergence theorem.

Theorem 4.8.8 *Let S_n be symmetric random walk with $S_0 = 0$ and let $T_1 = \min\{n : S_n = 1\}$.*

$$Es^{T_1} = \frac{1 - \sqrt{1 - s^2}}{s}$$

Inverting the generating function we find

$$P(T_1 = 2n - 1) = \frac{1}{2n - 1} \cdot \frac{(2n)!}{n!\,n!} 2^{-2n} \tag{4.8.1}$$

Proof We will use the exponential martingale $X_n = \exp(\theta S_n)/\phi(\theta)^n$ with $\theta > 0$. The remark after Theorem 4.8.7 implies $P_0(T_1 < \infty) = 1$.

$$\phi(\theta) = E\exp(\theta \xi_i) = (e^{\theta} + e^{-\theta})/2.$$

$\phi(0) = 1$, $\phi'(0) = 0$, and $\phi''(\theta) > 0$, so if $\theta > 0$, then $\phi(\theta) > 1$. This implies $X_{N \wedge n} \in [0, e^{\theta}]$ and it follows from the bounded convergence theorem that

$$1 = EX(T_1) \quad \text{and hence} \quad e^{-\theta} = E(\phi(\theta)^{-T_1})$$

To convert this into the formula for the generating function we set

$$\phi(\theta) = \frac{e^{\theta} + e^{-\theta}}{2} = 1/s.$$

Letting $x = e^{\theta}$ and doing some algebra, we want $x + x^{-1} = 2/s$ or

$$sx^2 - 2x + s = 0$$

The quadratic equation implies

$$x = \frac{2 \pm \sqrt{4 - 4s^2}}{2s} = \frac{1 \pm \sqrt{1 - s^2}}{s}$$

Since $Es^{T_1} = \sum_{k=1}^{\infty} s^k P(T_1 = k)$, we want the solution that is 0 when $s = 0$, which is $(1 - \sqrt{1-s^2})/s$.

To invert the generating function we we use Newton's binomial formula

$$(1+t)^a = 1 + \binom{a}{1}t + \binom{a}{2}t^2 + \binom{a}{3}t^3 + \cdots$$

where $\binom{x}{r} = x(x-1)\dots(x-r+1)/r!$. Taking $t = -s^2$ and $a = 1/2$, we have

$$\sqrt{1-s^2} = 1 - \binom{1/2}{1}s^2 + \binom{1/2}{2}s^4 - \binom{1/2}{3}s^6 + \cdots$$

$$\frac{1-\sqrt{1-s^2}}{s} = \binom{1/2}{1}s - \binom{1/2}{2}s^3 + \binom{1/2}{3}s^5 + \cdots$$

The coefficient of s^{2n-1} is

$$\begin{aligned}(-1)^{n-1}\frac{(1/2)(-1/2)\cdots(3-2n)/2}{n!} &= \frac{1\cdot 3\cdots(2n-3)}{n!}\cdot 2^{-n} \\ &= \frac{1}{2n-1}\frac{(2n)!}{n!n!}2^{-2n}\end{aligned}$$

which completes the proof. □

Theorem 4.8.9 (Asymmetric simple random walk) *refers to the special case in which* $P(\xi_i = 1) = p$ *and* $P(\xi_i = -1) = q \equiv 1 - p$ *with* $p \neq q$.

(a) If $\varphi(y) = \{(1-p)/p\}^y$, *then* $\varphi(S_n)$ *is a martingale.*

(b) If we let $T_z = \inf\{n : S_n = z\}$, *then for* $a < x < b$

$$P_x(T_a < T_b) = \frac{\varphi(b) - \varphi(x)}{\varphi(b) - \varphi(a)} \qquad P_x(T_b < T_a) = \frac{\varphi(x) - \varphi(a)}{\varphi(b) - \varphi(a)} \tag{4.8.2}$$

For the last two parts suppose $1/2 < p < 1$.

(c) If $a < 0$, *then* $P(\min_n S_n \le a) = P(T_a < \infty) = \{(1-p)/p\}^{-a}$.

(d) If $b > 0$, *then* $P(T_b < \infty) = 1$ *and* $ET_b = b/(2p-1)$.

Proof Since S_n and ξ_{n+1} are independent, Example 4.1.7 implies that on $\{S_n = m\}$,

$$\begin{aligned}E(\varphi(S_{n+1})|\mathcal{F}_n) &= p\cdot\left(\frac{1-p}{p}\right)^{m+1} + (1-p)\left(\frac{1-p}{p}\right)^{m-1} \\ &= \{1-p+p\}\left(\frac{1-p}{p}\right)^m = \varphi(S_n)\end{aligned}$$

which proves (a).

Let $N = T_a \wedge T_b$. Since $\varphi(S_{N\wedge n})$ is bounded, Theorem 4.8.2 implies

$$\varphi(x) = E\varphi(S_N) = P_x(T_a < T_b)\varphi(a) + [1 - P_x(T_a < T_a)]\varphi(b)$$

Rearranging gives the formula for $P_x(T_a < T_b)$, subtracting from 1 gives the one for $P_x(T_b < T_a)$.

Letting $b \to \infty$ and noting $\varphi(b) \to 0$ gives the result in (c), since $T_a < \infty$ if and only if $T_a < T_b$ for some b. To start to prove (d) we note that $\varphi(a) \to \infty$ as $a \to -\infty$, so $P(T_b < \infty) = 1$. For the second conclusion, we note that $X_n = S_n - (p-q)n$ is a martingale. Since $T_b \wedge n$ is a bounded stopping time, Theorem 4.4.1 implies

$$0 = E\left(S_{T_b \wedge n} - (p-q)(T_b \wedge n)\right)$$

Now $b \geq S_{T_b \wedge n} \geq \min_m S_m$ and (c) implies $E(\inf_m S_m) > -\infty$, so the dominated convergence theorem implies $ES_{T_b \wedge n} \to ES_{T_b}$ as $n \to \infty$. The monotone convergence theorem implies $E(T_b \wedge n) \uparrow ET_b$, so we have $b = (p-q)ET_b$. □

Example 4.8.10 A gambler is playing roulette and betting \$1 on black each time. The probability she wins \$1 is 18/38, and the probability she loses \$1 is 20/38. Calculate the probability that starting with \$20 she reaches \$40 before losing her money.

$(1-p)/p) = 20/18$, so using (4.8.2) we have

$$\begin{aligned} P_{20}(T_{40} < T_0) &= \frac{(10/9)^{20} - 1}{(10/9)^{40} - 1} \\ &= \frac{8.225 - 1}{67.655 - 1} = 0.1083 \end{aligned}$$

Exercises

4.8.1 Generalize Theorem 4.8.2 to show that if $L \leq M$ are stopping times and $Y_{M \wedge n}$ is a uniformly integrable submartingale, then $EY_L \leq EY_M$ and

$$Y_L \leq E(Y_M | \mathcal{F}_L)$$

4.8.2 If $X_n \geq 0$ is a supermartingale, then $P(\sup X_n > \lambda) \leq EX_0/\lambda$.

4.8.3 Let $S_n = \xi_1 + \cdots + \xi_n$ where the ξ_i are independent with $E\xi_i = 0$ and $\text{var}(X_i) = \sigma^2$. $S_n^2 - n\sigma^2$ is a martingale. Let $T = \min\{n : |S_n| > a\}$. Use Theorem 4.8.2 to show that $ET \geq a^2/\sigma^2$.

4.8.4 *Wald's second equation.* Let $S_n = \xi_1 + \cdots + \xi_n$, where the ξ_i are independent with $E\xi_i = 0$ and $\text{var}(\xi_i) = \sigma^2$. Use the martingale from the previous problem to show that if T is a stopping time with $ET < \infty$, then $ES_T^2 = \sigma^2 ET$.

4.8.5 *Variance of the time of gambler's ruin.* Let $\xi_1, \xi_2, \ldots$ be independent with $P(\xi_i = 1) = p$ and $P(\xi_i = -1) = q = 1 - p$, where $p < 1/2$. Let $S_n = S_0 + \xi_1 + \cdots + \xi_n$ and let $V_0 = \min\{n \geq 0 : S_n = 0\}$. Theorem 4.8.9 tells us that $E_x V_0 = x/(1-2p)$. The aim of this problem is to compute the variance of V_0. If we let $Y_i = \xi_i - (p-q)$ and note that $EY_i = 0$ and

$$\text{var}(Y_i) = \text{var}(X_i) = EX_i u^2 - (EX_i)^2$$

then it follows that $(S_n - (p-q)n)^2 - n(1-(p-q)^2)$ is a martingale. (a) Use this to conclude that when $S_0 = x$ the variance of V_0 is

$$x \cdot \frac{1 - (p-q)^2}{(q-p)^3}$$

(b) Why must the answer in (a) be of the form cx?

4.8.6 *Generating function of the time of gambler's ruin.* Continue with the set-up of the previous problem. (a) Use the exponential martingale and our stopping theorem to conclude that if $\theta \le 0$, then $e^{\theta x} = E_x(\phi(\theta)^{-V_0})$. (b) Let $0 < s < 1$. Solve the equation $\phi(\theta) = 1/s$, then use (a) to conclude

$$E_x(s^{V_0}) = \left(\frac{1 - \sqrt{1 - 4pqs^2}}{2ps}\right)^x$$

(c) Why must the answer in (b) be of the form $f(s)^x$?

4.8.7 Let S_n be a symmetric simple random walk starting at 0, and let $T = \inf\{n : S_n \notin (-a,a)\}$, where a is an integer. Find constants b and c so that $Y_n = S_n^4 - 6nS_n^2 + bn^2 + cn$ is a martingale, and use this to compute ET^2.

4.8.8 Let $S_n = \xi_1 + \cdots + \xi_n$ be a random walk. Suppose $\varphi(\theta_o) = E\exp(\theta_o \xi_1) = 1$ for some $\theta_o < 0$ and ξ_i is not constant. In this special case of the exponetial martingale $X_n = \exp(\theta_o S_n)$ is a martingale. Let $\tau = \inf\{n : S_n \notin (a,b)\}$ and $Y_n = X_{n \wedge T}$. Use Theorem 4.8.2 to conclude that $EX_\tau = 1$ and $P(S_\tau \le a) \le \exp(-\theta_o a)$.

4.8.9 Continuing with the set-up of the previous problem, suppose the ξ_i are integer valued with $P(\xi_i < -1) = 0$, $P(\xi_i = -1) > 0$, and $E\xi_i > 0$. Let $T = \inf\{n : S_n = a\}$ with $a < 0$. Use the martingale $X_n = \exp(\theta_o S_n)$ to conclude that $P(T < \infty) = \exp(-\theta_o a)$.

4.8.10 Consider a favorable game in which the payoffs are -1, 1, or 2 with probability 1/3 each. Use the results of the previous problem to compute the probability we ever go broke (i.e, our winnings W_n reach \$0) when we start with \$$i$.

4.8.11 Let S_n be the total assets of an insurance company at the end of year n. In year n, premiums totaling $c > 0$ are received and claims ζ_n are paid, where ζ_n is Normal(μ, σ^2) and $\mu < c$. To be precise, if $\xi_n = c - \zeta_n$, then $S_n = S_{n-1} + \xi_n$. The company is ruined if its assets drop to 0 or less. Show that if $S_0 > 0$ is nonrandom, then

$$P(\text{ ruin }) \le \exp(-2(c - \mu)S_0/\sigma^2)$$

4.9 Combinatorics of Simple Random Walk*

In the last section we proved some results for simple random walks using martingales. In this section we will delve deeper into their properties using combinatorial arguments. We will not be using martingales, but this section is a good transition to the study of Markov chains in the next chapter.

To facilitate discussion, we will think of the sequence $S_0, S_1, S_2, \ldots, S_n$ as being represented by a polygonal line with segments $(k-1, S_{k-1}) \to (k, S_k)$. A **path** is a polygonal line that is a possible outcome of simple random walk. To count the number of paths from (0,0) to (n,x), it is convenient to introduce a and b defined by: $a = (n+x)/2$ is the number of positive steps in the path and $b = (n-x)/2$ is the number of negative steps. Notice that

$n = a + b$ and $x = a - b$. If $-n \le x \le n$ and $n - x$ is even, the a and b defined above are nonnegative integers, and the number of paths from (0,0) to (n,x) is

$$N_{n,x} = \binom{n}{a} \tag{4.9.1}$$

Otherwise, the number of paths is 0.

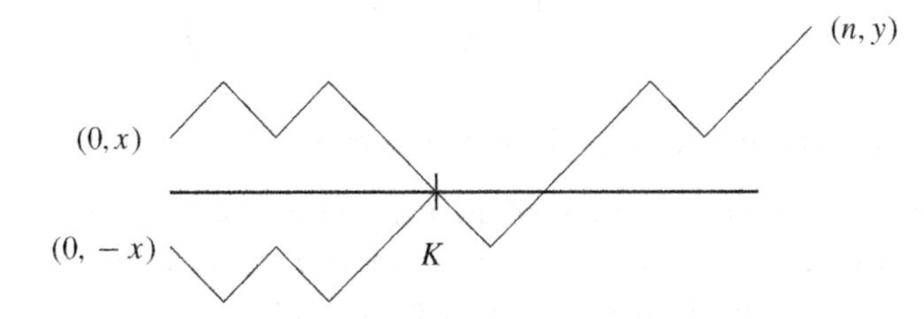

Figure 4.5 Reflection Principle

Theorem 4.9.1 (Reflection principle) *If $x, y > 0$, then the number of paths from $(0,x)$ to (n,y) that are 0 at some time is equal to the number of paths from $(0,-x)$ to (n,y).*

Proof Suppose $(0,s_0),(1,s_1),\ldots,(n,s_n)$ is a path from $(0,x)$ to (n,y). Let $K = \inf\{k : s_k = 0\}$. Let $s'_k = -s_k$ for $k \le K$, $s'_k = s_k$ for $K \le k \le n$. Then (k,s'_k), $0 \le k \le n$, is a path from $(0,-x)$ to (n,y). Conversely, if $(0,t_0),(1,t_1),\ldots,(n,t_n)$ is a path from $(0,-x)$ to (n,y), then it must cross 0. Let $K = \inf\{k : t_k = 0\}$. Let $t'_k = -t_k$ for $k \le K$, $t'_k = t_k$ for $K \le k \le n$. Then (k,t'_k), $0 \le k \le n$, is a path from $(0,-x)$ to (n,y) that is 0 at time K. The last two observations set up a one-to-one correspondence between the two classes of paths, so their numbers must be equal. □

From Theorem 4.9.1 we get a result first proved in 1878.

Theorem 4.9.2 (The Ballot Theorem) *Suppose that in an election candidate A gets α votes, and candidate B gets β votes where $\beta < \alpha$. The probability that throughout the counting A always leads B is $(\alpha - \beta)/(\alpha + \beta)$.*

Proof Let $x = \alpha - \beta$, $n = \alpha + \beta$. Clearly, there are as many such outcomes as there are paths from (1,1) to (n,x) that are never 0. The reflection principle implies that the number of paths from (1,1) to (n,x) that are 0 at some time the number of paths from (1,-1) to (n,x), so by (4.9.1) the number of paths from (1,1) to (n,x) that are never 0 is

$$\begin{aligned} N_{n-1,x-1} - N_{n-1,x+1} &= \binom{n-1}{\alpha-1} - \binom{n-1}{\alpha} \\ &= \frac{(n-1)!}{(\alpha-1)!\,(n-\alpha)!} - \frac{(n-1)!}{\alpha!\,(n-\alpha-1)!} \\ &= \frac{\alpha-(n-\alpha)}{n} \cdot \frac{n!}{\alpha!\,(n-\alpha)!} = \frac{\alpha-\beta}{\alpha+\beta} N_{n,x} \end{aligned}$$

since $n = \alpha + \beta$, this proves the desired result. □

Using the ballot theorem, we can compute the distribution of the time to hit 0 for simple random walk.

Lemma 4.9.3 $P(S_1 \neq 0, \ldots, S_{2n} \neq 0) = P(S_{2n} = 0)$.

Proof $P(S_1 > 0, \ldots, S_{2n} > 0) = \sum_{r=1}^{\infty} P(S_1 > 0, \ldots, S_{2n-1} > 0, S_{2n} = 2r)$. From the proof of Theorem 4.9.2, we see that the number of paths from (0,0) to $(2n, 2r)$ that are never 0 at positive times (= the number of paths from (1,1) to $(2n, 2r)$ that are never 0) is

$$N_{2n-1,2r-1} - N_{2n-1,2r+1}$$

If we let $p_{n,x} = P(S_n = x)$, then this implies

$$P(S_1 > 0, \ldots, S_{2n-1} > 0, S_{2n} = 2r) = \frac{1}{2}(p_{2n-1,2r-1} - p_{2n-1,2r+1})$$

Summing from $r = 1$ to ∞ gives

$$P(S_1 > 0, \ldots, S_{2n} > 0) = \frac{1}{2} p_{2n-1,1} = \frac{1}{2} P(S_{2n} = 0)$$

Symmetry implies $P(S_1 < 0, \ldots, S_{2n} < 0) = (1/2)P(S_{2n} = 0)$, and the proof is complete. □

Let $R = \inf\{m \geq 1 : S_m = 0\}$. Combining Lemma 4.9.3 with the central limit theorem for the binomial distribution, Theorem 3.1.2, gives

$$P(R > 2n) = P(S_{2n} = 0) \sim \pi^{-1/2} n^{-1/2} \tag{4.9.2}$$

Since $P(R > x)/\ P(|R| > x) = 1$, it follows from Theorem 3.8.8 that R is in the domain of attraction of the stable law with $\alpha = 1/2$ and $\kappa = 1$. This implies that if R_n is the time of the nth return to 0, then $R_n/n^2 \Rightarrow Y$, the indicated stable law. In Example 3.8.5, we considered $\tau = T_1$, where $T_x = \inf\{n : S_n = x\}$. Since $S_1 \in \{-1, 1\}$ and $T_1 =_d T_{-1}$, $R =_d 1 + T_1$, and it follows that $T_n/n^2 \Rightarrow Y$, the same stable law. In Example 8.1.12 we will use this observation to show that the limit has the same distribution as the hitting time of 1 for Brownian motion, which has a density given in (7.4.6).

From (4.9.2) we get

$$\begin{aligned} P(T_1 = 2n-1) = P(R = 2n) &= P(R > 2n-2) - P(R > 2n) \\ &= \frac{(2n-2)!}{(n-1)!\,(n-1)!} 2^{-2(n-1)} - \frac{(2n)!}{n!\,n!} 2^{-2n} \\ &= \frac{(2n)!}{n!\,n!} 2^{-2n} \left(\frac{n \cdot n}{(2n-1) \cdot 2n} \cdot 4 - 1 \right) \\ &= \frac{1}{2n-1} \frac{(2n)!}{n!\,n!} 2^{-2n} \sim \frac{1}{2} \pi^{-1/2} n^{-3/2} \end{aligned}$$

This completes our discussion of visits to 0. We turn now to the arcsine laws. The first one concerns

$$L_{2n} = \sup\{m \leq 2n : S_m = 0\}$$

It is remarkably easy to compute the distribution of L_{2n}.

Lemma 4.9.4 *Let* $u_{2m} = P(S_{2m} = 0)$. *Then* $P(L_{2n} = 2k) = u_{2k} u_{2n-2k}$.

Proof $P(L_{2n} = 2k) = P(S_{2k} = 0, S_{2k+1} \neq 0, \ldots, S_{2n} \neq 0)$, so the desired result follows from Lemma 4.9.3. □

Theorem 4.9.5 (Arcsine law for the last visit to 0) *For* $0 < a < b < 1$,

$$P(a \le L_{2n}/2n \le b) \to \int_a^b \pi^{-1}(x(1-x))^{-1/2}\,dx$$

To see the reason for the name, substitute $y = x^{1/2}$, $dy = (1/2)x^{-1/2}\,dx$ in the integral to obtain

$$\int_{\sqrt{a}}^{\sqrt{b}} \frac{2}{\pi}(1-y^2)^{-1/2}\,dy = \frac{2}{\pi}\{\arcsin(\sqrt{b}) - \arcsin(\sqrt{a})\}$$

Since L_{2n} is the time of the last zero before $2n$, it is surprising that the answer is symmetric about 1/2. The symmetry of the limit distribution implies

$$P(L_{2n}/2n \le 1/2) \to 1/2$$

In gambling terms, if two people were to bet \$1 on a coin flip every day of the year, then with probability 1/2, one of the players will be ahead from July 1 to the end of the year, an event that would undoubtedly cause the other player to complain about his bad luck.

Proof of Theorem 4.9.5. From the asymptotic formula for u_{2n}, it follows that if $k/n \to x$, then

$$nP(L_{2n} = 2k) \to \pi^{-1}(x(1-x))^{-1/2}$$

To get from this to the desired result, we let $2na_n$ = the smallest even integer $\ge 2na$, let $2nb_n$ = the largest even integer $\le 2nb$, and let $f_n(x) = nP(L_{2n} = k)$ for $2k/2n \le x < 2(k+1)/2n$, so we can write

$$P(a \le L_{2n}/2n \le b) = \sum_{k=na_n}^{nb_n} P(L_{2n} = 2k) = \int_{a_n}^{b_n+1/n} f_n(x)\,dx$$

Our first result implies that uniformly on compact sets

$$f_n(x) \to f(x) = \pi^{-1}(x(1-x))^{-1/2}$$

The uniformity of the convergence implies

$$\sup_{a_n \le x \le b_n+1/n} f_n(x) \to \sup_{a \le x \le b} f(x) < \infty$$

if $0 < a \le b < 1$, so the bounded convergence theorem gives

$$\int_{a_n}^{b_n+1/n} f_n(x)\,dx \to \int_a^b f(x)\,dx$$

□

The next result deals directly with the amount of time one player is ahead.

Theorem 4.9.6 (Arcsine law for time above 0) *Let π_{2n} be the number of segments $(k-1, S_{k-1}) \to (k, S_k)$ that lie above the axis (i.e., in $\{(x,y) : y \ge 0\}$), and let $u_m = P(S_m = 0)$.*

$$P(\pi_{2n} = 2k) = u_{2k}u_{2n-2k}$$

and consequently, if $0 < a < b < 1$

$$P(a \le \pi_{2n}/2n \le b) \to \int_a^b \pi^{-1}(x(1-x))^{-1/2}\,dx$$

Remark Since $\pi_{2n} =_d L_{2n}$, the second conclusion follows from the proof of Theorem 4.9.5. The reader should note that the limiting density $\pi^{-1}(x(1-x))^{-1/2}$ has a minimum at $x = 1/2$, and $\to \infty$ as $x \to 0$ or 1. An equal division of steps between the positive and negative side is therefore the least likely possibility, and completely one-sided divisions have the highest probability.

Proof Let $\beta_{2k,2n}$ denote the probability of interest. We will prove $\beta_{2k,2n} = u_{2k}u_{2n-2k}$ by induction. When $n = 1$, it is clear that

$$\beta_{0,2} = \beta_{2,2} = 1/2 = u_0 u_2$$

For a general n, first suppose $k = n$. From the proof of Lemma 4.9.3, we have

$$\begin{aligned}
\frac{1}{2}u_{2n} &= P(S_1 > 0, \ldots, S_{2n} > 0) \\
&= P(S_1 = 1, S_2 - S_1 \geq 0, \ldots, S_{2n} - S_1 \geq 0) \\
&= \frac{1}{2}P(S_1 \geq 0, \ldots, S_{2n-1} \geq 0) \\
&= \frac{1}{2}P(S_1 \geq 0, \ldots, S_{2n} \geq 0) = \frac{1}{2}\beta_{2n,2n}
\end{aligned}$$

The next to last equality follows from the observation that if $S_{2n-1} \geq 0$, then $S_{2n-1} \geq 1$, and hence $S_{2n} \geq 0$.

The last computation proves the result for $k = n$. Since $\beta_{0,2n} = \beta_{2n,2n}$, the result is also true when $k = 0$. Suppose now that $1 \leq k \leq n-1$. In this case, if R is the time of the first return to 0, then $R = 2m$ for some m with $0 < m < n$. Letting $f_{2m} = P(R = 2m)$ and breaking things up according to whether the first excursion was on the positive or negative side gives

$$\beta_{2k,2n} = \frac{1}{2}\sum_{m=1}^{k} f_{2m}\beta_{2k-2m,2n-2m} + \frac{1}{2}\sum_{m=1}^{n-k} f_{2m}\beta_{2k,2n-2m}$$

Using the induction hypothesis, it follows that

$$\beta_{2k,2n} = \frac{1}{2}u_{2n-2k}\sum_{m=1}^{k} f_{2m}u_{2k-2m} + \frac{1}{2}u_{2k}\sum_{m=1}^{n-k} f_{2m}u_{2n-2k-2m}$$

By considering the time of the first return to 0, we see

$$u_{2k} = \sum_{m=1}^{k} f_{2m}u_{2k-2m} \qquad u_{2n-2k} = \sum_{m=1}^{n-k} f_{2m}u_{2n-2k-2m}$$

and the desired result follows. □

Exercises

4.9.1 Let $a \in S$, $f_n = P_a(T_a = n)$, and $u_n = P_a(X_n = a)$. (i) Show that $u_n = \sum_{1 \leq m \leq n} f_m u_{n-m}$. (ii) Let $u(s) = \sum_{n \geq 0} u_n s^n$, $f(s) = \sum_{n \geq 1} f_n s^n$, and show $u(s) = 1/(1 - f(s))$.

5

Markov Chains

The main object of study in this chapter is (temporally homogeneous) Markov chains. These processes are important because the assumptions are satisfied in many examples and lead to a rich and detailed theory.

5.1 Examples

We begin with the case of countable state space. In this siutation the Markov property is that for any states $i_0, \dots i_{n-1}, i$, and j

$$P(X_{n+1} = j | X_n = i, X_{n-1} = i_{n-1}, \dots X_0 = i_0) = P(X_{n+1} = j | X_n = i)$$

In words, given the present state the rest of the past is irrelevant for predicting the future. In this section we will introduce a number of examples. In each case, we will not check the Markov property, but simply give the transition probability

$$p(i, j) = P(X_{n+1} = j | X_n = i)$$

and leave the rest to the reader. In the next section we will be more formal.

Example 5.1.1 (Random walk) Let $\xi_1, \xi_2, \dots \in \mathbf{Z}^d$ be independent with distribution μ. Let $X_n = X_0 + \xi_1 + \cdots + \xi_n$, where X_0 is constant. Then X_n is a Markov chain with transition probability.

$$p(i, j) = \mu(\{j - i\})$$

Example 5.1.2 (Branching processes) $S = \{0, 1, 2, \dots\}$

$$p(i, j) = P\left(\sum_{m=1}^{i} \xi_m = j\right)$$

where $\xi_1, \xi_2, \dots$ are i.i.d. nonnegative integer-valued random variables. In words, each of the i individuals at time n (or in generation n) gives birth to an independent and identically distributed number of offspring.

The first two chains are not good examples because, as we will see, they do not converge to equilibrium.

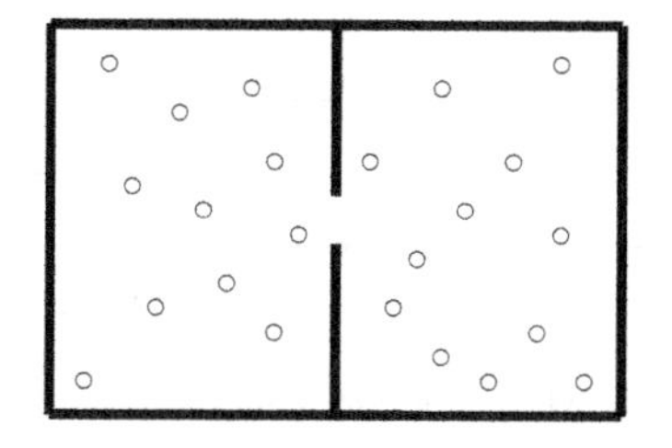

Figure 5.1 Physical motivation for the Ehrenfest chain.

Example 5.1.3 (Ehrenfest chain) $S = \{0, 1, \ldots, r\}$

$$p(k, k+1) = (r-k)/r$$
$$p(k, k-1) = k/r$$
$$p(i, j) = 0 \qquad \text{otherwise}$$

In words, there is a total of r balls in two urns; k in the first and $r-k$ in the second. We pick one of the r balls at random and move it to the other urn. Ehrenfest used this to model the division of air molecules between two chambers (of equal size and shape) that are connected by a small hole.

Example 5.1.4 (Wright–Fisher model) Thinking of a population of $N/2$ diploid individuals who have two copies of each of their chromosomes, or of N haploid individuals who have one copy, we consider a fixed population of N genes that can be one of two types: A or a. In the simplest version of this model the population at time $n+1$ is obtained by drawing with replacement from the population at time n. In this case, if we let X_n be the number of A alleles at time n, then X_n is a Markov chain with transition probability

$$p(i, j) = \binom{N}{j} \left(\frac{i}{N}\right)^j \left(1 - \frac{i}{N}\right)^{N-j}$$

since the right-hand side is the binomial distribution for N independent trials with success probability i/N.

In this model the states $x = 0$ and N that correspond to fixation of the population in the all a or all A states are **absorbing states**, that is, $p(x, x) = 1$. As we will see, this chain will eventually end up in state 0 or state N. To make this simple model more interesting, we introduce mutations. That is, an A that is drawn ends up being an a in the next generation with probability u, while an a that is drawn ends up being an A in the next generation with probability v. In this case, the probability an A is produced by a given draw is

$$\rho_i = \frac{i}{N}(1-u) + \frac{N-i}{N} v$$

but the transition probability still has the binomial form

$$p(i, j) = \binom{N}{j} (\rho_i)^j (1 - \rho_i)^{N-j}$$

If u and v are both positive, then 0 and N are no longer absorbing states, so the system can converge to an equilibrium distribution as time $t \to \infty$?

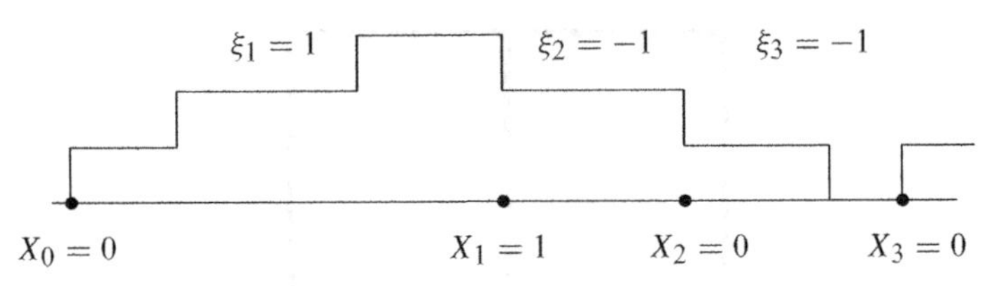

Figure 5.2 Realization of the M/G/1 queue. Black dots indicate the times at which the customers enter service.

Example 5.1.5 (M/G/1 queue) In this model, customers arrive according to a Poisson process with rate λ. (M is for Markov and refers to the fact that in a Poisson process the number of arrivals in disjoint time intervals is independent.) Each customer requires an independent amount of service with distribution F. (G is for general service distribution. 1 indicates that there is one server.) Let X_n be the number of customers waiting in the queue at the time the nth customer enters service. To be precise, when $X_0 = x$, the chain starts with x people waiting in line and customer 0 just beginning her service. To understand the definition, the picture in Figure 5.2 is useful:

To define our Markov chain X_n, let

$$a_k = \int_0^\infty e^{-\lambda t} \frac{(\lambda t)^k}{k!} \, dF(t)$$

be the probability that k customers arrive during a service time. Let $\xi_1, \xi_2, \ldots$ be i.i.d. with $P(\xi_i = k - 1) = a_k$. We think of ξ_i as the net number of customers to arrive during the ith service time, subtracting one for the customer who completed service. We define X_n by

$$X_{n+1} = (X_n + \xi_{n+1})^+ \tag{5.1.1}$$

The positive part only takes effect when $X_n = 0$ and $\xi_{n+1} = -1$ (e.g., $X_2 = 0$, $\xi_3 = -1$ in Figure 5.2) and reflects the fact that when the queue has size 0 and no one arrives during the service time, the next queue size is 0, since we do not start counting until the next customer arrives and then the queue length will be 0.

It is easy to see that the sequence defined in (5.1.1) is a Markov chain with transition probability

$$\begin{aligned} p(0,0) &= a_0 + a_1 \\ p(j, j-1+k) &= a_k \qquad \text{if } j \geq 1 \text{ or } k > 1 \end{aligned}$$

The formula for a_k is rather complicated, and its exact form is not important, so we will simplify things by assuming only that $a_k > 0$ for all $k \geq 0$ and $\sum_{k \geq 0} a_k = 1$.

Exercises

5.1.1 Let $\xi_1, \xi_2, \ldots$ be i.i.d. $\in \{1, 2, \ldots, N\}$ and taking each value with probability $1/N$. Show that $X_n = |\{\xi_1, \ldots, \xi_n\}|$ is a Markov chain and compute its transition probability.

5.1.2 Let $\xi_1, \xi_2, \ldots$ be i.i.d. $\in \{-1, 1\}$, taking each value with probability $1/2$. Let $S_0 = 0$, $S_n = \xi_1 + \cdots \xi_n$ and $X_n = \max\{S_m : 0 \le m \le n\}$. Show that X_n is not a Markov chain.

5.1.3 Let $\xi_0, \xi_1, \ldots$ be i.i.d. $\in \{H, T\}$, taking each value with probability $1/2$. Show that $X_n = (\xi_n, \xi_{n+1})$ is a Markov chain and compute its transition probability p. What is p^2?

5.1.4 **Brother-sister mating.** In this scheme, two animals are mated, and among their direct descendants two individuals of opposite sex are selected at random. These animals are mated and the process continues. Suppose each individual can be one of three genotypes AA, Aa, aa, and suppose that the type of the offspring is determined by selecting a letter from each parent. With these rules, the pair of genotypes in the nth generation is a Markov chain with six states:

$$AA, AA \quad AA, Aa \quad AA, aa \quad Aa, Aa \quad Aa, aa \quad aa, aa$$

Compute its transition probability.

5.1.5 **Bernoulli-Laplace model of diffusion.** Suppose two urns, which we will call left and right, have m balls each. b (which we will assume is $\le m$) balls are black, and $2m - b$ are white. At each time, we pick one ball from each urn and interchange them. Let the state at time n be the number of black balls in the left urn. Compute the transition probability.

5.1.6 Let $\theta, U_1, U_2, \ldots$ be independent and uniform on $(0, 1)$. Let $X_i = 1$ if $U_i \le \theta$, $= -1$ if $U_i > \theta$, and let $S_n = X_1 + \cdots + X_n$. In words, we first pick θ according to the uniform distribution and then flip a coin with probability θ of heads to generate a random walk. Compute $P(X_{n+1} = 1 | X_1, \ldots, X_n)$ and conclude S_n is a temporally inhomogeneous Markov chain. This is due to the fact that "S_n is a sufficient statistic for estimating θ."

5.2 Construction, Markov Properties

Let $(S, \mathcal{S})$ be a measurable space. This will be the state space for our Markov chain.

A function $p : S \times \mathcal{S} \to \mathbf{R}$ is said to be a **transition probability** if:

(i) For each $x \in S$, $A \to p(x, A)$ is a probability measure on $(S, \mathcal{S})$.

(ii) For each $A \in \mathcal{S}$, $x \to p(x, A)$ is a measurable function.

We say X_n is a Markov chain (w.r.t. $\mathcal{F}_n$) with transition probability p if

$$P(X_{n+1} \in B | \mathcal{F}_n) = p(X_n, B)$$

Given a transition probability p and an **initial distribution** μ on $(S, \mathcal{S})$, we can define a consistent set of finite dimensional distributions by

$$P(X_j \in B_j, 0 \le j \le n) = \int_{B_0} \mu(dx_0) \int_{B_1} p(x_0, dx_1) \cdots \int_{B_n} p(x_{n-1}, dx_n) \tag{5.2.1}$$

If we suppose that $(S, \mathcal{S})$ is nice, Kolmogorov's extenson theorem, Theorem 2.1.21, allows us to construct a probability measure P_μ on **sequence space**

$$(\Omega_o, \mathcal{F}_\infty) = (S^{\{0,1,\ldots\}}, \mathcal{S}^{\{0,1,\ldots\}})$$

so that the coordinate maps $X_n(\omega) = \omega_n$ have the desired distributions. Note that we have one set of very simple random variables, and a large family of measures, one for each initial condition. Of course, we only need one measure for each state since

$$P_\mu(A) = \int \mu(dx) P_x(A).$$

An advantage of building the chain on this canonical probability space is that we can define the shift operators by

$$\theta_n(\omega_0, \omega_1, \ldots) = (\omega_n, \omega_{n+1}, \ldots)$$

Our next step is to show

Theorem 5.2.1 *X_n is a Markov chain (with respect to $\mathcal{F}_n = \sigma(X_0, X_1, \ldots, X_n)$) with transition probability p. That is,*

$$P_\mu(X_{n+1} \in B | \mathcal{F}_n) = p(X_n, B)$$

Proof To prove this, we let $A = \{X_0 \in B_0, X_1 \in B_1, \ldots, X_n \in B_n\}$, $B_{n+1} = B$, and observe that using the definition of the integral, the definition of A, and the definition of P_μ

$$\begin{aligned}
\int_A 1_{(X_{n+1} \in B)} \, dP_\mu &= P_\mu(A, X_{n+1} \in B) \\
&= P_\mu(X_0 \in B_0, X_1 \in B_1, \ldots, X_n \in B_n, X_{n+1} \in B) \\
&= \int_{B_0} \mu(dx_0) \int_{B_1} p(x_0, dx_1) \cdots \int_{B_n} p(x_{n-1}, dx_n) \, p(x_n, B_{n+1})
\end{aligned}$$

We would like to assert that the last expression is

$$= \int_A p(X_n, B) \, dP_\mu$$

To do this, we begin by noting that

$$\int_{B_0} \mu(dx_0) \int_{B_1} p(x_0, dx_1) \cdots \int_{B_n} p(x_{n-1}, dx_n) \, 1_C(x_n) = \int_A 1_C(X_n) dP_\mu$$

Linearity implies that for simple functions,

$$\int_{B_0} \mu(dx_0) \int_{B_1} p(x_0, dx_1) \cdots \int_{B_n} p(x_{n-1}, dx_n) \, f(x_n) = \int_A f(X_n) dP_\mu$$

and the bounded convergence theorem implies that it is valid for bounded measurable f, e.g., $f(x) = p(x, B_{n+1})$.

The collection of sets for which

$$\int_A 1_{(X_{n+1} \in B)} \, dP_\mu = \int_A p(X_n, B) \, dP_\mu$$

holds is a λ-system, and the collection for which it has been proved is a π-system, so it follows from the $\pi - \lambda$ theorem, Theorem 2.1.6, that the equality is true for all $A \in \mathcal{F}_n$. This shows that

$$P(X_{n+1} \in B|\mathcal{F}_n) = p(X_n, B)$$

and proves the desired result. □

Our next small step is to show that if X_n has transition probability p, then for any bounded measurable f

$$E(f(X_{n+1})|\mathcal{F}_n) = \int p(X_n, dy) f(y) \tag{5.2.2}$$

The desired conclusion is a consequence of the next result, which will save us work in later proofs. Let $\mathcal{H} =$ the collection of bounded functions for which the identity holds.

Theorem 5.2.2 (Monotone class theorem) *Let $\mathcal{A}$ be a π-system that contains Ω and let $\mathcal{H}$ be a collection of real-valued functions that satisfies:*

(i) If $A \in \mathcal{A}$, then $1_A \in \mathcal{H}$.
(ii) If $f, g \in \mathcal{H}$, then $f + g$, and $cf \in \mathcal{H}$ for any real number c.
(iii) If $f_n \in \mathcal{H}$ are nonnegative and increase to a bounded function f, then $f \in \mathcal{H}$.

Then $\mathcal{H}$ contains all bounded functions measurable with respect to $\sigma(\mathcal{A})$.

Proof The assumption $\Omega \in \mathcal{A}$, (ii), and (iii) imply that $\mathcal{G} = \{A : 1_A \in \mathcal{H}\}$ is a λ-system, so by (i) and the $\pi - \lambda$ theorem, Theorem 2.1.6, $\mathcal{G} \supset \sigma(\mathcal{A})$. (ii) implies $\mathcal{H}$ contains all simple functions, and (iii) implies that $\mathcal{H}$ contains all bounded measurable functions. □

To extend (5.2.2), we observe that familiar properties of conditional expectation and (5.2.2) imply

$$\begin{aligned}
E\left(\prod_{m=0}^{n} f_m(X_m)\right) &= E\,E\left(\prod_{m=0}^{n} f_m(X_m)\middle|\mathcal{F}_{n-1}\right) \\
&= E\left(\prod_{m=0}^{n-1} f_m(X_m) E(f_n(X_n)|\mathcal{F}_{n-1})\right) \\
&= E\left(\prod_{m=0}^{n-1} f_m(X_m) \int p_{n-1}(X_{n-1}, dy) f_n(y)\right)
\end{aligned}$$

The last integral is a bounded measurable function of X_{n-1}, so it follows by induction that if μ is the distribution of X_0, then

$$\begin{aligned}
E\left(\prod_{m=0}^{n} f_m(X_m)\right) &= \int \mu(dx_0) f_0(x_0) \int p_0(x_0, dx_1) f_1(x_1) \\
&\quad \cdots \int p_{n-1}(x_{n-1}, dx_n) f_n(x_n)
\end{aligned} \tag{5.2.3}$$

Our next goal is to prove two extensions of the Markov property in which $\{X_{n+1} \in B\}$ is replaced by a bounded function of the future, $h(X_n, X_{n+1}, \ldots)$, and n is replaced by a stopping time N. These results, especially the second, will be the keys to developing the theory of Markov chains.

Theorem 5.2.3 (The Markov property) *Let $Y : \Omega_o \to \mathbf{R}$ be bounded and measurable.*

$$E_\mu(Y \circ \theta_m | \mathcal{F}_m) = E_{X_m} Y$$

Remark We denote the function by Y, a letter usually used for random variables, because that's exactly what Y is, a measurable function defined on our probability space Ω_o. Here the subscript μ on the left-hand side indicates that the conditional expectation is taken with respect to P_μ. The right-hand side is the function $\varphi(x) = E_x Y$ evaluated at $x = X_m$.

Proof We begin by proving the result in a special case and then use the $\pi-\lambda$ and monotone class theorems to get the general result. Let $A = \{\omega : \omega_0 \in A_0, \dots, \omega_m \in A_m\}$ and $g_0, \dots g_n$ be bounded and measurable. Applying (5.2.3) with $f_k = 1_{A_k}$ for $k < m$, $f_m = 1_{A_m} g_0$, and $f_k = g_{k-m}$ for $m < k \le m+n$ gives

$$\begin{aligned} E_\mu\left(\prod_{k=0}^n g_k(X_{m+k}); A\right) &= \int_{A_0} \mu(dx_0) \int_{A_1} p(x_0, dx_1) \cdots \int_{A_m} p(x_{m-1}, dx_m) \\ &\quad \cdot g_0(x_m) \int p(x_m, dx_{m+1}) g_1(x_{m+1}) \\ &\quad \cdots \int p(x_{m+n-1}, dx_{m+n}) g_n(x_{m+n}) \\ &= E_\mu\left(E_{X_m}\left(\prod_{k=0}^n g_k(X_k)\right); A\right) \end{aligned}$$

The collection of sets for which the last formula holds is a λ-system, and the collection for which it has been proved is a π-system, so using the $\pi - \lambda$ theorem, Theorem 2.1.6, shows that the last identity holds for all $A \in \mathcal{F}_m$.

Fix $A \in \mathcal{F}_m$ and let $\mathcal{H}$ be the collection of bounded measurable Y for which

$$(*) \qquad E_\mu(Y \circ \theta_m; A) = E_\mu(E_{X_m} Y; A)$$

The last computation shows that $(*)$ holds when

$$Y(\omega) = \prod_{0 \le k \le n} g_k(\omega_k)$$

To finish the proof, we will apply the monotone class theorem, Theorem 5.2.2. Let $\mathcal{A}$ be the collection of sets of the form $\{\omega : \omega_0 \in A_0, \dots, \omega_k \in A_k\}$. $\mathcal{A}$ is a π-system, so taking $g_k = 1_{A_k}$ shows (i) holds. $\mathcal{H}$ clearly has properties (ii) and (iii), so Theorem 5.2.2 implies that $\mathcal{H}$ contains the bounded functions measurable w.r.t $\sigma(\mathcal{A})$, and the proof is complete. □

As corollary of Theorem 5.2.3 we get:

Theorem 5.2.4 (Chapman-Kolmogorov equation)

$$P_x(X_{m+n} = z) = \sum_y P_x(X_m = y) P_y(X_n = z)$$

Intuitively, in order to go from x to z in $m+n$ steps we have to be at some y at time m, and the Markov property implies that for a given y the two parts of the journey are independent.

Proof $P_x(X_{n+m} = z) = E_x(P_x(X_{n+m} = z|\mathcal{F}_m)) = E_x(P_{X_m}(X_n = z))$ by the Markov property, Theorem 5.2.3, since $1_{(X_n=z)} \circ \theta_m = 1_{(X_{n+m}=z)}$. □

We are now ready for our second extension of the Markov property. Recall N is said to be a stopping time if $\{N = n\} \in \mathcal{F}_n$. As in Chapter 4, let

$$\mathcal{F}_N = \{A : A \cap \{N = n\} \in \mathcal{F}_n \text{ for all } n\}$$

be the information known at time N, and let

$$\theta_N \omega = \begin{cases} \theta_n \omega & \text{on } \{N = n\} \\ \Delta & \text{on } \{N = \infty\} \end{cases}$$

where Δ is an extra point that we add to Ω_o. In the next result and its applications, we will explicitly restrict our attention to $\{N < \infty\}$, so the reader does not have to worry about the second part of the definition of θ_N.

Theorem 5.2.5 (Strong Markov property) *Suppose that for each n, $Y_n : \Omega_0 \to \mathbf{R}$ is measurable and $|Y_n| \le M$ for all n. Then*

$$E_\mu(Y_N \circ \theta_N|\mathcal{F}_N) = E_{X_N} Y_N \text{ on } \{N < \infty\}$$

where the right-hand side is $\varphi(x,n) = E_x Y_n$ evaluated at $x = X_N$, $n = N$.

Proof Let $A \in \mathcal{F}_N$. Breaking things down according to the value of N.

$$E_\mu(Y_N \circ \theta_N; A \cap \{N < \infty\}) = \sum_{n=0}^{\infty} E_\mu(Y_n \circ \theta_n; A \cap \{N = n\})$$

Since $A \cap \{N = n\} \in \mathcal{F}_n$, using Theorem 5.2.3 now converts the right side into

$$\sum_{n=0}^{\infty} E_\mu(E_{X_n} Y_n; A \cap \{N = n\}) = E_\mu(E_{X_N} Y_N; A \cap \{N < \infty\})$$ □

Remark The reader should notice that the proof is trivial. All we do is break things down according to the value of N, replace N by n, apply the Markov property, and reverse the process. This is the standard technique for proving results about stopping times.

Our first application of the strong Markov property will be the key to developments in the next section. Let $T_y^0 = 0$, and for $k \ge 1$, let

$$T_y^k = \inf\{n > T_y^{k-1} : X_n = y\}$$

T_y^k is the time of the kth return to y. The reader should note that $T_y^1 > 0$, so any visit at time 0 does not count. We adopt this convention so that if we let $T_y = T_y^1$ and $\rho_{xy} = P_x(T_y < \infty)$, then:

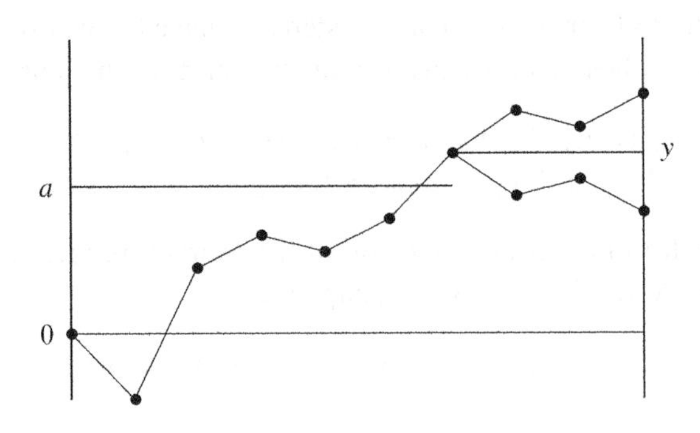

Figure 5.3 Proof by picture of the reflection principle.

Theorem 5.2.6 $P_x(T_y^k < \infty) = \rho_{xy}\rho_{yy}^{k-1}$.

Intuitively, in order to make k visits to y, we first have to go from x to y and then return $k-1$ times to y.

Proof When $k = 1$, the result is trivial, so we suppose $k \geq 2$. Let $Y(\omega) = 1$ if $\omega_n = y$ for some $n \geq 1$, $Y(\omega) = 0$ otherwise. If $N = T_y^{k-1}$, then $Y \circ \theta_N = 1$ if $T_y^k < \infty$. The strong Markov property, Theorem 5.2.5, implies

$$E_x(Y \circ \theta_N | \mathcal{F}_N) = E_{X_N} Y \quad \text{on } \{N < \infty\}$$

On $\{N < \infty\}$, $X_N = y$, so the right-hand side is $P_y(T_y < \infty) = \rho_{yy}$, and it follows that

$$\begin{aligned} P_x(T_y^k < \infty) &= E_x(Y \circ \theta_N; N < \infty) \\ &= E_x(E_x(Y \circ \theta_N | \mathcal{F}_N); N < \infty) \\ &= E_x(\rho_{yy}; N < \infty) = \rho_{yy} P_x(T_y^{k-1} < \infty) \end{aligned}$$

The result now follows by induction. □

The next example illustrates the use of Theorem 5.2.5, and explains why we want to allow the Y that we apply to the shifted path to depend on n.

Theorem 5.2.7 (Reflection principle) *Let $\xi_1, \xi_2, \ldots$ be independent and identically distributed with a distribution that is symmetric about 0. Let $S_n = \xi_1 + \cdots + \xi_n$. If $a > 0$, then*

$$P\left(\sup_{m \leq n} S_m \geq a\right) \leq 2P(S_n \geq a)$$

We do the proof in two steps because that is how formulas like this are derived in practice. First, one computes intuitively and then figures out how to extract the desired formula from Theorem 5.2.5.

Proof in words. First note that if Z has a distribution that is symmetric about 0, then

$$P(Z \geq 0) \geq P(Z > 0) + \frac{1}{2}P(Z = 0) = \frac{1}{2}$$

If we let $N = \inf\{m \le n : S_m > a\}$ (with $\inf \emptyset = \infty$), then on $\{N < \infty\}$, $S_n - S_N$ is independent of S_N and has $P(S_n - S_N \ge 0) \ge 1/2$. So

$$P(S_n > a) \ge \frac{1}{2} P(N \le n)$$

Proof Let $Y_m(\omega) = 1$ if $m \le n$ and $\omega_{n-m} \ge a$, $Y_m(\omega) = 0$ otherwise. The definition of Y_m is chosen so that $(Y_N \circ \theta_N)(\omega) = 1$ if $\omega_n \ge a$ (and hence $N \le n$), and $= 0$ otherwise. The strong Markov property implies

$$E_0(Y_N \circ \theta_N | \mathcal{F}_N) = E_{S_N} Y_N \quad \text{on } \{N < \infty\} = \{N \le n\}$$

To evaluate the right-hand side, we note that if $y > a$, then

$$E_y Y_m = P_y(S_{n-m} \ge a) \ge P_y(S_{n-m} \ge y) \ge 1/2$$

So integrating over $\{N \le n\}$ and using the definition of conditional expectation gives

$$\frac{1}{2} P(N \le n) \le E_0(E_0(Y_N \circ \theta_N | \mathcal{F}_N); N \le n) = E_0(Y_N \circ \theta_N; N \le n)$$

since $\{N \le n\} \in \mathcal{F}_N$. Recalling that $Y_N \circ \theta_N = 1_{\{S_n \ge a\}}$, the last quantity

$$= E_0(1_{\{S_n \ge a\}}; N \le n) = P_0(S_n \ge a)$$

since $\{S_n > a\} \subset \{N \le n\}$. □

Exercises

5.2.1 Let $A \in \sigma(X_0, \dots, X_n)$ and $B \in \sigma(X_n, X_{n+1}, \dots)$. Use the Markov property to show that for any initial distribution μ

$$P_\mu(A \cap B | X_n) = P_\mu(A|X_n) P_\mu(B|X_n)$$

In words, the past and future are conditionally independent given the present. Hint: Write the left-hand side as $E_\mu(E_\mu(1_A 1_B | \mathcal{F}_n)|X_n)$.

5.2.2 Let X_n be a Markov chain. Use Lévy's 0-1 law to show that if

$$P\left(\cup_{m=n+1}^{\infty} \{X_m \in B_m\} \middle| X_n\right) \ge \delta > 0 \quad \text{on } \{X_n \in A_n\}$$

then $P(\{X_n \in A_n \text{ i.o.}\} - \{X_n \in B_n \text{ i.o.}\}) = 0$.

5.2.3 A state a is called **absorbing** if $P_a(X_1 = a) = 1$. Let $D = \{X_n = a$ for some $n \ge 1\}$ and let $h(x) = P_x(D)$. Use the result of the previous exercise to conclude that $h(X_n) \to 0$ a.s. on D^c. Here a.s. means P_μ a.s. for any initial distribution μ.

We will suppose throughout the rest of the exercises that S is countable. Note that the times V_A and V_x of the first visit are inf *over $n \ge 0$, while the hitting times T_A and T_x are* inf *over $n \ge 1$.*

5.2.4 **First entrance decomposition.** Let $T_y = \inf\{n \ge 1 : X_n = y\}$. Show that

$$p^n(x, y) = \sum_{m=1}^{n} P_x(T_y = m) p^{n-m}(y, y)$$

5.2.5 Show that $\sum_{m=0}^{n} P_x(X_m = x) \geq \sum_{m=k}^{n+k} P_x(X_m = x)$.

5.2.6 Let $T_C = \inf\{n \geq 1 : X_n \in C\}$. Suppose that $S - C$ is finite and for each $x \in S - C$ $P_x(T_C < \infty) > 0$. Then there is an $N < \infty$ and $\epsilon > 0$ so that for all $y \in S - C$, $P_y(T_C > kN) \leq (1 - \epsilon)^k$.

5.2.7 **Exit distributions.** Let $V_C = \inf\{n \geq 0 : X_n \in C\}$ and let $h(x) = P_x(V_A < V_B)$. Suppose $A \cap B = \emptyset$, $S - (A \cup B)$ is finite, and $P_x(V_{A \cup B} < \infty) > 0$ for all $x \in S - (A \cup B)$. (i) Show that

$$(*) \qquad h(x) = \sum_y p(x,y)h(y) \quad \text{for } x \notin A \cup B$$

(ii) Show that if h satisfies $(*)$, then $h(X(n \wedge V_{A \cup B}))$ is a martingale. (iii) Use this and Exercise 5.2.6 to conclude that $h(x) = P_x(V_A < V_B)$ is the only solution of $(*)$ that is 1 on A and 0 on B.

5.2.8 Let X_n be a Markov chain with $S = \{0, 1, \ldots, N\}$ and suppose that X_n is a martingale. Let $V_x = \min\{n \geq 0 : X_n = x\}$ and suppose $P_x(V_0 \wedge V_N < \infty) > 0$ for all x. Show $P_x(V_N < V_0) = x/N$.

5.2.9 **Wright-Fisher model.** Suppose $S = \{0, 1, \ldots, N\}$ and consider

$$p(i,j) = \binom{N}{j}(i/N)^j(1 - i/N)^{N-j}$$

Show that this chain satisfies the hypotheses of Exercise 5.2.8.

5.2.10 In **brother-sister mating** described in Exercise 5.1.4, *AA*,*AA* and *aa*,*aa* are absorbing states. Show that the number of *A*'s in the pair is a martingale and use this to compute the probability of getting absorbed in *AA*,*AA* starting from each of the states.

5.2.11 **Exit Times.** Let $V_A = \inf\{n \geq 0 : X_n \in A\}$ and $g(x) = E_x V_A$. Suppose that $S - A$ is finite and for each $x \in S - A$, $P_x(V_A < \infty) > 0$. (i) Show that

$$(*) \qquad g(x) = 1 + \sum_y p(x,y)g(y) \quad \text{for } x \notin A$$

(ii) Show that if g satisfies $(*)$, $g(X(n \wedge V_A)) + n \wedge V_A$ is a martingale. (iii) Use this to conclude that $g(x) = E_x\tau_A$ is the only solution of $(*)$ that is 0 on A.

5.2.12 In this problem we imagine that you are bored and start flipping a coin to see how many tosses $N(HH)$ you need until you get two heads in a row. Let $\xi_0, \xi_1, \ldots$ be i.i.d. $\in \{H, T\}$, taking each value with probability 1/2, and let $X_n = (\xi_n, \xi_{n+1})$ be the Markov chain from Exercise 5.1.3. Use the result of the previous exercise to compute $g(x) =$ the expected time to reach state (H, H) starting from state x. The answer we want is

$$EN(HH) = 2 + (1/4)[g(H < H) + g(H,T) + g(T,H) + g(T,T)]$$

5.2.13 In this problem we consider the number of tosses $N(HHH)$ to get HHH, let X_n be the number of consecutive heads we have after n tosses. X_n is a Markov chain with transition probability

	0	**1**	**2**	**3**
0	1/2	1/2	0	0
1	1/2	0	1/2	0
2	1/2	0	0	1/2
3	0	0	0	1

Let T_3 be the time to reach state 3. Use Problem 5.2.11 to compute E_0T_3.

5.3 Recurrence and Transience

Let $T_y^0 = 0$, and for $k \geq 1$, let

$$T_y^k = \inf\{n > T_y^{k-1} : X_n = y\}$$

and recall (5.2.6)

$$\rho_{xy} = P_x(T_y < \infty)$$

A state y is said to be **recurrent** if $\rho_{yy} = 1$ and **transient** if $\rho_{yy} < 1$. If y is recurrent, Theorem 5.2.6 implies $P_y(T_y^k < \infty) = 1$ for all k, so $P_y(X_n = y \text{ i.o.}) = 1$.

If y is transient and we let $N(y) = \sum_{n=1}^{\infty} 1_{(X_n=y)}$ be the number of visits to y at positive times, then

$$\begin{aligned} E_x N(y) &= \sum_{k=1}^{\infty} P_x(N(y) \geq k) = \sum_{k=1}^{\infty} P_x(T_y^k < \infty) \\ &= \sum_{k=1}^{\infty} \rho_{xy}\rho_{yy}^{k-1} = \frac{\rho_{xy}}{1-\rho_{yy}} < \infty \end{aligned} \tag{5.3.1}$$

Combining the last computation with our result for recurrent states gives:

Theorem 5.3.1 *y is recurrent if and only if $E_y N(y) = \infty$.*

The next result shows that recurrence is contagious.

Theorem 5.3.2 *If x is recurrent and $\rho_{xy} > 0$, then y is recurrent and $\rho_{yx} = 1$.*

Proof We will first show $\rho_{yx} = 1$ by showing that if $\rho_{xy} > 0$ and $\rho_{yx} < 1$, then $\rho_{xx} < 1$. Let $K = \inf\{k : p^k(x,y) > 0\}$. There is a sequence $y_1, \ldots, y_{K-1}$ so that

$$p(x, y_1)p(y_1, y_2) \cdots p(y_{K-1}, y) > 0$$

Since K is minimal, $y_i \neq x$ for $1 \leq i \leq K-1$. If $\rho_{yx} < 1$, we have

$$P_x(T_x = \infty) \geq p(x, y_1)p(y_1, y_2) \cdots p(y_{K-1}, y)(1 - \rho_{yx}) > 0$$

a contradiction. So $\rho_{yx} = 1$.

To prove that y is recurrent, observe that $\rho_{yx} > 0$ implies there is an L so that $p^L(y,x) > 0$. Now

$$p^{L+n+K}(y,y) \geq p^L(y,x)p^n(x,x)p^K(x,y)$$

Summing over n, we see

$$\sum_{n=1}^{\infty} p^{L+n+K}(y,y) \geq p^L(y,x)p^K(x,y)\sum_{n=1}^{\infty} p^n(x,x) = \infty$$

so Theorem 5.3.1 implies y is recurrent. □

The next fact will help us identify recurrent states in examples. First we need two definitions. C is **closed** if $x \in C$ and $\rho_{xy} > 0$ implies $y \in C$. The name comes from the fact that if C is closed and $x \in C$, then $P_x(X_n \in C) = 1$ for all n. D is **irreducible** if $x, y \in D$ implies $\rho_{xy} > 0$.

Theorem 5.3.3 *Let C be a finite closed set. Then C contains a recurrent state. If C is irreducible then all states in C are recurrent.*

Proof In view of Theorem 5.3.2, it suffices to prove the first claim. Suppose it is false. Then for all $y \in C$, $\rho_{yy} < 1$ and $E_x N(y) = \rho_{xy}/(1-\rho_{yy})$, but this is ridiculous since it implies

$$\infty > \sum_{y\in C} E_x N(y) = \sum_{y\in C}\sum_{n=1}^{\infty} p^n(x,y) = \sum_{n=1}^{\infty}\sum_{y\in C} p^n(x,y) = \sum_{n=1}^{\infty} 1$$

The first inequality follows from the fact that C is finite and the last equality from the fact that C is closed. □

To illustrate the use of the last result consider:

Example 5.3.4 (A Seven-state chain) Consider the transition probability:

	1	**2**	**3**	**4**	**5**	**6**	**7**
1	.3	0	0	0	.7	0	0
2	.1	.2	.3	.4	0	0	0
3	0	0	.5	.5	0	0	0
4	0	0	0	.5	0	.5	0
5	.6	0	0	0	.4	0	0
6	0	0	0	.1	0	.1	.8
7	0	0	0	1	0	0	0

To identify the states that are recurrent and those that are transient, we begin by drawing a graph that will contain an arc from i to j if $p(i,j) > 0$ and $i \neq j$. We do not worry about drawing the self-loops corresponding to states with $p(i,i) > 0$, since such transitions cannot help the chain get somewhere new.

In the case under consideration, we draw arcs from $1 \to 5$, $2 \to 1$, $2 \to 3$, $2 \to 4$, $3 \to 4$, $4 \to 6$, $4 \to 7$, $5 \to 1$, $6 \to 4$, $6 \to 7$, $7 \to 4$.

(i) $\rho_{21} > 0$ and $\rho_{12} = 0$, so 2 must be transient, or we would contradict Theorem 5.3.2. Similarly, $\rho_{34} > 0$ and $\rho_{43} = 0$ so 3 must be transient

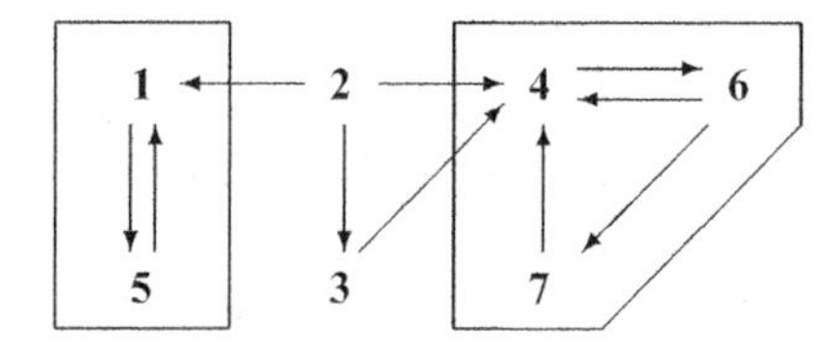

Figure 5.4 Graph for the seven state chain.

(ii) $\{1,5\}$ and $\{4,6,7\}$ are irreducible closed sets, so Theorem 5.3.3 implies these states are recurrent.

The last reasoning can be used to identify transient and recurrent states when S is finite, since for $x \in S$ either: (i) there is a y with $\rho_{xy} > 0$ and $\rho_{yx} = 0$ and x must be transient, or (ii) $\rho_{xy} > 0$ implies $\rho_{yx} > 0$. In case (ii), Exercise 5.3.2 implies $C_x = \{y : \rho_{xy} > 0\}$ is an irreducible closed set. (If $y, z \in C_x$, then $\rho_{yz} \geq \rho_{yx}\rho_{xz} > 0$. If $\rho_{yw} > 0$, then $\rho_{xw} \geq \rho_{xy}\rho_{yw} > 0$, so $w \in C_x$.) So Theorem 5.3.3 implies x is recurrent.

Example 5.3.4 motivates the following:

Theorem 5.3.5 (Decomposition theorem) *Let $R = \{x : \rho_{xx} = 1\}$ be the recurrent states of a Markov chain. R can be written as $\cup_i R_i$, where each R_i is closed and irreducible.*

Remark This result shows that for the study of recurrent states we can, without loss of generality, consider a single irreducible closed set.

Proof If $x \in R$, let $C_x = \{y : \rho_{xy} > 0\}$. By Theorem 5.3.2, $C_x \subset R$, and if $y \in C_x$, then $\rho_{yx} > 0$. From this it follows easily that either $C_x \cap C_y = \emptyset$ or $C_x = C_y$. To prove the last claim, suppose $C_x \cap C_y \neq \emptyset$. If $z \in C_x \cap C_y$, then $\rho_{xy} \geq \rho_{xz}\rho_{zy} > 0$, so if $w \in C_y$, we have $\rho_{xw} \geq \rho_{xy}\rho_{yw} > 0$ and it follows that $C_x \supset C_y$. Interchanging the roles of x and y gives $C_y \supset C_x$, and we have proved our claim. If we let R_i be a listing of the sets that appear as some C_x, we have the desired decomposition. □

The rest of this section is devoted to examples. Specifically, we concentrate on the question: How do we tell whether a state is recurrent or transient? Reasoning based on Theorem 5.3.2 works occasionally when S is infinite.

Example 5.3.6 (Branching process) If the probability of no children is positive, then $\rho_{k0} > 0$ and $\rho_{0k} = 0$ for $k \geq 1$, so Theorem 5.3.3 implies all states $k \geq 1$ are transient. The state 0 has $p(0,0) = 1$ and is recurrent. It is called an **absorbing state** to reflect the fact that once the chain enters 0, it remains there for all time.

If S is infinite and irreducible, all that Theorem 5.3.2 tells us is that either all the states are recurrent or all are transient, and we are left to figure out which case occurs.

Example 5.3.7 ($M/G/1$ queue) Let $\mu = \sum k a_k$ be the mean number of customers that arrive during one service time. We will now show that if $\mu > 1$, the chain is transient (i.e., all states are), but if $\mu \leq 1$, it is recurrent. For the case $\mu > 1$, we observe that if $\xi_1, \xi_2, \ldots$ are i.i.d. with $P(\xi_m = j) = a_{j+1}$ for $j \geq -1$ and $S_n = \xi_1 + \cdots + \xi_n$, then $X_0 + S_n$ and X_n behave the same until time $N = \inf\{n : X_0 + S_n = 0\}$. When $\mu > 1$, $E\xi_m = \mu - 1 > 0$, so $S_n \to \infty$ a.s., and $\inf S_n > -\infty$ a.s. It follows from the last observation that if x is large, $P_x(N < \infty) < 1$, and the chain is transient.

To deal with the case $\mu \le 1$, we observe that it follows from arguments in the last paragraph that $X_{n\wedge N}$ is a supermartingale. Let $T = \inf\{n : X_n \ge M\}$. Since $X_{n\wedge N}$ is a nonnegative supermartingale, using Theorem 4.8.4 at time $\tau = T \wedge N$, and observing $X_\tau \ge M$ on $\{T < N\}$, $X_\tau = 0$ on $\{N < T\}$ gives

$$x \ge M P_x(T < N)$$

Letting $M \to \infty$ shows $P_x(N < \infty) = 1$, so the chain is recurrent.

Remark There is another way of seeing that the $M/G/1$ queue is transient when $\mu > 1$. If we consider the customers that arrive during a person's service time to be her children, then we get a branching process. Results in Section 5.3 imply that when $\mu \le 1$, the branching process dies out with probability one (i.e., the queue becomes empty), so the chain is recurrent. When $\mu > 1$, Theorem 4.3.12 implies $P_x(T_0 < \infty) = \rho^x$, where ρ is the unique fixed point $\in (0,1)$ of the function $\varphi(\theta) = \sum_{k=0}^\infty a_k\theta^k$.

The next result encapsulates the techniques we used for birth and death chains and the $M/G/1$ queue.

Theorem 5.3.8 *Suppose S is irreducible, and $\varphi \ge 0$ with $E_x\varphi(X_1) \le \varphi(x)$ for $x \notin F$, a finite set, and $\varphi(x) \to \infty$ as $x \to \infty$, i.e., $\{x : \varphi(x) \le M\}$ is finite for any $M < \infty$, then the chain is recurrent.*

Proof Let $\tau = \inf\{n > 0 : X_n \in F\}$. Our assumptions imply that $Y_n = \varphi(X_{n\wedge\tau})$ is a supermartingale. Let $T_M = \inf\{n > 0 : X_n \in F \text{ or } \varphi(X_n) > M\}$. Since $\{x : \varphi(x) \le M\}$ is finite and the chain is irreducible, $T_M < \infty$ a.s. Using Theorem 4.8.4 now, we see that

$$\varphi(x) \ge E_x\varphi(X_{T_M}) \ge M P_x(T_M < \tau)$$

since $\varphi(X_{T_M}) \ge M$ when $T_M < \tau$. Letting $M \to \infty$, we see that $P_x(\tau < \infty) = 1$ for all $x \notin F$. So $P_y(X_n \in F \text{ i.o.}) = 1$ for all $y \in S$, and since F is finite, $P_y(X_n = z \text{ i.o.}) = 1$ for some $z \in F$. □

Example 5.3.9 (Birth and death chains on $\{0, 1, 2, \ldots\}$) Let

$$p(i, i+1) = p_i \quad p(i, i-1) = q_i \quad p(i, i) = r_i$$

where $q_0 = 0$. Let $N = \inf\{n : X_n = 0\}$. To analyze this example, we are going to define a function φ so that $\varphi(X_{N\wedge n})$ is a martingale. We start by setting $\varphi(0) = 0$ and $\varphi(1) = 1$. For the martingale property to hold when $X_n = k \ge 1$, we must have

$$\varphi(k) = p_k\varphi(k+1) + r_k\varphi(k) + q_k\varphi(k-1)$$

Using $r_k = 1 - (p_k + q_k)$, we can rewrite the last equation as

$$q_k(\varphi(k) - \varphi(k-1)) = p_k(\varphi(k+1) - \varphi(k))$$

$$\text{or} \quad \varphi(k+1) - \varphi(k) = \frac{q_k}{p_k}(\varphi(k) - \varphi(k-1))$$

Here and in what follows, we suppose that $p_k, q_k > 0$ for $k \geq 1$. Otherwise, the chain is not irreducible. Since $\varphi(1) - \varphi(0) = 1$, iterating the last result gives

$$\varphi(m+1) - \varphi(m) = \prod_{j=1}^{m} \frac{q_j}{p_j} \quad \text{for } m \geq 1$$

$$\varphi(n) = \sum_{m=0}^{n-1} \prod_{j=1}^{m} \frac{q_j}{p_j} \quad \text{for } n \geq 1$$

if we interpret the product as 1 when $m = 0$. Let $T_c = \inf\{n \geq 1 : X_n = c\}$. Now I claim that:

Theorem 5.3.10 *If $a < x < b$, then*

$$P_x(T_a < T_b) = \frac{\varphi(b) - \varphi(x)}{\varphi(b) - \varphi(a)} \qquad P_x(T_b < T_a) = \frac{\varphi(x) - \varphi(a)}{\varphi(b) - \varphi(a)}$$

Proof If we let $T = T_a \wedge T_b$, then $\varphi(X_{n \wedge T})$ is a bounded martingale, so $\varphi(x) = E_x \varphi(X_T)$ by Theorem 4.8.2. Since $X_T \in \{a, b\}$ a.s.,

$$\varphi(x) = \varphi(a) P_x(T_a < T_b) + \varphi(b)[1 - P_x(T_a < T_b)]$$

and solving gives the indicated formula. □

Remark The answer and the proof should remind the reader of our results for symmetric and asymetric simple random walk, which are special cases of this.

Letting $a = 0$ and $b = M$ in Theorem 5.3.10 gives

$$P_x(T_0 > T_M) = \varphi(x)/\varphi(M)$$

Letting $M \to \infty$ and observing that $T_M \geq M - x$, P_x a.s., we have proved:

Theorem 5.3.11 *0 is recurrent if and only if $\varphi(M) \to \infty$ as $M \to \infty$, i.e.,*

$$\varphi(\infty) \equiv \sum_{m=0}^{\infty} \prod_{j=1}^{m} \frac{q_j}{p_j} = \infty$$

If $\varphi(\infty) < \infty$, then $P_x(T_0 = \infty) = \varphi(x)/\varphi(\infty)$.

Exercises

5.3.1 Suppose y is recurrent and for $k \geq 0$, let $R_k = T_y^k$ be the time of the kth return to y, and for $k \geq 1$ let $r_k = R_k - R_{k-1}$ be the kth interarrival time. Use the strong Markov property to conclude that under P_y, the vectors $v_k = (r_k, X_{R_{k-1}}, \ldots, X_{R_k - 1})$, $k \geq 1$ are i.i.d.

5.3.2 Use the strong Markov property to show that $\rho_{xz} \geq \rho_{xy}\rho_{yz}$.

5.3.3 Show that in the Ehrenfest chain (Example 5.1.3) all states are recurrent.

5.3.4 To probe the boundary between recurrence and transience for birth and death chains, suppose $p_j = 1/2 + \epsilon_j$, where $\epsilon_j \sim Cj^{-\alpha}$ as $j \to \infty$, and $q_j = 1 - p_j$. Show that (i) if $\alpha > 1$ 0 is recurrent, (ii) if $\alpha < 1$ 0 is transient, (iii) if $\alpha = 1$, then 0 is transient if $C > 1/4$ and recurrent if $C < 1/4$.

5.3.5 Show that if we replace "$\varphi(x) \to \infty$" by "$\varphi(x) \to 0$" in the Theorem 5.3.8 and assume that $\varphi(x) > 0$ for $x \in F$, then we can conclude that the chain is transient.

5.3.6 Let X_n be a birth and death chain with $p_j - 1/2 \sim C/j$ as $j \to \infty$ and $q_j = 1 - p_j$. (i) Show that if we take $C < 1/4$, then we can pick $\alpha > 0$ so that $\varphi(x) = x^\alpha$ satisfies the hypotheses of Theorem 5.3.8. (ii) Show that when $C > 1/4$, we can take $\alpha < 0$ and apply Exercise 5.3.5.

5.3.7 f is said to be **superharmonic** if $f(x) \geq \sum_y p(x,y) f(y)$, or equivalently $f(X_n)$ is a supermartingale. Suppose p is irreducible. Show that p is recurrent if and only if every nonnegative superharmonic function is constant.

5.3.8 **M/M/∞ queue.** Consider a telephone system with an infinite number of lines. Let $X_n =$ the number of lines in use at time n, and suppose

$$X_{n+1} = \sum_{m=1}^{X_n} \xi_{n,m} + Y_{n+1}$$

where the $\xi_{n,m}$ are i.i.d. with $P(\xi_{n,m} = 1) = p$ and $P(\xi_{n,m} = 0) = 1-p$, and Y_n is an independent i.i.d. sequence of Poisson mean λ r.v.'s. In words, for each conversation we flip a coin with probability p of heads to see if it continues for another minute. Meanwhile, a Poisson mean λ number of conversations start between time n and $n+1$. Use Theorem 5.3.8 with $\varphi(x) = x$ to show that the chain is recurrent for any $p < 1$.

5.4 Recurrence of Random Walks Stararred Section

Throughout this section, S_n will be a random walk, i.e., $S_n = X_1 + \cdots + X_n$, where $X_1, X_2, \ldots$ are i.i.d. The number $x \in \mathbf{R}^d$ is said to be a **recurrent value** for the random walk S_n if for every $\epsilon > 0$, $P(\|S_n - x\| < \epsilon \text{ i.o.}) = 1$. Here $\|x\| = \sup |x_i|$. The reader will see the reason for this choice of norm in the proof of Lemma 5.4.6. The Hewitt-Savage 0-1 law, Theorem 2.5.4, implies that if the last probability is < 1, it is 0. Our first result shows that to know the set of recurrent values, it is enough to check $x = 0$. A number x is said to be a **possible value** of the random walk if for any $\epsilon > 0$, there is an n so that $P(\|S_n - x\| < \epsilon) > 0$.

Theorem 5.4.1 *The set $\mathcal{V}$ of recurrent values is either $\emptyset$ or a closed subgroup of $\mathbf{R}^d$. In the second case, $\mathcal{V} = \mathcal{U}$, the set of possible values.*

Proof Suppose $\mathcal{V} \neq \emptyset$. It is clear that $\mathcal{V}^c$ is open, so $\mathcal{V}$ is closed. To prove that $\mathcal{V}$ is a group, we will first show that

$(*)$ if $x \in \mathcal{U}$ and $y \in \mathcal{V}$, then $y - x \in \mathcal{V}$.

This statement has been formulated so that once it is established, the result follows easily. Let

$$p_{\delta,m}(z) = P(\|S_n - z\| \geq \delta \text{ for all } n \geq m)$$

If $y - x \notin \mathcal{V}$, there is an $\epsilon > 0$ and $m \geq 1$ so that $p_{2\epsilon,m}(y - x) > 0$. Since $x \in \mathcal{U}$, there is a k so that $P(\|S_k - x\| < \epsilon) > 0$. Since

$$P(\|S_n - S_k - (y - x)\| \geq 2\epsilon \text{ for all } n \geq k + m) = p_{2\epsilon,m}(y - x)$$

and is independent of $\{\|S_k - x\| < \epsilon\}$, it follows that

$$p_{\epsilon,m+k}(y) \geq P(\|S_k - x\| < \epsilon)p_{2\epsilon,m}(y - x) > 0$$

contradicting $y \in \mathcal{V}$, so $y - x \in \mathcal{V}$.

To conclude $\mathcal{V}$ is a group when $\mathcal{V} \neq \emptyset$, let $q, r \in \mathcal{V}$, and observe: (i) taking $x = y = r$ in $(*)$ shows $0 \in \mathcal{V}$, (ii) taking $x = r$, $y = 0$ shows $-r \in \mathcal{V}$, and (iii) taking $x = -r$, $y = q$ shows $q + r \in \mathcal{V}$. To prove that $\mathcal{V} = \mathcal{U}$ now, observe that if $u \in \mathcal{U}$ taking $x = u$, $y = 0$ shows $-u \in \mathcal{V}$ and since $\mathcal{V}$ is a group, it follows that $u \in \mathcal{V}$. □

If $\mathcal{V} = \emptyset$, the random walk is said to be **transient**, otherwise it is called **recurrent**. Before plunging into the technicalities needed to treat a general random walk, we begin by analyzing the special case Polya considered in 1921. Legend has it that Polya thought of this problem while wandering around in a park near Zürich when he noticed that he kept encountering the same young couple. History does not record what the young couple thought.

Example 5.4.2 (Simple random walk on $\mathbf{Z}^d$)

$$P(X_i = e_j) = P(X_i = -e_j) = 1/2d$$

for each of the d unit vectors e_j. To analyze this case, we begin with a result that is valid for any random walk. Let $\tau_0 = 0$ and $\tau_n = \inf\{m > \tau_{n-1} : S_m = 0\}$ be the time of the nth return to 0. From the strong Markov property, it follows that

$$P(\tau_n < \infty) = P(\tau_1 < \infty)^n$$

a fact that leads easily to:

Theorem 5.4.3 *For any random walk, the following are equivalent:*
(i) $P(\tau_1 < \infty) = 1$, *(ii)* $P(S_m = 0 \text{ i.o.}) = 1$, *and (iii)* $\sum_{m=0}^{\infty} P(S_m = 0) = \infty$.

Proof If $P(\tau_1 < \infty) = 1$, then $P(\tau_n < \infty) = 1$ for all n and $P(S_m = 0 \text{ i.o.}) = 1$. Let

$$V = \sum_{m=0}^{\infty} 1_{(S_m=0)} = \sum_{n=0}^{\infty} 1_{(\tau_n<\infty)}$$

be the number of visits to 0, counting the visit at time 0. Taking expected value and using Fubini's theorem to put the expected value inside the sum:

$$\begin{aligned} EV &= \sum_{m=0}^{\infty} P(S_m = 0) = \sum_{n=0}^{\infty} P(\tau_n < \infty) \\ &= \sum_{n=0}^{\infty} P(\tau_1 < \infty)^n = \frac{1}{1 - P(\tau_1 < \infty)} \end{aligned}$$

The second equality shows (ii) implies (iii), and in combination with the last two shows that if (i) is false, then (iii) is false (i.e., (iii) implies (i)). □

Theorem 5.4.4 *Simple random walk is recurrent in $d \le 2$ and transient in $d \ge 3$.*

To steal a joke from Kakutani (U.C.L.A. colloquium talk): "A drunk man will eventually find his way home but a drunk bird may get lost forever."

Proof Let $\rho_d(m) = P(S_m = 0)$. $\rho_d(m)$ is 0 if m is odd. From Theorem 3.1.3, we get $\rho_1(2n) \sim (\pi n)^{-1/2}$ as $n \to \infty$. This and Theorem 5.4.3 give the result in one dimension. Our next step is

Simple random walk is recurrent in two dimensions. Note that in order for $S_{2n} = 0$, we must for some $0 \le m \le n$ have m up steps, m down steps, $n - m$ to the left, and $n - m$ to the right, so

$$\begin{aligned}
\rho_2(2n) &= 4^{-2n} \sum_{m=0}^{n} \frac{2n!}{m!\, m!\, (n-m)!\, (n-m)!} \\
&= 4^{-2n} \binom{2n}{n} \sum_{m=0}^{n} \binom{n}{m} \binom{n}{n-m} = 4^{-2n} \binom{2n}{n}^2 = \rho_1(2n)^2
\end{aligned}$$

To see the next to last equality, consider choosing n students from a class with n boys and n girls and observe that for some $0 \le m \le n$ you must choose m boys and $n - m$ girls. Using the asymptotic formula $\rho_1(2n) \sim (\pi n)^{-1/2}$, we get $\rho_2(2n) \sim (\pi n)^{-1}$. Since $\sum n^{-1} = \infty$, the result follows from Theorem 5.4.3.

Remark For a direct proof of $\rho_2(2n) = \rho_1(2n)^2$, note that if T_n^1 and T_n^2 are independent, one-dimensional random walks, then T_n jumps from x to $x + (1,1)$, $x + (1,-1)$, $x + (-1,1)$, and $x + (-1,-1)$ with equal probability, so rotating T_n by 45 degrees and dividing by $\sqrt{2}$ gives S_n.

Simple random walk is transient in three dimensions. Intuitively, this holds, since the probability of being back at 0 after $2n$ steps is $\sim cn^{-3/2}$ and this is summable. We will not compute the probability exactly but will get an upper bound of the right order of magnitude. Again, since the number of steps in the directions $\pm e_i$ must be equal for $i = 1, 2, 3$

$$\begin{aligned}
\rho_3(2n) &= 6^{-2n} \sum_{j,k} \frac{(2n)!}{(j!\, k!\, (n-j-k)!)^2} \\
&= 2^{-2n} \binom{2n}{n} \sum_{j,k} \left(3^{-n} \frac{n!}{j!\, k!\, (n-j-k)!} \right)^2 \\
&\le 2^{-2n} \binom{2n}{n} \max_{j,k} 3^{-n} \frac{n!}{j!\, k!\, (n-j-k)!}
\end{aligned}$$

where in the last inequality we have used the fact that if $a_{j,k}$ are ≥ 0 and sum to 1, then $\sum_{j,k} a_{j,k}^2 \le \max_{j,k} a_{j,k}$. Our last step is to show

$$\max_{j,k} 3^{-n} \frac{n!}{j!\, k!\, (n-j-k)!} \le Cn^{-1}$$

To do this, we note that (a) if any of the numbers j, k, or $n-j-k$ is $< [n/3]$ increasing the smallest number and decreasing the largest number decreases the denominator (since $x(1-x)$ is maximized at 1/2), so the maximum occurs when all three numbers are as close as possible to $n/3$; (b) Stirling's formula implies

$$\frac{n!}{j!\,k!\,(n-j-k)!} \sim \frac{n^n}{j^j k^k (n-j-k)^{n-j-k}} \cdot \sqrt{\frac{n}{jk(n-j-k)}} \cdot \frac{1}{2\pi}$$

Taking j and k within 1 of $n/3$, the first term on the right is $\le C3^n$, and the desired result follows.

Simple random walk is transient in $d > 3$. Let $T_n = (S_n^1, S_n^2, S_n^3)$, $N(0) = 0$ and $N(n) = \inf\{m > N(n-1) : T_m \ne T_{N(n-1)}\}$. It is easy to see that $T_{N(n)}$ is a three-dimensional simple random walk. Since $T_{N(n)}$ returns infinitely often to 0 with probability 0 and the first three coordinates are constant in between the $N(n)$, S_n is transient. □

The rest of this section is devoted to proving the following facts about random walks:

- S_n is recurrent in $d = 1$ if $S_n/n \to 0$ in probability.
- S_n is recurrent in $d = 2$ if $S_n/n^{1/2} \Rightarrow$ a nondegenerate normal distribution.
- S_n is transient in $d \ge 3$ if it is "truly three dimensional."

To prove the last result, we will give a necessary and sufficient condition for recurrence.

The first step in deriving these results is to generalize Theorem 5.4.3.

Lemma 5.4.5 *If $\sum_{n=1}^\infty P(\|S_n\| < \epsilon) < \infty$, then $P(\|S_n\| < \epsilon$ i.o.$) = 0$.*
If $\sum_{n=1}^\infty P(\|S_n\| < \epsilon) = \infty$, then $P(\|S_n\| < 2\epsilon$ i.o.$) = 1$.

Proof The first conclusion follows from the Borel-Cantelli lemma. To prove the second, let $F = \{\|S_n\| < \epsilon \text{ i.o.}\}^c$. Breaking things down according to the last time $\|S_n\| < \epsilon$,

$$\begin{aligned} P(F) &= \sum_{m=0}^\infty P(\|S_m\| < \epsilon, \|S_n\| \ge \epsilon \text{ for all } n \ge m+1) \\ &\ge \sum_{m=0}^\infty P(\|S_m\| < \epsilon, \|S_n - S_m\| \ge 2\epsilon \text{ for all } n \ge m+1) \\ &= \sum_{m=0}^\infty P(\|S_m\| < \epsilon)\rho_{2\epsilon,1} \end{aligned}$$

where $\rho_{\delta,k} = P(\|S_n\| \ge \delta \text{ for all } n \ge k)$. Since $P(F) \le 1$, and

$$\sum_{m=0}^\infty P(\|S_m\| < \epsilon) = \infty$$

it follows that $\rho_{2\epsilon,1} = 0$. To extend this conclusion to $\rho_{2\epsilon,k}$ with $k \ge 2$, let

$$A_m = \{\|S_m\| < \epsilon, \|S_n\| \ge \epsilon \text{ for all } n \ge m+k\}$$

Since any ω can be in at most k of the A_m, repeating the argument gives

$$k \geq \sum_{m=0}^{\infty} P(A_m) \geq \sum_{m=0}^{\infty} P(\|S_m\| < \epsilon)\rho_{2\epsilon,k}$$

So $\rho_{2\epsilon,k} = P(\|S_n\| \geq 2\epsilon \text{ for all } j \geq k) = 0$, and since k is arbitrary, the desired conclusion follows. □

Our second step is to show that the convergence or divergence of the sums in Lemma 5.4.5 is independent of ϵ. The previous proof works for any norm. For the next one, we need $\|x\| = \sup_i |x_i|$.

Lemma 5.4.6 *Let m be an integer ≥ 2.*

$$\sum_{n=0}^{\infty} P(\|S_n\| < m\epsilon) \leq (2m)^d \sum_{n=0}^{\infty} P(\|S_n\| < \epsilon)$$

Proof We begin by observing

$$\sum_{n=0}^{\infty} P(\|S_n\| < m\epsilon) \leq \sum_{n=0}^{\infty} \sum_{k} P(S_n \in k\epsilon + [0,\epsilon)^d)$$

where the inner sum is over $k \in \{-m, \ldots, m-1\}^d$. If we let

$$T_k = \inf\{\ell \geq 0 : S_\ell \in k\epsilon + [0,\epsilon)^d\}$$

then breaking things down according to the value of T_k and using Fubini's theorem gives

$$\begin{aligned}
\sum_{n=0}^{\infty} P(S_n \in k\epsilon + [0,\epsilon)^d) &= \sum_{n=0}^{\infty} \sum_{\ell=0}^{n} P(S_n \in k\epsilon + [0,\epsilon)^d, T_k = \ell) \\
&\leq \sum_{\ell=0}^{\infty} \sum_{n=\ell}^{\infty} P(\|S_n - S_\ell\| < \epsilon, T_k = \ell)
\end{aligned}$$

Since $\{T_k = \ell\}$ and $\{\|S_n - S_\ell\| < \epsilon\}$ are independent, the last sum

$$= \sum_{m=0}^{\infty} P(T_k = m) \sum_{j=0}^{\infty} P(\|S_j\| < \epsilon) \leq \sum_{j=0}^{\infty} P(\|S_j\| < \epsilon)$$

Since there are $(2m)^d$ values of k in $\{-m, \ldots, m-1\}^d$, the proof is complete. □

Combining Lemmas 5.4.5 and 5.4.6 gives:

Theorem 5.4.7 *The convergence (resp. divergence) of $\sum_n P(\|S_n\| < \epsilon)$ for a single value of $\epsilon > 0$ is sufficient for transience (resp. recurrence).*

In $d = 1$, if $EX_i = \mu \neq 0$, then the strong law of large numbers implies $S_n/n \to \mu$ so $|S_n| \to \infty$ and S_n is transient. As a converse, we have:

Theorem 5.4.8 (Chung-Fuchs theorem) *Suppose $d = 1$. If the weak law of large numbers holds in the form $S_n/n \to 0$ in probability, then S_n is recurrent.*

Proof Let $u_n(x) = P(|S_n| < x)$ for $x > 0$. Lemma 5.4.6 implies

$$\sum_{n=0}^{\infty} u_n(1) \geq \frac{1}{2m} \sum_{n=0}^{\infty} u_n(m) \geq \frac{1}{2m} \sum_{n=0}^{Am} u_n(n/A)$$

for any $A < \infty$, since $u_n(x) \geq 0$ and is increasing in x. By hypothesis $u_n(n/A) \to 1$, so letting $m \to \infty$ and noticing the right-hand side is $A/2$ times the average of the first Am terms

$$\sum_{n=0}^{\infty} u_n(1) \geq A/2$$

Since A is arbitrary, the sum must be ∞, and the desired conclusion follows from Theorem 5.4.7. □

Theorem 5.4.9 *If S_n is a random walk in $\mathbf{R}^2$ and $S_n/n^{1/2} \Rightarrow$ a nondegenerate normal distribution, then S_n is recurrent.*

Remark The conclusion is also true if the limit is degenerate, but in that case the random walk is essentially one- (or zero-) dimensional, and the result follows from the Chung-Fuchs theorem.

Proof Let $u(n,m) = P(\|S_n\| < m)$. Lemma 5.4.6 implies

$$\sum_{n=0}^{\infty} u(n,1) \geq (4m^2)^{-1} \sum_{n=0}^{\infty} u(n,m)$$

If $m/\sqrt{n} \to c$, then

$$u(n,m) \to \int_{[-c,c]^2} n(x)\, dx$$

where $n(x)$ is the density of the limiting normal distribution. If we use $\rho(c)$ to denote the right-hand side and let $n = [\theta m^2]$, it follows that $u([\theta m^2],m) \to \rho(\theta^{-1/2})$. If we write

$$m^{-2} \sum_{n=0}^{\infty} u(n,m) = \int_0^{\infty} u([\theta m^2],m)\, d\theta$$

let $m \to \infty$, and use Fatou's lemma, we get

$$\liminf_{m\to\infty} (4m^2)^{-1} \sum_{n=0}^{\infty} u(n,m) \geq 4^{-1} \int_0^{\infty} \rho(\theta^{-1/2})\, d\theta$$

Since the normal density is positive and continuous at 0

$$\rho(c) = \int_{[-c,c]^2} n(x)\, dx \sim n(0)(2c)^2$$

as $c \to 0$. So $\rho(\theta^{-1/2}) \sim 4n(0)/\theta$ as $\theta \to \infty$, the integral diverges, and backtracking to the first inequality in the proof it follows that $\sum_{n=0}^{\infty} u(n,1) = \infty$, proving the result. □

We come now to the promised necessary and sufficient condition for recurrence. Here $\phi = E \exp(it \cdot X_j)$ is the ch.f. of one step of the random walk.

Theorem 5.4.10 *Let $\delta > 0$. S_n is recurrent if and only if*

$$\int_{(-\delta,\delta)^d} Re \frac{1}{1-\varphi(y)}\, dy = \infty$$

We will prove a weaker result:

Theorem 5.4.11 *Let $\delta > 0$. S_n is recurrent if and only if*

$$\sup_{r<1} \int_{(-\delta,\delta)^d} Re \frac{1}{1-r\varphi(y)}\, dy = \infty$$

Remark Half of the work needed to get the first result from the second is trivial.

$$0 \le \mathrm{Re} \frac{1}{1-r\varphi(y)} \to \mathrm{Re} \frac{1}{1-\varphi(y)} \qquad \text{as } r \to 1$$

so Fatou's lemma shows that if the integral is infinite, the walk is recurrent. The other direction is rather difficult: the second result is in Chung and Fuchs (1951), but a proof of the first result had to wait for Ornstein (1969) and Stone (1969) to solve the problem independently. Their proofs use a trick to reduce to the case where the increments have a density and then a second trick to deal with that case, so we will not give the details here. The reader can consult either of the sources cited or Port and Stone (1969), where the result is demonstrated for random walks on Abelian groups.

Proof The first ingredient in the solution is the:

Lemma 5.4.12 (Parseval relation) *Let μ and ν be probability measures on $\mathbf{R}^d$ with ch.f.'s φ and ψ.*

$$\int \psi(t)\,\mu(dt) = \int \varphi(x)\,\nu(dx)$$

Proof Since $e^{it\cdot x}$ is bounded, Fubini's theorem implies

$$\int \psi(t)\mu(dt) = \int\int e^{itx}\nu(dx)\mu(dt) = \int\int e^{itx}\mu(dt)\nu(dx) = \int \varphi(x)\nu(dx) \qquad \square$$

Our second ingredient is a little calculus.

Lemma 5.4.13 *If $|x| \le \pi/3$, then $1 - \cos x \ge x^2/4$.*

Proof It suffices to prove the result for $x > 0$. If $z \le \pi/3$, then $\cos z \ge 1/2$,

$$\sin y = \int_0^y \cos z\, dz \ge \frac{y}{2}$$

$$1 - \cos x = \int_0^x \sin y\, dy \ge \int_0^x \frac{y}{2}\, dy = \frac{x^2}{4}$$

which proves the desired result. $\square$

From Example 3.3.7, we see that the density

$$\frac{\delta - |x|}{\delta^2} \quad \text{when} \quad |x| \le \delta, \qquad 0 \quad \text{otherwise}$$

has ch.f. $2(1 - \cos \delta t)/(\delta t)^2$. Let μ_n denote the distribution of S_n. Using Lemma 5.4.13 (note $\pi/3 \geq 1$) and then Lemma 5.4.12, we have

$$P(\|S_n\| < 1/\delta) \leq 4^d \int \prod_{i=1}^{d} \frac{1 - \cos(\delta t_i)}{(\delta t_i)^2} \mu_n(dt)$$

$$= 2^d \int_{(-\delta,\delta)^d} \prod_{i=1}^{d} \frac{\delta - |x_i|}{\delta^2} \varphi^n(x)\, dx$$

Our next step is to sum from 0 to ∞. To be able to interchange the sum and the integral, we first multiply by r^n, where $r < 1$.

$$\sum_{n=0}^{\infty} r^n P(\|S_n\| < 1/\delta) \leq 2^d \int_{(-\delta,\delta)^d} \prod_{i=1}^{d} \frac{\delta - |x_i|}{\delta^2} \frac{1}{1 - r\varphi(x)} dx$$

Symmetry dictates that the integral on the right is real, so we can take the real part without affecting its value. Letting $r \uparrow 1$ and using $(\delta - |x|)/\delta \leq 1$

$$\sum_{n=0}^{\infty} P(\|S_n\| < 1/\delta) \leq \left(\frac{2}{\delta}\right)^d \sup_{r<1} \int_{(-\delta,\delta)^d} \text{Re} \frac{1}{1 - r\varphi(x)} dx$$

and using Theorem 5.4.7 gives half of Theorem 5.4.11.

To prove the other direction, we begin by noting that Example 3.3.15 shows that the density $(1 - \cos(x/\delta))/\pi x^2/\delta$ has ch.f. $1 - |\delta t|$ when $|t| \leq 1/\delta$, 0 otherwise. Using $1 \geq \prod_{i=1}^{d}(1 - |\delta x_i|)$ and then Lemma 5.4.12,

$$P(\|S_n\| < 1/\delta) \geq \int_{(-1/\delta,1/\delta)^d} \prod_{i=1}^{d}(1 - |\delta x_i|)\, \mu_n(dx)$$

$$= \int \prod_{i=1}^{d} \frac{1 - \cos(t_i/\delta)}{\pi t_i^2/\delta} \varphi^n(t)\, dt$$

Multiplying by r^n and summing gives

$$\sum_{n=0}^{\infty} r^n P(\|S_n\| < 1/\delta) \geq \int \prod_{i=1}^{d} \frac{1 - \cos(t_i/\delta)}{\pi t_i^2/\delta} \frac{1}{1 - r\varphi(t)} dt$$

The last integral is real, so its value is unaffected if we integrate only the real part of the integrand. If we do this and apply Lemma 5.4.13, we get

$$\sum_{n=0}^{\infty} r^n P(\|S_n\| < 1/\delta) \geq (4\pi\delta)^{-d} \int_{(-\delta,\delta)^d} \text{Re} \frac{1}{1 - r\varphi(t)} dt$$

Letting $r \uparrow 1$ and using Theorem 5.4.7 now completes the proof of Theorem 5.4.11. □

We will now consider some examples. Our goal in $d = 1$ and $d = 2$ is to convince you that the conditions in Theorems 5.4.8 and 5.4.9 are close to the best possible.

d = 1. Consider the symmetric stable laws that have ch.f. $\varphi(t) = \exp(-|t|^\alpha)$. To avoid using facts that we have not proved, we will obtain our conclusions from Theorem 5.4.11. It is not hard to use that form of the criterion in this case, since

$$1 - r\varphi(t) \downarrow 1 - \exp(-|t|^\alpha) \quad \text{as } r \uparrow 1$$
$$1 - \exp(-|t|^\alpha) \sim |t|^\alpha \quad \text{as } t \to 0$$

From this, it follows that the corresponding random walk is transient for $\alpha < 1$ and recurrent for $\alpha \geq 1$. The case $\alpha > 1$ is covered by Theorem 5.4.8, since these random walks have mean 0. The result for $\alpha = 1$ is new because the Cauchy distribution does not satisfy $S_n/n \to 0$ in probability. The random walks with $\alpha < 1$ are interesting because Exercise 5.4.1 implies

$$-\infty = \liminf S_n < \limsup S_n = \infty$$

but $P(|S_n| < M \text{ i.o.}) = 0$ for any $M < \infty$.

Remark The stable law examples are misleading in one respect. Shepp (1964) has proved that recurrent random walks may have arbitrarily large tails. To be precise, given a function $\epsilon(x) \downarrow 0$ as $x \uparrow \infty$, there is a recurrent random walk with $P(|X_1| \geq x) \geq \epsilon(x)$ for large x.

d = 2. Let $\alpha < 2$, and let $\varphi(t) = \exp(-|t|^\alpha)$, where $|t| = (t_1^2 + t_2^2)^{1/2}$. φ is the characteristic function of a random vector (X_1, X_2) that has two nice properties:

(i) the distribution of (X_1, X_2) is invariant under rotations,

(ii) X_1 and X_2 have symmetric stable laws with index α.

Again, $1 - r\varphi(t) \downarrow 1 - \exp(-|t|^\alpha)$ as $r \uparrow 1$ and $1 - \exp(-|t|^\alpha) \sim |t|^\alpha$ as $t \to 0$. Changing to polar coordinates and noticing

$$2\pi \int_0^\delta dx\, x\, x^{-\alpha} < \infty$$

when $1 - \alpha > -1$ shows the random walks with ch.f. $\exp(-|t|^\alpha)$, $\alpha < 2$ are transient. When $p < \alpha$, we have $E|X_1|^p < \infty$ by Exercise 3.8.4, so these examples show that Theorem 5.4.9 is reasonably sharp.

d ≥ 3. The integral $\int_0^\delta dx\, x^{d-1} x^{-2} < \infty$, so if a random walk is recurrent in $d \geq 3$, its ch.f. must $\to 1$ faster than t^2. In Exercise 3.3.16, we observed that (in one dimension) if $\varphi(r) = 1 + o(r^2)$, then $\varphi(r) \equiv 1$. By considering $\varphi(r\theta)$, where r is real and θ is a fixed vector, the last conclusion generalizes easily to $\mathbf{R}^d$, $d > 1$ and suggests that once we exclude walks that stay on a plane through 0, no three-dimensional random walks are recurrent.

A random walk in $\mathbf{R}^3$ is **truly three-dimensional** if the distribution of X_1 has $P(X_1 \cdot \theta \neq 0) > 0$ for all $\theta \neq 0$.

Theorem 5.4.14 *No truly three-dimensional random walk is recurrent.*

Proof We will deduce the result from Theorem 5.4.11. We begin with some arithmetic. If z is complex, the conjugate of $1 - z$ is $1 - \bar{z}$, so

$$\frac{1}{1-z} = \frac{1-\bar{z}}{|1-z|^2} \quad \text{and} \quad \mathrm{Re}\,\frac{1}{1-z} = \frac{\mathrm{Re}\,(1-z)}{|1-z|^2}$$

If $z = a + bi$ with $a \le 1$, then using the previous formula and dropping the b^2 from the denominator

$$\text{Re}\,\frac{1}{1-z} = \frac{1-a}{(1-a)^2+b^2} \le \frac{1}{1-a}$$

Taking $z = r\phi(t)$ and supposing for the second inequality that $0 \le \text{Re}\,\phi(t) \le 1$, we have

(a) $$\text{Re}\,\frac{1}{1-r\varphi(t)} \le \frac{1}{\text{Re}\,(1-r\varphi(t))} \le \frac{1}{\text{Re}\,(1-\varphi(t))}$$

The last calculation shows that it is enough to estimate

$$\text{Re}\,(1-\varphi(t)) = \int \{1-\cos(x\cdot t)\}\mu(dx) \ge \int_{|x\cdot t|<\pi/3} \frac{|x\cdot t|^2}{4}\,\mu(dx)$$

by Lemma 5.4.13. Writing $t = \rho\theta$, where $\theta \in S = \{x : |x| = 1\}$ gives

(b) $$\text{Re}\,(1-\varphi(\rho\theta)) \ge \frac{\rho^2}{4}\int_{|x\cdot\theta|<\pi/3\rho} |x\cdot\theta|^2\mu(dx)$$

Fatou's lemma implies that if we let $\rho \to 0$ and $\theta(\rho) \to \theta$, then

(c) $$\liminf_{\rho\to 0} \int_{|x\cdot\theta(\rho)|<\pi/3\rho} |x\cdot\theta(\rho)|^2\mu(dx) \ge \int |x\cdot\theta|^2\mu(dx) > 0$$

I claim this implies that for $\rho < \rho_0$

(d) $$\inf_{\theta\in S} \int_{|x\cdot\theta|<\pi/3\rho} |x\cdot\theta|^2\mu(dx) = C > 0$$

To get the last conclusion, observe that if it is false, then for $\rho = 1/n$ there is a θ_n so that

$$\int_{|x\cdot\theta_n|<n\pi/3} |x\cdot\theta_n|^2\mu(dx) \le 1/n$$

All the θ_n lie in S, a compact set, so if we pick a convergent subsequence we contradict (c). Combining (b) and (d) gives

$$\text{Re}\,(1-\varphi(\rho\theta)) \ge C\rho^2/4$$

Using the last result and (a) then changing to polar coordinates, we see that if δ is small (so $\text{Re}\,\phi(y) \ge 0$ on $(-\delta,\delta)^d$)

$$\begin{aligned}\int_{(-\delta,\delta)^d} \text{Re}\,\frac{1}{1-r\phi(y)}\,dy &\le \int_0^{\delta\sqrt{d}} d\rho\,\rho^{d-1}\int d\theta\,\frac{1}{\text{Re}\,(1-\phi(\rho\theta))}\\ &\le C'\int_0^1 d\rho\,\rho^{d-3} < \infty\end{aligned}$$

when $d > 2$, so the desired result follows from Theorem 5.4.11. □

Remark The analysis becomes much simpler when we consider random walks on $\mathbf{Z}^d$. The inversion formula given in Exercise 3.3.2 implies

$$P(S_n = 0) = (2\pi)^{-d}\int_{(-\pi,\pi)^d} \varphi^n(t)\,dt$$

Multiplying by r^n and summing gives

$$\sum_{n=0}^{\infty} r^n P(S_n = 0) = (2\pi)^{-d} \int_{(-\pi,\pi)^d} \frac{1}{1 - r\varphi(t)}\, dt$$

In the case of simple random walk in $d = 3$, $\phi(t) = \frac{1}{3}\sum_{j=1}^{3} \cos t_j$ is real.

$$\frac{1}{1 - r\phi(t)} \uparrow \frac{1}{1 - \phi(t)} \quad \text{when } \phi(t) > 0$$
$$0 \le \frac{1}{1 - r\phi(t)} \le 1 \quad \text{when } \phi(t) \le 0$$

So, using the monotone and bounded convergence theorems

$$\sum_{n=0}^{\infty} P(S_n = 0) = (2\pi)^{-3} \int_{(-\pi,\pi)^3} \left(1 - \frac{1}{3}\sum_{i=1}^{3} \cos x_i\right)^{-1} dx$$

This integral was first evaluated by Watson in 1939 in terms of elliptic integrals, which could be found in tables. Glasser and Zucker (1977) showed that it was

$$(\sqrt{6}/32\pi^3)\Gamma(1/24)\Gamma(5/24)\Gamma(7/24)\Gamma(11/24) = 1.51638606\ldots$$

so it follows from (5.3.1) that if $\beta_3 = P_0(T_0 < \infty)$, then

$$\sum_{n=0}^{\infty} P_0(S_n = 0) = \frac{1}{1 - \beta_3}$$

and hence $\beta_3 = 0.34053733\ldots$ For numerical results in $4 \le d \le 9$, see Kondo and Hara (1987).

Exercises

5.4.1 For a random walk on $\mathbf{R}$, there are only four possibilities, one of which has probability one.
(i) $S_n = 0$ for all n.
(ii) $S_n \to \infty$.
(iii) $S_n \to -\infty$.
(iv) $-\infty = \liminf S_n < \limsup S_n = \infty$.

5.4.2 **Ladder variables.** Let $\alpha(\omega) = \inf\{n : \omega_1 + \cdots + \omega_n > 0\}$, where $\inf \emptyset = \infty$, and set $\alpha(\Delta) = \infty$. Let $\alpha_0 = 0$ and let

$$\alpha_k(\omega) = \alpha_{k-1}(\omega) + \alpha(\theta^{\alpha_{k-1}}\omega)$$

for $k \ge 1$. At time α_k, the random walk is at a record high value.
(i) If $P(\alpha < \infty) < 1$, then $P(\sup S_n < \infty) = 1$.
(ii) If $P(\alpha < \infty) = 1$, then $P(\sup S_n = \infty) = 1$.

5.4.3 Suppose f is superharmonic on $\mathbf{R}^d$, i.e.,

$$f(x) \geq \frac{1}{|B(x,r)|} \int_{B(x,r)} f(y)\, dy$$

Let $\xi_1, \xi_2, \ldots$ be i.i.d. uniform on $B(0,1)$, and define S_n by $S_n = S_{n-1} + \xi_n$ for $n \geq 1$ and $S_0 = x$. (i) Show that $X_n = f(S_n)$ is a supermartingale. (ii) Use this to conclude that in $d \leq 2$, nonnegative superharmonic functions must be constant. The example $f(x) = |x|^{2-d}$ shows this is false in $d > 2$.

5.4.4 Suppose h is harmonic on $\mathbf{R}^d$, i.e.,

$$h(x) = \frac{1}{|B(x,r)|} \int_{B(x,r)} f(y)\, dy$$

Let $\xi_1, \xi_2, \ldots$ be i.i.d. uniform on $B(0,1)$, and define S_n by $S_n = S_{n-1} + \xi_n$ for $n \geq 1$ and $S_0 = x$. (i) Show that $X_n = f(S_n)$ is a martingale. (ii) Use this to conclude that in any dimension, bounded harmonic functions must be constant. The Hewitt-Savage 0-1 law will be useful here.

5.5 Stationary Measures

A measure μ is said to be a **stationary measure** if

$$\sum_x \mu(x)p(x,y) = \mu(y)$$

The last equation says $P_\mu(X_1 = y) = \mu(y)$. Using the Markov property and induction, it follows that $P_\mu(X_n = y) = \mu(y)$ for all $n \geq 1$. If μ is a probability measure, we call μ a **stationary distribution**, and it represents a possible equilibrium for the chain. That is, if X_0 has distribution μ, then so does X_n for all $n \geq 1$. If we stretch our imagination a little, we can also apply this interpretation when μ is an infinite measure. (When the total mass is finite, we can divide by $\mu(S)$ to get a stationary distribution.) Before getting into the theory, we consider some examples.

Example 5.5.1 (Random walk) $S = \mathbf{Z}^d$. $p(x,y) = f(y-x)$, where $f(z) \geq 0$ and $\sum f(z) = 1$. In this case, $\mu(x) \equiv 1$ is a stationary measure, since

$$\sum_x p(x,y) = \sum_x f(y-x) = 1$$

A transition probability that has $\sum_x p(x,y) = 1$ is called **doubly stochastic**. This is obviously a necessary and sufficient condition for $\mu(x) \equiv 1$ to be a stationary measure.

Example 5.5.2 (Asymmetric simple random walk) $S = \mathbf{Z}$.

$$p(x,x+1) = p \qquad p(x,x-1) = q = 1-p$$

By the last example, $\mu(x) \equiv 1$ is a stationary measure. When $p \neq q$, $\mu(x) = (p/q)^x$ is a second one. To check this, we observe that

$$\begin{aligned}\sum_x \mu(x)p(x,y) &= \mu(y+1)p(y+1,y) + \mu(y-1)p(y-1,y) \\ &= (p/q)^{y+1}q + (p/q)^{y-1}p = (p/q)^y[p+q] = (p/q)^y\end{aligned}$$

To simplify the computations in the next two examples, we introdcue a concept that is stronger than being a stationary measure. μ satisfies the **detailed balance condition** if

$$\mu(x)p(x,y) = \mu(y)p(y,x) \tag{5.5.1}$$

Summing over x gives

$$\sum_x \mu(x)p(x,y) = \mu(y)$$

(5.5.1) asserts that the amount of mass that moves from x to y in one jump is exactly the same as the amount that moves from y to x. A measure μ that satisfies (5.5.1) is said to be a **reversible measure**. Theorem 5.5.5 will explain the term reversible.

Example 5.5.3 (The Ehrenfest chain) $S = \{0, 1, \dots, r\}$

$$p(k,k+1) = (r-k)/r \qquad p(k,k-1) = k/r$$

In this case, $\mu(x) = 2^{-r}\binom{r}{x}$ is a stationary distribution. One can check this without pencil and paper by observing that μ corresponds to flipping r coins to determine which urn each ball is to be placed in, and the transitions of the chain correspond to picking a coin at random and turning it over. Alternatively, you can check that

$$\begin{aligned}\mu(k+1)p(k+1,k) &= 2^{-r}\frac{r!}{(k+1)!\,(r-k-1)!}\cdot\frac{k+1}{r} = \frac{(r-1)!}{k!\,(r-k-1)!}\\ &= 2^{-r}\frac{r!}{k!\,(r-k)!}\cdot\frac{r-k}{r} = \mu(k-1)p(k-1,k) = \mu(k).\end{aligned}$$

Example 5.5.4 (Birth and death chains) $S = \{0, 1, 2, \dots\}$

$$p(x,x+1) = p_x \quad p(x,x) = r_x \quad p(x,x-1) = q_x$$

with $q_0 = 0$ and $p(i,j) = 0$ otherwise. In this case, there is the measure

$$\mu(x) = \prod_{k=1}^{x} \frac{p_{k-1}}{q_k}$$

which satisfies detailed balance:

$$\mu(x)p(x,x+1) = p_x \prod_{k=1}^{x} \frac{p_{k-1}}{q_k} = \mu(x+1)p(x+1,x)$$

The next result explains the name "reversible."

Theorem 5.5.5 *Let μ be a stationary measure and suppose X_0 has "distribution" μ. Then $Y_m = X_{n-m}$, $0 \le m \le n$ is a Markov chain with initial measure μ and transition probability*

$$q(x,y) = \mu(y)p(y,x)/\mu(x)$$

q is called the **dual transition probability**. *If μ is a reversible measure, then $q = p$.*

Proof

$$\begin{aligned}P(Y_{m+1} = y|Y_m = x) &= P(X_{n-m-1} = y|X_{n-m} = x) \\ &= \frac{P(X_{n-m-1} = y, X_{n-m} = x)}{P(X_{n-m} = x)} \\ &= \frac{P(X_{n-m} = x|X_{n-m-1} = y)P(X_{n-m-1} = y)}{P(X_{n-m} = x)} \\ &= \frac{\mu(y)p(y,x)}{\mu(x)}\end{aligned}$$

which proves the result. □

While many chains have reversible measure, most do not. If there is a pair of states with $p(x,y) > 0$ and $p(y,x) = 0$, then it is impossible to have $\mu(p(x,y) = \pi(y)p(y,x)$. The $M/G/1$ queue also has this problem. The next result gives a necessary and sufficient condition for a chain to have a reversible measure.

Theorem 5.5.6 (Kolmogorov's cycle condition) *Suppose p is irreducible. A necessary and sufficient condition for the existence of a reversible measure is that (i) $p(x,y) > 0$ implies $p(y,x) > 0$, and (ii) for any loop $x_0, x_1, \ldots, x_n = x_0$ with $\prod_{1 \le i \le n} p(x_i, x_{i-1}) > 0$,*

$$\prod_{i=1}^{n} \frac{p(x_{i-1}, x_i)}{p(x_i, x_{i-1})} = 1$$

Proof To prove the necessity of this condition, we note that irreducibility implies that any stationary measure has $\mu(x) > 0$ for all x, so (5.5.1) implies (i) holds. To check (ii), note that (5.5.1) implies that for the sequences considered here

$$\prod_{i=1}^{n} \frac{p(x_{i-1}, x_i)}{p(x_i, x_{i-1})} = \prod_{i=1}^{n} \frac{\mu(x_i)}{\mu(x_{i-1})} = 1$$

To prove sufficiency, fix $a \in S$, set $\mu(a) = 1$, and if $x_0 = a, x_1, \ldots, x_n = x$ is a sequence with $\prod_{1 \le i \le n} p(x_i, x_{i-1}) > 0$ (irreducibility implies such a sequence will exist), we let

$$\mu(x) = \prod_{i=1}^{n} \frac{p(x_{i-1}, x_i)}{p(x_i, x_{i-1})}$$

The cycle condition guarantees that the last definition is independent of the path. To check (5.5.1) now, observe that if $p(y,x) > 0$, then adding $x_{n+1} = y$ to the end of a path to x we have

$$\mu(x)\frac{p(x,y)}{p(y,x)} = \mu(y)$$ □

Only special chains have reversible measures, but as the next result shows, many Markov chains have stationary measures.

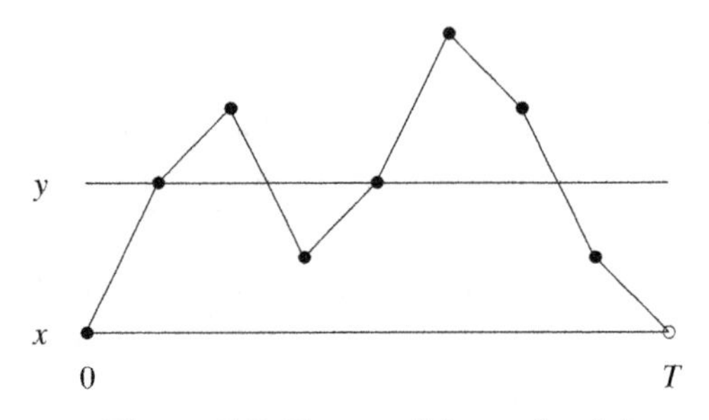

Figure 5.5 Picture of the cycle trick.

Theorem 5.5.7 *Let x be a recurrent state, and let $T = \inf\{n \geq 1 : X_n = x\}$. Then*

$$\mu_x(y) = E_x\left(\sum_{n=0}^{T-1} 1_{\{X_n=y\}}\right) = \sum_{n=0}^{\infty} P_x(X_n = y, T > n)$$

defines a stationary measure.

Proof This is called the "cycle trick." The proof in words is simple. $\mu_x(y)$ is the expected number of visits to y in $\{0, \ldots, T-1\}$. $\mu_x p(y) \equiv \sum \mu_x(z)p(z,y)$ is the expected number of visits to y in $\{1, \ldots, T\}$, which is $= \mu_x(y)$, since $X_T = X_0 = x$.

To translate this intuition into a proof, let $\bar{p}_n(x,y) = P_x(X_n = y, T > n)$ and use Fubini's theorem to get

$$\sum_y \mu_x(y)p(y,z) = \sum_{n=0}^{\infty}\sum_y \bar{p}_n(x,y)p(y,z)$$

Case 1. $z \neq x$.

$$\begin{aligned}\sum_y \bar{p}_n(x,y)p(y,z) &= \sum_y P_x(X_n = y, T > n, X_{n+1} = z) \\ &= P_x(T > n+1, X_{n+1} = z) = \bar{p}_{n+1}(x,z)\end{aligned}$$

so $\sum_{n=0}^{\infty}\sum_y \bar{p}_n(x,y)p(y,z) = \sum_{n=0}^{\infty} \bar{p}_{n+1}(x,z) = \mu_x(z)$ since $\bar{p}_0(x,z) = 0$.

Case 2. $z = x$.

$$\sum_y \bar{p}_n(x,y)p(y,x) = \sum_y P_x(X_n = y, T > n, X_{n+1} = x) = P_x(T = n+1)$$

so $\sum_{n=0}^{\infty}\sum_y \bar{p}_n(x,y)p(y,x) = \sum_{n=0}^{\infty} P_x(T = n+1) = 1 = \mu_x(x)$, since by definition $P_x(T = 0) = 0$. □

Remark If x is transient, then we have $\mu_x p(z) \leq \mu_x(z)$ with equality for all $z \neq x$.

Technical Note To show that we are not cheating, we should prove that $\mu_x(y) < \infty$ for all y. First, observe that $\mu_x p = \mu_x$ implies $\mu_x p^n = \mu_x$ for all $n \geq 1$, and $\mu_x(x) = 1$, so if $p^n(y,x) > 0$, then $\mu_x(y) < \infty$. Since the last result is true for all n, we see that $\mu_x(y) < \infty$ whenever $\rho_{yx} > 0$, but this is good enough. By Theorem 5.3.2, when x is recurrent, $\rho_{xy} > 0$ implies $\rho_{yx} > 0$, and it follows from the argument above that $\mu_x(y) < \infty$. If $\rho_{xy} = 0$, then $\mu_x(y) = 0$.

Example 5.5.8 (Renewal chain) $S = \{0, 1, 2, \dots\}$, $f_k \ge 0$, and $\sum_{k=1}^{\infty} f_k = 1$.

$$\begin{aligned} p(0, j) &= f_{j+1} && \text{for } j \ge 0 \\ p(i, i-1) &= 1 && \text{for } i \ge 1 \\ p(i, j) &= 0 && \text{otherwise} \end{aligned}$$

To explain the definition, let $\xi_1, \xi_2, \dots$ be i.i.d. with $P(\xi_m = j) = f_j$, let $T_0 = i_0$, and for $k \ge 1$ let $T_k = T_{k-1} + \xi_k$. T_k is the time of the kth arrival in a renewal process that has its first arrival at time i_0. Let

$$Y_m = \begin{cases} 1 & \text{if } m \in \{T_0, T_1, T_2, \dots\} \\ 0 & \text{otherwise} \end{cases}$$

and let $X_n = \inf\{m - n : m \ge n, Y_m = 1\}$. $Y_m = 1$ if a renewal occurs at time m, and X_n is the amount of time until the first renewal $\ge n$.

An example should help clarify the definition:

Y_n	0	0	0	1	0	0	1	1	0	0	0	0	1
X_n	3	2	1	0	2	1	0	0	4	3	2	1	0

It is clear that if $X_n = i > 0$, then $X_{n+1} = i - 1$. When $X_n = 0$, the next ξ will be equal to k with probability f_k. In this case, the next positive value of Y is at time $n + k$, so $X_{n+1} = k - 1$.

$P_0(T_0 < \infty) = 1$, so 0 is always recurrent. If $f_K > 0$ but $f_k = 0$ for $k > K$, the chain is not irreducible on $\{0, 1, 2, \dots\}$ but it is on $\{0, 1, \dots K - 1\}$. To compute the stationary distribution, we use the cycle trick starting at 0. The chain will hit 0 if and only if the first jump is to a point $k \ge j$, so $\mu_0(j) = P(\xi > j)$. $E_0 T_0 = \sum_{j=0}^{\infty} P(\xi > j) = E\xi$. If $E\xi < \infty$, then $\pi(j) = P(\xi > j)/E\xi$ is a stationary distribution.

Theorem 5.5.7 allows us to construct a stationary measure for each closed set of recurrent states. Conversely, we have:

Theorem 5.5.9 *If p is irreducible and recurrent (i.e., all states are) then the stationary measure is unique up to constant multiples.*

Proof Let ν be a stationary measure and let $a \in S$.

$$\nu(z) = \sum_y \nu(y) p(y, z) = \nu(a) p(a, z) + \sum_{y \ne a} \nu(y) p(y, z)$$

Using the last identity to replace $\nu(y)$ on the right-hand side,

$$\begin{aligned} \nu(z) &= \nu(a) p(a, z) + \sum_{y \ne a} \nu(a) p(a, y) p(y, z) \\ &\quad + \sum_{x \ne a} \sum_{y \ne a} \nu(x) p(x, y) p(y, z) \\ &= \nu(a) P_a(X_1 = z) + \nu(a) P_a(X_1 \ne a, X_2 = z) \\ &\quad + P_\nu(X_0 \ne a, X_1 \ne a, X_2 = z) \end{aligned}$$

Continuing in the obvious way, we get

$$\nu(z) = \nu(a) \sum_{m=1}^{n} P_a(X_k \neq a, 1 \leq k < m, X_m = z) \\ + P_\nu(X_j \neq a, 0 \leq j < n, X_n = z)$$

The last term is ≥ 0. Letting $n \to \infty$ gives $\nu(z) \geq \nu(a)\mu_a(z)$, where μ_a is the measure defined in Theorem 5.5.7 for $x = a$.

It is tempting to claim that recurrence implies

$$P_\nu(X_j \neq a, 0 \leq j < n) \to 0$$

but ν may be an infinite measure, so we need another approach. It follows from Theorem 5.5.7 that μ_a is a stationary measure with $\mu_a(a) = 1$. (Here we are summing from 1 to T rather than from 0 to $T - 1$.) To turn the $\geq$ in the last equation into $=$, we observe

$$\nu(a) = \sum_x \nu(x)p^n(x,a) \geq \nu(a) \sum_x \mu_a(x)p^n(x,a) = \nu(a)\mu_a(a) = \nu(a)$$

Since $\nu(x) \geq \nu(a)\mu_a(x)$ and the left- and right-hand sides are equal, we must have $\nu(x) = \nu(a)\mu_a(x)$ whenever $p^n(x,a) > 0$. Since p is irreducible, it follows that $\nu(x) = \nu(a)\mu_a(x)$ for all $x \in S$, and the proof is complete. □

Theorems 5.5.7 and 5.5.9 make a good team. The first result gives us a formula for a stationary distribution we call μ_x, and the second shows it is unique up to constant multiples. Together they allow us to derive a lot of formulas.

Having examined the existence and uniqueness of stationary measures, we turn our attention now to **stationary distributions**, i.e., probability measures π with $\pi p = \pi$. Stationary measures may exist for transient chains, e.g., random walks in $d \geq 3$, but:

Theorem 5.5.10 *If there is a stationary distribution, then all states y that have $\pi(y) > 0$ are recurrent.*

Proof Since $\pi p^n = \pi$, Fubini's theorem implies

$$\sum_x \pi(x) \sum_{n=1}^{\infty} p^n(x,y) = \sum_{n=1}^{\infty} \pi(y) = \infty$$

when $\pi(y) > 0$. Using Theorem 5.3.1 now gives

$$\infty = \sum_x \pi(x) \frac{\rho_{xy}}{1 - \rho_{yy}} \leq \frac{1}{1 - \rho_{yy}}$$

since $\rho_{xy} \leq 1$ and π is a probability measure. So $\rho_{yy} = 1$. □

Theorem 5.5.11 *If p is irreducible and has stationary distribution π, then*

$$\pi(x) = 1/E_x T_x$$

Remark Recycling Chung's quote regarding Theorem 4.6.9, we note that the proof will make $\pi(x) = 1/E_x T_x$ obvious, but it seems incredible that

$$\sum_x \frac{1}{E_x T_x} p(x,y) = \frac{1}{E_y T_y}$$

Proof Irreducibility implies $\pi(x) > 0$, so all states are recurrent by Theorem 5.5.10. From Theorem 5.5.7,

$$\mu_x(y) = \sum_{n=0}^{\infty} P_x(X_n = y, T_x > n)$$

defines a stationary measure with $\mu_x(x) = 1$, and Fubini's theorem implies

$$\sum_y \mu_x(y) = \sum_{n=0}^{\infty} P_x(T_x > n) = E_x T_x$$

By Theorem 5.5.9, the stationary measure is unique up to constant multiples, so $\pi(x) = \mu_x(x)/E_x T_x$. Since $\mu_x(x) = 1$ by definition, the desired result follows. □

If a state x has $E_x T_x < \infty$, it is said to be **positive recurrent**. A recurrent state with $E_x T_x = \infty$ is said to be **null recurrent**. Theorem 5.6.1 will explain these names. The next result helps us identify positive recurrent states.

Theorem 5.5.12 *If p is irreducible, then the following are equivalent:*
(i) Some x is positive recurrent.
(ii) There is a stationary distribution.
(iii) All states are positive recurrent.

This result shows that being positive recurrent is a **class property**. If it holds for one state in an irreducible set, then it is true for all.

Proof *(i) implies (ii).* If x is positive recurrent, then

$$\pi(y) = \sum_{n=0}^{\infty} P_x(X_n = y, T_x > n)/E_x T_x$$

defines a stationary distribution.

(ii) implies (iii). Theorem 5.5.11 implies $\pi(y) = 1/E_y T_y$, and irreducibility tells us $\pi(y) > 0$ for all y, so $E_y T_y < \infty$.

(iii) implies (i). Trivial. □

Example 5.5.13 **Birth and death chains** have a stationary distribution if and only if

$$\sum_x \prod_{k=1}^{x} \frac{p_{k-1}}{q_k} < \infty$$

By Theorem 5.3.11, the chain is recurrent if and only if

$$\sum_{m=0}^{\infty} \prod_{j=1}^{m} \frac{q_j}{p_j} = \infty$$

Example 5.5.14 ($M/G/1$ queue) Let $\mu = \sum k a_k$ be the mean number of customers that arrive during one service time. In Example 5.3.7, we showed that the chain is recurrent if and only if $\mu \le 1$. We will now show that the chain is positive recurrent if and only if $\mu < 1$. First, suppose that $\mu < 1$. When $X_n > 0$, the chain behaves like a random walk that has jumps with mean $\mu - 1$, so if $N = \inf\{n \ge 0 : X_n = 0\}$, then $X_{N \wedge n} - (\mu - 1)(N \wedge n)$ is a martingale. If $X_0 = x > 0$, then the martingale property implies

$$x = E_x X_{N \wedge n} + (1 - \mu) E_x(N \wedge n) \ge (1 - \mu) E_x(N \wedge n)$$

since $X_{N \wedge n} \ge 0$, and it follows that $E_x N \le x/(1 - \mu)$.

To prove that there is equality, observe that X_n decreases by at most one each time and for $x \ge 1$, $E_x T_{x-1} = E_1 T_0$, so $E_x N = cx$. To identify the constant, observe that

$$E_1 N = 1 + \sum_{k=0}^{\infty} a_k E_k N$$

so $c = 1 + \mu c$ and $c = 1/(1 - \mu)$. If $X_0 = 0$ then $p(0,0) = a_0 + a_1$ and $p(0, k - 1) = a_k$ for $k \ge 2$. By considering what happens on the first jump, we see that (the first term may look wrong, but recall $k - 1 = 0$ when $k = 1$)

$$E_0 T_0 = 1 + \sum_{k=1}^{\infty} a_k \frac{k-1}{1-\mu} = 1 + \frac{\mu - (1 - a_0)}{1 - \mu} = \frac{a_0}{1 - \mu} < \infty$$

This shows that the chain is positive recurrent if $\mu < 1$. To prove the converse, observe that the arguments above show that if $E_0 T_0 < \infty$, then $E_k N < \infty$ for all k, $E_k N = ck$, and $c = 1/(1 - \mu)$, which is impossible if $\mu \ge 1$.

The last result when combined with Theorem 5.5.7 and 5.5.11 allows us to conclude that the stationary distribution has $\pi(0) = (1 - \mu)/a_0$. This may not seem like much, but the equations in $\pi p = \pi$ are:

$$\begin{aligned}
\pi(0) &= \pi(0)(a_0 + a_1) + \pi(1) a_0 \\
\pi(1) &= \pi(0) a_2 + \pi(1) a_1 + \pi(2) a_0 \\
\pi(2) &= \pi(0) a_3 + \pi(1) a_2 + \pi(2) a_1 + \pi(3) a_0
\end{aligned}$$

or, in general, for $j \ge 1$

$$\pi(j) = \sum_{i=0}^{j+1} \pi(i) a_{j+1-i}$$

The equations have a "triangular" form, so knowing $\pi(0)$, we can solve for $\pi(1), \pi(2), \ldots$ The first expression,

$$\pi(1) = \pi(0)(1 - (a_0 + a_1))/a_0$$

is simple, but the formulas get progressively messier, and there is no nice closed form solution.

To close the section, we will give a self-contained proof of:

Theorem 5.5.15 *If p is irreducible and has a stationary distribution π, then any other stationary measure is a multiple of π.*

Remark This result is a consequence of Theorems 5.5.10 and 5.5.9, but we find the method of proof amusing.

Proof Since p is irreducible, $\pi(x) > 0$ for all x. Let φ be a concave function that is bounded on $(0,\infty)$, e.g., $\varphi(x) = x/(x+1)$. Define the **entropy** of μ by

$$\mathcal{E}(\mu) = \sum_y \varphi\left(\frac{\mu(y)}{\pi(y)}\right)\pi(y)$$

The reason for the name will become clear during the proof.

$$\mathcal{E}(\mu p) = \sum_y \varphi\left(\sum_x \frac{\mu(x)p(x,y)}{\pi(y)}\right)\pi(y) = \sum_y \varphi\left(\sum_x \frac{\mu(x)}{\pi(x)} \cdot \frac{\pi(x)p(x,y)}{\pi(y)}\right)\pi(y)$$
$$\geq \sum_y \sum_x \varphi\left(\frac{\mu(x)}{\pi(x)}\right)\frac{\pi(x)p(x,y)}{\pi(y)}\pi(y)$$

since φ is concave, and $\nu(x) = \pi(x)p(x,y)/\pi(y)$ is a probability distribution. Since the $\pi(y)$'s cancel and $\sum_y p(x,y) = 1$, the last expression $= \mathcal{E}(\mu)$, and we have shown $\mathcal{E}(\mu p) \geq \mathcal{E}(\mu)$, i.e., the entropy of an arbitrary initial measure μ is increased by an application of p.

If $p(x,y) > 0$ for all x and y, and $\mu p = \mu$, it follows that $\mu(x)/\pi(x)$ must be constant, for otherwise there would be strict inequality in the application of Jensen's inequality. To get from the last special case to the general result, observe that if p is irreducible

$$\bar{p}(x,y) = \sum_{n=1}^{\infty} 2^{-n} p^n(x,y) > 0 \quad \text{for all } x, y$$

and $\mu p = \mu$ implies $\mu \bar{p} = \mu$. □

Exercises

5.5.1 Find the stationary distribution for the Bernoulli-Laplace model of diffusion from Exercise 5.1.5.

5.5.2 Let $w_{xy} = P_x(T_y < T_x)$. Show that $\mu_x(y) = w_{xy}/w_{yx}$.

5.5.3 Show that if p is irreducible and recurrent, then

$$\mu_x(y)\mu_y(z) = \mu_x(z)$$

5.5.4 Suppose p is irreducible and positive recurrent. Then $E_x T_y < \infty$ for all x, y.

5.5.5 Suppose p is irreducible and has a stationary measure μ with $\sum_x \mu(x) = \infty$. Then p is not positive recurrent.

5.5.6 Use Theorems 5.5.7 and 5.5.9 to show that for simple random walk, (i) the expected number of visits to k between successive visits to 0 is 1 for all k, and (ii) if we start from k, the expected number of visits to k before hitting 0 is $2k$.

5.5.7 Let $\xi_1, \xi_2, \ldots$ be i.i.d. with $P(\xi_m = k) = a_{k+1}$ for $k \geq -1$, let $S_n = x + \xi_1 + \cdots + \xi_n$, where $x \geq 0$, and let

$$X_n = S_n + \left(\min_{m \leq n} S_m \right)^-$$

(5.1.1) shows that X_n has the same distribution as the $M/G/1$ queue starting from $X_0 = x$. Use this representation to conclude that if $\mu = \sum k a_k < 1$, then as $n \to \infty$

$$\frac{1}{n} |\{m \leq n : X_{m-1} = 0, \xi_m = -1\}| \to (1 - \mu) \quad \text{a.s.}$$

and hence $\pi(0) = (1 - \mu)/a_0$ as proved earlier.

5.5.8 $M/M/\infty$ **queue**. In this chain, introduced in Exercise 5.3.8,

$$X_{n+1} = \sum_{m=1}^{X_n} \xi_{n,m} + Y_{n+1}$$

where $\xi_{n,m}$ are i.i.d. Bernoulli with mean p and Y_{n+1} is an independent Poisson mean λ. Compute the distribution at time 1 when $X_0 =$ Poissson(μ) and use the result to find the stationary distribution.

5.5.9 Let $X_n \geq 0$ be a Markov chain and suppose $E_x X_1 \leq x - \epsilon$ for $x > K$, where $\epsilon > 0$. Let $Y_n = X_n + n\epsilon$ and $\tau = \inf\{n : X_n \leq K\}$. $Y_{n \wedge \tau}$ is a positive supermartingale and the optional stopping theorem implies $E_x \tau \leq x/\epsilon$. With assumptions about the behavior of the chain starting from points $x \leq K$, this leads to conditions for positive recurrence

5.6 Asymptotic Behavior

The first topic in this section is to investigate the asymptotic behavior of $p^n(x, y)$. If y is transient, $\sum_n p^n(x, y) < \infty$, so $p^n(x, y) \to 0$ as $n \to \infty$. To deal with the recurrent states, we let

$$N_n(y) = \sum_{m=1}^{n} 1_{\{X_m = y\}}$$

be the number of visits to y by time n.

Theorem 5.6.1 *Suppose y is recurrent. For any $x \in S$, as $n \to \infty$*

$$\frac{N_n(y)}{n} \to \frac{1}{E_y T_y} 1_{\{T_y < \infty\}} \quad P_x\text{-a.s.}$$

Here $1/\infty = 0$.

Proof Suppose first that we start at y. Let $R(k) = \min\{n \geq 1 : N_n(y) = k\}$ = the time of the kth return to y. Let $t_k = R(k) - R(k-1)$, where $R(0) = 0$. Since we have assumed $X_0 = y$, $t_1, t_2, \ldots$ are i.i.d. and the strong law of large numbers implies

$$R(k)/k \to E_y T_y \quad P_y\text{-a.s.}$$

Since $R(N_n(y)) \le n < R(N_n(y)+1)$,

$$\frac{R(N_n(y))}{N_n(y)} \le \frac{n}{N_n(y)} < \frac{R(N_n(y)+1)}{N_n(y)+1} \cdot \frac{N_n(y)+1}{N_n(y)}$$

Letting $n \to \infty$, and recalling $N_n(y) \to \infty$ a.s. since y is recurrent, we have

$$\frac{n}{N_n(y)} \to E_y T_y \quad P_y\text{-a.s.}$$

To generalize now to $x \neq y$, observe that if $T_y = \infty$, then $N_n(y) = 0$ for all n and hence

$$N_n(y)/n \to 0 \quad \text{on } \{T_y = \infty\}$$

The strong Markov property implies that conditional on $\{T_y < \infty\}$, $t_2, t_3, \ldots$ are i.i.d. and have $P_x(t_k = n) = P_y(T_y = n)$, so

$$R(k)/k = t_1/k + (t_2 + \cdots + t_k)/k \to 0 + E_y T_y \quad P_x\text{-a.s.}$$

Repeating the proof for the case $x = y$ shows

$$N_n(y)/n \to 1/E_y T_y \quad P_x\text{-a.s. on } \{T_y < \infty\}$$

and combining this with the result for $\{T_y = \infty\}$ completes the proof. □

Remark Theorem 5.6.1 should help explain the terms positive and null recurrent. If we start from x, then in the first case the asymptotic fraction of time spent at x is positive and in the second case it is 0.

Since $0 \le N_n(y)/n \le 1$, it follows from the bounded convergence theorem that $E_x N_n(y)/n \to E_x(1_{\{T_y<\infty\}}/E_y T_y)$, so

$$\frac{1}{n}\sum_{m=1}^{n} p^m(x,y) \to \rho_{xy}/E_y T_y \tag{5.6.1}$$

The last result was proved for recurrent y but also holds for transient y, since in that case, $E_y T_y = \infty$, and the limit is 0, since $\sum_m p^m(x,y) < \infty$.

(5.6.1) shows that the sequence $p^n(x,y)$ always converges in the Cesaro sense. The next example shows that $p^n(x,y)$ need not converge.

Example 5.6.2

$$p = \begin{pmatrix} 0 & 1 \\ 1 & 0 \end{pmatrix} \quad p^2 = \begin{pmatrix} 1 & 0 \\ 0 & 1 \end{pmatrix} \quad p^3 = p, \quad p^4 = p^2, \ldots$$

A similar problem also occurs in the Ehrenfest chain. In that case, if X_0 is even, then X_1 is odd, X_2 is even, ... so $p^n(x,x) = 0$ unless n is even. It is easy to construct examples with $p^n(x,x) = 0$ unless n is a multiple of 3 or 17 or ...

Theorem 5.6.6 will show that this "periodicity" is the only thing that can prevent the convergence of the $p^n(x,y)$. First, we need a definition and two preliminary results. Let x be a recurrent state, let $I_x = \{n \ge 1 : p^n(x,x) > 0\}$, and let d_x be the greatest common divisor of I_x. d_x is called the **period** of x.

Example 5.6.3 (Triangle and square) Consider the transition matrix:

	−2	−1	0	1	2	3
−2	0	0	1	0	0	0
−1	1	0	0	0	0	0
0	0	0.5	0	0.5	0	0
1	0	0	0	0	1	0
2	0	0	0	0	0	1
3	0	0	1	0	0	0

In words, from 0 we are equally likely to go to 1 or -1. From -1 we go with probability one to -2 and then back to 0, from 1 we go to 2 then to 3 and back to 0. The name refers to the fact that $0 \to -1 \to -2 \to 0$ is a triangle and $0 \to 1 \to 2 \to 3 \to 0$ is a square.

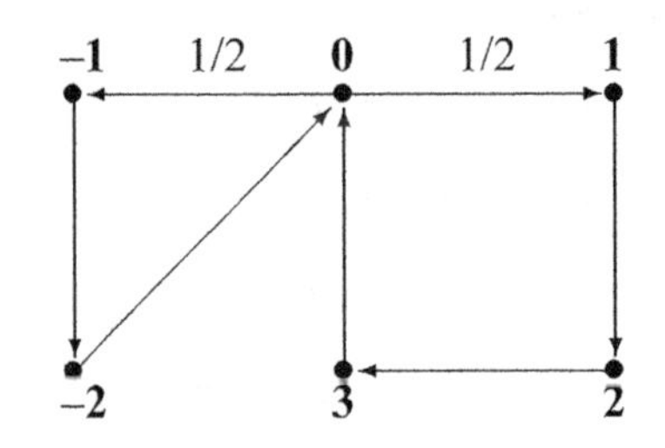

Clearly, $p^3(0,0) > 0$ and $p^4(0,0) > 0$, so $3, 4 \in I_0$ and hence $d_0 = 1$.

The next result says that the period is a class property. In particular, we can use it to conclude that in the last example all states have period 1.

Lemma 5.6.4 *If $\rho_{xy} > 0$, then $d_y = d_x$.*

Proof Let K and L be such that $p^K(x,y) > 0$ and $p^L(y,x) > 0$. (x is recurrent, so $\rho_{yx} > 0$.)

$$p^{K+L}(y,y) \geq p^L(y,x)p^K(x,y) > 0$$

so d_y divides $K + L$, abbreviated $d_y|(K + L)$. Let n be such that $p^n(x,x) > 0$.

$$p^{K+n+L}(y,y) \geq p^L(y,x)p^n(x,x)p^K(x,y) > 0$$

so $d_y|(K + n + L)$, and hence $d_y|n$. Since $n \in I_x$ is arbitrary, $d_y|d_x$. Interchanging the roles of y and x gives $d_x|d_y$, and hence $d_x = d_y$. □

If a chain is irreducible and $d_x = 1$, it is said to be aperiodic. The easiest way to check this is to find a state with $p(x,x) > 0$. The $M/G/1$ queue has $a_k > 0$ for all $k \geq 0$, so it has this property. The renewal chain is aperiodic if $g.c.d.\{k : f_k > 0\} = 1$.

Lemma 5.6.5 *If $d_x = 1$, then $p^m(x,x) > 0$ for $m \geq m_0$.*

Proof by example Consider the renewal chain (Example 5.5.8) with $f_5 = f_{12} = 1/2$. $5, 12 \in I_0$. Since $m, n \in I_0$ implies $m + n \in I_0$

$$I_0 = \{5, 10, 12, 15, 17, 20, 22, 24, 25, 27, 29, 30, 32,$$
$$34, 35, 36, 37, 39, 40, 41, 42, 43, \ldots\}$$

To check this, note that 5 gives rise to 10=5+5 and 17=5+12, 10 to 15 and 22, 12 to 17 and 24, etc. Once we have five consecutive numbers in I_0, here 39–43, we have all the rest.

Proof We begin by observing that it enough to show that I_x will contain two consecutive integers: k and $k+1$. For then it will contain $2k, 2k+1, 2k+2, 3k, 3k+1, 3k+2, 3k+3$, and so on until we have k consecutive integers and then we $(k-1)k, (k-1)k+1, \dots (k-1)k+k-1$ and we have the rest.

To show that there are two consecutive integers, we cheat and use a fact from number theory: if the greatest common divisor of a set I_x is 1, then there are integers $i_1, \dots i_m \in I_x$ and (positive or negative) integer coefficients c_i so that $c_1 i_1 + \cdots + c_m i_m = 1$. Let $a_i = c_i^+$ and $b_i = (c_i)^-$. In words, the a_i are the positive coefficients and the b_i are -1 times the negative coefficients. Rearranging the last equation gives

$$a_1 i_1 + \cdots + a_m i_m = (b_1 i_1 + \cdots + b_m i_m) + 1$$

and we have found our two consecutive integers in I_x. In the numerical example $5 \cdot 5 - 2 \cdot 12 = 1$, which gives us consecutive integers 24 and 25. □

Theorem 5.6.6 (Convergence theorem) *Suppose p is irreducible, aperiodic (i.e., all states have $d_x = 1$), and has stationary distribution π. Then, as $n \to \infty$, $p^n(x,y) \to \pi(y)$.*

Proof Let $S^2 = S \times S$. Define a transition probability $\bar{p}$ on $S \times S$ by

$$\bar{p}((x_1, y_1), (x_2, y_2)) = p(x_1, x_2) p(y_1, y_2)$$

i.e., each coordinate moves independently. Our first step is to check that $\bar{p}$ is irreducible. This may seem like a silly thing to do first, but this is the only step that requires aperiodicity. Since p is irreducible, there are K, L, so that $p^K(x_1, x_2) > 0$ and $p^L(y_1, y_2) > 0$. From Lemma 5.6.5 it follows that if M is large, $p^{L+M}(x_2, x_2) > 0$ and $p^{K+M}(y_2, y_2) > 0$, so

$$\bar{p}^{K+L+M}((x_1, y_1), (x_2, y_2)) > 0$$

Our second step is to observe that since the two coordinates are independent, $\bar{\pi}(a,b) = \pi(a)\pi(b)$ defines a stationary distribution for $\bar{p}$, and Theorem 5.5.10 implies that for $\bar{p}$ all states are recurrent. Let (X_n, Y_n) denote the chain on $S \times S$, and let T be the first time that this chain hits the diagonal $\{(y,y) : y \in S\}$. Let $T_{(x,x)}$ be the hitting time of (x,x). Since $\bar{p}$ is irreducible and recurrent, $T_{(x,x)} < \infty$ a.s. and hence $T < \infty$ a.s. The final step is to observe that on $\{T \le n\}$, the two coordinates X_n and Y_n have the same distribution. By considering the time and place of the first intersection and then using the Markov property,

$$\begin{aligned}
P(X_n = y, T \le n) &= \sum_{m=1}^{n} \sum_x P(T = m, X_m = x, X_n = y) \\
&= \sum_{m=1}^{n} \sum_x P(T = m, X_m = x) P(X_n = y | X_m = x) \\
&= \sum_{m=1}^{n} \sum_x P(T = m, Y_m = x) P(Y_n = y | Y_m = x) \\
&= P(Y_n = y, T \le n)
\end{aligned}$$

To finish up, we observe that

$$
\begin{aligned}
P(X_n = y) &= P(Y_n = y, T \le n) + P(X_n = y, T > n) \\
&\le P(Y_n = y) + P(X_n = y, T > n)
\end{aligned}
$$

and similarly, $P(Y_n = y) \le P(X_n = y) + P(Y_n = y, T > n)$. So

$$|P(X_n = y) - P(Y_n = y)| \le P(X_n = y, T > n) + P(Y_n = y, T > n)$$

and summing over y gives

$$\sum_y |P(X_n = y) - P(Y_n = y)| \le 2P(T > n)$$

If we let $X_0 = x$ and let Y_0 have the stationary distribution π, then Y_n has distribution π, and it follows that

$$\sum_y |p^n(x, y) - \pi(y)| \le 2P(T > n) \to 0$$

proving the desired result. If we recall the definition of the total variation distance given in Section 3.6, the last conclusion can be written as

$$\|p^n(x, \cdot) - \pi(\cdot)\| \le P(T > n) \to 0$$

□

At first glance, it may seem strange to prove the convergence theorem by running independent copies of the chain. An approach that is slightly more complicated but explains better what is happening is to define

$$
q((x_1, y_1), (x_2, y_2)) = \begin{cases} p(x_1, x_2) p(y_1, y_2) & \text{if } x_1 \ne y_1 \\ p(x_1, x_2) & \text{if } x_1 = y_1, x_2 = y_2 \\ 0 & \text{otherwise} \end{cases}
$$

In words, the two coordinates move independently until they hit and then move together. It is easy to see from the definition that each coordinate is a copy of the original process. If T' is the hitting time of the diagonal for the new chain (X'_n, Y'_n), then $X'_n = Y'_n$ on $T' \le n$, so it is clear that

$$\sum_y |P(X'_n = y) - P(Y'_n = y)| \le 2\, P(X'_n \ne Y'_n) = 2P(T' > n)$$

On the other hand, T and T' have the same distribution, so $P(T' > n) \to 0$, and the conclusion follows as before. The technique used in the last proof is called **coupling**. Generally, this term refers to building two sequences X_n and Y_n on the same space to conclude that X_n converges in distribution by showing $P(X_n \ne Y_n) \to 0$, or more generally, that for some metric ρ, $\rho(X_n, Y_n) \to 0$ in probability.

Exercises

5.6.1 Suppose $S = \{0, 1\}$ and

$$p = \begin{pmatrix} 1-\alpha & \alpha \\ \beta & 1-\beta \end{pmatrix}$$

Use induction to show that

$$P_\mu(X_n = 0) = \frac{\beta}{\alpha+\beta} + (1-\alpha-\beta)^n \left\{ \mu(0) - \frac{\beta}{\alpha+\beta} \right\}$$

5.6.2 Show that if S is finite and p is irreducible and aperiodic, then there is an m so that $p^m(x,y) > 0$ for all x, y.

5.6.3 Show that if S is finite, p is irreducible and aperiodic, and T is the coupling time defined in the proof of Theorem 5.6.6, then $P(T > n) \le Cr^n$ for some $r < 1$ and $C < \infty$. So the convergence to equilibrium occurs exponentially rapidly in this case. Hint: First, consider the case in which $p(x,y) > 0$ for all x and y and reduce the general case to this one by looking at a power of p.

5.6.4 For any transition matrix p, define

$$\alpha_n = \sup_{i,j} \frac{1}{2} \sum_k |p^n(i,k) - p^n(j,k)|$$

The $1/2$ is there because for any i and j we can define r.v.'s X and Y so that $P(X = k) = p^n(i,k)$, $P(Y = k) = p^n(j,k)$, and

$$P(X \neq Y) = (1/2) \sum_k |p^n(i,k) - p^n(j,k)|$$

Show that $\alpha_{m+n} \le \alpha_n \alpha_m$. Here you may find the coupling interpretation may help you from getting lost in the algebra. Using Lemma 2.7.1, we can conclude that

$$\frac{1}{n} \log \alpha_n \to \inf_{m \ge 1} \frac{1}{m} \log \alpha_m$$

so if $\alpha_m < 1$ for some m, it approaches 0 exponentially fast.

5.6.5 **Strong law for additive functionals.** Suppose p is irreducible and has stationary distribution π. Let f be a function that has $\sum |f(y)|\pi(y) < \infty$. Let T_x^k be the time of the kth return to x. (i) Show that

$$V_k^f = f(X(T_x^k)) + \cdots + f(X(T_x^{k+1} - 1)), \quad k \ge 1 \text{ are i.i.d.}$$

with $E|V_k^f| < \infty$. (ii) Let $K_n = \inf\{k : T_x^k \ge n\}$ and show that

$$\frac{1}{n} \sum_{m=1}^{K_n} V_m^f \to \frac{EV_1^f}{E_x T_x^1} = \sum f(y)\pi(y) \quad P_\mu - \text{a.s.}$$

(iii) Show that $\max_{1 \le m \le n} V_m^{|f|}/n \to 0$ and conclude

$$\frac{1}{n} \sum_{m=1}^{n} f(X_m) \to \sum_y f(y)\pi(y) \quad P_\mu - \text{a.s.}$$

for any initial distribution μ.

5.6.6 **Central limit theorem for additive functionals.** Suppose in addition to the conditions in the Exercise 5.6.5 that $\sum f(y)\pi(y) = 0$, and $E_x(V_k^{|f|})^2 < \infty$. (i) Use the random index central limit theorem (Exercise 3.4.6) to conclude that for any initial distribution μ

$$\frac{1}{\sqrt{n}} \sum_{m=1}^{K_n} V_m^f \Rightarrow c\chi \quad \text{under } P_\mu$$

(ii) Show that $\max_{1 \le m \le n} V_m^{|f|}/\sqrt{n} \to 0$ in probability and conclude

$$\frac{1}{\sqrt{n}} \sum_{m=1}^{n} f(X_m) \Rightarrow c\chi \quad \text{under } P_\mu$$

5.6.7 **Ratio Limit Theorems.** Theorem 5.6.1 does not say much in the null recurrent case. To get a more informative limit theorem, suppose that y is recurrent and m is the (unique up to constant multiples) stationary measure on $C_y = \{z : \rho_{yz} > 0\}$. Let $N_n(z) = |\{m \le n : X_n = z\}|$. Break up the path at successive returns to y and show that $N_n(z)/N_n(y) \to m(z)/m(y)$ P_x-a.s. for all $x, z \in C_y$. Note that $n \to N_n(z)$ is increasing, so this is much easier than the previous problem.

5.6.8 We got (5.6.1) from Theorem 5.6.1 by taking expected value. This does not work for the ratio in the previous exercise, so we need another approach. Suppose $z \ne y$. (i) Let $\bar{p}_n(x,z) = P_x(X_n = z, T_y > n)$ and decompose $p^m(x,z)$ according to the value of $J = \sup\{j \in [1,m) : X_j = y\}$ to get

$$\sum_{m=1}^{n} p^m(x,z) = \sum_{m=1}^{n} \bar{p}_m(x,z) + \sum_{j=1}^{n-1} p^j(x,y) \sum_{k=1}^{n-j} \bar{p}_k(y,z)$$

(ii) Show that

$$\sum_{m=1}^{n} p^m(x,z) \Big/ \sum_{m=1}^{n} p^m(x,y) \to \frac{m(z)}{m(y)}$$

5.7 Periodicity, Tail σ-Field*

Lemma 5.7.1 *Suppose p is irreducible, recurrent, and all states have period d. Fix $x \in S$, and for each $y \in S$, let $K_y = \{n \ge 1 : p^n(x,y) > 0\}$. (i) There is an $r_y \in \{0, 1, \dots, d-1\}$ so that if $n \in K_y$, then $n = r_y$ mod d, i.e., the difference $n - r_y$ is a multiple of d. (ii) Let $S_r = \{y : r_y = r\}$ for $0 \le r < d$. If $y \in S_i$, $z \in S_j$, and $p^n(y,z) > 0$, then $n = (j - i)$ mod d. (iii) $S_0, S_1, \dots, S_{d-1}$ are irreducible classes for p^d, and all states have period 1.*

Proof (i) Let $m(y)$ be such that $p^{m(y)}(y,x) > 0$. If $n \in K_y$, then $p^{n+m(y)}(x,x)$ is positive, so $d|(n+m)$. Let $r_y = (d - m(y))$ mod d. (ii) Let m, n be such that $p^n(y,z)$, $p^m(x,y) > 0$. Since $p^{n+m}(x,z) > 0$, it follows from (i) that $n + m = j$ mod d. Since $m = i$ mod d, the result follows. The irreducibility in (iii) follows immediately from (ii). The aperiodicity follows from the definition of the period as the g.c.d. $\{x : p^n(x,x) > 0\}$. □

A partition of the state space $S_0, S_1, \ldots, S_{d-1}$ satisfying (ii) in Lemma 5.7.1 is called a **cyclic decomposition** of the state space. Except for the choice of the set to put first, it is unique. (Pick an $x \in S$. It lies in some S_j, but once the value of j is known, irreducibility and (ii) allow us to calculate all the sets.)

Exercise 5.7.1 Find the decomposition for the Markov chain with transition probability

	1	**2**	**3**	**4**	**5**	**6**	**7**
1	0	0	0	.5	.5	0	0
2	.3	0	0	0	0	0	.7
3	0	0	0	0	0	0	1
4	0	0	1	0	0	0	0
5	0	0	1	0	0	0	0
6	0	1	0	0	0	0	0
7	0	0	0	.4	0	.6	0

Theorem 5.7.2 (Convergence theorem, periodic case) *Suppose p is irreducible, has a stationary distribution π, and all states have period d. Let $x \in S$, and let $S_0, S_1, \ldots, S_{d-1}$ be the cyclic decomposition of the state space with $x \in S_0$. If $y \in S_r$, then*

$$\lim_{m\to\infty} p^{md+r}(x,y) = \pi(y)d$$

Proof If $y \in S_0$, then using (iii) in Lemma 5.7.1 and applying Theorem 5.6.6 to p^d shows

$$\lim_{m\to\infty} p^{md}(x,y) \text{ exists}$$

To identify the limit, we note that (5.6.1) implies

$$\frac{1}{n}\sum_{m=1}^{n} p^m(x,y) \to \pi(y)$$

and (ii) of Lemma 5.7.1 implies $p^m(x,y) = 0$ unless $d|m$, so the limit in the first display must be $\pi(y)d$. If $y \in S_r$ with $1 \le r < d$, then

$$p^{md+r}(x,y) = \sum_{z\in S_r} p^r(x,z)p^{md}(z,y)$$

Since $y, z \in S_r$ it follows from the first case in the proof that $p^{md}(z,y) \to \pi(y)d$ as $m \to \infty$. $p^{md}(z,y) \le 1$, and $\sum_z p^r(x,z) = 1$, so the result follows from the dominated convergence theorem. □

Let $\mathcal{F}'_n = \sigma(X_{n+1}, X_{n+2}, \ldots)$ and $\mathcal{T} = \cap_n \mathcal{F}'_n$ be the tail σ-field. The next result is due to Orey. The proof we give is from Blackwell and Freedman (1964).

Theorem 5.7.3 *Suppose p is irreducible, recurrent, and all states have period d, $\mathcal{T} = \sigma(\{X_0 \in S_r\} : 0 \le r < d)$.*

Remark To be precise, if μ is any initial distribution and $A \in \mathcal{T}$, then there is an r so that $A = \{X_0 \in S_r\}$ P_μ-a.s.

Proof We build up to the general result in three steps.

Case 1. Suppose $P(X_0 = x) = 1$. Let $T_0 = 0$, and for $n \geq 1$, let $T_n = \inf\{m > T_{n-1} : X_m = x\}$ be the time of the nth return to x. Let

$$V_n = (X(T_{n-1}), \ldots, X(T_n - 1))$$

The vectors V_n are i.i.d. by Exercise 5.3.1, and the tail σ-field is contained in the exchangeable field of the V_n, so the Hewitt-Savage 0-1 law (Theorem 2.5.4, proved there for r.v.'s taking values in a general measurable space) implies that $\mathcal{T}$ is trivial in this case.

Case 2. Suppose that the initial distribution is concentrated on one cyclic class, say S_0. If $A \in \mathcal{T}$ then $P_x(A) \in \{0,1\}$ for each x by case 1. If $P_x(A) = 0$ for all $x \in S_0$, then $P_\mu(A) = 0$. Suppose $P_y(A) > 0$, and hence $= 1$, for some $y \in S_0$. Let $z \in S_0$. Since p^d is irreducible and aperiodic on S_0, there is an n so that $p^n(z,y) > 0$ and $p^n(y,y) > 0$. If we write $1_A = 1_B \circ \theta_n$ then the Markov property implies

$$1 = P_y(A) = E_y(E_y(1_B \circ \theta_n | \mathcal{F}_n)) = E_y(E_{X_n} 1_B)$$

so $P_y(B) = 1$. Another application of the Markov property gives

$$P_z(A) = E_z(E_{X_n} 1_B) \geq p^n(z,y) > 0$$

so $P_z(A) = 1$, and since $z \in S_0$ is arbitrary, $P_\mu(A) = 1$.

General Case. From case 2, we see that $P(A|X_0 = y) \equiv 1$ or $\equiv 0$ on each cyclic class. This implies that either $\{X_0 \in S_r\} \subset A$ or $\{X_0 \in S_r\} \cap A = \emptyset$ P_μ a.s. Conversely, it is clear that $\{X_0 \in S_r\} = \{X_{nd} \in S_r \text{ i.o.}\} \in \mathcal{T}$, and the proof is complete. □

The next result will help us identify the tail σ-field in transient examples.

Theorem 5.7.4 *Suppose X_0 has initial distribution μ. The equations*

$$h(X_n, n) = E_\mu(Z|\mathcal{F}_n) \quad \text{and} \quad Z = \lim_{n\to\infty} h(X_n, n)$$

set up a 1-1 correspondence between bounded $Z \in \mathcal{T}$ and bounded **space-time harmonic functions**, *i.e., bounded $h : S \times \{0,1,\ldots\} \to \mathbf{R}$, so that $h(X_n, n)$ is a martingale.*

Proof Let $Z \in \mathcal{T}$, write $Z = Y_n \circ \theta_n$, and let $h(x,n) = E_x Y_n$.

$$E_\mu(Z|\mathcal{F}_n) = E_\mu(Y_n \circ \theta_n | \mathcal{F}_n) = h(X_n, n)$$

by the Markov property, so $h(X_n, n)$ is a martingale. Conversely, if $h(X_n, n)$ is a bounded martingale, using Theorems 4.2.11 and 4.6.7 shows $h(X_n, n) \to Z \in \mathcal{T}$ as $n \to \infty$, and $h(X_n, n) = E_\mu(Z|\mathcal{F}_n)$. □

Exercise 5.7.2 A random variable Z with $Z = Z \circ \theta$, and hence $= Z \circ \theta_n$ for all n, is called **invariant**. Show there is a 1-1 correspondence between bounded invariant random variables and bounded harmonic functions. We will have more to say about invariant r.v.'s in Section 7.1.

Example 5.7.5 (Simple random walk in d dimensions) We begin by constructing a coupling for this process. Let $i_1, i_2, \ldots$ be i.i.d. uniform on $\{1, \ldots, d\}$. Let $\xi_1, \xi_2, \ldots$ and

$\eta_1, \eta_2, \ldots$ be i.i.d. uniform on $\{-1, 1\}$. Let e_j be the jth unit vector. Construct a coupled pair of d-dimensional simple random walks by

$$X_n = X_{n-1} + e(i_n)\xi_n$$

$$Y_n = \begin{cases} Y_{n-1} + e(i_n)\xi_n & \text{if } X^{i_n}_{n-1} = Y^{i_n}_{n-1} \\ Y_{n-1} + e(i_n)\eta_n & \text{if } X^{i_n}_{n-1} \neq Y^{i_n}_{n-1} \end{cases}$$

In words, the coordinate that changes is always the same in the two walks, and once they agree in one coordinate, future movements in that direction are the same. It is easy to see that if $X^i_0 - Y^i_0$ is even for $1 \le i \le d$, then the two random walks will hit with probability one.

Let $L_0 = \{z \in \mathbf{Z}^d : z^1 + \cdots + z^d \text{ is even }\}$ and $L_1 = \mathbf{Z}^d - L_0$. Although we have only defined the notion for the recurrent case, it should be clear that L_0, L_1 is the cyclic decomposition of the state space for simple random walk. If $S_n \in L_i$, then $S_{n+1} \in L_{1-i}$ and p^2 is irreducible on each L_i. To couple two random walks starting from $x, y \in L_i$, let them run independently until the first time all the coordinate differences are even, and then use the last coupling. In the remaining case, $x \in L_0$, $y \in L_1$ coupling is impossible.

The next result should explain our interest in coupling two d-dimensional simple random walks.

Theorem 5.7.6 *For d-dimensional simple random walk,*

$$\mathcal{T} = \sigma(\{X_0 \in L_i\}, i = 0, 1)$$

Proof Let $x, y \in L_i$, and let X_n, Y_n be a realization of the coupling defined above for $X_0 = x$ and $Y_0 = y$. Let $h(x,n)$ be a bounded space-time harmonic function. The martingale property implies $h(x,0) = E_x h(X_n, n)$. If $|h| \le C$, it follows from the coupling that

$$|h(x,0) - h(y,0)| = |Eh(X_n,n) - Eh(Y_n,n)| \le 2CP(X_n \neq Y_n) \to 0$$

so $h(x,0)$ is constant on L_0 and L_1. Applying the last result to $h'(x,m) = h(x, n+m)$, we see that $h(x,n) = a^i_n$ on L_i. The martingale property implies $a^i_n = a^{1-i}_{n+1}$, and the desired result follows from Theorem 5.7.4. □

Example 5.7.7 (Ornstein's coupling) Let $p(x,y) = f(y - x)$ be the transition probability for an irreducible aperiodic random walk on $\mathbf{Z}$. To prove that the tail σ-field is trivial, pick M large enough so that the random walk generated by the probability distribution $f_M(x)$ with $f_M(x) = c_M f(x)$ for $|x| \le M$ and $f_M(x) = 0$ for $|x| > M$ is irreducible and aperiodic. Let $Z_1, Z_2, \ldots$ be i.i.d. with distribution f and let $W_1, W_2, \ldots$ be i.i.d. with distribution f_M. Let $X_n = X_{n-1} + Z_n$ for $n \ge 1$. If $X_{n-1} = Y_{n-1}$, we set $X_n = Y_n$. Otherwise, we let

$$Y_n = \begin{cases} Y_{n-1} + Z_n & \text{if } |Z_n| > m \\ Y_{n-1} + W_n & \text{if } |Z_n| \le m \end{cases}$$

In words, the big jumps are taken in parallel and the small jumps are independent. The recurrence of one-dimensional random walks with mean 0 implies $P(X_n \neq Y_n) \to 0$. Repeating the proof of Theorem 5.7.6, we see that $\mathcal{T}$ is trivial.

The tail σ-field in Theorem 5.7.6 is essentially the same as in Theorem 5.7.3. To get a more interesting $\mathcal{T}$, we look at:

Example 5.7.8 (Random walk on a tree) To facilitate definitions, we will consider the system as a random walk on a group with three generators a, b, c that have $a^2 = b^2 = c^2 = e$, the identity element. To form the random walk, let $\xi_1, \xi_2, \ldots$ be i.i.d. with $P(\xi_n = x) = 1/3$ for $x = a, b, c$, and let $X_n = X_{n-1}\xi_n$. (This is equivalent to a random walk on the tree in which each vertex has degree 3 but the algebraic formulation is convenient for computations.) Let L_n be the length of the word X_n when it has been reduced as much as possible, with $L_n = 0$ if $X_n = e$. The reduction can be done as we go along. If the last letter of X_{n-1} is the same as ξ_n, we erase it, otherwise we add the new letter. It is easy to see that L_n is a Markov chain with a transition probability that has $p(0, 1) = 1$ and

$$p(j, j-1) = 1/3 \qquad p(j, j+1) = 2/3 \quad \text{for } j \geq 1$$

As $n \to \infty$, $L_n \to \infty$. From this, it follows easily that the word X_n has a limit in the sense that the ith letter X_n^i stays the same for large n. Let X_∞ be the limiting word, i.e., $X_\infty^i = \lim X_n^i$. $\mathcal{T} \supset \sigma(X_\infty^i, i \geq 1)$, but it is easy to see that this is not all. If $S_0 =$ the words of even length, and $S_1 = S_0^c$, then $X_n \in S_i$ implies $X_{n+1} \in S_{1-i}$, so $\{X_0 \in S_0\} \in \mathcal{T}$. Can the reader prove that we have now found all of $\mathcal{T}$? As Fermat once said, "I have a proof but it won't fit in the margin."

Remark This time the solution does not involve elliptic curves but uses "h-paths." See Furstenburg (1970) or decode the following: "Condition on the exit point (the infinite word). Then the resulting RW is an h-process, which moves closer to the boundary with probability 2/3 and farther with probability 1/3 (1/6 each to the two possibilities). Two such random walks couple, provided they have same parity." The quote is from Robin Pemantle, who says he consulted Itai Benajamini and Yuval Peres.

5.8 General State Space*

In this section, we will generalize the results from countable state space to a collection of Markov chains with uncountable state space called Harris chains. The developments here are motivated by three ideas. First, the proofs for countable state space if there is one point in the state space that the chain hits with probability one. (Think, for example, about the construction of the stationary measure via the cycle trick.) Second, a recurrent Harris chain can be modified to contain such a point. Third, the collection of Harris chains is a comfortable level of generality; broad enough to contain a large number of interesting examples, yet restrictive enough to allow for a rich theory.

We say that a Markov chain X_n is a **Harris chain** if we can find sets $A, B \in \mathcal{S}$, a function q with $q(x, y) \geq \epsilon > 0$ for $x \in A$, $y \in B$, and a probability measure ρ concentrated on B so that:

(i) If $\tau_A = \inf\{n \geq 0 : X_n \in A\}$, then $P_z(\tau_A < \infty) > 0$ for all $z \in S$.

(ii) If $x \in A$ and $C \subset B$, then $p(x, C) \geq \int_C q(x, y)\, \rho(dy)$.

To explain the definition we turn to some examples:

Example 5.8.1 (Countable state space) If S is countable and there is a point a with $\rho_{xa} > 0$ for all x (a condition slightly weaker than irreducibility), then we can take $A = \{a\}$, $B = \{b\}$, where b is any state with $p(a,b) > 0$, $\mu = \delta_b$ the point mass at b, and $q(a,b) = p(a,b)$.

Example 5.8.2 (Chains with continuous densities) Suppose $X_n \in \mathbf{R}^d$ is a Markov chain with a transition probability that has $p(x,dy) = p(x,y)\,dy$, where $(x,y) \to p(x,y)$ is continuous. Pick (x_0,y_0) so that $p(x_0,y_0) > 0$. Let A and B be open sets around x_0 and y_0 that are small enough so that $p(x,y) \geq \epsilon > 0$ on $A \times B$. If we let $\rho(C) = |B \cap C|/|B|$, where $|B|$ is the Lebesgue measure of B, then (ii) holds. If (i) holds, then X_n is a Harris chain.

For concrete examples, consider:

(a) **Diffusion processes** are a large class of examples that lie outside the scope of this book, but are too important to ignore. When things are nice, specifically, if the generator of X has Hölder continuous coefficients satisfying suitable growth conditions, see the Appendix of Dynkin (1965), then $P_x(X_1 \in dy) = p(x,y)\,dy$, and p satisfies the previously noted conditions.

(b) The **discrete Ornstein-Uhlenbeck process**. Let $\xi_1, \xi_2, \ldots$ be i.i.d. standard normals and let $V_n = \theta V_{n-1} + \xi_n$. The Ornstein-Uhlenbeck process is a diffusion process $\{V_t, t \in [0,\infty)\}$ that models the velocity of a particle suspended in a liquid. See, e.g., Breiman (1968) section 16.1. Looking at V_t at integer times (and dividing by a constant to make the variance 1) gives a Markov chain with the indicated distributions.

Example 5.8.3 (GI/G/1 queue, or storage model) Let $\xi_1, \xi_2, \ldots$ be i.i.d. and define W_n inductively by $W_n = (W_{n-1} + \xi_n)^+$. If $P(\xi_n < 0) > 0$, then we can take $A = B = \{0\}$ and (i) and (ii) hold. To explain the first name in the title, consider a queueing system in which customers arrive at times of a renewal process, i.e., at times $0 = T_0 < T_1 < T_2 \ldots$ with $\zeta_n = T_n - T_{n-1}$, $n \geq 1$ i.i.d. Let η_n, $n \geq 0$, be the amount of service time the nth customer requires and let $\xi_n = \eta_{n-1} - \zeta_n$. I claim that W_n is the amount of time the nth customer has to wait to enter service. To see this, notice that the $(n-1)$th customer adds η_{n-1} to the server's workload, and if the server is busy at all times in $[T_{n-1}, T_n)$, he reduces his workload by ζ_n. If $W_{n-1} + \eta_{n-1} < \zeta_n$, then the server has enough time to finish his work and the next arriving customer will find an empty queue.

The second name in the title refers to the fact that W_n can be used to model the contents of a storage facility. For an intuitive description, consider water reservoirs. We assume that rain storms occur at times of a renewal process $\{T_n : n \geq 1\}$, that the nth rainstorm contributes an amount of water η_n, and that water is consumed at constant rate c. If we let $\zeta_n = T_n - T_{n-1}$ as before, and $\xi_n = \eta_{n-1} - c\zeta_n$, then W_n gives the amount of water in the reservoir just before the nth rainstorm.

History Lesson. Doeblin was the first to prove results for Markov chains on general state space. He supposed that there was an n so that $p^n(x,C) \geq \epsilon\rho(C)$ for all $x \in S$ and $C \subset S$. See Doob (1953), section V.5, for an account of his results. Harris (1956) generalized Doeblin's result by observing that it was enough to have a set A so that (i) holds and the chain viewed on A ($Y_k = X(T_A^k)$, where $T_A^k = \inf\{n > T_A^{k-1} : X_n \in A\}$ and $T_A^0 = 0$) satisfies Doeblin's condition. Our formulation, as well as most of the proofs in this section, follows Athreya and Ney (1978). For a nice description of the "traditional approach," see Revuz (1984).

Given a Harris chain on $(S,\mathcal{S})$, we will construct a Markov chain $\bar{X}_n$ with transition probability $\bar{p}$ on $(\bar{S},\bar{\mathcal{S}})$, where $\bar{S} = S \cup \{\alpha\}$ and $\bar{\mathcal{S}} = \{B,\ B\cup\{\alpha\} : B \in \mathcal{S}\}$. The aim, as advertised earlier, is to manufacture a point α that the process hits with probability 1 in the recurrent case.

$$\begin{array}{ll} \text{If } x \in S - A & \bar{p}(x,C) = p(x,C) \text{ for } C \in \mathcal{S} \\ \text{If } x \in A & \bar{p}(x,\{\alpha\}) = \epsilon \\ & \bar{p}(x,C) = p(x,C) - \epsilon\rho(C) \text{ for } C \in \mathcal{S} \\ \text{If } x = \alpha & \bar{p}(\alpha,D) = \int \rho(dx)\bar{p}(x,D) \text{ for } D \in \bar{\mathcal{S}} \end{array}$$

Intuitively, $\bar{X}_n = \alpha$ corresponds to X_n being distributed on B according to ρ. Here, and in what follows, we will reserve A and B for the special sets that occur in the definition and use C and D for generic elements of $\mathcal{S}$. We will often simplify notation by writing $\bar{p}(x,\alpha)$ instead of $\bar{p}(x,\{\alpha\})$, $\mu(\alpha)$ instead of $\mu(\{\alpha\})$, etc.

Our next step is to prove three technical lemmas that will help us develop the theory. Define a transition probability v by

$$v(x,\{x\}) = 1 \quad \text{if} \quad x \in S \qquad v(\alpha,C) = \rho(C)$$

In words, V leaves mass in S alone but returns the mass at α to S and distributes it according to ρ.

Lemma 5.8.4 *$v\bar{p} = \bar{p}$ and $\bar{p}v = p$.*

Proof Before giving the proof, we would like to remind the reader that measures multiply the transition probability on the left, i.e., in the first case we want to show $\mu v\bar{p} = \mu\bar{p}$. If we first make a transition according to v and then one according to $\bar{p}$, this amounts to one transition according to $\bar{p}$, since only mass at α is affected by v and

$$\bar{p}(\alpha,D) = \int \rho(dx)\bar{p}(x,D)$$

The second equality also follows easily from the definition. In words, if $\bar{p}$ acts first and then v, then v returns the mass at α to where it came from. □

From Lemma 5.8.4, it follows easily that we have:

Lemma 5.8.5 *Let Y_n be an inhomogeneous Markov chain with $p_{2k} = v$ and $p_{2k+1} = \bar{p}$. Then $\bar{X}_n = Y_{2n}$ is a Markov chain with transition probability $\bar{p}$ and $X_n = Y_{2n+1}$ is a Markov chain with transition probability p.*

Lemma 5.8.5 shows that there is an intimate relationship between the asymptotic behavior of X_n and of $\bar{X}_n$. To quantify this, we need a definition. If f is a bounded measurable function on S, let $\bar{f} = vf$, i.e., $\bar{f}(x) = f(x)$ for $x \in S$ and $\bar{f}(\alpha) = \int f\,d\rho$.

Lemma 5.8.6 *If μ is a probability measure on $(S,\mathcal{S})$, then*

$$E_\mu f(X_n) = E_\mu \bar{f}(\bar{X}_n)$$

Proof Observe that if X_n and $\bar{X}_n$ are constructed as in Lemma 5.8.5, and $P(\bar{X}_0 \in S) = 1$, then $X_0 = \bar{X}_0$ and X_n is obtained from $\bar{X}_n$ by making a transition according to v. □

The last three lemmas will allow us to obtain results for X_n from those for $\bar{X}_n$. We turn now to the task of generalizing the results from countable state space to $\bar{X}_n$. Before developing the theory, we will give one last example that explains why some of the statements are messy.

Example 5.8.7 (Perverted O.U. process) Take the discrete O.U. process from part (b) of Example 5.8.2 and modify the transition probability at the integers $x \geq 2$ so that

$$p(x, \{x+1\}) = 1 - x^{-2}$$
$$p(x, A) = x^{-2}|A| \quad \text{for } A \subset (0, 1)$$

p is the transition probability of a Harris chain, but

$$P_2(X_n = n + 2 \text{ for all } n) > 0$$

I can sympathize with the reader who thinks that such chains will not arise "in applications," but it seems easier (and better) to adapt the theory to include them than to modify the assumptions to exclude them.

5.8.1 Recurrence and Transience

We begin with the dichotomy between recurrence and transience. Let $R = \inf\{n \geq 1 : \bar{X}_n = \alpha\}$. If $P_\alpha(R < \infty) = 1$, then we call the chain **recurrent**, otherwise we call it **transient**. Let $R_1 = R$ and for $k \geq 2$, let $R_k = \inf\{n > R_{k-1} : \bar{X}_n = \alpha\}$ be the time of the kth return to α. The strong Markov property implies $P_\alpha(R_k < \infty) = P_\alpha(R < \infty)^k$, so $P_\alpha(\bar{X}_n = \alpha \text{ i.o.}) = 1$ in the recurrent case and $= 0$ in the transient case.

The next result generalizes Lemma 5.3.2. (If x is recurrent and $\rho_{xy} > 0$, then y is recurrent and $\rho_{yx} = 1$.)

Theorem 5.8.8 *Let $\lambda(C) = \sum_{n=1}^{\infty} 2^{-n} \bar{p}^n(\alpha, C)$. In the recurrent case, if $\lambda(C) > 0$, then $P_\alpha(\bar{X}_n \in C$ i.o.$) = 1$. For λ-a.e. x, $P_x(R < \infty) = 1$.*

Proof The first conclusion follows from Exercise 4.6.4. For the second let $D = \{x : P_x(R < \infty) < 1\}$ and observe that if $p^n(\alpha, D) > 0$ for some n, then

$$P_\alpha(\bar{X}_m = \alpha \text{ i.o.}) \leq \int \bar{p}^n(\alpha, dx) P_x(R < \infty) < 1$$ □

Remark Example 5.8.7 shows that we cannot expect to have $P_x(R < \infty) = 1$ for all x. To see that even when the state space is countable, we need not hit every point starting from α (see Exercise 5.8.2).

5.8.2 Stationary Measures

Theorem 5.8.9 *In the recurrent case, there is a σ-finite stationary measure $\bar{\mu} \ll \lambda$.*

Proof Let $R = \inf\{n \geq 1 : \bar{X}_n = \alpha\}$, and let

$$\bar{\mu}(C) = E_\alpha\left(\sum_{n=0}^{R-1} 1_{\{\bar{X}_n \in C\}}\right) = \sum_{n=0}^{\infty} P_\alpha(\bar{X}_n \in C, R > n)$$

Repeating the proof of Theorem 5.5.7 shows that $\bar{\mu}\bar{p} = \bar{\mu}$. If we let $\mu = \bar{\mu}v$, then it follows from Lemma 5.8.4 that $\bar{\mu}v\, p = \bar{\mu}\bar{p}v = \bar{\mu}v$, so $\mu\, p = \mu$.

To prove that it is σ-finite, let $G_{k,\delta} = \{x : \bar{p}^k(x,\alpha) \geq \delta\}$. Let $T_1 = \inf\{n : X(T_1) \in G_k\}$ and for $m \geq 2$, let $X(T_m) = \inf\{n \geq T_{m-1}\}$, and let $M = \sup\{m : T_m < R\}$. By assumption, $EM < 1/\delta$. In between T_m and T_{m-1} there can be at most k visits to $G_{k,\delta}$. The fact that $\bar{\mu} \ll \lambda$ is clear by comparing the definitions. □

To investigate uniqueness of the stationary measure, we begin with:

Lemma 5.8.10 *If ν is a σ-finite stationary measure for p, then $\nu(A) < \infty$ and $\bar{\nu} = \nu\bar{p}$ is a stationary measure for $\bar{p}$ with $\bar{\nu}(\alpha) < \infty$.*

Proof We will first show that $\nu(A) < \infty$. If $\nu(A) = \infty$, then part (ii) of the definition implies $\nu(C) = \infty$ for all sets C with $\rho(C) > 0$. If $B = \cup_i B_i$ with $\nu(B_i) < \infty$, then $\rho(B_i) = 0$ by the last observation and $\rho(B) = 0$ by countable subadditivity, a contradiction. So $\nu(A) < \infty$ and $\bar{\nu}(\alpha) = \nu\bar{p}(\alpha) = \epsilon\nu(A) < \infty$. Using the fact that $\nu\, p = \nu$, we find

$$\nu\bar{p}(C) = \nu(C) - \epsilon\nu(A)\rho(B \cap C)$$

the last subtraction being well-defined since $\nu(A) < \infty$, and it follows that $\bar{\nu}v = \nu$. To check $\bar{\nu}\bar{p} = \bar{\nu}$, we observe that Lemma 5.8.4 and the last result imply $\bar{\nu}\bar{p} = \bar{\nu}v\bar{p} = \nu\bar{p} = \bar{\nu}$. □

Theorem 5.8.11 *Suppose p is recurrent. If ν is a σ-finite stationary measure, then $\nu = \bar{\nu}(\alpha)\mu$, where μ is the measure constructed in the proof of Theorem 5.8.9.*

Proof By Lemma 5.8.10, it suffices to prove that if $\bar{\nu}$ is a stationary measure for $\bar{p}$ with $\bar{\nu}(\alpha) < \infty$, then $\bar{\nu} = \bar{\nu}(\alpha)\bar{\mu}$. Repeating the proof of Theorem 5.5.9 with $a = \alpha$, it is easy to show that $\bar{\nu}(C) \geq \bar{\nu}(\alpha)\bar{\mu}(C)$. Continuing to compute as in that proof:

$$\bar{\nu}(\alpha) = \int \bar{\nu}(dx)\bar{p}^n(x,\alpha) \geq \bar{\nu}(\alpha)\int \bar{\mu}(dx)\bar{p}^n(x,\alpha) = \bar{\nu}(\alpha)\bar{\mu}(\alpha) = \bar{\nu}(\alpha)$$

Let $S_n = \{x : p^n(x,\alpha) > 0\}$. By assumption, $\cup_n S_n = S$. If $\bar{\nu}(D) > \bar{\nu}(\alpha)\bar{\mu}(D)$ for some D, then $\bar{\nu}(D \cap S_n) > \bar{\nu}(\alpha)\bar{\mu}(D \cap S_n)$, and it follows that $\bar{\nu}(\alpha) > \bar{\nu}(\alpha)$ a contradiction. □

5.8.3 Convergence Theorem

We say that a recurrent Harris chain X_n is **aperiodic** if g.c.d. $\{n \geq 1 : p^n(\alpha,\alpha) > 0\} = 1$. This occurs, for example, if we can take $A = B$ in the definition for then $p(\alpha,\alpha) > 0$.

Theorem 5.8.12 *Let X_n be an aperiodic recurrent Harris chain with stationary distribution π. If $P_x(R < \infty) = 1$, then as $n \to \infty$,*

$$\|p^n(x,\cdot) - \pi(\cdot)\| \to 0$$

Note. Here $\|\ \|$ denotes the total variation distance between the measures. Lemma 5.8.8 guarantees that π a.e. x satisfies the hypothesis.

Proof In view of Lemma 5.8.6, it suffices to prove the result for $\bar{p}$. We begin by observing that the existence of a stationary probability measure and the uniqueness result in Theorem 5.8.11 imply that the measure constructed in Theorem 5.8.9 has $E_\alpha R = \bar{\mu}(S) < \infty$. As in the proof of Theorem 5.6.6, we let X_n and Y_n be independent copies of the chain with initial distributions δ_x and π, respectively, and let $\tau = \inf\{n \geq 0 : X_n = Y_n = \alpha\}$. For $m \geq 0$, let S_m (resp. T_m) be the times at which X_n (resp. Y_n) visit α for the $(m+1)$th time. $S_m - T_m$ is a random walk with mean 0 steps, so $M = \inf\{m \geq 1 : S_m = T_m\} < \infty$ a.s., and it follows that this is true for τ as well. The computations in the proof of Theorem 5.6.6 show $|P(X_n \in C) - P(Y_n \in C)| \leq P(\tau > n)$. Since this is true for all C, $\|p^n(x, \cdot) - \pi(\cdot)\| \leq P(\tau > n)$, and the proof is complete. □

5.8.4 GI/G/1 Queue

For the rest of the section, we will concentrate on the $GI/G/1$ queue. Let $\xi_1, \xi_2, \ldots$ be i.i.d., let $W_n = (W_{n-1} + \xi_n)^+$, and let $S_n = \xi_1 + \cdots + \xi_n$. Recall $\xi_n = \eta_{n-1} - \zeta_n$, where the η's are service times, ζ's are the interarrival times, and suppose $E\xi_n < 0$ so that Exercise 5.8.5 implies there is a stationary distribution.

Explicit formulas for the distribution of M are in general difficult to obtain. However, this can be done if either the arrival or service distribution is exponential. One reason for this is:

Lemma 5.8.13 *Suppose $X, Y \geq 0$ are independent and $P(X > x) = e^{-\lambda x}$. Show that $P(X - Y > x) = ae^{-\lambda x}$, where $a = P(X - Y > 0)$.*

Proof Let F be the distribution of Y

$$\begin{aligned} P(X - Y > x) &= \int_0^\infty dF(y) e^{-\lambda(x+y)} \\ &= e^{-\lambda x} \int_0^\infty dF(y) e^{-\lambda(y)} = e^{-\lambda x} P(X - Y > 0) \end{aligned}$$

where the last equality follows from setting $x = 0$ in the first one. □

Example 5.8.14 (Exponential service time) Suppose $P(\eta_n > x) = e^{-\beta x}$ and $E\zeta_n > E\eta_n$. Let $T = \inf\{n : S_n > 0\}$ and $L = S_T$, setting $L = -\infty$ if $T = \infty$. The lack of memory property of the exponential distribution implies that $P(L > x) = re^{-\beta x}$, where $r = P(T < \infty)$. To compute the distribution of the maximum, M, let $T_1 = T$ and let $T_k = \inf\{n > T_{k-1} : S_n > S_{T_{k-1}}\}$ for $k \geq 2$. The strong Markov property implies that if $T_k < \infty$, then $S(T_{k+1}) - S(T_k) =_d L$ and is independent of $S(T_k)$. Using this and breaking things down according to the value of $K = \inf\{k : L_{k+1} = -\infty\}$, we see that for $x > 0$ the density function

$$P(M = x) = \sum_{k=1}^\infty r^k (1 - r) e^{-\beta x} \beta^k x^{k-1}/(k-1)! = \beta r(1 - r) e^{-\beta x(1-r)}$$

To complete the calculation, we need to calculate r. To do this, let

$$\phi(\theta) = E\exp(\theta\xi_n) = E\exp(\theta\eta_{n-1}) E\exp(-\theta\zeta_n)$$

which is finite for $0 < \theta < \beta$, since $\zeta_n \ge 0$ and η_{n-1} has an exponential distribution. It is easy to see that

$$\phi'(0) = E\xi_n < 0 \qquad \lim_{\theta\uparrow\beta} \phi(\theta) = \infty$$

so there is a $\theta \in (0, \beta)$ with $\phi(\theta) = 1$. Example 4.2.3 implies $\exp(\theta S_n)$ is a martingale. Theorem 4.4.1 implies $1 = E\exp(\theta S_{T\wedge n})$. Letting $n \to \infty$ and noting that $(S_n | T = n)$ has an exponential distribution and $S_n \to -\infty$ on $\{T = \infty\}$, we have

$$1 = r\int_0^\infty e^{\theta x}\beta e^{-\beta x}\,dx = \frac{r\beta}{\beta - \theta}$$

Example 5.8.15 (Poisson arrivals) Suppose $P(\zeta_n > x) = e^{-\alpha x}$ and $E\zeta_n > E\eta_n$. Let $\bar{S}_n = -S_n$. Reversing time as in (ii) of Exercise 5.8.6, we see (for $n \ge 1$)

$$P\left(\max_{0\le k<n} \bar{S}_k < \bar{S}_n \in A\right) = P\left(\min_{1\le k\le n} \bar{S}_k > 0, \bar{S}_n \in A\right)$$

Let $\psi_n(A)$ be the common value of the last two expressions and let $\psi(A) = \sum_{n\ge 0}\psi_n(A)$. $\psi_n(A)$ is the probability the random walk reaches a new maximum (or ladder height, see Exercise 5.4.2 in A at time n, so $\psi(A)$ is the number of ladder points in A with $\psi(\{0\}) = 1$. Letting the random walk take one more step

$$P\left(\min_{1\le k\le n} \bar{S}_k > 0, \bar{S}_{n+1} \le x\right) = \int F(x - z)\,d\psi_n(z)$$

The last identity is valid for $n = 0$ if we interpret the left-hand side as $F(x)$. Let $\tau = \inf\{n \ge 1 : \bar{S}_n \le 0\}$ and $x \le 0$. Integrating by parts on the right-hand side and then summing over $n \ge 0$ gives

$$\begin{aligned} P(\bar{S}_\tau \le x) &= \sum_{n=0}^\infty P\left(\min_{1\le k\le n} \bar{S}_k > 0, \bar{S}_{n+1} \le x\right) \\ &= \int_{y\le x} \psi[0, x - y]\,dF(y) \end{aligned} \tag{5.8.1}$$

The limit $y \le x$ comes from the fact that $\psi((-\infty, 0)) = 0$.

Let $\bar{\xi}_n = \bar{S}_n - \bar{S}_{n-1} = -\xi_n$. Exercise 5.8.13 implies $P(\bar{\xi}_n > x) = ae^{-\alpha x}$. Let $\bar{T} = \inf\{n : \bar{S}_n > 0\}$. $E\bar{\xi}_n > 0$, so $P(\bar{T} < \infty) = 1$. Let $J = \bar{S}_T$. As in the previous example, $P(J > x) = e^{-\alpha x}$. Let $V_n = J_1 + \cdots + J_n$. V_n is a rate α Poisson process, so $\psi[0, x - y] = 1 + \alpha(x - y)$ for $x - y \ge 0$. Using (5.8.1) now and integrating by parts gives

$$\begin{aligned} P(\bar{S}_\tau \le x) &= \int_{y\le x} (1 + \alpha(x - y))\,dF(y) \\ &= F(x) + \alpha\int_{-\infty}^x F(y)\,dy \qquad \text{for } x \le 0 \end{aligned} \tag{5.8.2}$$

Since $P(\bar{S}_n = 0) = 0$ for $n \ge 1$, $-\bar{S}_\tau$ has the same distribution as S_T, where $T = \inf\{n : S_n > 0\}$. Combining this with part (ii) of Exercise 5.8.6 gives a "formula" for $P(M > x)$. Straightforward but somewhat tedious calculations show that if $B(s) = E\exp(-s\eta_n)$, then

$$E\exp(-sM) = \frac{(1 - \alpha\cdot E\eta)s}{s - \alpha + \alpha B(s)}$$

a result known as the **Pollaczek-Khintchine formula**. The computations we omitted can be found in Billingsley (1979) on p. 277 or several times in Feller, Vol. II (1971).

Exercises

5.8.1 $\bar{X}_n$ is recurrent if and only if $\sum_{n=1}^{\infty} \bar{p}^n(\alpha,\alpha) = \infty$.

5.8.2 If X_n is a recurrent Harris chain on a countable state space, then S can only have one irreducible set of recurrent states but may have a nonempty set of transient states. For a concrete example, consider a branching process in which the probability of no children $p_0 > 0$ and set $A = B = \{0\}$.

5.8.3 Use Exercise 5.8.1 and imitate the proof of Theorem 5.5.10 to show that a Harris chain with a stationary distribution must be recurrent.

5.8.4 Suppose X_n is a recurrent Harris chain. Show that if (A', B') is another pair satisfying the conditions of the definition, then Theorem 5.8.8 implies $P_\alpha(\bar{X}_n \in A' \text{ i.o.}) = 1$, so the recurrence or transience does not depend on the choice of (A, B).

5.8.5 In the $GI/G/1$ queue, the waiting time W_n and the random walk $S_n = X_0 + \xi_1 + \cdots + \xi_n$ agree until $N = \inf\{n : S_n < 0\}$, and at this time $W_N = 0$. Use this observation as we did in Example 5.3.7 to show that Example 5.8.3 is transient when $E\xi_n > 0$, recurrent when $E\xi_n \le 0$, and there is a stationary distribution when $E\xi_n < 0$.

5.8.6 Let $m_n = \min(S_0, S_1, \ldots, S_n)$, where S_n is the random walk defined previously. (i) Show that $S_n - m_n =_d W_n$. (ii) Let $\xi'_m = \xi_{n+1-m}$ for $1 \le m \le n$. Show that $S_n - m_n = \max(S'_0, S'_1, \ldots, S'_n)$. (iii) Conclude that as $n \to \infty$ we have $W_n \Rightarrow M \equiv \max(S'_0, S'_1, S'_2, \ldots)$.

5.8.7 In the discrete O.U. process, X_{n+1} is normal with mean θX_n and variance 1. What happens to the recurrence and transience if instead Y_{n+1} is normal with mean 0 and variance $\beta^2 |Y_n|$?

6

Ergodic Theorems

$X_n, n \geq 0$, is said to be a stationary sequence if for each $k \geq 1$, it has the same distribution as the shifted sequence $X_{n+k}, n \geq 0$. The basic fact about these sequences, called the ergodic theorem, is that if $E|f(X_0)| < \infty$, then

$$\lim_{n\to\infty} \frac{1}{n} \sum_{m=0}^{n-1} f(X_m) \quad \text{exists a.s.}$$

If X_n is ergodic (a generalization of the notion of irreducibility for Markov chains), then the limit is $Ef(X_0)$. Sections 6.1 and 6.2 develop the theory needed to prove the ergodic theorem. In Section 6.3 we apply the ergodic theorem to study the recurrence of random walks with increments that are stationary sequences finding remarkable generalizations of the i.i.d. case. In Section 6.4 we prove a subadditive ergodic theorem. As the examples in Sections 6.4 and 6.5 should indicate, this is a useful generalization of the ergodic theorem.

6.1 Definitions and Examples

$X_0, X_1, \dots$ is said to be a **stationary sequence** if for every k, the shifted sequence $\{X_{k+n}, n \geq 0\}$ has the same distribution, i.e., for each m, $(X_0, \dots, X_m)$ and $(X_k, \dots, X_{k+m})$ have the same distribution. We begin by giving four examples that will be our constant companions.

Example 6.1.1 $X_0, X_1, \dots$ are i.i.d.

Example 6.1.2 Let X_n be a Markov chain with transition probability $p(x, A)$ and stationary distribution π, i.e., $\pi(A) = \int \pi(dx)\, p(x, A)$. If X_0 has distribution π, then $X_0, X_1, \dots$ is a stationary sequence. A special case to keep in mind for counterexamples is the chain with state space $S = \{0, 1\}$ and transition probability $p(x, \{1-x\}) = 1$. In this case, the stationary distribution has $\pi(0) = \pi(1) = 1/2$ and $(X_0, X_1, \dots) = (0, 1, 0, 1, \dots)$ or $(1, 0, 1, 0, \dots)$ with probability $1/2$ each.

Example 6.1.3 (Rotation of the circle) Let $\Omega = [0, 1)$, $\mathcal{F} =$ Borel subsets, $P =$ Lebesgue measure. Let $\theta \in (0, 1)$, and for $n \geq 0$, let $X_n(\omega) = (\omega + n\theta) \bmod 1$, where $x \bmod 1 = x - [x]$, $[x]$ being the greatest integer $\leq x$. To see the reason for the name, map $[0, 1)$ into $\mathbf{C}$ by $x \to \exp(2\pi i x)$. This example is a special case of the last one. Let $p(x, \{y\}) = 1$ if $y = (x + \theta) \bmod 1$.

Our last "example" contains all other examples as a special case,

Example 6.1.4 Let $(\Omega, \mathcal{F}, P)$ be a probability space. A measurable map $\varphi : \Omega \to \Omega$ is said to be **measure preserving** if $P(\varphi^{-1}A) = P(A)$ for all $A \in \mathcal{F}$. Let φ^n be the nth iterate of φ defined inductively by $\varphi^n = \varphi(\varphi^{n-1})$ for $n \geq 1$, where $\varphi^0(\omega) = \omega$. We claim that if $X \in \mathcal{F}$, then $X_n(\omega) = X(\varphi^n\omega)$ defines a stationary sequence. To check this, let $B \in \mathcal{R}^{n+1}$ and $A = \{\omega : (X_0(\omega), \ldots, X_n(\omega)) \in B\}$. Then

$$P((X_k, \ldots, X_{k+n}) \in B) = P(\varphi^k\omega \in A) = P(\omega \in A) = P((X_0, \ldots, X_n) \in B)$$

The last example is more than an important example. In fact, it is the only example! If $Y_0, Y_1, \ldots$ is a stationary sequence taking values in a nice space, Kolmogorov's extension theorem, Theorem A.3.1, allows us to construct a measure P on sequence space $(S^{\{0,1,\ldots\}}, \mathcal{S}^{\{0,1,\ldots\}})$ so that the sequence $X_n(\omega) = \omega_n$ has the same distribution as that of $\{Y_n, n \geq 0\}$. If we let φ be the shift operator, i.e., $\varphi(\omega_0, \omega_1, \ldots) = (\omega_1, \omega_2, \ldots)$, and let $X(\omega) = \omega_0$, then φ is measure preserving and $X_n(\omega) = X(\varphi^n\omega)$.

In view of these observations, it suffices to give our definitions and prove our results in the setting of Example 6.1.4. Thus, our basic set up consists of

$(\Omega, \mathcal{F}, P)$	a probability space
φ	a map that preserves P
$X_n(\omega) = X(\varphi^n\omega)$	where X is a random variable

We will now give some important definitions. Here and in what follows we assume φ is measure-preserving. A set $A \in \mathcal{F}$ is said to be **invariant** if $\varphi^{-1}A = A$. (Here, as usual, two sets are considered to be equal if their symmetric difference has probability 0.) Let $\mathcal{I}$ be the collection of invariant events. In Exercise 6.1.1 you will prove that $\mathcal{I}$ is a σ-field.

A measure-preserving transformation on $(\Omega, \mathcal{F}, P)$ is said to be **ergodic** if $\mathcal{I}$ is trivial, i.e., for every $A \in \mathcal{I}$, $P(A) \in \{0, 1\}$. If φ is not ergodic, then the space can be split into two sets A and A^c, each having positive measure so that $\varphi(A) = A$ and $\varphi(A^c) = A^c$. In words, φ is not irreducible.

To investigate further the meaning of ergodicity, we return to our examples, renumbering them because the new focus is on checking ergodicity.

Example 6.1.5 (i.i.d. sequence) We begin by observing that if $\Omega = \mathbf{R}^{\{0,1,\ldots\}}$ and φ is the shift operator, then an invariant set A has $\{\omega : \omega \in A\} = \{\omega : \varphi\omega \in A\} \in \sigma(X_1, X_2, \ldots)$. Iterating gives

$$A \in \cap_{n=1}^{\infty} \sigma(X_n, X_{n+1}, \ldots) = \mathcal{T}, \quad \text{the tail } \sigma\text{-field}$$

so $\mathcal{I} \subset \mathcal{T}$. For an i.i.d. sequence, Kolmogorov's 0-1 law implies $\mathcal{T}$ is trivial, so $\mathcal{I}$ is trivial and the sequence is ergodic (i.e., when the corresponding measure is put on sequence space $\Omega = \mathbf{R}^{\{0,1,2,\ldots\}}$ the shift is).

Example 6.1.6 (Markov chains) Suppose the state space S is countable and the stationary distribution has $\pi(x) > 0$ for all $x \in S$. By Theorems 5.5.10 and 5.3.5, all states are recurrent, and we can write $S = \cup R_i$, where the R_i are disjoint irreducible closed sets. If $X_0 \in R_i$, then with probability one, $X_n \in R_i$ for all $n \geq 1$ so $\{\omega : X_0(\omega) \in R_i\} \in \mathcal{I}$. The last observation shows that if the Markov chain is not irreducible, then the sequence is not ergodic. To prove the converse, observe that if $A \in \mathcal{I}$, $1_A \circ \theta_n = 1_A$, where

$\theta_n(\omega_0, \omega_1, \ldots) = (\omega_n, \omega_{n+1}, \ldots)$. So if we let $\mathcal{F}_n = \sigma(X_0, \ldots, X_n)$, the shift invariance of 1_A and the Markov property imply

$$E_\pi(1_A|\mathcal{F}_n) = E_\pi(1_A \circ \theta_n|\mathcal{F}_n) = h(X_n)$$

where $h(x) = E_x 1_A$. Lévy's 0-1 law implies that the left-hand side converges to 1_A as $n \to \infty$. If X_n is irreducible and recurrent, then for any $y \in S$, the right-hand side $= h(y)$ i.o., so either $h(x) \equiv 0$ or $h(x) \equiv 1$, and $P_\pi(A) \in \{0, 1\}$. This example also shows that $\mathcal{I}$ and $\mathcal{T}$ may be different. When the transition probability p is irreducible, $\mathcal{I}$ is trivial, but if all the states have period $d > 1$, $\mathcal{T}$ is not. In Theorem 5.7.3, we showed that if $S_0, \ldots, S_{d-1}$ is the cyclic decomposition of S, then $\mathcal{T} = \sigma(\{X_0 \in S_r\} : 0 \le r < d)$.

Example 6.1.7 Rotation of the circle is not ergodic if $\theta = m/n$, where $m < n$ are positive integers. If B is a Borel subset of $[0, 1/n)$ and

$$A = \cup_{k=0}^{n-1}(B + k/n)$$

then A is invariant. Conversely, if θ is irrational, then φ is ergodic. To prove this, we need a fact from Fourier analysis. If f is a measurable function on $[0, 1)$ with $\int f^2(x)\,dx < \infty$, then f can be written as $f(x) = \sum_k c_k e^{2\pi ikx}$, where the equality is in the sense that as $K \to \infty$

$$\sum_{k=-K}^{K} c_k e^{2\pi ikx} \to f(x) \text{ in } L^2[0, 1)$$

and this is possible for only one choice of the coefficients $c_k = \int f(x)e^{-2\pi ikx}\,dx$. Now

$$f(\varphi(x)) = \sum_k c_k e^{2\pi ik(x+\theta)} = \sum_k (c_k e^{2\pi ik\theta})e^{2\pi ikx}$$

The uniqueness of the coefficients c_k implies that $f(\varphi(x)) = f(x)$ if and only if $c_k(e^{2\pi ik\theta} - 1) = 0$. If θ is irrational, this implies $c_k = 0$ for $k \ne 0$, so f is constant. Applying the last result to $f = 1_A$ with $A \in \mathcal{I}$ shows that $A = \emptyset$ or $[0, 1)$ a.s.

Exercises

6.1.1 Show that the class of invariant events $\mathcal{I}$ is a σ-field, and $X \in \mathcal{I}$ if and only if X is **invariant**, i.e., $X \circ \varphi = X$ a.s.

6.1.2 Some authors call A **almost invariant** if $P(A \Delta \varphi^{-1}(A)) = 0$ and call C **invariant in the strict sense** if $C = \varphi^{-1}(C)$. (i) Let A be any set, let $B = \cup_{n=0}^{\infty} \varphi^{-n}(A)$. Show $\varphi^{-1}(B) \subset B$. (ii) Let B be any set with $\varphi^{-1}(B) \subset B$ and let $C = \cap_{n=0}^{\infty} \varphi^{-n}(B)$. Show that $\varphi^{-1}(C) = C$. (iii) Show that A is almost invariant if and only if there is a C invariant in the strict sense with $P(A \Delta C) = 0$.

6.1.3 *A direct proof of ergodicity of rotation of the circle.* (i) Show that if θ is irrational, $x_n = n\theta \bmod 1$ is dense in [0,1). Hint: All the x_n are distinct, so for any $N < \infty$, $|x_n - x_m| \le 1/N$ for some $m < n \le N$. (ii) Use Exercise A.2.1 to show that if A is a Borel set with $|A| > 0$, then for any $\delta > 0$ there is an interval $J = [a, b)$ so that $|A \cap J| > (1 - \delta)|J|$. (iii) Combine this with (i) to conclude $P(A) = 1$.

6.1.4 Any stationary sequence $\{X_n\ ,\ n \geq 0\}$ can be embedded in a two-sided stationary sequence $\{Y_n : n \in \mathbf{Z}\}$.

6.1.5 If $X_0, X_1, \ldots$ is a stationary sequence and $g : \mathbf{R}^{\{0,1,\ldots\}} \to \mathbf{R}$ is measurable, then $Y_k = g(X_k, X_{k+1}, \ldots)$ is a stationary sequence. If X_n is ergodic then so is Y_n

6.1.6 **Independent blocks**. Let $X_1, X_2, \ldots$ be a stationary sequence. Let $n < \infty$ and let $Y_1, Y_2, \ldots$ be a sequence so that $(Y_{nk+1}, \ldots, Y_{n(k+1)})$, $k \geq 0$ are i.i.d. and $(Y_1, \ldots, Y_n) = (X_1, \ldots, X_n)$. Finally, let ν be uniformly distributed on $\{1, 2, \ldots, n\}$, independent of Y, and let $Z_m = Y_{\nu+m}$ for $m \geq 1$. Show that Z is stationary and ergodic.

6.1.7 **Continued fractions.** Let $\varphi(x) = 1/x - [1/x]$ for $x \in (0,1)$ and $A(x) = [1/x]$, where $[1/x] =$ the largest integer $\leq 1/x$. $a_n = A(\varphi^n x)$, $n = 0, 1, 2, \ldots$ gives the continued fraction representation of x, i.e.,

$$x = 1/(a_0 + 1/(a_1 + 1/(a_2 + 1/\ldots)))$$

Show that φ preserves $\mu(A) = \frac{1}{\log 2} \int_A \frac{dx}{1+x}$ for $A \subset (0,1)$.

6.2 Birkhoff's Ergodic Theorem

Throughout this section, φ is a measure-preserving transformation on $(\Omega, \mathcal{F}, P)$. See Example 6.1.4 for details. We begin by proving a result that is usually referred to as:

Theorem 6.2.1 (The ergodic theorem) *For any $X \in L^1$,*

$$\frac{1}{n} \sum_{m=0}^{n-1} X(\varphi^m \omega) \to E(X|\mathcal{I}) \quad \textit{a.s. and in } L^1$$

This result due to Birkhoff (1931) is sometimes called the pointwise or individual ergodic theorem because of the a.s. convergence in the conclusion. When the sequence is ergodic, the limit is the mean EX. In this case, if we take $X = 1_A$, it follows that the asymptotic fraction of time $\varphi^m \in A$ is $P(A)$.

The proof we give is based on an odd integration inequality due to Yosida and Kakutani (1939). We follow Garsia (1965). The proof is not intuitive, but none of the steps are difficult.

Lemma 6.2.2 (Maximal ergodic lemma) *Let $X_j(\omega) = X(\varphi^j \omega)$, $S_k(\omega) = X_0(\omega) + \cdots + X_{k-1}(\omega)$, and $M_k(\omega) = \max(0, S_1(\omega), \ldots, S_k(\omega))$. Then $E(X; M_k > 0) \geq 0$.*

Proof If $j \leq k$, then $M_k(\varphi\omega) \geq S_j(\varphi\omega)$, so adding $X(\omega)$ gives

$$X(\omega) + M_k(\varphi\omega) \geq X(\omega) + S_j(\varphi\omega) = S_{j+1}(\omega)$$

and rearranging we have

$$X(\omega) \geq S_{j+1}(\omega) - M_k(\varphi\omega) \text{ for } j = 1, \ldots, k$$

Trivially, $X(\omega) \geq S_1(\omega) - M_k(\varphi\omega)$, since $S_1(\omega) = X(\omega)$ and $M_k(\varphi\omega) \geq 0$. Therefore

$$E(X(\omega); M_k > 0) \geq \int_{\{M_k>0\}} \max(S_1(\omega), \ldots, S_k(\omega)) - M_k(\varphi\omega)\, dP$$
$$= \int_{\{M_k>0\}} M_k(\omega) - M_k(\varphi\omega)\, dP$$

Now $M_k(\omega) = 0$ and $M_k(\varphi\omega) \geq 0$ on $\{M_k > 0\}^c$, so the last expression is

$$\geq \int M_k(\omega) - M_k(\varphi\omega)\, dP = 0$$

since φ is measure preserving. □

Proof of Theorem 6.2.1. $E(X|\mathcal{I})$ is invariant under φ (see Exercise 6.1.1), so letting $X' = X - E(X|\mathcal{I})$ we can assume without loss of generality that $E(X|\mathcal{I}) = 0$. Let $\bar{X} = \limsup S_n/n$, let $\epsilon > 0$, and let $D = \{\omega : \bar{X}(\omega) > \epsilon\}$. Our goal is to prove that $P(D) = 0$. $\bar{X}(\varphi\omega) = \bar{X}(\omega)$, so $D \in \mathcal{I}$. Let

$$X^*(\omega) = (X(\omega) - \epsilon)1_D(\omega) \qquad S_n^*(\omega) = X^*(\omega) + \cdots + X^*(\varphi^{n-1}\omega)$$
$$M_n^*(\omega) = \max(0, S_1^*(\omega), \ldots, S_n^*(\omega)) \qquad F_n = \{M_n^* > 0\}$$
$$F = \cup_n F_n = \left\{ \sup_{k\geq 1} S_k^*/k > 0 \right\}$$

Since $X^*(\omega) = (X(\omega) - \epsilon)1_D(\omega)$ and $D = \{\limsup S_k/k > \epsilon\}$, it follows that

$$F = \left\{ \sup_{k\geq 1} S_k/k > \epsilon \right\} \cap D = D$$

Lemma 6.2.2 implies that $E(X^*; F_n) \geq 0$. Since $E|X^*| \leq E|X| + \epsilon < \infty$, the dominated convergence theorem implies $E(X^*; F_n) \to E(X^*; F)$, and it follows that $E(X^*; F) \geq 0$. The last conclusion looks innocent, but $F = D \in \mathcal{I}$, so it implies

$$0 \leq E(X^*; D) = E(X - \epsilon; D) = E(E(X|\mathcal{I}); D) - \epsilon P(D) = -\epsilon P(D)$$

since $E(X|\mathcal{I}) = 0$. The last inequality implies that

$$0 = P(D) = P(\limsup S_n/n > \epsilon)$$

and since $\epsilon > 0$ is arbitrary, it follows that $\limsup S_n/n \leq 0$. Applying the last result to $-X$ shows that $S_n/n \to 0$ a.s.

The clever part of the proof is over and the rest is routine. To prove that convergence occurs in L^1, let

$$X'_M(\omega) = X(\omega)1_{(|X(\omega)|\leq M)} \quad \text{and} \quad X''_M(\omega) = X(\omega) - X'_M(\omega)$$

The part of the ergodic theorem we have proved implies

$$\frac{1}{n}\sum_{m=0}^{n-1} X'_M(\varphi^m\omega) \to E(X'_M|\mathcal{I}) \quad \text{a.s.}$$

Since X'_M is bounded, the bounded convergence theorem implies

$$E\left|\frac{1}{n}\sum_{m=0}^{n-1} X'_M(\varphi^m\omega) - E(X'_M|\mathcal{I})\right| \to 0$$

To handle X''_M, we observe

$$E\left|\frac{1}{n}\sum_{m=0}^{n-1} X''_M(\varphi^m\omega)\right| \le \frac{1}{n}\sum_{m=0}^{n-1} E|X''_M(\varphi^m\omega)| = E|X''_M|$$

and $E|E(X''_M|\mathcal{I})| \le EE(|X''_M||\mathcal{I}) = E|X''_M|$. So

$$E\left|\frac{1}{n}\sum_{m=0}^{n-1} X''_M(\varphi^m\omega) - E(X''_M|\mathcal{I})\right| \le 2E|X''_M|$$

and it follows that

$$\limsup_{n\to\infty} E\left|\frac{1}{n}\sum_{m=0}^{n-1} X(\varphi^m\omega) - E(X|\mathcal{I})\right| \le 2E|X''_M|$$

As $M \to \infty$, $E|X''_M| \to 0$ by the dominated convergence theorem, which completes the proof. □

Our next step is to see what Theorem 6.2.2 says about our examples.

Example 6.2.3 (i.i.d. sequences) Since $\mathcal{I}$ is trivial, the ergodic theorem implies that

$$\frac{1}{n}\sum_{m=0}^{n-1} X_m \to EX_0 \quad \text{a.s. and in } L^1$$

The a.s. convergence is the strong law of large numbers.

Remark We can prove the L^1 convergence in the law of large numbers without invoking the ergodic theorem. To do this, note that

$$\frac{1}{n}\sum_{m=1}^{n} X_m^+ \to EX^+ \quad \text{a.s.} \qquad E\left(\frac{1}{n}\sum_{m=1}^{n} X_m^+\right) = EX^+$$

and use Theorem 4.6.3 to conclude that $\frac{1}{n}\sum_{m=1}^{n} X_m^+ \to EX^+$ in L^1. Similar results for the negative part and the triangle inequality now give the desired result.

Example 6.2.4 (Markov chains) Let X_n be an irreducible Markov chain on a countable state space that has a stationary distribution π. Let f be a function with

$$\sum_x |f(x)|\pi(x) < \infty$$

In Example 6.1.6, we showed that $\mathcal{I}$ is trivial, so applying the ergodic theorem to $f(X_0(\omega))$ gives

$$\frac{1}{n}\sum_{m=0}^{n-1} f(X_m) \to \sum_x f(x)\pi(x) \quad \text{a.s. and in } L^1$$

For another proof of the almost sure convergence, see Exercise 5.6.5.

Example 6.2.5 (Rotation of the circle) $\Omega = [0,1)$ $\varphi(\omega) = (\omega + \theta) \bmod 1$. Suppose that $\theta \in (0,1)$ is irrational so that by a result in Section 7.1 $\mathcal{I}$ is trivial. If we set $X(\omega) = 1_A(\omega)$, with A a Borel subset of [0,1), then the ergodic theorem implies

$$\frac{1}{n}\sum_{m=0}^{n-1} 1_{(\varphi^m\omega\in A)} \to |A| \quad \text{a.s.}$$

where $|A|$ denotes the Lebesgue measure of A. The last result for $\omega = 0$ is usually called **Weyl's equidistribution theorem**, although Bohl and Sierpinski should also get credit. For the history and a nonprobabilistic proof, see Hardy and Wright (1959), pp. 390–393.

To recover the number theoretic result, we will now show that:

Theorem 6.2.6 *If $A = [a,b)$, then the exceptional set is $\emptyset$.*

Proof Let $A_k = [a + 1/k, b - 1/k)$. If $b - a > 2/k$, the ergodic theorem implies

$$\frac{1}{n}\sum_{m=0}^{n-1} 1_{A_k}(\varphi^m\omega) \to b - a - \frac{2}{k}$$

for $\omega \in \Omega_k$ with $P(\Omega_k) = 1$. Let $G = \cap\Omega_k$, where the intersection is over integers k with $b-a > 2/k$. $P(G) = 1$, so G is dense in [0,1). If $x \in [0,1)$ and $\omega_k \in G$ with $|\omega_k - x| < 1/k$, then $\varphi^m\omega_k \in A_k$ implies $\varphi^m x \in A$, so

$$\liminf_{n\to\infty} \frac{1}{n}\sum_{m=0}^{n-1} 1_A(\varphi^m x) \geq b - a - \frac{2}{k}$$

for all large enough k. Noting that k is arbitrary and applying similar reasoning to A^c shows

$$\frac{1}{n}\sum_{m=0}^{n-1} 1_A(\varphi^m x) \to b - a$$

□

Example 6.2.7 (Benford's law) As Gelfand first observed, the equidistribution theorem says something interesting about 2^m. Let $\theta = \log_{10} 2$, $1 \leq k \leq 9$, and $A_k = [\log_{10} k, \log_{10}(k+1))$, where $\log_{10} y$ is the logarithm of y to the base 10. Taking $x = 0$ in the last result, we have

$$\frac{1}{n}\sum_{m=0}^{n-1} 1_A(\varphi^m 0) \to \log_{10}\left(\frac{k+1}{k}\right)$$

A little thought reveals that the first digit of 2^m is k if and only if $m\theta \bmod 1 \in A_k$. The numerical values of the limiting probabilities are

1	2	3	4	5	6	7	8	9
.3010	.1761	.1249	.0969	.0792	.0669	.0580	.0512	.0458

The limit distribution on $\{1, \ldots, 9\}$ is called Benford's (1938) law, although it was discovered by Newcomb (1881). As Raimi (1976) explains, in many tables the observed frequency with which k appears as a first digit is approximately $\log_{10}((k+1)/k)$. Some of the many examples that are supposed to follow Benford's law are: census populations of 3,259 counties, 308 numbers from Reader's Digest, areas of 335 rivers, 342 addresses of *American Men of Science*. The next table compares the percentages of the observations in the first five categories to Benford's law:

	1	2	3	4	5
Census	33.9	20.4	14.2	8.1	7.2
Reader's Digest	33.4	18.5	12.4	7.5	7.1
Rivers	31.0	16.4	10.7	11.3	7.2
Benford's Law	30.1	17.6	12.5	9.7	7.9
Addresses	28.9	19.2	12.6	8.8	8.5

The fits are far from perfect, but in each case Benford's law matches the general shape of the observed distribution. The IRS and other government agencies use Benford's law to detect fraud. When records are made up, the first digit distribution does not match Benford's law.

Exercises

6.2.1 Show that if $X \in L^p$ with $p > 1$, then the convergence in Theorem 6.2.1 occurs in L^p.

6.2.2 (i) Show that if $g_n(\omega) \to g(\omega)$ a.s. and $E(\sup_k |g_k(\omega)|) < \infty$, then

$$\lim_{n\to\infty} \frac{1}{n} \sum_{m=0}^{n-1} g_m(\varphi^m \omega) = E(g|\mathcal{I}) \quad \text{a.s.}$$

(ii) Show that if we suppose only that $g_n \to g$ in L^1, we get L^1 convergence.

6.2.3 **Wiener's maximal inequality.** Let $X_j(\omega) = X(\varphi^j \omega)$, $S_k(\omega) = X_0(\omega) + \cdots + X_{k-1}(\omega)$, $A_k(\omega) = S_k(\omega)/k$, and $D_k = \max(A_1, \ldots, A_k)$. Use Lemma 6.2.2 to show that if $\alpha > 0$, then

$$P(D_k > \alpha) \le \alpha^{-1} E|X|$$

6.3 Recurrence

In this section, we will study the recurrence properties of stationary sequences. Our first result is an application of the ergodic theorem. Let $X_1, X_2, \ldots$ be a stationary sequence taking values in $\mathbf{R}^d$, let $S_k = X_1 + \cdots + X_k$, let $A = \{S_k \neq 0 \text{ for all } k \ge 1\}$, and let $R_n = |\{S_1, \ldots, S_n\}|$ be the number of points visited at time n. Kesten, Spitzer, and Whitman, see Spitzer (1964), p. 40, proved the next result when the X_i are i.i.d. In that case, $\mathcal{I}$ is trivial, so the limit is $P(A)$.

Theorem 6.3.1 *As $n \to \infty$, $R_n/n \to E(1_A|\mathcal{I})$ a.s.*

Proof Suppose $X_1, X_2, \ldots$ are constructed on $(\mathbf{R}^d)^{\{0,1,\ldots\}}$ with $X_n(\omega) = \omega_n$, and let φ be the shift operator. It is clear that

$$R_n \geq \sum_{m=1}^{n} 1_A(\varphi^m \omega)$$

since the right-hand side $= |\{m : 1 \leq m \leq n, S_\ell \neq S_m \text{ for all } \ell > m\}|$. Using the ergodic theorem now gives

$$\liminf_{n\to\infty} R_n/n \geq E(1_A|\mathcal{I}) \quad \text{a.s.}$$

To prove the opposite inequality, let $A_k = \{S_1 \neq 0, S_2 \neq 0, \ldots, S_k \neq 0\}$. It is clear that

$$R_n \leq k + \sum_{m=1}^{n-k} 1_{A_k}(\varphi^m \omega)$$

since the sum on the right-hand side $= |\{m : 1 \leq m \leq n-k, S_\ell \neq S_m \text{ for } m < \ell \leq m+k\}|$. Using the ergodic theorem now gives

$$\limsup_{n\to\infty} R_n/n \leq E(1_{A_k}|\mathcal{I})$$

As $k \uparrow \infty$, $A_k \downarrow A$, so the monotone convergence theorem for conditional expectations, (c) in Theorem 4.1.9, implies

$$E(1_{A_k}|\mathcal{I}) \downarrow E(1_A|\mathcal{I}) \quad \text{as } k \uparrow \infty$$

and the proof is complete. □

From Theorem 6.3.1, we get a result about the recurrence of random walks with stationary increments that is (for integer valued random walks) a generalization of a result of Chung-Fuchs, Theorem 5.4.8.

Theorem 6.3.2 *Let $X_1, X_2, \ldots$ be a stationary sequence taking values in $\mathbf{Z}$ with $E|X_i| < \infty$. Let $S_n = X_1 + \cdots + X_n$, and let $A = \{S_1 \neq 0, S_2 \neq 0, \ldots\}$. (i) If $E(X_1|\mathcal{I}) = 0$, then $P(A) = 0$. (ii) If $P(A) = 0$, then $P(S_n = 0 \text{ i.o.}) = 1$.*

Remark In words, mean zero implies recurrence. The condition $E(X_1|\mathcal{I}) = 0$ is needed to rule out trivial examples that have mean 0 but are a combination of a sequence with positive and negative means, e.g., $P(X_n = 1 \text{ for all } n) = P(X_n = -1 \text{ for all } n) = 1/2$.

Proof If $E(X_1|\mathcal{I}) = 0$, then the ergodic theorem implies $S_n/n \to 0$ a.s. Now

$$\limsup_{n\to\infty}\left(\max_{1\leq k\leq n} |S_k|/n\right) = \limsup_{n\to\infty}\left(\max_{K\leq k\leq n} |S_k|/n\right) \leq \left(\max_{k\geq K} |S_k|/k\right)$$

for any K and the right-hand side $\downarrow 0$ as $K \uparrow \infty$. The last conclusion leads easily to

$$\lim_{n\to\infty}\left(\max_{1\leq k\leq n} |S_k|\right)\Big/ n = 0$$

Since $R_n \leq 1 + 2\max_{1\leq k\leq n} |S_k|$, it follows that $R_n/n \to 0$ and Theorem 6.3.1 implies $P(A) = 0$.

Let $F_j = \{S_i \neq 0 \text{ for } 1 \leq i < j, S_j = 0\}$ and

$$G_{j,k} = \{S_{j+i} - S_j \neq 0 \text{ for } 1 \leq i < k,\ S_{j+k} - S_j = 0\}.$$

$P(A) = 0$ implies that $\sum P(F_k) = 1$. Stationarity implies $P(G_{j,k}) = P(F_k)$, and for fixed j the $G_{j,k}$ are disjoint, so $\cup_k G_{j,k} = \Omega$ a.s. It follows that

$$\sum_k P(F_j \cap G_{j,k}) = P(F_j) \quad \text{and} \quad \sum_{j,k} P(F_j \cap G_{j,k}) = 1$$

On $F_j \cap G_{j,k}$, $S_j = 0$ and $S_{j+k} = 0$, so we have shown $P(S_n = 0 \text{ at least two times }) = 1$. Repeating the last argument shows $P(S_n = 0 \text{ at least } k \text{ times}) = 1$ for all k, and the proof is complete. □

Extending the reasoning in the proof of part (ii) of Theorem 6.3.2 gives a result of Kac (1947b). Let $X_0, X_1, \ldots$ be a stationary sequence taking values in $(S, \mathcal{S})$. Let $A \in \mathcal{S}$, let $T_0 = 0$, and for $n \geq 1$, let $T_n = \inf\{m > T_{n-1} : X_m \in A\}$ be the time of the nth return to A.

Theorem 6.3.3 *If $P(X_n \in A$ at least once$) = 1$, then under $P(\cdot|X_0 \in A)$, $t_n = T_n - T_{n-1}$ is a stationary sequence with $E(T_1|X_0 \in A) = 1/P(X_0 \in A)$.*

Remark If X_n is an irreducible Markov chain on a countable state space S starting from its stationary distribution π, and $A = \{x\}$, then Theorem 6.3.3 says $E_x T_x = 1/\pi(x)$, which is Theorem 5.5.11. Theorem 6.3.3 extends that result to an arbitrary $A \subset S$ and drops the assumption that X_n is a Markov chain.

Proof We first show that under $P(\cdot|X_0 \in A)$, $t_1, t_2, \ldots$ is stationary. To cut down on ...'s, we will only show that

$$P(t_1 = m, t_2 = n|X_0 \in A) = P(t_2 = m, t_3 = n|X_0 \in A)$$

It will be clear that the same proof works for any finite-dimensional distribution. Our first step is to extend $\{X_n, n \geq 0\}$ to a two-sided stationary sequence $\{X_n, n \in \mathbf{Z}\}$ using Theorem 6.1.4. Let $C_k = \{X_{-1} \notin A, \ldots, X_{-k+1} \notin A, X_{-k} \in A\}$.

$$\left(\cup_{k=1}^K C_k\right)^c = \{X_k \notin A \text{ for } -K \leq k \leq -1\}$$

The last event has the same probability as $\{X_k \notin A \text{ for } 1 \leq k \leq K\}$, so letting $K \to \infty$, we get $P\left(\cup_{k=1}^\infty C_k\right) = 1$. To prove the desired stationarity, we let $I_{j,k} = \{i \in [j,k] : X_i \in A\}$ and observe that

$$\begin{aligned} P(t_2 = m, t_3 = n, X_0 \in A) &= \sum_{\ell=1}^\infty P(X_0 \in A, t_1 = \ell, t_2 = m, t_3 = n) \\ &= \sum_{\ell=1}^\infty P(C_\ell, X_0 \in A, t_1 = m, t_2 = n) \end{aligned}$$

To complete the proof, we compute

$$E(t_1|X_0 \in A) = \sum_{k=1}^{\infty} P(t_1 \ge k|X_0 \in A) = P(X_0 \in A)^{-1} \sum_{k=1}^{\infty} P(t_1 \ge k, X_0 \in A)$$
$$= P(X_0 \in A)^{-1} \sum_{k=1}^{\infty} P(C_k) = 1/P(X_0 \in A)$$

since the C_k are disjoint and their union has probability 1. □

Exercises

6.3.1 Let $g_n = P(S_1 \neq 0, \ldots, S_n \neq 0)$ for $n \ge 1$ and $g_0 = 1$. Show that

$$ER_n = \sum_{m=1}^{n} g_{m-1}.$$

6.3.2 Imitate the proof of (i) in Theorem 6.3.2 to show that if we assume $P(X_i > 1) = 0$, $EX_i > 0$, and the sequence X_i is ergodic in addition to the hypotheses of Theorem 6.3.2, then $P(A) = EX_i$.

6.3.3 Show that if $P(X_n \in A \text{ at least once}) = 1$ and $A \cap B = \emptyset$, then

$$E\left(\sum_{1 \le m \le T_1} 1_{(X_m \in B)} \middle| X_0 \in A \right) = \frac{P(X_0 \in B)}{P(X_0 \in A)}$$

When $A = \{x\}$ and X_n is a Markov chain, this is the "cycle trick" for defining a stationary measure. See Theorem 5.5.7.

6.3.4 Consider the special case in which $X_n \in \{0, 1\}$, and let $\bar{P} = P(\cdot|X_0 = 1)$. Here $A = \{1\}$ and so $T_1 = \inf\{m > 0 : X_m = 1\}$. Show $P(T_1 = n) = \bar{P}(T_1 \ge n)/\bar{E}T_1$. When $t_1, t_2, \ldots$ are i.i.d., this reduces to the formula for the first waiting time in a stationary renewal process.

6.4 A Subadditive Ergodic Theorem

In this section we will prove Liggett's (1985) version of Kingman's (1968):

Theorem 6.4.1 (Subadditive ergodic theorem) *Suppose $X_{m,n}$, $0 \le m < n$ satisfy:*

(i) $X_{0,m} + X_{m,n} \ge X_{0,n}$

(ii) $\{X_{nk,(n+1)k}, n \ge 1\}$ *is a stationary sequence for each* k.

(iii) The distribution of $\{X_{m,m+k}, k \ge 1\}$ *does not depend on* m.

(iv) $EX^+_{0,1} < \infty$ *and for each* n, $EX_{0,n} \ge \gamma_0 n$, *where* $\gamma_0 > -\infty$.

Then

(a) $\lim_{n\to\infty} EX_{0,n}/n = \inf_m EX_{0,m}/m \equiv \gamma$

(b) $X = \lim_{n\to\infty} X_{0,n}/n$ *exists a.s. and in* L^1, *so* $EX = \gamma$.

(c) If all the stationary sequences in (ii) are ergodic, then $X = \gamma$ *a.s.*

Remark Kingman assumed (iv), but instead of (i)–(iii) he assumed that $X_{\ell,m}+X_{m,n} \geq X_{\ell,n}$ for all $\ell < m < n$ and that the distribution of $\{X_{m+k,n+k}, 0 \leq m < n\}$ does not depend on k. In the last application in the next section, these stronger conditions do not hold.

Before giving the proof, which is somewhat lengthy, we will consider several examples for motivation. Since the validity of (ii) and (iii) in each case is clear, we will only check (i) and (iv). The first example shows that Theorem 6.4.1 contains Birkhoff's ergodic theorem as a special case.

Example 6.4.2 (Stationary sequences) Suppose $\xi_1, \xi_2, \ldots$ is a stationary sequence with $E|\xi_k| < \infty$, and let $X_{m,n} = \xi_{m+1} + \cdots + \xi_n$. Then $X_{0,n} = X_{0,m} + X_{m,n}$, and (iv) holds.

Example 6.4.3 (Range of random walk) Suppose $\xi_1, \xi_2, \ldots$ is a stationary sequence and let $S_n = \xi_1 + \cdots + \xi_n$. Let $X_{m,n} = |\{S_{m+1}, \ldots, S_n\}|$. It is clear that $X_{0,m} + X_{m,n} \geq X_{0,n}$. $0 \leq X_{0,n} \leq n$, so (iv) holds. Applying Theorem 6.4.1 now gives $X_{0,n}/n \to X$ a.s. and in L^1, but it does not tell us what the limit is.

Example 6.4.4 (Longest common subsequences) Given are ergodic stationary sequences $X_1, X_2, X_3, \ldots$ and Y_1, Y_2, Y_3. Let $L_{m,n} = \max\{K : X_{i_k} = Y_{j_k}$ for $1 \leq k \leq K$, where $m < i_1 < i_2 \ldots < i_K \leq n$ and $m < j_1 < j_2 \ldots < j_K \leq n\}$. It is clear that

$$L_{0,m} + L_{m,n} \leq L_{0,n}$$

so $X_{m,n} = -L_{m,n}$ is subadditive. $0 \leq L_{0,n} \leq n$ so (iv) holds. Applying Theorem 6.4.1 now, we conclude that

$$L_{0,n}/n \to \gamma = \sup_{m\geq 1} E(L_{0,m}/m)$$

The examples above should provide enough motivation for now. In the next section, we will give four more applications of Theorem 6.4.1.

Proof of Theorem 6.4.1. There are four steps. The first, second, and fourth date back to Kingman (1968). The half dozen proofs of subadditive ergodic theorems that exist all do the crucial third step in a different way. Here we use the approach of S. Leventhal (1988), who in turn based his proof on Katznelson and Weiss (1982).

Step 1. The first thing to check is that $E|X_{0,n}| \leq Cn$. To do this, we note that (i) implies $X^+_{0,m} + X^+_{m,n} \geq X^+_{0,n}$. Repeatedly using the last inequality and invoking (iii) gives $EX^+_{0,n} \leq nEX^+_{0,1} < \infty$. Since $|x| = 2x^+ - x$, it follows from (iv) that

$$E|X_{0,n}| \leq 2EX^+_{0,n} - EX_{0,n} \leq Cn < \infty$$

Let $a_n = EX_{0,n}$. (i) and (iii) imply that

$$a_m + a_{n-m} \geq a_n \tag{6.4.1}$$

From this, it follows easily that

$$a_n/n \to \inf_{m\geq 1} a_m/m \equiv \gamma \tag{6.4.2}$$

To prove this, we observe that the liminf is clearly $\geq \gamma$, so all we have to show is that the limsup $\leq a_m/m$ for any m. The last fact is easy, for if we write $n = km + \ell$ with $0 \leq \ell < m$, then repeated use of (6.4.1) gives $a_n \leq ka_m + a_\ell$. Dividing by $n = km + \ell$ gives

$$\frac{a_n}{n} \leq \frac{km}{km+\ell} \cdot \frac{a_m}{m} + \frac{a_\ell}{n}$$

Letting $n \to \infty$ and recalling $0 \leq \ell < m$ gives (6.4.2) and proves (a) in Theorem 6.4.1.

Step 2. Making repeated use of (i), we get

$$X_{0,n} \leq X_{0,km} + X_{km,n}$$
$$X_{0,n} \leq X_{0,(k-1)m} + X_{(k-1)m,km} + X_{km,n}$$

and so on until the first term on the right is $X_{0,m}$. Dividing by $n = km + \ell$ then gives

$$\frac{X_{0,n}}{n} \leq \frac{k}{km+\ell} \cdot \frac{X_{0,m} + \cdots + X_{(k-1)m,km}}{k} + \frac{X_{km,n}}{n} \tag{6.4.3}$$

Using (ii) and the ergodic theorem now gives that

$$\frac{X_{0,m} + \cdots + X_{(k-1)m,km}}{k} \to A_m \quad \text{a.s. and in } L^1$$

where $A_m = E(X_{0,m}|\mathcal{I}_m)$ and the subscript indicates that $\mathcal{I}_m$ is the shift invariant σ-field for the sequence $X_{(k-1)m,km}$, $k \geq 1$. The exact formula for the limit is not important, but we will need to know later that $EA_m = EX_{0,m}$.

If we fix ℓ and let $\epsilon > 0$, then (iii) implies

$$\sum_{k=1}^{\infty} P(X_{km,km+\ell} > (km+\ell)\epsilon) \leq \sum_{k=1}^{\infty} P(X_{0,\ell} > k\epsilon) < \infty$$

since $EX_{0,\ell}^+ < \infty$ by the result at the beginning of Step 1. The last two observations imply

$$\overline{X} \equiv \limsup_{n\to\infty} X_{0,n}/n \leq A_m/m \tag{6.4.4}$$

Taking expected values now gives $E\overline{X} \leq E(X_{0,m}/m)$, and taking the infimum over m, we have $E\overline{X} \leq \gamma$. Note that if all the stationary sequences in (ii) are ergodic, we have $\overline{X} \leq \gamma$.

Remark If (i)–(iii) hold, $EX_{0,1}^+ < \infty$, and $\inf EX_{0,m}/m = -\infty$, then it follows from the last argument that as $X_{0,n}/n \to -\infty$ a.s. as $n \to \infty$.

Step 3. The next step is to let

$$\underline{X} = \liminf_{n\to\infty} X_{0,n}/n$$

and show that $E\underline{X} \geq \gamma$. Since $\infty > EX_{0,1} \geq \gamma \geq \gamma_0 > -\infty$, and we have shown in Step 2 that $E\overline{X} \leq \gamma$, it will follow that $\underline{X} = \overline{X}$, i.e., the limit of $X_{0,n}/n$ exists a.s. Let

$$\underline{X}_m = \liminf_{n\to\infty} X_{m,m+n}/n$$

(i) implies

$$X_{0,m+n} \le X_{0,m} + X_{m,m+n}$$

Dividing both sides by n and letting $n \to \infty$ gives $\underline{X} \le \underline{X}_m$ a.s. However, (iii) implies that $\underline{X}_m$ and $\underline{X}$ have the same distribution, so $\underline{X} = \underline{X}_m$ a.s.

Let $\epsilon > 0$ and let $Z = \epsilon + (\underline{X} \vee -M)$. Since $\underline{X} \le \overline{X}$ and $E\overline{X} \le \gamma < \infty$ by Step 2, $E|Z| < \infty$. Let

$$Y_{m,n} = X_{m,n} - (n-m)Z$$

Y satisfies (i)–(iv), since $Z_{m,n} = -(n-m)Z$ does, and has

$$\underline{Y} \equiv \liminf_{n\to\infty} Y_{0,n}/n \le -\epsilon \tag{6.4.5}$$

Let $T_m = \min\{n \ge 1 : Y_{m,m+n} \le 0\}$. (iii) implies $T_m =_d T_0$ and

$$E(Y_{m,m+1}; T_m > N) = E(Y_{0,1}; T_0 > N)$$

(6.4.5) implies that $P(T_0 < \infty) = 1$, so we can pick N large enough so that

$$E(Y_{0,1}; T_0 > N) \le \epsilon$$

Let

$$S_m = \begin{cases} T_m & \text{on } \{T_m \le N\} \\ 1 & \text{on } \{T_m > N\} \end{cases}$$

This is not a stopping time but there is nothing special about stopping times for a stationary sequence! Let

$$\xi_m = \begin{cases} 0 & \text{on } \{T_m \le N\} \\ Y_{m,m+1} & \text{on } \{T_m > N\} \end{cases}$$

Since $Y(m, m+T_m) \le 0$ always and we have $S_m = 1$, $Y_{m,m+1} > 0$ on $\{T_m > N\}$, we have $Y(m, m+S_m) \le \xi_m$ and $\xi_m \ge 0$. Let $R_0 = 0$, and for $k \ge 1$, let $R_k = R_{k-1} + S(R_{k-1})$. Let $K = \max\{k : R_k \le n\}$. From (i), it follows that

$$Y(0,n) \le Y(R_0, R_1) + \cdots + Y(R_{K-1}, R_K) + Y(R_K, n)$$

Since $\xi_m \ge 0$ and $n - R_K \le N$, the last quantity is

$$\le \sum_{m=0}^{n-1} \xi_m + \sum_{j=1}^{N} |Y_{n-j,n-j+1}|$$

Here we have used (i) on $Y(R_K, n)$. Dividing both sides by n, taking expected values, and letting $n \to \infty$ gives

$$\limsup_{n\to\infty} EY_{0,n}/n \le E\xi_0 \le E(Y_{0,1}; T_0 > N) \le \epsilon$$

It follows from (a) and the definition of $Y_{0,n}$ that

$$\gamma = \lim_{n\to\infty} EX_{0,n}/n \le 2\epsilon + E(\underline{X} \vee -M)$$

Since $\epsilon > 0$ and M are arbitrary, it follows that $E\underline{X} \ge \gamma$ and Step 3 is complete.

Step 4. It only remains to prove convergence in L^1. Let $\Gamma_m = A_m/m$ be the limit in (6.4.4), recall $E\Gamma_m = E(X_{0,m}/m)$, and let $\Gamma = \inf \Gamma_m$. Observing that $|z| = 2z^+ - z$ (consider two cases $z \geq 0$ and $z < 0$), we can write

$$E|X_{0,n}/n - \Gamma| = 2E(X_{0,n}/n - \Gamma)^+ - E(X_{0,n}/n - \Gamma) \leq 2E(X_{0,n}/n - \Gamma)^+$$

since

$$E(X_{0,n}/n) \geq \gamma = \inf E\Gamma_m \geq E\Gamma$$

Using the trivial inequality $(x+y)^+ \leq x^+ + y^+$ and noticing $\Gamma_m \geq \Gamma$ now gives

$$E(X_{0,n}/n - \Gamma)^+ \leq E(X_{0,n}/n - \Gamma_m)^+ + E(\Gamma_m - \Gamma)$$

Now $E\Gamma_m \to \gamma$ as $m \to \infty$ and $E\Gamma \geq E\bar{X} \geq E\underline{X} \geq \gamma$ by steps 2 and 3, so $E\Gamma = \gamma$, and it follows that $E(\Gamma_m - \Gamma)$ is small if m is large. To bound the other term, observe that (i) implies

$$\begin{aligned} E(X_{0,n}/n - \Gamma_m)^+ \leq\ & E\left(\frac{X(0,m) + \cdots + X((k-1)m, km)}{km + \ell} - \Gamma_m\right)^+ \\ & + E\left(\frac{X(km,n)}{n}\right)^+ \end{aligned}$$

The second term $= E(X^+_{0,\ell}/n) \to 0$ as $n \to \infty$. For the first, we observe $y^+ \leq |y|$, and the ergodic theorem implies

$$E\left|\frac{X(0,m) + \cdots + X((k-1)m, km)}{k} - \Gamma_m\right| \to 0$$

so the proof of Theorem 6.4.1 is complete. □

6.5 Applications

In this section, we will give three applications of our subadditive ergodic theorem, 6.4.1. These examples are independent of each other and can be read in any order. In Example 6.5.5, we encounter situations to which Liggett's version applies but Kingman's version does not.

Example 6.5.1 (Products of random matrices) Suppose $A_1, A_2, \ldots$ is a stationary sequence of $k \times k$ matrices with positive entries and let

$$\alpha_{m,n}(i,j) = (A_{m+1} \cdots A_n)(i,j),$$

i.e., the entry in row i of column j of the product. It is clear that

$$\alpha_{0,m}(1,1)\alpha_{m,n}(1,1) \leq \alpha_{0,n}(1,1)$$

so if we let $X_{m,n} = -\log \alpha_{m,n}(1,1)$, then $X_{0,m} + X_{m,n} \geq X_{0,n}$. To check (iv), we observe that

$$\prod_{m=1}^{n} A_m(1,1) \leq \alpha_{0,n}(1,1) \leq k^{n-1} \prod_{m=1}^{n} \left(\sup_{i,j} A_m(i,j)\right)$$

or taking logs

$$-\sum_{m=1}^{n} \log A_m(1,1) \geq X_{0,n} \geq -(n \log k) - \sum_{m=1}^{n} \log \left(\sup_{i,j} A_m(i,j) \right)$$

So if $E \log A_m(1,1) > -\infty$, then $EX_{0,1}^+ < \infty$, and if

$$E \log \left(\sup_{i,j} A_m(i,j) \right) < \infty$$

then $EX_{0,n}^- \leq \gamma_0 n$. If we observe that

$$P \left(\log \left(\sup_{i,j} A_m(i,j) \right) \geq x \right) \leq \sum_{i,j} P \left(\log A_m(i,j) \geq x \right)$$

we see that it is enough to assume that

$$(*) \qquad E|\log A_m(i,j)| < \infty \quad \text{for all } i,j$$

When $(*)$ holds, applying Theorem 6.4.1 gives $X_{0,n}/n \to X$ a.s. Using the strict positivity of the entries, it is easy to improve that result to

$$\frac{1}{n} \log \alpha_{0,n}(i,j) \to -X \quad \text{a.s. for all } i,j \tag{6.5.1}$$

a result first proved by Furstenberg and Kesten (1960). □

An alternative approach is to let

$$\|A\| = \max_i \sum_j |A(i,j)| = \max\{\|xA\|_1 : \|x\|_1 = 1\},$$

where $(xA)_j = \sum_i x_i A(i,j)$ and $\|x\|_1 = |x_1| + \cdots + |x_k|$. It is clear that $\|AB\| \leq \|A\| \cdot \|B\|$, so if we let

$$\beta_{m,n} = \|A_{m+1} \cdots A_n\|$$

and $Y_{m,n} = \log \beta_{m,n}$, then $Y_{m,n}$ is subadditive. It is easy to use the subadditinve ergodic theorem to conclude

$$\frac{1}{n} \log \|A_{m+1} \cdots A_n\| \to -X \quad \text{a.s.}$$

where X is the limit of $X_{0,n}/n$. From this we see that

$$\sup_{m \geq 1} (E \log \alpha_{0,m})/m = -X = \inf_{m \geq 1} (E \log \beta_{0,m})/m$$

Example 6.5.2 (Increasing sequences in random permutations) Let π be a permutation of $\{1, 2, \ldots, n\}$ and let $\ell(\pi)$ be the length of the longest increasing sequence in π. That is, the largest k for which there are integers $i_1 < i_2 \ldots < i_k$ so that $\pi(i_1) < \pi(i_2) < \ldots < \pi(i_k)$. Hammersley (1970) attacked this problem by putting a rate one Poisson process in the plane, and for $s < t \in [0, \infty)$, letting $Y_{s,t}$ denote the length of the longest increasing path lying in the square $R_{s,t}$ with vertices (s,s), (s,t), (t,t), and (t,s). That is, the largest k for which there are points (x_i, y_i) in the Poisson process with $s < x_1 < \ldots < x_k < t$ and

$s < y_1 < \ldots < y_k < t$. It is clear that $Y_{0,m} + Y_{m,n} \le Y_{0,n}$. Applying Theorem 6.4.1 to $-Y_{0,n}$ shows

$$Y_{0,n}/n \to \gamma \equiv \sup_{m \ge 1} EY_{0,m}/m \quad \text{a.s.}$$

For each k, $Y_{nk,(n+1)k}$, $n \ge 0$ is i.i.d., so the limit is constant. We will show that $\gamma < \infty$ in Exercise 6.5.4.

To get from the result about the Poisson process back to the random permutation problem, let $\tau(n)$ be the smallest value of t for which there are n points in $R_{0,t}$. Let the n points in $R_{0,\tau(n)}$ be written as (x_i, y_i), where $0 < x_1 < x_2 \ldots < x_n \le \tau(n)$ and let π_n be the unique permutation of $\{1,2,\ldots,n\}$ so that $y_{\pi_n(1)} < y_{\pi_n(2)} \ldots < y_{\pi_n(n)}$. It is clear that $Y_{0,\tau(n)} = \ell(\pi_n)$. An easy argument shows:

Lemma 6.5.3 $\tau(n)/\sqrt{n} \to 1$ *a.s.*

Proof Let S_n be the number of points in $R_{0,\sqrt{n}}$. $S_n - S_{n-1}$ are independent Poisson r.v.'s with mean 1, so the strong law of large numbers implies $S_n/n \to 1$ a.s. If $\epsilon > 0$, then for large n, $S_{n(1-\epsilon)} < n < S_{n(1+\epsilon)}$ and hence $\sqrt{(1-\epsilon)n} \le \tau(n) \le \sqrt{(1+\epsilon)n}$. □

It follows from Lemma 6.5.3 and the monotonicity of $m \to Y_{0,m}$ that

$$n^{-1/2}\ell(\pi_n) \to \gamma \quad \text{a.s.}$$

Hammersley (1970) has a proof that $\pi/2 \le \gamma \le e$, and Kingman (1973) shows that $1.59 < \gamma < 2.49$. See Exercises 6.5.3 and 6.5.4. Subsequent work on the random permutation problem, see Logan and Shepp (1977) and Vershik and Kerov (1977), has shown that $\gamma = 2$.

Example 6.5.4 (First passage percolation) Consider $\mathbf{Z}^d$ as a graph with edges connecting each $x, y \in \mathbf{Z}^d$ with $|x - y| = 1$. Assign an independent nonnegative random variable $\tau(e)$ to each edge that represents the time required to traverse the edge going in either direction. If e is the edge connecting x and y, let $\tau(x,y) = \tau(y,x) = \tau(e)$. If $x_0 = x, x_1, \ldots, x_n = y$ is a path from x to y, i.e., a sequence with $|x_m - x_{m-1}| = 1$ for $1 \le m \le n$, we define the **travel time** for the path to be $\tau(x_0,x_1) + \cdots + \tau(x_{n-1},x_n)$. Define the **passage time** from x to y, $t(x,y) =$ the infimum of the travel times over all paths from x to y. Let $z \in \mathbf{Z}^d$ and let $X_{m,n} = t(mu, nu)$, where $u = (1,0,\ldots,0)$.

Clearly $X_{0,m} + X_{m,n} \ge X_{0,n}$. $X_{0,n} \ge 0$, so if $E\tau(x,y) < \infty$, then (iv) holds, and Theorem 6.4.1 implies that $X_{0,n}/n \to X$ a.s. To see that the limit is constant, enumerate the edges in some order $e_1, e_2, \ldots$ and observe that X is measurable with respect to the tail σ-field of the i.i.d. sequence $\tau(e_1), \tau(e_2), \ldots$

It is not hard to see that the assumption of finite first moment can be weakened. If τ has distribution F with

$$(*) \qquad \int_0^\infty (1 - F(x))^{2d}\, dx < \infty$$

i.e., the minimum of $2d$ independent copies has finite mean, then by finding $2d$ disjoint paths from 0 to $u = (1,0,\ldots,0)$, one concludes that $E\tau(0,u) < \infty$ and (6.1) can be applied.

The condition $(*)$ is also necessary for $X_{0,n}/n$ to converge to a finite limit. If $(*)$ fails and Y_n is the minimum of $t(e)$ over all the edges from v, then

$$\limsup_{n\to\infty} X_{0,n}/n \geq \limsup_{n\to\infty} Y_n/n = \infty \quad \text{a.s.}$$

Example 6.5.5 (Age-dependent branching processes) This is a variation of the branching process introduced in which each individual lives for an amount of time with distribution F before producing k offspring with probability p_k. The description of the process is completed by supposing that the process starts with one individual in generation 0 who is born at time 0, and when this particle dies, its offspring start independent copies of the original process.

Suppose $p_0 = 0$, let $X_{0,m}$ be the birth time of the first member of generation m, and let $X_{m,n}$ be the time lag necessary for that individual to have an offspring in generation n. In case of ties, pick an individual at random from those in generation m born at time $X_{0,m}$. It is clear that $X_{0,n} \leq X_{0,m} + X_{m,n}$. Since $X_{0,n} \geq 0$, (iv) holds if we assume F has finite mean. Applying Theorem 6.4.1 now, it follows that

$$X_{0,n}/n \to \gamma \quad \text{a.s.}$$

The limit is constant because the sequences $\{X_{nk,(n+1)k}, n \geq 0\}$ are i.i.d.

Remark The inequality $X_{\ell,m} + X_{m,n} \geq X_{\ell,n}$ is false when $\ell > 0$, because if we call i_m the individual that determines the value of $X_{m,n}$ for $n > m$, then i_m may not be a descendant of i_ℓ.

As usual, one has to use other methods to identify the constant. Let $t_1, t_2, \ldots$ be i.i.d. with distribution F, let $T_n = t_1 + \cdots + t_n$, and $\mu = \sum k p_k$. Let $Z_n(an)$ be the number of individuals in generation n born by time an. Each individual in generation n has probability $P(T_n \leq an)$ to be born by time an, and the times are independent of the offspring numbers so

$$EZ_n(an) = EE(Z_n(an)|Z_n) = E(Z_n P(T_n \leq an)) = \mu^n P(T_n \leq an)$$

By results in Section 2.6, $n^{-1} \log P(T_n \leq an) \to -c(a)$ as $n \to \infty$. If $\log \mu - c(a) < 0$, then Chebyshev's inequality and the Borel-Cantelli lemma imply $P(Z_n(an) \geq 1 \text{ i.o.}) = 0$. Conversely, if $EZ_n(an) > 1$ for some n, then we can define a supercritical branching process Y_m that consists of the offspring in generation mn that are descendants of individuals in Y_{m-1} in generation $(m-1)n$ that are born less than an units of time after their parents. This shows that with positive probability, $X_{0,mn} \leq mna$ for all m. Combining the last two observations with the fact that $c(a)$ is strictly increasing gives

$$\gamma = \inf\{a : \log \mu - c(a) > 0\}$$

The last result is from Biggins (1977). See his (1978) and (1979) papers for extensions and refinements.

Exercises

6.5.1 To show that the convergence in (a) of Theorem 6.4.1 may occur arbitrarily slowly, let $X_{m,m+k} = f(k) \geq 0$, where $f(k)/k$ is decreasing, and check that $X_{m,m+k}$ is subadddtive.

6.5.2 Consider the longest common subsequence problem, Example 6.4.4 when $X_1, X_2, \dots$ and $Y_1, Y_2, \dots$ are i.i.d. and take the values 0 and 1 with probability 1/2 each. (a) Compute EL_1 and $EL_2/2$ to get lower bounds on γ. (b) Show $\gamma < 1$ by computing the expected number of i and j sequences of length $K = an$ with the desired property. Chvatal and Sankoff (1975) have shown $0.727273 \le \gamma \le 0.866595$.

6.5.3 Given a rate one Poisson process in $[0, \infty) \times [0, \infty)$, let (X_1, Y_1) be the point that minimizes $x + y$. Let (X_2, Y_2) be the point in $[X_1, \infty) \times [Y_1, \infty)$ that minimizes $x + y$, and so on. Use this construction to show that in Example 6.5.2 $\gamma \ge (8/\pi)^{1/2} > 1.59$.

6.5.4 Let π_n be a random permutation of $\{1, \dots, n\}$ and let J_k^n be the number of subsets of $\{1, \dots n\}$ of size k so that the associated $\pi_n(j)$ form an increasing subsequence. Compute EJ_k^n and take $k \sim \alpha n^{1/2}$ to conclude that in Example 6.5.2 $\gamma \le e$.

6.5.5 Let $\varphi(\theta) = E \exp(-\theta t_i)$ and

$$Y_n = (\mu\varphi(\theta))^{-n} \sum_{i=1}^{Z_n} \exp(-\theta T_n(i))$$

where the sum is over individuals in generation n and $T_n(i)$ is the ith person's birth time. Show that Y_n is a nonnegative martingale and use this to conclude that if $\exp(-\theta a)/\mu\varphi(\theta) > 1$, then $P(X_{0,n} \le an) \to 0$. A little thought reveals that this bound is the same as the answer in the last exercise.

7

Brownian Motion

Brownian motion is a process of tremendous practical and theoretical significance. It originated (a) as a model of the phenomenon observed by Robert Brown in 1828 that "pollen grains suspended in water perform a continual swarming motion," and (b) in Bachelier's (1900) work as a model of the stock market. These are just two of many systems that Brownian motion has been used to model. On the theoretical side, Brownian motion is a Gaussian Markov process with stationary independent increments. It lies in the intersection of three important classes of processes and is a fundamental example in each theory.

The first part of this chapter develops properties of Brownian motion. In Section 7.1 we define Brownian motion and investigate continuity properties of its paths. In Section 7.2 we prove the Markov property and a related 0-1 law. In Section 7.3 we define stopping times and prove the strong Markov property. In Section 7.4, we use the strong Markov prperty to investigate properties of the paths of Brownian motion. In Section 7.5 we introduce some martingales associated with Brownian motion and use them to obtain information about exit distribution and times. In Section 7.6 we prove three versions of Itô's formula.

7.1 Definition and Construction

A one-dimensional **Brownian motion** is a real-valued process B_t, $t \geq 0$ that has the following properties:

(a) If $t_0 < t_1 < \ldots < t_n$, then $B(t_0), B(t_1) - B(t_0), \ldots, B(t_n) - B(t_{n-1})$ are independent.

(b) If $s, t \geq 0$, then

$$P(B(s+t) - B(s) \in A) = \int_A (2\pi t)^{-1/2} \exp(-x^2/2t)\, dx$$

(c) With probability one, $t \to B_t$ is continuous.

(a) says that B_t has independent increments. (b) says that the increment $B(s+t) - B(s)$ has a normal distribution with mean 0 and variance t. (c) is self-explanatory.

Thinking of Brown's pollen grain (c) is certainly reasonable. (a) and (b) can be justified by noting that the movement of the pollen grain is due to the net effect of the bombardment of millions of water molecules, so by the central limit theorem, the displacement in any one interval should have a normal distribution, and the displacements in two disjoint intervals should be independent.

Two immediate consequences of the definition that will be useful many times are:

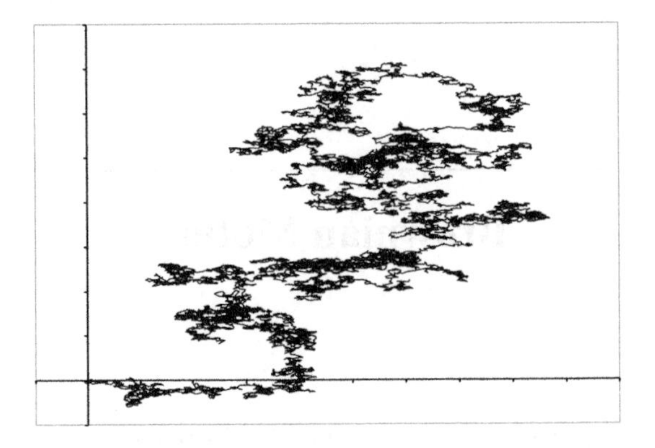

Figure 7.1 Simulation of two dimensional Brownian motion.

Translation invariance $\{B_t - B_0, t \geq 0\}$ is independent of B_0 and has the same distribution as a Brownian motion with $B_0 = 0$.

Proof Let $\mathcal{A}_1 = \sigma(B_0)$ and $\mathcal{A}_2$ be the events of the form

$$\{B(t_1) - B(t_0) \in A_1, \dots, B(t_n) - B(t_{n-1}) \in A_n\}.$$

The $\mathcal{A}_i$ are π-systems that are independent, so the desired result follows from the $\pi - \lambda$ theorem 2.1.6. □

The Brownian scaling relation If $B_0 = 0$, then for any $t > 0$,

$$\{B_{st}, s \geq 0\} \stackrel{d}{=} \{t^{1/2} B_s, s \geq 0\} \tag{7.1.1}$$

To be precise, the two families of r.v.'s have the same finite dimensional distributions, i.e., if $s_1 < \ldots < s_n$, then

$$(B_{s_1 t}, \dots, B_{s_n t}) \stackrel{d}{=} (t^{1/2} B_{s_1}, \dots t^{1/2} B_{s_n})$$

Proof To check this when $n = 1$, we note that $t^{1/2}$ times a normal with mean 0 and variance s is a normal with mean 0 and variance st. The result for $n > 1$ follows from independent increments. □

A second equivalent definition of Brownian motion starting from $B_0 = 0$, that we will occasionally find useful is that B_t, $t \geq 0$, is a real-valued process satisfying

(a′) $B(t)$ is a **Gaussian process** (i.e., all its finite dimensional distributions are multivariate normal).

(b′) $EB_s = 0$ and $EB_s B_t = s \wedge t$.

(c′) With probability one, $t \to B_t$ is continuous.

It is easy to see that (a) and (b) imply (a′). To get (b′) from (a) and (b), suppose $s < t$ and write

$$EB_s B_t = E(B_s^2) + E(B_s(B_t - B_s)) = s$$

The converse is even easier. (a′) and (b′) specify the finite dimensional distributions of B_t, which by the last calculation must agree with the ones defined in (a) and (b).

The first question that must be addressed in any treatment of Brownian motion is, "Is there a process with these properties?" The answer is "Yes," of course, or this chapter would not exist. For pedagogical reasons, we will pursue an approach that leads to a dead end and then retreat a little to rectify the difficulty. Fix an $x \in \mathbf{R}$ and for each $0 < t_1 < \ldots < t_n$, define a measure on $\mathbf{R}^n$ by

$$\mu_{x,t_1,\ldots,t_n}(A_1 \times \ldots \times A_n) = \int_{A_1} dx_1 \cdots \int_{A_n} dx_n \prod_{m=1}^{n} p_{t_m - t_{m-1}}(x_{m-1}, x_m) \tag{7.1.2}$$

where $A_i \in \mathcal{R}$, $x_0 = x$, $t_0 = 0$, and

$$p_t(a,b) = (2\pi t)^{-1/2} \exp(-(b-a)^2/2t)$$

From this formula, it is easy to see that for fixed x the family μ is a consistent set of finite dimensional distributions (f.d.d.'s), that is, if $\{s_1, \ldots, s_{n-1}\} \subset \{t_1, \ldots, t_n\}$ and $t_j \notin \{s_1, \ldots, s_{n-1}\}$, then

$$\mu_{x,s_1,\ldots,s_{n-1}}(A_1 \times \cdots \times A_{n-1}) = \mu_{x,t_1,\ldots,t_n}(A_1 \times \cdots \times A_{j-1} \times \mathbf{R} \times A_j \times \cdots \times A_{n-1})$$

This is clear when $j = n$. To check the equality when $1 \le j < n$, it is enough to show that

$$\int p_{t_j - t_{j-1}}(x,y) p_{t_{j+1} - t_j}(y,z)\, dy = p_{t_{j+1} - t_{j-1}}(x,z)$$

By translation invariance, we can without loss of generality assume $x = 0$, but all this says is that the sum of independent normals with mean 0 and variances $t_j - t_{j-1}$ and $t_{j+1} - t_j$ has a normal distribution with mean 0 and variance $t_{j+1} - t_{j-1}$.

With the consistency of f.d.d.'s verified, we get our first construction of Brownian motion:

Theorem 7.1.1 *Let $\Omega_o = \{$functions $\omega : [0,\infty) \to \mathbf{R}\}$ and $\mathcal{F}_o$ be the σ-field generated by the finite dimensional sets $\{\omega : \omega(t_i) \in A_i$ for $1 \le i \le n\}$, where $A_i \in \mathcal{R}$. For each $x \in \mathbf{R}$, there is a unique probability measure ν_x on $(\Omega_o, \mathcal{F}_o)$ so that $\nu_x\{\omega : \omega(0) = x\} = 1$ and when $0 < t_1 < \ldots < t_n$*

$$\nu_x\{\omega : \omega(t_i) \in A_i\} = \mu_{x,t_1,\ldots,t_n}(A_1 \times \cdots \times A_n) \tag{7.1.3}$$

This follows from a generalization of Kolmogorov's extension theorem, (7.1) in Appendix 3. We will not bother with the details, since at this point we are at the dead end referred to earlier. If $C = \{\omega : t \to \omega(t)$ is continuous$\}$, then $C \notin \mathcal{F}_o$, that is, C is not a measurable set. The easiest way of proving $C \notin \mathcal{F}_o$ is to do Exercise 7.1.4.

This problem is easy to solve. Let $\mathbf{Q}_2 = \{m2^{-n} : m,n \ge 0\}$ be the **dyadic rationals**. If $\Omega_q = \{\omega : \mathbf{Q}_2 \to \mathbf{R}\}$ and $\mathcal{F}_q$ is the σ-field generated by the finite dimensional sets, then enumerating the rationals $q_1, q_2, \ldots$ and applying Kolmogorov's extension theorem shows that we can construct a probability ν_x on $(\Omega_q, \mathcal{F}_q)$ so that $\nu_x\{\omega : \omega(0) = x\} = 1$ and (7.1.3) holds when the $t_i \in \mathbf{Q}_2$. To extend B_t to a process defined on $[0,\infty)$, we will show:

Theorem 7.1.2 *Let $T < \infty$ and $x \in \mathbf{R}$. ν_x assigns probability one to paths $\omega : \mathbf{Q}_2 \to \mathbf{R}$ that are uniformly continuous on $\mathbf{Q}_2 \cap [0,T]$.*

Remark It will take quite a bit of work to prove Theorem 7.1.2. Before taking on that task, we will attend to the last measure theoretic detail: We tidy things up by moving our probability measures to $(C, \mathcal{C})$, where $C = \{\text{continuous } \omega : [0,\infty) \to \mathbf{R}\}$ and $\mathcal{C}$ is the σ-field generated by the coordinate maps $t \to \omega(t)$. To do this, we observe that the map ψ that takes a uniformly continuous point in Ω_q to its unique continuous extension in C is measurable, and we set

$$P_x = \nu_x \circ \psi^{-1}$$

Our construction guarantees that $B_t(\omega) = \omega_t$ has the right finite dimensional distributions for $t \in \mathbf{Q}_2$. Continuity of paths and a simple limiting argument show that this is true when $t \in [0,\infty)$. Finally, the reader should note that, as in the case of Markov chains, we have one set of random variables $B_t(\omega) = \omega(t)$, and a family of probability measures P_x, $x \in \mathbf{R}$, so that under P_x, B_t is a Brownian motion with $P_x(B_0 = x) = 1$.

Proof By translation invariance and scaling (7.1.1), we can without loss of generality suppose $B_0 = 0$ and prove the result for $T = 1$. In this case, part (b) of the definition and the scaling relation imply

$$E_0(|B_t - B_s|)^4 = E_0|B_{t-s}|^4 = C(t-s)^2$$

where $C = E_0|B_1|^4 < \infty$. From the last observation, we get the desired uniform continuity by using the following result due to Kolmogorov. Thanks to Robin Pemantle for simplifying the proof and to Timo Seppäläinen for correcting the simplification.

Theorem 7.1.3 *Suppose $E|X_s - X_t|^\beta \le K|t-s|^{1+\alpha}$, where $\alpha, \beta > 0$. If $\gamma < \alpha/\beta$, then with probability one there is a constant $C(\omega)$ so that*

$$|X(q) - X(r)| \le C|q-r|^\gamma \quad \text{for all } q, r \in \mathbf{Q}_2 \cap [0,1]$$

Proof Let $G_n = \{|X(i/2^n) - X((i-1)/2^n)| \le 2^{-\gamma n} \text{ for all } 0 < i \le 2^n\}$. Chebyshev's inequality implies $P(|Y| > a) \le a^{-\beta} E|Y|^\beta$, so if we let $\lambda = \alpha - \beta\gamma > 0$, then

$$P(G_n^c) \le 2^n \cdot 2^{n\beta\gamma} \cdot E|X(j2^{-n}) - X(i2^{-n})|^\beta = K2^{-n\lambda}$$

Lemma 7.1.4 *On $H_N = \cap_{n=N}^\infty G_n$ we have*

$$|X(q) - X(r)| \le \frac{3}{1-2^{-\gamma}}|q-r|^\gamma$$

for $q, r \in \mathbf{Q}_2 \cap [0,1]$ with $|q-r| < 2^{-N}$.

Proof of Lemma 7.1.4 Let $q, r \in \mathbf{Q}_2 \cap [0,1]$ with $0 < r - q < 2^{-N}$. Let $I_i^k = [(i-1)/2^k, i/2^k)$ and let m be the smallest value of k for which q and r are in different intervals I_i^k. Since they were in the same interval on level $k-1$, $q \in I_i^m$ and $r \in I_{i+1}^m$. We can write

$$r = i2^{-m} + 2^{-r(1)} + \cdots + 2^{-r(\ell)}$$
$$q = i2^{-m} - 2^{-q(1)} - \cdots - 2^{-q(k)}$$

where $N < r(1) < \cdots < r(\ell)$ and $N < q(1) < \cdots < q(k)$. To see this, note that as we further subdivide, r will lie in the left or right half. When it is in the right half we add another term.

On H_N

$$|X(q) - X((i-1)2^{-m})| \le \sum_{h=1}^{k} (2^{-q(h)})^\gamma \le \sum_{h=m}^{\infty} (2^{-\gamma})^h = \frac{2^{-\gamma m}}{1-2^{-\gamma}}$$

$$|X(r) - X(i2^{-m})| \le \frac{2^{-\gamma m}}{1-2^{-\gamma}}$$

Combining the last three inequalities with $2^{-m} \le |q-r|$ and $1 - 2^{-\gamma} > 1$ completes the proof of Lemma 7.1.4. □

To prove Theorem 7.1.3 now, we note that

$$P(H_N^c) \le \sum_{n=N}^{\infty} P(G_n^c) \le K \sum_{n=N}^{\infty} 2^{-n\lambda} = K 2^{-N\lambda}/(1-2^{-\lambda})$$

Since $\sum_{N=1}^{\infty} P(H_N^c) < \infty$, the Borel-Cantelli lemma, Theorem 2.3.1, implies

$$|X(q) - X(r)| \le A|q-r|^\gamma \quad \text{for } q,r \in \mathbf{Q}_2 \text{ with } |q-r| < \delta(\omega).$$

To extend this to $q, r \in \mathbf{Q}_2 \cap [0,1]$, let $s_0 = q < s_1 < \ldots < s_n = r$ with $|s_i - s_{i-1}| < \delta(\omega)$ and use the triangle inequality to conclude $|X(q) - X(r)| \le C(\omega)|q-r|^\gamma$, where $C(\omega) = 1 + \delta(\omega)^{-1}$. □

The scaling relation, (7.1.1), implies

$$E|B_t - B_s|^{2m} = C_m |t-s|^m \quad \text{where } C_m = E|B_1|^{2m}$$

so using Theorem 7.1.3 with $\beta = 2m$, $\alpha = m - 1$ and letting $m \to \infty$ gives a result of Wiener (1923).

Theorem 7.1.5 *Brownian paths are Hölder continuous for any exponent $\gamma < 1/2$.*

It is easy to show:

Theorem 7.1.6 *With probability one, Brownian paths are not Lipschitz continuous (and hence not differentiable) at any point.*

Remark The nondifferentiability of Brownian paths was discovered by Paley, Wiener, and Zygmund (1933). Paley died in 1933 at the age of 26 in a skiing accident while the paper was in press. The proof we are about to give is due to Dvoretsky, Erdös, and Kakutani (1961).

Proof Fix a constant $C < \infty$ and let $A_n = \{\omega :$ there is an $s \in [0,1]$ so that $|B_t - B_s| \le C|t-s|$ when $|t-s| \le 3/n\}$. For $1 \le k \le n-2$, let

$$Y_{k,n} = \max\left\{\left|B\left(\frac{k+j}{n}\right) - B\left(\frac{k+j-1}{n}\right)\right| : j = 0,1,2\right\}$$

$$B_n = \{\text{ at least one } Y_{k,n} \le 5C/n\}$$

The triangle inequality implies $A_n \subset B_n$. The worst case is $s = 1$. We pick $k = n - 2$ and observe

$$\left|B\left(\frac{n-3}{n}\right) - B\left(\frac{n-2}{n}\right)\right| \le \left|B\left(\frac{n-3}{n}\right) - B(1)\right| + \left|B(1) - B\left(\frac{n-2}{n}\right)\right|$$
$$\le C(3/n + 2/n)$$

Using $A_n \subset B_n$ and the scaling relation (7.1.1) now gives

$$P(A_n) \le P(B_n) \le nP(|B(1/n)| \le 5C/n)^3 = nP(|B(1)| \le 5C/n^{1/2})^3$$
$$\le n\{(10C/n^{1/2}) \cdot (2\pi)^{-1/2}\}^3$$

since $\exp(-x^2/2) \le 1$. Letting $n \to \infty$ shows $P(A_n) \to 0$. Noticing $n \to A_n$ is increasing shows $P(A_n) = 0$ for all n and completes the proof. □

Remark Let $\mathcal{H}_\gamma(\omega)$ be the set of times at which the path $\omega \in C$ is Hölder continuous of order γ. Theorem 7.1.5 shows that $P(\mathcal{H}_\gamma = [0,\infty)) = 1$ for $\gamma < 1/2$. Exercise 7.1.5 shows that $P(\mathcal{H}_\gamma = \emptyset) = 1$ for $\gamma > 1/2$. Exercise 7.2.4 shows $P(t \in \mathcal{H}_{1/2}) = 0$ for each t, but B. Davis (1983) has shown $P(\mathcal{H}_{1/2} \ne \emptyset) = 1$. Perkins (1983) has computed the Hausdorff dimension of

$$\left\{t \in (0,1) : \limsup_{h \downarrow 0} \frac{|B_{t+h} - B_t|}{h^{1/2}} \le c\right\}$$

Multidimensional Brownian Motion

All of the results in this section have been for one-dimensional Brownian motion. To define a d-dimensional Brownian motion starting at $x \in \mathbf{R}^d$ we let $B_t^1, \dots B_t^d$ be independent Brownian motions with $B_0^i = x_i$. As in the case $d = 1$, these are realized as probability measures P_x on $(C, \mathcal{C})$ where $C = \{\text{continuous } \omega : [0,\infty) \to \mathbf{R}^d\}$ and $\mathcal{C}$ is the σ-field generated by the coordinate maps. Since the coordinates are independent, it is easy to see that the finite dimensional distributions satisfy (7.1.2) with transition probability

$$p_t(x,y) = (2\pi t)^{-d/2} \exp(-|y-x|^2/2t) \tag{7.1.4}$$

Exercises

7.1.1 Given $s < t$ find $P(B(s) > 0, B(t) > 0)$. Hint: write the probability in terms of $B(s)$ and $B(t) - B(s)$.

7.1.2 Find $E(B_1^2 B_2 B_3)$.

7.1.3 Let $W = \int_0^t B_s\, ds$. Find EW and EW^2. What is the distribution of W?

7.1.4 $A \in \mathcal{F}_o$ if and only if there is a sequence of times $t_1, t_2, \dots$ in $[0,\infty)$ and a $B \in \mathcal{R}^{\{1,2,\dots\}}$ so that $A = \{\omega : (\omega(t_1), \omega(t_2), \dots) \in B\}$. In words, all events in $\mathcal{F}_o$ depend on only countably many coordinates.

7.1.5 Looking at the proof of Theorem 7.1.6 carefully shows that if $\gamma > 5/6$, then B_t is not Hölder continuous with exponent γ at any point in [0,1]. Show, by considering k increments instead of 3, that the last conclusion is true for all $\gamma > 1/2 + 1/k$.

7.1.6 Fix t and let $\Delta_{m,n} = B(tm2^{-n}) - B(t(m-1)2^{-n})$. Compute

$$E\Big(\sum_{m\le 2^n} \Delta_{m,n}^2 - t\Big)^2$$

and use Borel-Cantelli to conclude that $\sum_{m\le 2^n} \Delta_{m,n}^2 \to t$ a.s. as $n \to \infty$.

Remark The last result is true if we consider a sequence of partitions $\Pi_1 \subset \Pi_2 \subset \ldots$ with mesh $\to 0$. See Freedman (1971a) pp. 42–46. However, the true quadratic variation, defined as the sup over all partitions, is ∞.

7.2 Markov Property, Blumenthal's 0-1 Law

Intuitively, the Markov property says "if $s \ge 0$ then $B(t+s) - B(s)$, $t \ge 0$ is a Brownian motion that is independent of what happened before time s." The first step in making this into a precise statement is to explain what we mean by "what happened before time s." The first thing that comes to mind is

$$\mathcal{F}_s^o = \sigma(B_r : r \le s)$$

For reasons that will become clear as we go along, it is convenient to replace $\mathcal{F}_s^o$ by

$$\mathcal{F}_s^+ = \cap_{t>s} \mathcal{F}_t^o$$

The fields $\mathcal{F}_s^+$ are nicer because they are **right continuous**:

$$\cap_{t>s} \mathcal{F}_t^+ = \cap_{t>s} \left(\cap_{u>t} \mathcal{F}_u^o\right) = \cap_{u>s} \mathcal{F}_u^o = \mathcal{F}_s^+$$

In words, the $\mathcal{F}_s^+$ allow us an "infinitesimal peek at the future," i.e., $A \in \mathcal{F}_s^+$ if it is in $\mathcal{F}_{s+\epsilon}^o$ for any $\epsilon > 0$. If $f(u) > 0$ for all $u > 0$, then in $d = 1$ the random variable

$$\limsup_{t \downarrow s} \frac{B_t - B_s}{f(t-s)}$$

is measurable with respect to $\mathcal{F}_s^+$ but not $\mathcal{F}_s^o$. We will see that there are no interesting examples, i.e., $\mathcal{F}_s^+$ and $\mathcal{F}_s^o$ are the same (up to null sets).

To state the Markov property, we need some notation. Recall that we have a family of measures P_x, $x \in \mathbf{R}^d$, on $(C, \mathcal{C})$ so that under P_x, $B_t(\omega) = \omega(t)$ is a Brownian motion starting at x. For $s \ge 0$, we define the **shift transformation** $\theta_s : C \to C$ by

$$(\theta_s \omega)(t) = \omega(s+t) \quad \text{for } t \ge 0$$

In words, we cut off the part of the path before time s and then shift the path so that time s becomes time 0.

Theorem 7.2.1 (Markov property) *If $s \ge 0$ and Y is bounded and $\mathcal{C}$ measurable, then for all $x \in \mathbf{R}^d$*

$$E_x(Y \circ \theta_s | \mathcal{F}_s^+) = E_{B_s} Y$$

where the right-hand side is the function $\varphi(x) = E_x Y$ evaluated at $x = B_s$.

Proof By the definition of conditional expectation, what we need to show is that

$$E_x(Y \circ \theta_s; A) = E_x(E_{B_s} Y; A) \quad \text{for all } A \in \mathcal{F}_s^+ \tag{7.2.1}$$

We will begin by proving the result for a carefully chosen special case and then use the monotone class theorem (MCT) to get the general case. Suppose $Y(\omega) = \prod_{1 \le m \le n} f_m(\omega(t_m))$, where $0 < t_1 < \ldots < t_n$ and the f_m are bounded and measurable. Let $0 < h < t_1$, let $0 < s_1 \ldots < s_k \le s + h$, and let $A = \{\omega : \omega(s_j) \in A_j, 1 \le j \le k\}$, where $A_j \in \mathcal{R}$ for $1 \le j \le k$. From the definition of Brownian motion, it follows that

$$\begin{aligned} E_x(Y \circ \theta_s; A) = \int_{A_1} dx_1\, p_{s_1}(x, x_1) \int_{A_2} dx_2\, p_{s_2 - s_1}(x_1, x_2) \cdots \\ \int_{A_k} dx_k\, p_{s_k - s_{k-1}}(x_{k-1}, x_k) \int dy\, p_{s+h-s_k}(x_k, y) \varphi(y, h) \end{aligned}$$

where

$$\varphi(y, h) = \int dy_1\, p_{t_1 - h}(y, y_1) f_1(y_1) \ldots \int dy_n\, p_{t_n - t_{n-1}}(y_{n-1}, y_n) f_n(y_n)$$

For more details, see the proof of (5.2.3), which applies without change here. Using that identity on the right-hand side, we have

$$E_x(Y \circ \theta_s; A) = E_x(\varphi(B_{s+h}, h); A) \tag{7.2.2}$$

The last equality holds for all finite dimensional sets A, so the $\pi - \lambda$ theorem, Theorem 2.1.6, implies that it is valid for all $A \in \mathcal{F}_{s+h}^o \supset \mathcal{F}_s^+$.

It is easy to see by induction on n that

$$\begin{aligned} \psi(y_1) = f_1(y_1) \int dy_2\, p_{t_2 - t_1}(y_1, y_2) f_2(y_2) \\ \ldots \int dy_n\, p_{t_n - t_{n-1}}(y_{n-1}, y_n) f_n(y_n) \end{aligned}$$

is bounded and measurable. Letting $h \downarrow 0$ and using the dominated convergence theorem shows that if $x_h \to x$, then

$$\phi(x_h, h) = \int dy_1\, p_{t_1 - h}(x_h, y_1) \psi(y_1) \to \phi(x, 0)$$

as $h \downarrow 0$. Using (7.2.2) and the bounded convergence theorem now gives

$$E_x(Y \circ \theta_s; A) = E_x(\varphi(B_s, 0); A)$$

for all $A \in \mathcal{F}_s^+$. This shows that (7.2.1) holds for $Y = \prod_{1 \le m \le n} f_m(\omega(t_m))$ and the f_m are bounded and measurable.

The desired conclusion now follows from the monotone class theorem, 5.2.2. Let $\mathcal{H} =$ the collection of bounded functions for which (7.2.1) holds. $\mathcal{H}$ clearly has properties (ii) and (iii). Let $\mathcal{A}$ be the collection of sets of the form $\{\omega : \omega(t_j) \in A_j\}$, where $A_j \in \mathcal{R}$. The special case treated here shows (i) holds and the desired conclusion follows. □

The reader will see many applications of the Markov property later, so we turn our attention now to a "triviality" that has surprising consequences. Since

$$E_x(Y \circ \theta_s | \mathcal{F}_s^+) = E_{B(s)} Y \in \mathcal{F}_s^o$$

it follows from Theorem 4.1.12 that

$$E_x(Y \circ \theta_s | \mathcal{F}_s^+) = E_x(Y \circ \theta_s | \mathcal{F}_s^o)$$

From the last equation, it is a short step to:

Theorem 7.2.2 *If $Z \in \mathcal{C}$ is bounded, then for all $s \geq 0$ and $x \in \mathbf{R}^d$,*

$$E_x(Z | \mathcal{F}_s^+) = E_x(Z | \mathcal{F}_s^o)$$

Proof As in the proof of Theorem 7.2.1, it suffices to prove the result when

$$Z = \prod_{m=1}^{n} f_m(B(t_m))$$

and the f_m are bounded and measurable. In this case, Z can be written as $X(Y \circ \theta_s)$, where $X \in \mathcal{F}_s^o$ and Y is $\mathcal{C}$ measurable, so

$$E_x(Z | \mathcal{F}_s^+) = X E_x(Y \circ \theta_s | \mathcal{F}_s^+) = X E_{B_s} Y \in \mathcal{F}_s^o$$

and the proof is complete. □

If we let $Z \in \mathcal{F}_s^+$, then Theorem 7.2.2 implies $Z = E_x(Z | \mathcal{F}_s^o) \in \mathcal{F}_s^o$, so the two σ-fields are the same up to null sets. At first glance, this conclusion is not exciting. The fun starts when we take $s = 0$ in Theorem 7.2.2 to get:

Theorem 7.2.3 (Blumenthal's 0-1 law) *If $A \in \mathcal{F}_0^+$, then for all $x \in \mathbf{R}^d$,*

$$P_x(A) \in \{0, 1\}.$$

Proof Using $A \in \mathcal{F}_0^+$, Theorem 7.2.2, and $\mathcal{F}_0^o = \sigma(B_0)$ is trivial under P_x gives

$$1_A = E_x(1_A | \mathcal{F}_0^+) = E_x(1_A | \mathcal{F}_0^o) = P_x(A) \quad P_x \text{ a.s.}$$

This shows that the indicator function 1_A is a.s. equal to the number $P_x(A)$, and the result follows. □

In words, the last result says that the **germ field**, $\mathcal{F}_0^+$, is trivial. This result is very useful in studying the local behavior of Brownian paths. For the rest of the section we restrict our attention to $d = 1$.

Theorem 7.2.4 *If $\tau = \inf\{t \geq 0 : B_t > 0\}$, then $P_0(\tau = 0) = 1$.*

Proof $P_0(\tau \leq t) \geq P_0(B_t > 0) = 1/2$, since the normal distribution is symmetric about 0. Letting $t \downarrow 0$, we conclude

$$P_0(\tau = 0) = \lim_{t \downarrow 0} P_0(\tau \leq t) \geq 1/2$$

so it follows from Theorem 7.2.3 that $P_0(\tau = 0) = 1$. □

Once Brownian motion must hit $(0,\infty)$ immediately starting from 0, it must also hit $(-\infty,0)$ immediately. Since $t \to B_t$ is continuous, this forces:

Theorem 7.2.5 *If* $T_0 = \inf\{t > 0 : B_t = 0\}$ *then* $P_0(T_0 = 0) = 1$.

Theorem 7.2.3 concerns the behavior of B_t as $t \to 0$. By using a trick, we can use this result to get information about the behavior as $t \to \infty$.

Theorem 7.2.6 *If B_t is a Brownian motion starting at 0, then so is the process defined by $X_0 = 0$ and $X_t = tB(1/t)$ for $t > 0$.*

Proof Here we will check the second definition of Brownian motion. To do this, we note: (i) If $0 < t_1 < \ldots < t_n$, then $(X(t_1), \ldots, X(t_n))$ has a multivariate normal distribution with mean 0. (ii) $EX_s = 0$ and if $s < t$, then

$$E(X_sX_t) = stE(B(1/s)B(1/t)) = s$$

For (iii) we note that X is clearly continuous at $t \neq 0$.

To handle $t = 0$, we begin by observing that the strong law of large numbers implies $B_n/n \to 0$ as $n \to \infty$ through the integers. To handle values in between integers, we note that Kolmogorov's inequality, Theorem 2.5.5, implies

$$P\left(\sup_{0<k\le 2^m} |B(n+k2^{-m}) - B_n| > n^{2/3}\right) \le n^{-4/3}E(B_{n+1} - B_n)^2$$

Letting $m \to \infty$, we have

$$P\left(\sup_{u\in[n,n+1]} |B_u - B_n| > n^{2/3}\right) \le n^{-4/3}$$

Since $\sum_n n^{-4/3} < \infty$, the Borel-Cantelli lemma implies $B_u/u \to 0$ as $u \to \infty$. Taking $u = 1/t$, we have $X_t \to 0$ as $t \to 0$. □

Theorem 7.2.6 allows us to relate the behavior of B_t as $t \to \infty$ and as $t \to 0$. Combining this idea with Blumenthal's 0-1 law leads to a very useful result. Let

$$\mathcal{F}'_t = \sigma(B_s : s \ge t) = \text{ the future at time } t$$
$$\mathcal{T} = \cap_{t\ge 0}\mathcal{F}'_t = \text{ the tail } \sigma\text{-field.}$$

Theorem 7.2.7 *If $A \in \mathcal{T}$, then either $P_x(A) \equiv 0$ or $P_x(A) \equiv 1$.*

Remark Notice that this is stronger than the conclusion of Blumenthal's 0-1 law. The examples $A = \{\omega : \omega(0) \in D\}$ show that for A in the germ σ-field $\mathcal{F}_0^+$, the value of $P_x(A)$, $1_D(x)$ in this case, may depend on x.

Proof Since the tail σ-field of B is the same as the germ σ-field for X, it follows that $P_0(A) \in \{0,1\}$. To improve this to the conclusion given, observe that $A \in \mathcal{F}'_1$, so 1_A can be written as $1_D \circ \theta_1$. Applying the Markov property gives

$$P_x(A) = E_x(1_D \circ \theta_1) = E_x(E_x(1_D \circ \theta_1|\mathcal{F}_1)) = E_x(E_{B_1}1_D)$$
$$= \int (2\pi)^{-1/2}\exp(-(y-x)^2/2)P_y(D)\,dy$$

Taking $x = 0$, we see that if $P_0(A) = 0$, then $P_y(D) = 0$ for a.e. y with respect to Lebesgue measure, and using the formula again shows $P_x(A) = 0$ for all x. To handle the case $P_0(A) = 1$, observe that $A^c \in \mathcal{T}$ and $P_0(A^c) = 0$, so the last result implies $P_x(A^c) = 0$ for all x. □

The next result is a typical application of Theorem 7.2.7.

Theorem 7.2.8 *Let B_t be a one-dimensional Brownian motion starting at 0 then with probability 1,*

$$\limsup_{t\to\infty} B_t/\sqrt{t} = \infty \qquad \liminf_{t\to\infty} B_t/\sqrt{t} = -\infty$$

Proof Let $K < \infty$. By Exercise 2.3.1 and scaling

$$P_0(B_n/\sqrt{n} \geq K \text{ i.o.}) \geq \limsup_{n\to\infty} P_0(B_n \geq K\sqrt{n}) = P_0(B_1 \geq K) > 0$$

so the 0–1 law in Theorem 7.2.7 implies the probability is 1. Since K is arbitrary, this proves the first result. The second one follows from symmetry. □

From Theorem 7.2.8, translation invariance, and the continuity of Brownian paths it follows that we have:

Theorem 7.2.9 *Let B_t be a one-dimensional Brownian motion and let $A = \cap_n \{B_t = 0$ for some $t \geq n\}$. Then $P_x(A) = 1$ for all x.*

In words, one-dimensional Brownian motion is recurrent. For any starting point x, it will return to 0 "infinitely often," i.e., there is a sequence of times $t_n \uparrow \infty$ so that $B_{t_n} = 0$. We have to be careful with the interpretation of the phrase in quotes, since starting from 0, B_t will hit 0 infinitely many times by time $\epsilon > 0$.

Last rites With our discussion of Blumenthal's 0-1 law complete, the distinction between $\mathcal{F}_s^+$ and $\mathcal{F}_s^o$ is no longer important, so we will make one final improvement in our σ-fields and remove the superscripts. Let

$$\begin{aligned}
\mathcal{N}_x &= \{A : A \subset D \text{ with } P_x(D) = 0\} \\
\mathcal{F}_s^x &= \sigma(\mathcal{F}_s^+ \cup \mathcal{N}_x) \\
\mathcal{F}_s &= \cap_x \mathcal{F}_s^x
\end{aligned}$$

$\mathcal{N}_x$ are the **null sets** and $\mathcal{F}_s^x$ are the completed σ-fields for P_x. Since we do not want the filtration to depend on the initial state, we take the intersection of all the σ-fields. The reader should note that it follows from the definition that the $\mathcal{F}_s$ are right continuous.

Exercises

7.2.1 Let $T_0 = \inf\{s > 0 : B_s = 0\}$ and let $R = \inf\{t > 1 : B_t = 0\}$. R is for right or return. Use the Markov property at time 1 to get

$$P_x(R > 1 + t) = \int p_1(x, y) P_y(T_0 > t)\, dy \tag{7.2.3}$$

7.2.2 Let $T_0 = \inf\{s > 0 : B_s = 0\}$ and let $L = \sup\{t \le 1 : B_t = 0\}$. L is for left or last. Use the Markov property at time $0 < t < 1$ to conclude

$$P_0(L \le t) = \int p_t(0,y)P_y(T_0 > 1-t)\,dy \tag{7.2.4}$$

7.2.3 If $a < b$, then with probability one a is the limit of local maximum of B_t in (a,b). So the set of local maxima of B_t is almost surely a dense set. However, unlike the zero set it is countable.

7.2.4 (i) Suppose $f(t) > 0$ for all $t > 0$. Use Theorem 7.2.3 to conclude that $\limsup_{t\downarrow 0} B(t)/f(t) = c$, P_0 a.s., where $c \in [0,\infty]$ is a constant. (ii) Show that if $f(t) = \sqrt{t}$, then $c = \infty$, so with probability one Brownian paths are not Hölder continuous of order $1/2$ at 0.

7.3 Stopping Times, Strong Markov Property

Generalizing the definition from discrete time, we call a random variable S taking values in $[0,\infty]$ a **stopping time** if for all $t \ge 0$, $\{S < t\} \in \mathcal{F}_t$. In the last definition, we have obviously made a choice between $\{S < t\}$ and $\{S \le t\}$. This makes a big difference in discrete time but none in continuous time (for a right continuous filtration $\mathcal{F}_t$) :

If $\{S \le t\} \in \mathcal{F}_t$ then $\{S < t\} = \cup_n\{S \le t - 1/n\} \in \mathcal{F}_t$.

If $\{S < t\} \in \mathcal{F}_t$ then $\{S \le t\} = \cap_n\{S < t + 1/n\} \in \mathcal{F}_t$.

The first conclusion requires only that $t \to \mathcal{F}_t$ is increasing. The second relies on the fact that $t \to \mathcal{F}_t$ is right continuous. Theorem 7.3.2 and 7.3.3 show that when checking something is a stopping time, it is nice to know that the two definitions are equivalent.

Theorem 7.3.1 *If G is an open set and $T = \inf\{t \ge 0 : B_t \in G\}$, then T is a stopping time.*

Proof Since G is open and $t \to B_t$ is continuous, $\{T < t\} = \cup_{q<t}\{B_q \in G\}$, where the union is over all rational q, so $\{T < t\} \in \mathcal{F}_t$. Here we need to use the rationals to get a countable union, and hence a measurable set. □

Theorem 7.3.2 *If T_n is a sequence of stopping times and $T_n \downarrow T$, then T is a stopping time.*

Proof $\{T < t\} = \cup_n\{T_n < t\}$. □

Theorem 7.3.3 *If T_n is a sequence of stopping times and $T_n \uparrow T$, then T is a stopping time.*

Proof $\{T \le t\} = \cap_n\{T_n \le t\}$. □

Theorem 7.3.4 *If K is a closed set and $T = \inf\{t \ge 0 : B_t \in K\}$, then T is a stopping time.*

Proof Let $B(x,r) = \{y : |y - x| < r\}$, let $G_n = \cup_{x\in K} B(x,1/n)$, and let $T_n = \inf\{t \ge 0 : B_t \in G_n\}$. Since G_n is open, it follows from Theorem 7.3.1 that T_n is a stopping time. I claim that as $n \uparrow \infty$, $T_n \uparrow T$. To prove this, notice that $T \ge T_n$ for all n, so $\lim T_n \le T$. To prove $T \le \lim T_n$, we can suppose that $T_n \uparrow t < \infty$. Since $B(T_n) \in \bar{G}_n$ for all n and $B(T_n) \to B(t)$, it follows that $B(t) \in K$ and $T \le t$. □

Theorems 7.3.1 and 7.3.4 will take care of all the hitting times we will consider. Using 7.3.2 with the second result, one can show that the hitting time of a union of closed sets is a stopping time. It is known that the hitting time of a Borel set is a stopping time but proving that is very difficult.

We will need the next result in the proof of the strong Markov property.

Theorem 7.3.5 *Let S be a stopping time and let $S_n = ([2^n S] + 1)/2^n$, where $[x]$ = the largest integer $\leq x$. That is,*

$$S_n = (m+1)2^{-n} \text{ if } m2^{-n} \leq S < (m+1)2^{-n}$$

In words, we stop at the first time of the form $k2^{-n}$ after S (i.e., $> S$). From the verbal description, it should be clear that S_n is a stopping time.

Proof If $m/2^{-n} << (m+1)/2^{-n}$, then $\{S_n < t\} = \{S \leq m/2^n\} \in \mathcal{F}_{m/2^n} \subset \mathcal{F}_t$. It is impossible to have $S_n = 0$. □

Our next goal is to state and prove the strong Markov property. To do this, we need to generalize two definitions from Section 4.1. Given a nonnegative random variable $S(\omega)$ we define the random shift θ_S, which "cuts off the part of ω before $S(\omega)$ and then shifts the path so that time $S(\omega)$ becomes time 0." In symbols, we set

$$(\theta_S\omega)(t) = \begin{cases} \omega(S(\omega)+t) & \text{on } \{S < \infty\} \\ \Delta & \text{on } \{S = \infty\} \end{cases}$$

where Δ is an extra point we add to C. As in Section 6.3, we will usually explicitly restrict our attention to $\{S < \infty\}$, so the reader does not have to worry about the second half of the definition.

The second quantity $\mathcal{F}_S$, "the information known at time S," is a little more subtle. Imitating the discrete time definition from Section 4.1, we let

$$\mathcal{F}_S = \{A : A \cap \{S \leq t\} \in \mathcal{F}_t \text{ for all } t \geq 0\}$$

In words, this makes the reasonable demand that the part of A that lies in $\{S \leq t\}$ should be measurable with respect to the information available at time t. Again we have made a choice between $\leq t$ and $< t$, but as in the case of stopping times, this makes no difference, and it is useful to know that the two definitions are equivalent.

Remark When $\mathcal{F}_t$ is right continuous, the last definition is unchanged if we replace $\{S \leq t\}$ by $\{S < t\}$.

We will now derive some properties of $\mathcal{F}_S$. The first is intuitively obvious: at a later time we have more information.

Theorem 7.3.6 *If $S \leq T$ are stopping times, then $\mathcal{F}_S \subset \mathcal{F}_T$.*

Proof If $A \in \mathcal{F}_S$, then $A \cap \{T \leq t\} = (A \cap \{S \leq t\}) \cap \{T \leq t\} \in \mathcal{F}_t$. □

Theorem 7.3.7 *If $T_n \downarrow T$ are stopping times, then $\mathcal{F}_T = \cap \mathcal{F}(T_n)$.*

Proof Theorem 7.3.6 implies $\mathcal{F}(T_n) \supset \mathcal{F}_T$ for all n. To prove the other inclusion, let $A \in \cap \mathcal{F}(T_n)$. Since $A \cap \{T_n < t\} \in \mathcal{F}_t$ and $T_n \downarrow T$, it follows that $A \cap \{T < t\} \in \mathcal{F}_t$. □

The next result is obvious but takes a little work.

Theorem 7.3.8 *$B_S \in \mathcal{F}_S$, i.e., the value of B_S is measurable with respect to the information known at time S!*

Proof To prove this, let $S_n = ([2^n S] + 1)/2^n$ be the stopping times defined in Theorem 7.3.5. If A is a Borel set.

$$\{B(S_n) \in A\} = \cup_{m=1}^{\infty} \{S_n = m/2^n, B(m/2^n) \in A\} \in \mathcal{F}_{S_n}$$

Now let $n \to \infty$ and use Theorem 7.3.7. □

We are now ready to state the strong Markov property, which says that the Markov property holds at stopping times. It is interesting that the notion of Brownian motion dates to the the very beginning of the twentieth century, but the first proofs of the strong Markov property were given independently by Hunt (1956), and Dynkin and Yushkevich (1956). Hunt writes "Although mathematicians use this extended Markoff property, at least as a heuristic principle, I have nowhere found it discussed with rigor."

Theorem 7.3.9 (Strong Markov property) *Let $(s,\omega) \to Y_s(\omega)$ be bounded and $\mathcal{R} \times \mathcal{C}$ measurable. If S is a stopping time, then for all $x \in \mathbf{R}^d$*

$$E_x(Y_S \circ \theta_S | \mathcal{F}_S) = E_{B(S)} Y_S \text{ on } \{S < \infty\}$$

where the right-hand side is the function $\varphi(x,t) = E_x Y_t$ evaluated at $x = B(S)$, $t = S$.

Remark The only facts about Brownian motion used here are that (i) it is a Markov process, and (ii) if f is bounded and continuous, then $x \to E_x f(B_t)$ is continuous. In Markov process theory (ii) is called the Feller property. Hunt's proof only applies to Brownian motion, and Dynkin and Yushkevich proved the result in this generality.

Proof We first prove the result under the assumption that there is a sequence of times $t_n \uparrow \infty$, so that $P_x(S < \infty) = \sum P_x(S = t_n)$. In this case, the proof is basically the same as the proof of Theorem 5.2.5. We break things down according to the value of S, apply the Markov property, and put the pieces back together. If we let $Z_n = Y_{t_n}(\omega)$ and $A \in \mathcal{F}_S$, then

$$E_x(Y_S \circ \theta_S; A \cap \{S < \infty\}) = \sum_{n=1}^{\infty} E_x(Z_n \circ \theta_{t_n}; A \cap \{S = t_n\})$$

Now if $A \in \mathcal{F}_S$, $A \cap \{S = t_n\} = (A \cap \{S \le t_n\}) - (A \cap \{S \le t_{n-1}\}) \in \mathcal{F}_{t_n}$, so it follows from the Markov property that the sum is

$$= \sum_{n=1}^{\infty} E_x(E_{B(t_n)} Z_n; A \cap \{S = t_n\}) = E_x(E_{B(S)} Y_S; A \cap \{S < \infty\})$$

To prove the result in general, we let $S_n = ([2^n S] + 1)/2^n$ be the stopping time defined in Exercise 7.3.5. To be able to let $n \to \infty$, we restrict our attention to Y's of the form

$$Y_s(\omega) = f_0(s) \prod_{m=1}^{n} f_m(\omega(t_m)) \tag{7.3.1}$$

where $0 < t_1 < \ldots < t_n$ and $f_0, \ldots, f_n$ are bounded and continuous. If f is bounded and continuous, then the dominated convergence theorem implies that

$$x \to \int dy\, p_t(x,y) f(y)$$

is continuous. From this and induction, it follows that

$$\varphi(x,s) = E_x Y_s = f_0(s) \int dy_1\, p_{t_1}(x,y_1) f_1(y_1)$$
$$\ldots \int dy_n\, p_{t_n - t_{n-1}}(y_{n-1}, y_n) f_n(y_n)$$

is bounded and continuous.

Having assembled the necessary ingredients, we can now complete the proof. Let $A \in \mathcal{F}_S$. Since $S \leq S_n$, Theorem 7.3.6 implies $A \in \mathcal{F}(S_n)$. Applying the special case proved here to S_n and observing that $\{S_n < \infty\} = \{S < \infty\}$ gives

$$E_x(Y_{S_n} \circ \theta_{S_n}; A \cap \{S < \infty\}) = E_x(\varphi(B(S_n), S_n); A \cap \{S < \infty\})$$

Now, as $n \to \infty$, $S_n \downarrow S$, $B(S_n) \to B(S)$, $\varphi(B(S_n), S_n) \to \varphi(B(S), S)$ and

$$Y_{S_n} \circ \theta_{S_n} \to Y_S \circ \theta_S$$

so the bounded convergence theorem implies that the result holds when Y has the form given in (7.3.1).

To complete the proof now, we will apply the monotone class theorem. As in the proof of Theorem 7.2.1, we let $\mathcal{H}$ be the collection of Y for which

$$E_x(Y_S \circ \theta_S; A) = E_x(E_{B(S)} Y_S; A) \quad \text{for all } A \in \mathcal{F}_S$$

and it is easy to see that (ii) and (iii) hold. This time, however, we take $\mathcal{A}$ to be the sets of the form $A = G_0 \times \{\omega : \omega(s_j) \in G_j, 1 \leq j \leq k\}$, where the G_j are open sets. To verify (i), we note that if $K_j = G_j^c$ and $f_j^n(x) = 1 \wedge n\rho(x, K_j)$, where $\rho(x, K) = \inf\{|x - y| : y \in K\}$, then f_j^n are continuous functions with $f_j^n \uparrow 1_{G_j}$ as $n \uparrow \infty$. The facts that

$$Y_s^n(\omega) = f_0^n(s) \prod_{j=1}^{k} f_j^n(\omega(s_j)) \in \mathcal{H}$$

and (iii) holds for $\mathcal{H}$ imply that $1_A \in \mathcal{H}$. This verifies (i) in the monotone class theorem and completes the proof. □

Exercises

7.3.1 Let A be an F_σ, that is, a countable union of closed sets. Show that $T_A = \inf\{t : B_t \in A\}$ is a stopping time.

7.3.2 If S and T are stopping times, then $S \wedge T = \min\{S, T\}$, $S \vee T = \max\{S, T\}$, and $S + T$ are also stopping times. In particular, if $t \geq 0$, then $S \wedge t$, $S \vee t$, and $S + t$ are stopping times.

7.3.3 Let T_n be a sequence of stopping times. Show that

$$\sup_n T_n, \quad \inf_n T_n, \quad \limsup_n T_n, \quad \liminf_n T_n$$

are stopping times.

7.3.4 Let S be a stopping time, let $A \in \mathcal{F}_S$, and let $R = S$ on A and $R = \infty$ on A^c. Show that R is a stopping time.

7.3.5 Let S and T be stopping times. $\{S < T\}$, $\{S > T\}$, and $\{S = T\}$ are in $\mathcal{F}_S$ (and in $\mathcal{F}_T$).

7.3.6 Let S and T be stopping times. $\mathcal{F}_S \cap \mathcal{F}_T = \mathcal{F}_{S \wedge T}$.

7.4 Path Properties

In this section, we will use the strong Markov property to derive properties of the zero set $\{t : B_t = 0\}$, the hitting times $T_a = \inf\{t : B_t = a\}$, and $\max_{0 \le s \le t} B_s$ for one-dimensional Brownian motion.

7.4.1 Zeros of Brownian Motion

Let $R_t = \inf\{u > t : B_u = 0\}$ and let $T_0 = \inf\{u > 0 : B_u = 0\}$. Now Theorem 7.2.9 implies $P_x(R_t < \infty) = 1$, so $B(R_t) = 0$ and the strong Markov property and Theorem 7.2.5 imply

$$P_x(T_0 \circ \theta_{R_t} > 0 | \mathcal{F}_{R_t}) = P_0(T_0 > 0) = 0$$

Taking expected value of the last equation, we see that

$$P_x(T_0 \circ \theta_{R_t} > 0 \text{ for some rational } t) = 0$$

From this, it follows that if a point $u \in \mathcal{Z}(\omega) \equiv \{t : B_t(\omega) = 0\}$ is isolated on the left (i.e., there is a rational $t < u$ so that $(t, u) \cap \mathcal{Z}(\omega) = \emptyset$), then it is, with probability one,

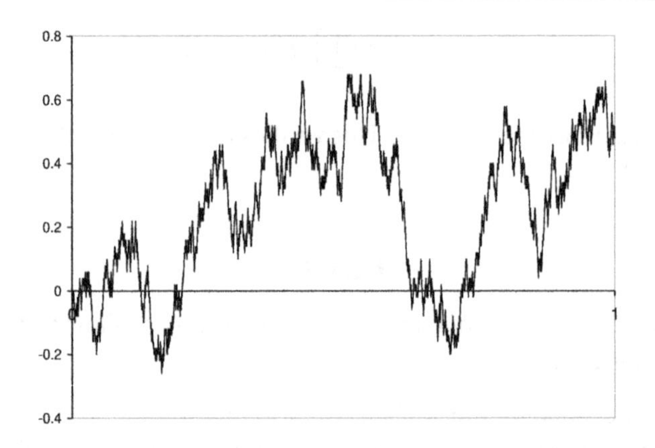

Figure 7.2 Simulation one-dimensional Brownian motion.

a decreasing limit of points in $\mathcal{Z}(\omega)$. This shows that the closed set $\mathcal{Z}(\omega)$ has no isolated points and hence must be uncountable. For the last step, see Hewitt and Stromberg (1965), p. 72.

If we let $|\mathcal{Z}(\omega)|$ denote the Lebesgue measure of $\mathcal{Z}(\omega)$, then Fubini's theorem implies

$$E_x(|\mathcal{Z}(\omega)| \cap [0,T]) = \int_0^T P_x(B_t = 0)\, dt = 0$$

So $\mathcal{Z}(\omega)$ is a set of measure zero.

The last four observations show that $\mathcal{Z}$ is like the Cantor set that is obtained by removing $(1/3, 2/3)$ from $[0,1]$ and then repeatedly removing the middle third from the intervals that remain. The Cantor set is bigger, however. Its Hausdorff dimension is $\log 2/\log 3$, while $\mathcal{Z}$ has dimension 1/2.

7.4.2 Hitting Times

Theorem 7.4.1 *Under P_0, $\{T_a, a \geq 0\}$ has stationary independent increments.*

Proof The first step is to notice that if $0 < a < b$, then

$$T_b \circ \theta_{T_a} = T_b - T_a,$$

so if f is bounded and measurable, the strong Markov property, 7.3.9, and translation invariance imply

$$\begin{aligned} E_0\left(f(T_b - T_a)\,\middle|\mathcal{F}_{T_a}\right) &= E_0\left(f(T_b) \circ \theta_{T_a}\,\middle|\mathcal{F}_{T_a}\right) \\ &= E_a f(T_b) = E_0 f(T_{b-a}) \end{aligned}$$

To show that the increments are independent, let $a_0 < a_1 \ldots < a_n$, let f_i, $1 \leq i \leq n$ be bounded and measurable, and let $F_i = f_i(T_{a_i} - T_{a_{i-1}})$. Conditioning on $\mathcal{F}_{T_{a_{n-1}}}$ and using the preceding calculation, we have

$$E_0\left(\prod_{i=1}^{n} F_i\right) = E_0\left(\prod_{i=1}^{n-1} F_i \cdot E_0(F_n|\mathcal{F}_{T_{a_{n-1}}})\right) = E_0\left(\prod_{i=1}^{n-1} F_i\right) E_0 F_n$$

By induction, it follows that $E_0 \prod_{i=1}^{n} F_i = \prod_{i=1}^{n} E_0 F_i$, which implies the desired conclusion. □

The scaling relation (7.1.1) implies

$$T_a \stackrel{d}{=} a^2 T_1 \tag{7.4.1}$$

Combining Theorem 7.4.1 and (7.4.1), we see that $t_k = T_k - T_{k-1}$ are i.i.d. and

$$\frac{t_1 + \cdots + t_n}{n^2} \to T_1$$

so using Theorem 3.8.8, we see that T_a has a stable law. Since we are dividing by n^2 and $T_a \geq 0$, the index $\alpha = 1/2$ and the skewness parameter $\kappa = 1$, see (3.8.11).

Without knowing the theory mentioned in the previous paragraph, it is easy to determine the Laplace transform

$$\varphi_a(\lambda) = E_0 \exp(-\lambda T_a) \quad \text{for } a \geq 0$$

and reach the same conclusion. To do this, we start by observing that Theorem 7.4.1 implies

$$\varphi_x(\lambda)\varphi_y(\lambda) = \varphi_{x+y}(\lambda).$$

It follows easily from this that

$$\varphi_a(\lambda) = \exp(-ac(\lambda)) \tag{7.4.2}$$

Proof Let $c(\lambda) = -\log \varphi_1(\lambda)$, so (7.4.2) holds when $a = 1$. Using the previous identity with $x = y = 2^{-m}$ and induction gives the result for $a = 2^{-m}$, $m \geq 1$. Then, letting $x = k2^{-m}$ and $y = 2^{-m}$, we get the result for $a = (k+1)2^{-m}$ with $k \geq 1$. Finally, to extend to $a \in [0, \infty)$, note that $a \to \phi_a(\lambda)$ is decreasing. □

To identify $c(\lambda)$, we observe that (7.4.1) implies

$$E \exp(-T_a) = E \exp(-a^2 T_1)$$

so $ac(1) = c(a^2)$, i.e., $c(\lambda) = c(1)\sqrt{\lambda}$. Since all of our arguments also apply to σB_t we cannot hope to compute $c(1)$. Theorem 7.5.7 will show

$$E_0(\exp(-\lambda T_a)) = \exp(-a\sqrt{2\lambda}) \tag{7.4.3}$$

Our next goal is to compute the distribution of the hitting times T_a. This application of the strong Markov property shows why we want to allow the function Y that we apply to the shifted path to depend on the stopping time S.

Example 7.4.2 (Reflection principle) Let $a > 0$ and let $T_a = \inf\{t : B_t = a\}$. Then

$$P_0(T_a < t) = 2P_0(B_t \geq a) \tag{7.4.4}$$

Intuitive proof. We observe that if B_s hits a at some time $s < t$, then the strong Markov property implies that $B_t - B(T_a)$ is independent of what happened before time T_a. The symmetry of the normal distribution and $P_a(B_u = a) = 0$ for $u > 0$ then imply

$$P_0(T_a < t, B_t > a) = \frac{1}{2} P_0(T_a < t) \tag{7.4.5}$$

Rearranging the last equation and using $\{B_t > a\} \subset \{T_a < t\}$ gives

$$P_0(T_a < t) = 2P_0(T_a < t, B_t > a) = 2P_0(B_t > a)$$

Proof To make the intuitive proof rigorous, we only have to prove (7.4.5). To extract this from the strong Markov property, Theorem 7.3.9, we let

$$Y_s(\omega) = \begin{cases} 1 & \text{if } s < t,\ \omega(t-s) > a \\ 0 & \text{otherwise} \end{cases}$$

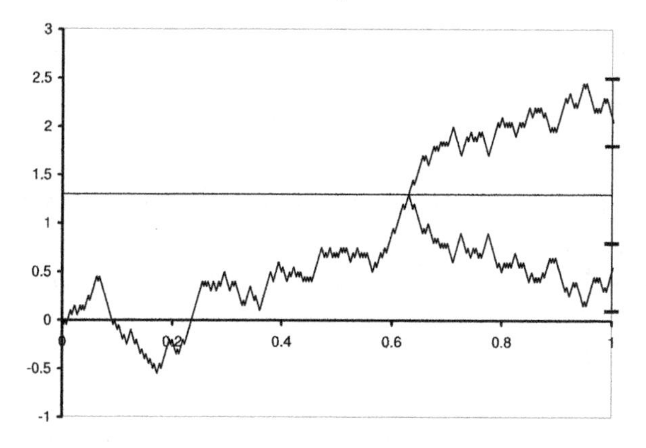

Figure 7.3 Proof by picture of the reflection principle.

We do this so that if we let $S = \inf\{s < t : B_s = a\}$ with $\inf \emptyset = \infty$, then

$$Y_S(\theta_S \omega) = \begin{cases} 1 & \text{if } S < t,\, B_t > a \\ 0 & \text{otherwise} \end{cases}$$

and the strong Markov property implies

$$E_0(Y_S \circ \theta_S | \mathcal{F}_S) = \varphi(B_S, S) \quad \text{on } \{S < \infty\} = \{T_a < t\}$$

where $\varphi(x, s) = E_x Y_s$. $B_S = a$ on $\{S < \infty\}$ and $\varphi(a, s) = 1/2$ if $s < t$, so taking expected values gives

$$\begin{aligned} P_0(T_a < t, B_t \geq a) &= E_0(Y_S \circ \theta_S; S < \infty) \\ &= E_0(E_0(Y_S \circ \theta_S | \mathcal{F}_S); S < \infty) = E_0(1/2; T_a < t) \end{aligned}$$

which proves (7.4.5). □

Using (7.4.4), we can compute the probability density of T_a. We begin by noting that

$$P(T_a \leq t) = 2\, P_0(B_t \geq a) = 2 \int_a^\infty (2\pi t)^{-1/2} \exp(-x^2/2t) dx$$

then change variables $x = (t^{1/2}a)/s^{1/2}$ to get

$$\begin{aligned} P_0(T_a \leq t) &= 2 \int_t^0 (2\pi t)^{-1/2} \exp(-a^2/2s) \left(-t^{1/2}a/2s^{3/2}\right) ds \\ &= \int_0^t (2\pi s^3)^{-1/2} a \exp(-a^2/2s)\, ds \end{aligned} \tag{7.4.6}$$

Using the last formula, we can compute:

Example 7.4.3 The distribution of $L = \sup\{t \le 1 : B_t = 0\}$. By (7.2.4),

$$\begin{aligned} P_0(L \le s) &= \int_{-\infty}^{\infty} p_s(0,x)P_x(T_0 > 1-s)\,dx \\ &= 2\int_0^{\infty} (2\pi s)^{-1/2}\exp(-x^2/2s)\int_{1-s}^{\infty}(2\pi r^3)^{-1/2}x\exp(-x^2/2r)\,dr\,dx \\ &= \frac{1}{\pi}\int_{1-s}^{\infty}(sr^3)^{-1/2}\int_0^{\infty} x\exp(-x^2(r+s)/2rs)\,dx\,dr \\ &= \frac{1}{\pi}\int_{1-s}^{\infty}(sr^3)^{-1/2} rs/(r+s)\,dr \end{aligned}$$

Our next step is to let $t = s/(r+s)$ to convert the integral over $r \in [1-s,\infty)$ into one over $t \in [0,s]$. $dt = -s/(r+s)^2 dr$, so to make the calculations easier we first rewrite the integral as

$$= \frac{1}{\pi}\int_{1-s}^{\infty}\left(\frac{(r+s)^2}{rs}\right)^{1/2}\frac{s}{(r+s)^2}\,dr$$

and then change variables to get

$$P_0(L \le s) = \frac{1}{\pi}\int_0^s (t(1-t))^{-1/2}\,dt = \frac{2}{\pi}\arcsin(\sqrt{s}) \tag{7.4.7}$$

Since L is the time of the last zero before time 1, it is remarkable that the density is symmetric about 1/2 and blows up at 0. The arcsin may remind the reader of the limit theorem for $L_{2n} = \sup\{m \le 2n : S_m = 0\}$ given in Theorem 4.9.5. We will see in Section 8.6 that our new result is a consequence of the old one.

Exercises

7.4.1 Let $B_t = (B_t^1, B_t^2)$ be a two-dimensional Brownian motion starting from 0. Let $T_a = \inf\{t : B_t^2 = a\}$. Show that (i) $B^1(T_a)$, $a \ge 0$ has independent increments and $B^1(T_a) =_d aB^1(T_1)$. Use this to conclude that $B^1(T_a)$ has a Cauchy distirubtion.

7.4.2 Use (7.2.3) to show that $R = \inf\{t > 1 : B_t = 0\}$ has probability density

$$P_0(R = 1+t) = 1/(\pi t^{1/2}(1+t))$$

7.4.3 (a) Generalize the proof of (7.4.5) to conclude that if $u < v \le a$, then

$$P_0(T_a < t, u < B_t < v) = P_0(2a - v < B_t < 2a - u) \tag{7.4.8}$$

(b) Let (u, v) shrink down to x in (7.4.8) to conclude that for $a < x$

$$P_0(T_a < t, B_t = x) = p_t(0, 2a - x)$$

i.e., the (subprobability) density for B_t on $\{T_a < t\}$. (c) Use the fact that $\{T_a < t\} = \{M_t > a\}$ and differentiate with respect to a gives the joint density

$$f_{(M_t,B_t)}(a,x) = \frac{2(2a-x)}{\sqrt{2\pi t^3}}e^{-(2a-x)^2/2t}$$

7.4.4 Let $A_{s,t}$ be the event Brownian motion has at least one zero in $[s,t]$. Show that $P_0(A_{s,t}) = (2/\pi)\arccos(\sqrt{s/t})$. Hint: To get started note that

$$P_0(A_{s,t}) = 2\int_0^\infty (2\pi s)^{-1/2} e^{-x^2/2} P_x(T_0 \le t - s)\, dx$$

and use the formula for the density function of T_x. At the end, you may need some trig to convert arctan into arccos.

Remark Using the previous exercise, it is easy to see that if $s \ge 1$ then $P(A_{s,s+h}) \le Ch^{1/2}$. Thus if we divide the interval $[1,2]$ into n equal pieces, then expected number of intervals that contains zeros is $\le Cn^{-1/2}$. It follows that the Hausdorff dimension of the zero set is at most 1/2.

7.5 Martingales

At the end of Section 5.7, we used martingales to study the hitting times of random walks. The same methods can be used on Brownian motion once we prove:

Theorem 7.5.1 *Let X_t be a right continuous martingale adapted to a right continuous filtration. If T is a bounded stopping time, then $EX_T = EX_0$.*

Proof Let n be an integer so that $P(T \le n-1) = 1$. As in the proof of the strong Markov property, let $T_m = ([2^m T]+1)/2^m$. $Y_k^m = X(k2^{-m})$ is a martingale with respect to $\mathcal{F}_k^m = \mathcal{F}(k2^{-m})$ and $S_m = 2^m T_m$ is a stopping time for $(Y_k^m, \mathcal{F}_k^m)$, so by Exercise 4.4.4

$$X(T_m) = Y_{S_m}^m = E(Y_{n2^m}^m | \mathcal{F}_{S_m}^m) = E(X_n | \mathcal{F}(T_m))$$

As $m \uparrow \infty$, $X(T_m) \to X(T)$ by right continuity and $\mathcal{F}(T_m) \downarrow \mathcal{F}(T)$ by Theorem 7.3.7, so it follows from Theorem 4.7.3 that

$$X(T) = E(X_n | \mathcal{F}(T))$$

Taking expected values gives $EX(T) = EX_n = EX_0$, since X_n is a martingale. □

Theorem 7.5.2 *B_t is a martingale w.r.t. the σ-fields $\mathcal{F}_t$ defined in Section 8.2.*

Note: We will use these σ-fields in all of the martingale results but will not mention them explicitly in the statements.

Proof The Markov property implies that

$$E_x(B_t | \mathcal{F}_s) = E_{B_s}(B_{t-s}) = B_s$$

since symmetry implies $E_y B_u = y$ for all $u \ge 0$. □

From Theorem 7.5.2, it follows immediately that we have:

Theorem 7.5.3 *If $a < x < b$, then $P_x(T_a < T_b) = (b-x)/(b-a)$.*

Proof Let $T = T_a \wedge T_b$. Theorem 7.2.8 implies that $T < \infty$ a.s. Using Theorems 7.5.1 and 7.5.2, it follows that $x = E_x B(T \wedge t)$. Letting $t \to \infty$ and using the bounded convergence theorem, it follows that

$$x = aP_x(T_a < T_b) + b(1 - P_x(T_a < T_b))$$

Solving for $P_x(T_a < T_b)$ now gives the desired result. □

Theorem 7.5.4 $B_t^2 - t$ *is a martingale.*

Proof Writing $B_t^2 = (B_s + B_t - B_s)^2$, we have

$$\begin{aligned} E_x(B_t^2|\mathcal{F}_s) &= E_x(B_s^2 + 2B_s(B_t - B_s) + (B_t - B_s)^2|\mathcal{F}_s) \\ &= B_s^2 + 2B_s E_x(B_t - B_s|\mathcal{F}_s) + E_x((B_t - B_s)^2|\mathcal{F}_s) \\ &= B_s^2 + 0 + (t - s) \end{aligned}$$

since $B_t - B_s$ is independent of $\mathcal{F}_s$ and has mean 0 and variance $t - s$. □

Theorem 7.5.5 *Let* $T = \inf\{t : B_t \notin (a,b)\}$, *where* $a < 0 < b$.

$$E_0 T = -ab$$

Proof Theorem 7.5.1 and 7.5.4 imply $E_0(B^2(T \wedge t)) = E_0(T \wedge t))$. Letting $t \to \infty$ and using the monotone convergence theorem gives $E_0(T \wedge t) \uparrow E_0 T$. Using the bounded convergence theorem and Theorem 7.5.3, we have

$$E_0 B^2(T \wedge t) \to E_0 B_T^2 = a^2 \frac{b}{b-a} + b^2 \frac{-a}{b-a} = ab\frac{a-b}{b-a} = -ab$$ □

Theorem 7.5.6 $\exp(\theta B_t - (\theta^2 t/2))$ *is a martingale.*

Proof Bringing $\exp(\theta B_s)$ outside

$$\begin{aligned} E_x(\exp(\theta B_t)|\mathcal{F}_s) &= \exp(\theta B_s) E(\exp(\theta(B_t - B_s))|\mathcal{F}_s) \\ &= \exp(\theta B_s)\exp(\theta^2(t-s)/2) \end{aligned}$$

since $B_t - B_s$ is independent of $\mathcal{F}_s$ and has a normal distribution with mean 0 and variance $t - s$. □

Theorem 7.5.7 *If* $T_a = \inf\{t : B_t = a\}$, *then* $E_0 \exp(-\lambda T_a) = \exp(-a\sqrt{2\lambda})$.

Proof Theorem 7.5.1 and 7.5.6 imply that $1 = E_0 \exp(\theta B(T \wedge t) - \theta^2(T_a \wedge t)/2)$. Taking $\theta = \sqrt{2\lambda}$, letting $t \to \infty$ and using the bounded convergence theorem gives $1 = E_0 \exp(a\sqrt{2\lambda} - \lambda T_a)$. □

Theorem 7.5.8 *If* $u(t,x)$ *is a polynomial in* t *and* x *with*

$$\frac{\partial u}{\partial t} + \frac{1}{2}\frac{\partial^2 u}{\partial x^2} = 0 \tag{7.5.1}$$

then $u(t, B_t)$ *is a martingale.*

Proof Let $p_t(x,y) = (2\pi)^{-1/2} t^{-1/2} \exp(-(y-x)^2/2t)$. The first step is to check that p_t satisfies the heat equation: $\partial p_t/\partial t = (1/2)\partial^2 p_t/\partial y^2$.

$$\begin{aligned} \frac{\partial p}{\partial t} &= -\frac{1}{2} t^{-1} p_t + \frac{(y-x)^2}{2t^2} p_t \\ \frac{\partial p}{\partial y} &= -\frac{y-x}{t} p_t \\ \frac{\partial^2 p}{\partial y^2} &= -\frac{1}{t} p_t + \frac{(y-x)^2}{t^2} p_t \end{aligned}$$

Interchanging $\partial/\partial t$ and $\int$, and using the heat equation

$$\begin{aligned}\frac{\partial}{\partial t}E_x u(t,B_t) &= \int \frac{\partial}{\partial t}(p_t(x,y)u(t,y))\,dy \\ &= \int \frac{1}{2}\frac{\partial}{\partial y^2}p_t(x,y)u(t,y) + p_t(x,y)\frac{\partial}{\partial t}u(t,y)\,dy\end{aligned}$$

Integrating by parts twice the previous

$$= \int p_t(x,y)\left(\frac{\partial}{\partial t} + \frac{1}{2}\frac{\partial}{\partial y^2}\right)u(t,y)\,dy = 0$$

Since $u(t,y)$ is a polynomial, there is no question about the convergence of integrals and there is no contribution from the boundary terms when we integrate by parts.

At this point, we have shown that $t \to E_x u(t,B_t)$ is constant. To check the martingale property, we need to show that

$$E(u(t,B_t)|\mathcal{F}_s) = u(s,B_s)$$

To do this, we note that $v(r,x) = u(s+r,x)$ satisfies $\partial v/\partial r = (1/2)\partial v^2/\partial x^2$, so using the Markov property

$$\begin{aligned}E(u(t,B_t)|\mathcal{F}_s) &= E(v(t-s,B_{t-s})\circ\theta_s|\mathcal{F}_s) \\ &= E_{B(s)}v(t-s,B_{t-s}) = v(0,B_s) = u(s,B_s)\end{aligned}$$

where in the next to last step we have used the previously proved constancy of the expected value. □

Examples of polynomials that satisfy (7.5.1) are $x, x^2-t, x^3-3tx, x^4-6x^2t+3t^2\dots$ The result can also be extended to $\exp(\theta x - \theta^2 t/2)$. The only place we used the fact that $u(t,x)$ was a polynomial was in checking the convergence of the integral to justify integration by parts.

Theorem 7.5.9 *If* $T = \inf\{t : B_t \notin (-a,a)\}$, *then* $ET^2 = 5a^4/3$.

Proof Theorem 7.5.1 implies

$$E(B(T\wedge t)^4 - 6(T\wedge t)B(T\wedge t)^2) = -3E(T\wedge t)^2.$$

From Theorem 7.5.5, we know that $ET = a^2 < \infty$. Letting $t \to \infty$, using the dominated convergence theorem on the left-hand side, and the monotone convergence theorem on the right gives

$$a^4 - 6a^2 ET = -3E(T^2)$$

Plugging in $ET = a^2$ gives the desired result. □

Exercises

7.5.1 Let $T = \inf\{B_t \notin (-a,a)\}$. Show that

$$E_0\exp(-\lambda T) = 1/\cosh(a\sqrt{2\lambda}).$$

7.5.2 The point of this exercise is to get information about the amount of time it takes Brownian motion with drift $-b$, $X_t \equiv B_t - bt$ to hit level a. Let $\tau = \inf\{t : B_t = a + bt\}$, where $a > 0$. (i) Use the martingale $\exp(\theta B_t - \theta^2 t/2)$ with $\theta = b + (b^2 + 2\lambda)^{1/2}$ to show

$$E_0 \exp(-\lambda\tau) = \exp(-a\{b + (b^2 + 2\lambda)^{1/2}\})$$

Letting $\lambda \to 0$ gives $P_0(\tau < \infty) = \exp(-2ab)$.

7.5.3 Let $\sigma = \inf\{t : B_t \notin (a,b)\}$ and let $\lambda > 0$. Use the strong Markov property to show

$$E_x \exp(-\lambda T_a) = E_x(e^{-\lambda\sigma}; T_a < T_b) + E_x(e^{-\lambda\sigma}; T_b < T_a) E_b \exp(-\lambda T_a)$$

(ii) Interchange the roles of a and b to get a second equation, use Theorem 7.5.7, and solve to get

$$E_x(e^{-\lambda\sigma}; T_a < T_b) = \sinh(\sqrt{2\lambda}(b - x))/\sinh(\sqrt{2\lambda}(b - a))$$
$$E_x(e^{-\lambda\sigma}; T_b < T_a) = \sinh(\sqrt{2\lambda}(x - a))/\sinh(\sqrt{2\lambda}(b - a))$$

7.5.4 If $T = \inf\{t : B_t \notin (a,b)\}$, where $a < 0 < b$ and $a \neq -b$, then T and B_T^2 are not independent, so we cannot calculate ET^2 as we did in the proof of Theorem 7.5.9. Use the Cauchy-Schwarz inequality to estimate $E(TB_T^2)$ and conclude (i) $ET^2 \leq 4\, E(B_T^4)$ and (ii) $EB_T^4 \leq 36ET^2$.

7.5.5 Find a martingale of the form $B_t^6 - c_1 t B_t^4 + c_2 t^2 B_t^2 - c_3 t^3$ and use it to compute the third moment of $T = \inf\{t : B_t \notin (-a,a)\}$.

7.5.6 Show that $(1+t)^{-1/2}\exp(B_t^2/2(1+t))$ is a martingale and use this to conclude that $\limsup_{t\to\infty} B_t/((1+t)\log(1+t))^{1/2} \leq 1$ a.s.

7.6 Itô's Formula*

Our goal in this section is to prove three versions of Itô's formula. The first one, for one-dimensional Brownian motion is:

Theorem 7.6.1 *Suppose that $f \in C^2$, i.e., it has two continuous derivatives. With probability one, for all $t \geq 0$,*

$$f(B_t) - f(B_0) = \int_0^t f'(B_s)\, dB_s + \frac{1}{2}\int_0^t f''(B_s)\, ds \tag{7.6.1}$$

Of course, as part of the derivation we have to explain the meaning of the first term on the right-hand side. In contrast, if A_s is continuous and has bounded variation on each finite interval, and $f \in C^1$:

$$f(A_t) - f(A_0) = \int_0^t f'(A_s)\, dA_s$$

where the Riemann-Steiltjes integral on the right-hand side is

$$\lim_{n\to\infty} \sum_{i=1}^{k(n)} f'(A_{t_{i-1}^n})[A(t_i^n) - A(t_{i-1}^n)]$$

over a sequence of partitions $0 = t_0^n < t_1^n < \ldots < t_{k(n)}^n = t$ with mesh $\max_i t_i^n - t_{i-1}^n \to 0$.

Proof To derive (7.6.1), we let $t_i^n = ti/2^n$ for $0 \le i \le k(n) = 2^n$. From calculus we know that for any a and b there is a $c(a,b)$ in between a and b such that

$$f(b) - f(a) = (b-a)f'(a) + \frac{1}{2}(b-a)^2 f''(c(a,b)) \tag{7.6.2}$$

Using the calculus fact, it follows that

$$\begin{aligned} f(B_t) - f(B_0) &= \sum_i f(B_{t_{i+1}^n}) - f(B_{t_i^n}) \\ &= \sum_i f'(B_{t_i^n})(B_{t_{i+1}^n} - B_{t_i^n}) \\ &\quad + \frac{1}{2}\sum_i g_i^n(\omega)(B_{t_{i+1}^n} - B_{t_i^n})^2 \end{aligned} \tag{7.6.3}$$

where $g_i^n(\omega) = f''(c(B_{t_i^n}, B_{t_{i+1}^n}))$. Comparing (7.6.3) with (7.6.1), it becomes clear that we want to show

$$\sum_i f'(B_{t_i^n})(B_{t_{i+1}^n} - B_{t_i^n}) \to \int_0^t f'(B_s)\,dB_s \tag{7.6.4}$$

$$\frac{1}{2}\sum_i g_i(\omega)(B_{t_{i+1}^n} - B_{t_i^n})^2 \to \frac{1}{2}\int_0^t f''(B_s)\,ds \tag{7.6.5}$$

in probability as $n \to \infty$.

We will prove the result first under the assumption that $|f'(x)|, |f''(x)| \le K$. Let

$$I_m^1 \equiv \sum_i f'(B_{t_i^m})(B_{t_{i+1}^m} - B_{t_i^m})$$

In order to compare I_m^1 and I_n^1 with $n \ge m$, we want to further subdivide the intervals of length $t2^{-m}$ to be of length $t2^{-n}$. To leave the sum unchanged, we have to redefine things. Let $i(n,m,j) = [2^m j/2^n]/2^m$ and $H_j^m = f'(B_{i(n,m,j)})$. If we let $K_j^n = f'(B_{t_i^n})$, then we have

$$I_m^1 - I_n^1 = \sum_j (H_j^m - K_j^n)(B_{t_{j+1}^n} - B_{t_j^n})$$

Squaring, we have

$$\begin{aligned} (I_m^1 - I_n^1)^2 &= \sum_j (H_j^m - K_j^n)^2 (B_{t_{j+1}^n} - B_{t_j^n})^2 \\ &\quad + 2\sum_{j<k} (H_j^m - K_j^n)(H_k^m - K_k^n)(B_{t_{j+1}^n} - B_{t_j^n})(B_{t_{k+1}^n} - B_{t_k^n}) \end{aligned}$$

The next step is to take expected value. The expected value of the second sum vanishes because if we take expected value of of the j,k term with respect to $\mathcal{F}_{t_k^n}$, the first three terms can be taken outside the expected value

$$(H_j^m - K_j^n)(H_k^m - K_k^n)(B_{t_{j+1}^n} - B_{t_j^n})E(B_{t_{k+1}^n} - B_{t_k^n}|\mathcal{F}_{t_k^n}) = 0$$

Conditioning on $\mathcal{F}_{t_j^n}$, in the first sum we have

$$E(I_m^1 - I_n^1)^2 = \sum_j E(H_j^m - K_j^n)^2 \cdot 2^{-n} t \tag{7.6.6}$$

Since $|i(n,m,j)/2^m - j/2^n| \le 1/2^m$ and $|f''(x)| \le K$, we have

$$\sup_{j \le 2^n} |H_j^m - K_j^n| \le K \sup\{|B_s - B_r| : 0 \le r \le s \le t, |s - r| \le 2^{-m}\}$$

Since Brownian motion is uniformly continuous on $[0,t]$ and $E \sup_{s \le t} |B_s|^2 < \infty$, it follows from the dominated convergence theorem that

$$\lim_{m,n \to \infty} E(I_m^1 - I_n^1)^2 = 0$$

This shows that I_m^1 is a Cauchy sequence in L^2, so there is an I_∞^1 so that $I_m^1 \to I_\infty^1$ in L^2. The limit defines $\int_0^t f'(B_s)\, dB_s$ as a limit of approximating Riemann sums.

To prove (7.6.5), we let $G_s^n = g_i^n(\omega) = f''(c(B_{t_i^n}, B_{t_{i+1}^n}))$ when $s \in (t_i^n, t_{i+1}^n]$ and let

$$A_s^n = \sum_{t_{i+1}^n \le s} (B_{t_{i+1}^n} - B_{t_i^n})^2$$

so that

$$\sum_i g_i^n(\omega)(B_{t_{i+1}^n} - B_{t_i^n})^2 = \int_0^t G_s^n dA_s^n$$

and what we want to show is

$$\int_0^t G_s^n dA_s^n \to \int_0^t f''(B_s)\, ds \tag{7.6.7}$$

To do this, we begin by observing that continuity of f'' implies that if $s_n \to s$ as $n \to \infty$, then $G_{s_n}^n \to f''(B_s)$, while Exercise 7.1.6 implies that A_s^n converges almost surely to s. At this point, we can fix ω and deduce (7.6.7) from the following simple result:

Lemma 7.6.2 *If (i) measures μ_n on $[0,t]$ converge weakly to μ_∞, a finite measure, and (ii) g_n is a sequence of functions with $|g_n| \le K$ that have the property that whenever $s_n \in [0,t] \to s$ we have $g_n(s_n) \to g(s)$, then as $n \to \infty$*

$$\int g_n d\mu_n \to \int g\, d\mu_\infty$$

Proof By letting $\mu_n'(A) = \mu_n(A)/\mu_n([0,t])$, we can assume that all the μ_n are probability measures. A standard construction (see Theorem 3.2.8) shows that there is a sequence of random variables X_n with distribution μ_n so that $X_n \to X_\infty$ a.s. as $n \to \infty$ The convergence of g_n to g implies $g_n(X_n) \to g(X_\infty)$, so the result follows from the bounded convergence theorem. □

Lemma 7.6.2 is the last piece in the proof of Theorem under the additional assumptions that $|f'(x)|, |f''(x)| \le M$. Tracing back through the proof, we see that Lemma 7.6.2 implies

(7.6.7), which in turn completes the proof of (7.6.5). So adding (7.6.4) and using (7.6.3) gives that

$$f(B_t) - f(B_0) = \int_0^t f'(B_s)\,dB_s + \frac{1}{2}\int_0^t f''(B_s)\,ds \quad \text{a.s.}$$

To remove the assumed bounds on the derivatives, we let M be large and define f_M with $f_M = f$ on $[-M, M]$ and $|f_M'|, |f_M''| \le M$. The formula holds for f_M but on $\{\sup_{s \le t} |B_s| \le M\}$ there is no difference in the formula between using f_M and f. □

Example 7.6.3 Taking $f = x^2$ in (7.6.1), we have

$$B_t^2 - B_0^2 = \int_0^t 2B_s\,dB_s + t$$

so if $B_0 = 0$, we have

$$\int_0^t 2B_s\,dB_s = B_t^2 - t$$

In contrast to the calculus formula $\int_0^a 2x\,dx = a^2$.

Direct proof. Using the subdivision we introduced previously, we can write

$$B_t^2 = \left(\sum_{i=0}^{2^n-1} (B(t_{i+1}^n - B(t_i^n) \right)^2$$
$$+ \sum_{i=0}^{2^n-1} \sum_{j=0}^{2^n-1} (B(t_{i+1}^n - B(t_i^n)(B(t_{j+1}^n - B(t_j^n)$$

If we sum over $j < i$, this becomes

$$\sum_{i=0}^{2^n-1} B(t^n - i)(B(t_{i+1}^n - B(t_i^n)) \to \int_0^t B_s\,dB_s$$

We get the same limit for the sum over $i < j$. When we sum over $i = j$

$$\sum_{i=0}^{2^n-1} (B(t_{i+1}^n - B(t_i^n)^2 \to t$$

To give one reason why the answer cannot be $B_t^2/2$, we note that:

Theorem 7.6.4 *If g is continuous and $E\int_0^t |g(B_s)|^2\,ds < \infty$, then $\int_0^t g(B_s)\,dB_s$ is a continuous martingale.*

If g grows too fast at ∞, then the integrability condition might not hold. To deal with this, we introduce a weaker concept: M_t is a local martingale if there is an increasing sequence of stopping times $T_n \uparrow \infty$ so that $M(t \wedge T_n)$ is a martingale. If g is continuous, then the stochastic integral is always a local martingale.

Proof We first prove the result assuming $|g| \le K$. Let $t_i^n = ti/2^n$ for $0 \le i \le 2^n$, let $J = \max\{i : t_{i+1}^n \le s\}$, and let

$$I_n^1(s) = \sum_{i=0}^{J} g(B_{t_i^n})(B_{t_{i+1}^n} - B_{t_i^n}) + g(B_{t_{J+1}^n})(B_s - B_{t_{J+1}^n})$$

It is easy to check that $I_n^1(s)$, $s \le t$ is a continuous martingale with

$$EI_n^1(t)^2 = \sum_{0 \le i < 2^n} E[g(B_{t_i^n})^2] \cdot t2^{-n}.$$

The L^2 maximal inequality implies that

$$E\left(\sup_{s \le t} |I_m^1(s) - I_n^1(s)|^2\right) \le 4E(I_m^1 - I_n^1)^2$$

The right-hand side $\to 0$ by the previous proof, so $I_m^1(s) \to I_\infty^1(s)$ uniformly on $[0,t]$. Let $Y \in \mathcal{F}_r$ with $EY^2 < \infty$. Since $I_n^1(s)$ is a martingale, if $r < s < t$, we have

$$E[Y\{I_n^1(s) - I_n^1(r)\}] = 0$$

From this and the Cauchy-Schwarz inequality we conclude that

$$E[Y\{I_\infty^1(s) - I_\infty^1(r)\}] = 0$$

Since this holds for $Y = 1_A$ with $A \in \mathcal{F}_r$, we have proved the desired result under the assumption $|g| \le K$.

Approximating g by $g_M = g$ on $[-M, M]$ and g_M bounded as before and stopping at time $T_M = \inf\{t : |B_t| \ge M\}$, we can conclude that if g is continuous, then

$$\int_0^{s \wedge T_M} g(B_r)\, dB_r$$

is a continuous martingale with

$$E\left(\int_0^{s \wedge T_M} f'(B_r)\, dB_r\right)^2 = E\int_0^{s \wedge T_M} f'(B_r)^2\, dr$$

Letting $M \to \infty$ it follows that if $E\int_0^t |f'(B_s)|^2\, ds < \infty$, then $\int_0^t f'(B_s)\, dB_s$ is a continuous martingale. □

Our second version of Itô's formula allows f to depend on t as well.

Theorem 7.6.5 *Suppose $f \in C^2$, i.e., it has continuous partial derivatives up to order two. Then with probability 1, for all $t \ge 0$*

$$f(t, B_t) - f(0, B_0) = \int_0^t \frac{\partial f}{\partial t}(s, B_s)\, ds + \int_0^t \frac{\partial f}{\partial x}(s, B_s)\, dB_s + \frac{1}{2}\int_0^t \frac{\partial^2 f}{\partial x^2}(s, B_s)\, ds \tag{7.6.8}$$

Proof Perhaps the simplest way to give a completely rigorous proof of this result (and the next one) is to prove the result when f is a polynomial in t and x and then note that for any $f \in C^2$ and $M < \infty$, we can find polynomials ϕ_n so that ϕ_n and all of its partial derivatives up to order two converge to those of f uniformly on $[0,t] \times [-M, M]$. Here, we will content ourselves to explain why this result is true.

Expanding to second order and being a little less precise about the error term

$$f(t,b) - f(s,a) \approx \frac{\partial f}{\partial t}(s,a)(t-s) + \frac{\partial f}{\partial x}(s,a)(b-a)$$
$$+ \frac{1}{2}\frac{\partial^2 f}{\partial x^2}(s,a)(b-a)^2 + \frac{\partial^2 f}{\partial t \partial x}(s,a)(t-s)(b-a) + \frac{1}{2}\frac{\partial^2 f}{\partial t^2}(s,a)(t-s)^2$$

Let t be a fixed positive number and let $t_k^n = tk/2^n$. As before, we first prove the result under the assumption that all of the first- and second-order partial derivatives are bounded by K. The details of the convergence of the first three terms to their limits is similar to the previous proof. To see that the fifth term vanishes in the limit, use the fact that $t_{i+1}^n - t_i^n = 2^{-n}$ to conclude that

$$\sum_i \left| \frac{\partial^2 f}{\partial x^2}(t_i^n, B(t_i^n)) \right| (t_{i+1}^n - t_i^n)^2 \leq Kt2^{-n} \to 0$$

To treat the fourth term, we first consider

$$V_n^{1,1}(t) = \sum_i (t_{i+1}^n - t_i^n)|B(t_{i+1}^n) - B(t_i^n)|$$

$EV_n^{1,1}(t) = \sum_i C_1(t_{i+1}^n - t_i^n)^{3/2}$, where $C_1 = E\chi^{1/2}$ and χ is a standard normal. Arguing as before,

$$E\sum_i \left| \frac{\partial^2 f}{\partial x^2}(t_i^n, B(t_i^n)) \right| (t_{i+1}^n - t_i^n)|B(t_{i+1}^n) - B(t_i^n)| \leq KC_1 t \cdot 2^{-n/2} \to 0$$

Using Markov's inequality now, we see that if $\epsilon > 0$

$$P\left(\sum_i \left| \frac{\partial^2 f}{\partial x^2}(t_i^n, B(t_i^n)) \right| (t_{i+1}^n - t_i^n)|B(t_{i+1}^n) - B(t_i^n)| > \epsilon \right) \leq KC_1 t 2^{-n/2}/\epsilon \to 0$$

and hence the sum converges to 0 in probability.

To remove the assumption that the partial derivatives are bounded, let $T_M = \min\{t : |B_t| > M\}$ and use the previous calculation to conclude that the formula holds up to time $T_M \wedge t$. □

Example 7.6.6 Using (7.6.8), we can prove Theorem 7.5.8: if $u(t,x)$ is a polynomial in t and x with $\partial u/\partial t + (1/2)\partial^2 u/\partial x^2 = 0$, then Itô's formula implies

$$u(t, B_t) - u(0, B_0) = \int_0^t \frac{\partial u}{\partial x}(s, B_s)\, dB_s$$

Since $\partial u/\partial x$ is a polynomial, it satisfies the integrability condition in Theorem 7.6.4; $u(t, B_t)$ is a martingale. To get a new conclusion, let $u(t,x) = \exp(\mu t + \sigma x)$. Exponential Brownian motion $X_t = u(t, B_t)$ is often used as a model of the price of a stock. Since

$$\frac{\partial u}{\partial t} = \mu u \qquad \frac{\partial u}{\partial x} = \sigma u \qquad \frac{\partial^2 u}{\partial x^2} = \sigma^2 u$$

Itô's formula gives

$$X_t - X_0 = \int_0^t \left(\mu - \frac{\sigma^2}{2}\right) X_s\, ds + \int_0^t \sigma X_s\, dB_s$$

From this we see that if $\mu = \sigma^2/2$, then X_t is a martingale, but this formula gives more information about how X_t behaves.

Our third and final version of Itô's formula is for multi-dimensional Brownian motion. To simplify notation we write $D_i f = \partial f/\partial x_i$ and $D_{ij} = \partial^2 f/\partial x_i \partial x_j$.

Theorem 7.6.7 *Suppose $f \in C^2$. Then with probability 1, for all $t \geq 0$*

$$\begin{aligned} f(t, B_t) - f(0, B_0) = &\int_0^t \frac{\partial f}{\partial t}(s, B_s)\, ds + \sum_{i=1}^d \int_0^t D_i f(B_s)\, dB_s^i \\ &+ \frac{1}{2} \sum_{i=1}^d \int_0^t D_{ii} f(B_s)\, ds \end{aligned} \tag{7.6.9}$$

Proof Let $x_0 = t$ and write $D_0 f = \partial f/\partial x_0$. Expanding to second order and again not worrying about the error term

$$\begin{aligned} f(y) - f(x) \approx &\sum_i D_i f(x)(y_i - x_i) \\ &+ \frac{1}{2} \sum_i D_{ii} f(x)(y_i - x_i)^2 + \sum_{i<j} D_{ij} f(x)(y_i - x_i)(y_j - x_j) \end{aligned}$$

Let t be a fixed positive number and let $t_i^n ti/2^n$. The details of the convergence of the first two terms to their limits are similar to the previous two proofs. To handle the third term, we let

$$F_{i,j}(t_k^n) = D_{ij} f(B(t_k^n))$$

and note that independence of coordinates implies

$$\begin{aligned} &E(F_{i,j}(t_k^n)(B^i(t_{k+1}^n) - B^i(t_k^n)) \cdot (B^j(t_{k+1}^n) - B^j(t_k^n))|\mathcal{F}_{t_k^n}) \\ &\quad = F_{i,j}(t_k^n) E((B^i(t_{k+1}^n) - B^i(t_k^n)) \cdot (B^j(t_{k+1}^n) - B^j(t_k^n))|\mathcal{F}_{t_k^n}) = 0 \end{aligned}$$

Fix $i < j$ and let $G_k = F_{i,j}(t_k^n)(B^i(t_{k+1}^n) - B^i(t_k^n)) \cdot (B^j(t_{k+1}^n) - B^j(t_k^n))$. The last calculation shows that

$$E\left(\sum_k G_k\right) = 0$$

To compute the variance of the sum, we note that

$$E\left(\sum_k G_k\right)^2 = E \sum_k G_k^2 + 2E \sum_{k<\ell} G_k G_\ell$$

Again we first complete the proof under the assumption that $|\partial^2 f/\partial x_i \partial x_j| \leq C$. In this case, conditioning on $\mathcal{F}_{t_k^n}$ in order to bring $F_{i,j}^2(t_k^n)$ outside the first term,

$$E(F_{i,j}^2(t_k^n)(B^i(t_{k+1}^n) - B^i(t_k^n))^2 \cdot (B^j(t_{k+1}^n) - B^j(t_k^n))^2 | \mathcal{F}_{t_k^n})$$
$$= F_{i,j}^2(t_k^n) E((B^i(t_{k+1}^n) - B^i(t_k^n))^2 \cdot (B^j(t_{k+1}^n) - B^j(t_k^n))^2 | \mathcal{F}_{t_k^n}) = F_{i,j}^2(t_k^n)(t_{k+1}^n - t_k^n)^2$$

and it follows that

$$E \sum_k G_k^2 \leq K^2 \sum_k (t_{k+1}^n - t_k^n)^2 \leq K^2 t 2^{-n} \to 0$$

To handle the cross terms, we condition on $\mathcal{F}_{t_\ell^n}$ to get

$$E(G_k G_\ell | \mathcal{F}_{t_k^n}) = G_k F_{i,j}(t_\ell^n) E((B^i(t_{\ell+1}^n) - B^i(t_\ell^n))(B^j(t_{\ell+1}^n) - B^j(t_\ell^n)) | \mathcal{F}_{t_\ell^n}) = 0$$

At this point, we have shown $E\left(\sum_k G_k\right)^2 \to 0$ and the desired result follows from Markov's inequality. The assumption that the second derivatives are bounded can be removed as before. □

Exercises

7.6.1 Suppose $f \in C^2$. Show that $f(B_t)$ is a martingale if and only if $f(x) - a + bx$.

7.6.2 Show that $B_t^3 - 3tB_t$ and $B_t^3 - \int_0^t 3B_s\,ds$ are a martingale.

7.6.3 Let $\beta_{2k}(t) = E_0 B_t^{2k}$. Use Itô's formula to relate $\beta_{2k}(t)$ to $\beta_{2k-2}(t)$ and use this relationship to derive a formula for $\beta_{2k}(t)$.

7.6.4 Apply Itô's formula to $|B_t|$.

7.6.5 Apply Itô's formula to $|B_t|^2$. Use this to conclude that $E_0|B_t|^2 = td$.

8

Applications to Random Walk

The key to the developments in the section is that we can embed sums of mean zero independent random variables (see Section 8.1) and martingales (see Section 8.2) into Brownian motion and derive central limit theorems. These results lead in Section 8.3 to CLTs for stationary sequences and in Section 8.4 to the convergence of rescaled empirical distributions to Brownian Bridge. Finally, in Section 8.5 we use the embedding to prove a law of the iterated logarithm.

8.1 Donsker's Theorem

In this section we will prove Donsker's (1951) theorem. The key to its proof is:

Theorem 8.1.1 (Skorokhod's representation theorem) *If $EX = 0$ and $EX^2 < \infty$, then there is a stopping time T for Brownian motion so that $B_T =_d X$ and $ET = EX^2$.*

Remark The Brownian motion in the statement and all the Brownian motions in this section have $B_0 = 0$.

Proof Suppose first that X is supported on $\{a,b\}$, where $a < 0 < b$. Since $EX = 0$, we must have

$$P(X = a) = \frac{b}{b-a} \qquad P(X = b) = \frac{-a}{b-a}$$

If we let $T = T_{a,b} = \inf\{t : B_t \notin (a,b)\}$, then Theorem 7.5.3 implies $B_T =_d X$ and Theorem 7.5.5 tells us that

$$ET = -ab = EB_T^2$$

To treat the general case, we will write $F(x) = P(X \le x)$ as a mixture of two point distributions with mean 0. Let

$$c = \int_{-\infty}^{0} (-u)\, dF(u) = \int_0^{\infty} v\, dF(v)$$

If φ is bounded and $\varphi(0) = 0$, then using the two formulas for c

$$
\begin{aligned}
c\int \varphi(x)\,dF(x) &= \left(\int_0^\infty \varphi(v)\,dF(v)\right)\int_{-\infty}^0 (-u)dF(u) \\
&\quad + \left(\int_{-\infty}^0 \varphi(u)\,dF(u)\right)\int_0^\infty v\,dF(v) \\
&= \int_0^\infty dF(v)\int_{-\infty}^0 dF(u)\,(v\varphi(u) - u\varphi(v))
\end{aligned}
$$

So we have

$$
\int \varphi(x)\,dF(x) = c^{-1}\int_0^\infty dF(v)\int_{-\infty}^0 dF(u)(v-u)\left\{\frac{v}{v-u}\varphi(u) + \frac{-u}{v-u}\varphi(v)\right\}
$$

The last equation gives the desired mixture. If we let $(U,V) \in \mathbf{R}^2$ have

$$
\begin{aligned}
&P\{(U,V) = (0,0)\} = F(\{0\}) \\
&P((U,V) \in A) = c^{-1}\iint_{(u,v)\in A} dF(u)\,dF(v)\,(v-u)
\end{aligned}
\tag{8.1.1}
$$

for $A \subset (-\infty,0) \times (0,\infty)$ and define probability measures by $\mu_{0,0}(\{0\}) = 1$ and

$$
\mu_{u,v}(\{u\}) = \frac{v}{v-u} \qquad \mu_{u,v}(\{v\}) = \frac{-u}{v-u} \qquad \text{for } u < 0 < v
$$

then

$$
\int \varphi(x)\,dF(x) = E\left(\int \varphi(x)\,\mu_{U,V}(dx)\right)
$$

We proved the last formula when $\varphi(0) = 0$, but it is easy to see that it is true in general. Letting $\varphi \equiv 1$ in the last equation shows that the measure defined in (8.1.1) has total mass 1.

From these calculations it follows that if we have (U,V) with distribution given in (8.1.1) and an independent Brownian motion defined on the same space, then $B(T_{U,V}) =_d X$. Sticklers for detail will notice that $T_{U,V}$ is not a stopping time for B_t, since (U,V) is independent of the Brownian motion. This is not a serious problem, since if we condition on $U = u$ and $V = v$, then $T_{u,v}$ is a stopping time, and this is good enough for all the following calculations. For instance, to compute $E(T_{U,V})$ we observe

$$
E(T_{U,V}) = E\{E(T_{U,V}|(U,V))\} = E(-UV)
$$

by Theorem 7.5.5, (8.1.1) implies

$$
\begin{aligned}
E(-UV) &= \int_{-\infty}^0 dF(u)(-u)\int_0^\infty dF(v)v(v-u)c^{-1} \\
&= \int_{-\infty}^0 dF(u)(-u)\left\{-u + \int_0^\infty dF(v)c^{-1}v^2\right\}
\end{aligned}
$$

since

$$
c = \int_0^\infty v\,dF(v) = \int_{-\infty}^0 (-u)\,dF(u)
$$

Using the second expression for c now gives

$$E(T_{U,V}) = E(-UV) = \int_{-\infty}^{0} u^2 dF(u) + \int_{0}^{\infty} v^2 dF(v) = EX^2 \qquad \square$$

From Theorem 8.1.1, it is only a small step to:

Theorem 8.1.2 *Let $X_1, X_2, \ldots$ be i.i.d. with a distribution F, which has mean 0 and variance 1, and let $S_n = X_1 + \cdots + X_n$. There is a sequence of stopping times $T_0 = 0, T_1, T_2, \ldots$ such that $S_n =_d B(T_n)$ and $T_n - T_{n-1}$ are independent and identically distributed.*

Proof Let $(U_1, V_1), (U_2, V_2), \ldots$ be i.i.d. and have distribution given in (8.1.1) and let B_t be an independent Brownian motion. Let $T_0 = 0$, and for $n \geq 1$, let

$$T_n = \inf\{t \geq T_{n-1} : B_t - B(T_{n-1}) \notin (U_n, V_n)\} \qquad \square$$

As a corollary of Theorem 8.1.2, we get:

Theorem 8.1.3 (Central limit theorem) *Under the hypotheses of Theorem 8.1.2, $S_n/\sqrt{n} \Rightarrow \chi$, where χ has the standard normal distribution.*

Proof If we let $W_n(t) = B(nt)/\sqrt{n} =_d B_t$ by Brownian scaling, then

$$S_n/\sqrt{n} \stackrel{d}{=} B(T_n)/\sqrt{n} = W_n(T_n/n)$$

The weak law of large numbers implies that $T_n/n \to 1$ in probability. It should be clear from this that $S_n/\sqrt{n} \Rightarrow B_1$. To fill in the details, let $\epsilon > 0$, pick δ so that

$$P(|B_t - B_1| > \epsilon \text{ for some } t \in (1-\delta, 1+\delta)) < \epsilon/2$$

then pick N large enough so that for $n \geq N$, $P(|T_n/n - 1| > \delta) < \epsilon/2$. The last two estimates imply that for $n \geq N$

$$P(|W_n(T_n/n) - W_n(1)| > \epsilon) < \epsilon$$

Since ϵ is arbitrary, it follows that $W_n(T_n/n) - W_n(1) \to 0$ in probability. Applying the converging together lemma, Lemma 3.2.13, with $X_n = W_n(1)$ and $Z_n = W_n(T_n/n)$, the desired result follows. $\square$

Let $\mathbf{N}$ be the nonnegative integers and let

$$S(u) = \begin{cases} S_k & \text{if } u = k \in \mathbf{N} \\ \text{linear on } [k, k+1] & \text{for } k \in \mathbf{N} \end{cases}$$

Later in this section, we will prove:

Theorem 8.1.4 (Donsker's theorem) *Under the hypotheses of Theorem 8.1.2,*

$$S(n\cdot)/\sqrt{n} \Rightarrow B(\cdot),$$

i.e., the associated measures on $C[0,1]$ converge weakly.

To motivate ourselves for the proof, we will begin by extracting several corollaries. The key to each one is a consequence of the following result, which follows from Theorem 3.10.1.

Theorem 8.1.5 *If $\psi : C[0,1] \to \mathbf{R}$ has the property that it is continuous P_0-a.s., then*

$$\psi(S(n\cdot)/\sqrt{n}) \Rightarrow \psi(B(\cdot))$$

Example 8.1.6 (Maxima) Let $\psi(\omega) = \max\{\omega(t) : 0 \le t \le 1\}$. $\psi : C[0,1] \to \mathbf{R}$ is continuous, so Theorem 8.1.5 implies

$$\max_{0 \le m \le n} S_m/\sqrt{n} \Rightarrow M_1 \equiv \max_{0 \le t \le 1} B_t$$

To complete the picture, we observe that by (7.4.4) the distribution of the right-hand side is

$$P_0(M_1 \ge a) = P_0(T_a \le 1) = 2P_0(B_1 \ge a)$$

Example 8.1.7 (Last 0 before time *n*) Let $\psi(\omega) = \sup\{t \le 1 : \omega(t) = 0\}$. This time, ψ is not continuous, for if ω_ϵ with $\omega_\epsilon(0) = 0$ is piecewise linear with slope 1 on $[0, 1/3 + \epsilon]$, -1 on $[1/3 + \epsilon, 2/3]$, and slope 1 on $[2/3, 1]$, then $\psi(\omega_0) = 2/3$ but $\psi(\omega_\epsilon) = 0$ for $\epsilon > 0$.

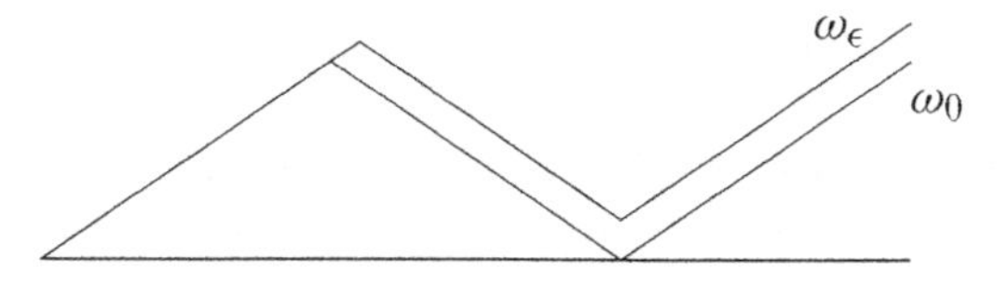

It is easy to see that if $\psi(\omega) < 1$ and $\omega(t)$ has positive and negative values in each interval $(\psi(\omega) - \delta, \psi(\omega) + \delta)$, then ψ is continuous at ω. By arguments in Subection 8.4.1, the last set has P_0 measure 1. (If the zero at $\psi(\omega)$ was isolated on the left, it would not be isolated on the right.) It follows that

$$\sup\{m \le n : S_{m-1} \cdot S_m \le 0\}/n \Rightarrow L = \sup\{t \le 1 : B_t = 0\}$$

The distribution of L is given in (7.4.7). The last result shows that the arcsine law, Theorem 4.9.5, proved for simple random walks holds when the mean is 0 and variance is finite.

Example 8.1.8 (Occupation times of half-lines) Let

$$\psi(\omega) = |\{t \in [0,1] : \omega(t) > a\}|.$$

The point $\omega \equiv a$ shows that ψ is not continuous, but it is easy to see that ψ is continuous at paths ω with $|\{t \in [0,1] : \omega(t) = a\}| = 0$. Fubini's theorem implies that

$$E_0|\{t \in [0,1] : B_t = a\}| = \int_0^1 P_0(B_t = a)\, dt = 0$$

so ψ is continuous P_0-a.s. Using Theorem 8.1.5, we now get that

$$|\{u \le n : S(u) > a\sqrt{n}\}|/n \Rightarrow |\{t \in [0,1] : B_t > a\}|$$

As we will now show, with a little work, one can convert this into the more natural result

$$|\{m \le n : S_m > a\sqrt{n}\}|/n \Rightarrow |\{t \in [0,1] : B_t > a\}|$$

Proof Application of Theorem 8.1.5 gives that for any a,

$$|\{t \in [0,1] : S(nt) > a\sqrt{n}\}| \Rightarrow |\{t \in [0,1] : B_t > a\}|$$

To convert this into a result about $|\{m \le n : S_m > a\sqrt{n}\}|$, we note that on $\{\max_{m\le n} |X_m| \le \epsilon\sqrt{n}\}$, which by Chebyshev's inequality has a probability $\to 1$, we have

$$|\{t \in [0,1] : S(nt) > (a+\epsilon)\sqrt{n}\}| \le \frac{1}{n}|\{m \le n : S_m > a\sqrt{n}\}|$$
$$\le |\{t \in [0,1] : S(nt) > (a-\epsilon)\sqrt{n}\}|$$

Combining this with the first conclusion of the proof and using the fact that $b \to |\{t \in [0,1] : B_t > b\}|$ is continuous at $b = a$ with probability one, one arrives easily at the desired conclusion. □

To compute the distribution of $|\{t \in [0,1] : B_t > 0\}|$, observe that we proved in Theorem 4.9.6 that if S_n is symmetric simple random walk, then the left-hand side converges to the arcsine law, so the right-hand side has that distribution and is the limit for any random walk with mean 0 and finite variance. The last argument uses an idea called the "invariance principle" that originated with Erdös and Kac (1946, 1947): The asymptotic behavior of functionals of S_n should be the same as long as the central limit theorem applies.

Proof of Theorem 8.1.4 To simplify the proof and prepare for generalizations in the next section, let $X_{n,m}$, $1 \le m \le n$, be a triangular array of random variables, $S_{n,m} = X_{n,1} + \cdots + X_{n,m}$ and suppose $S_{n,m} = B(\tau^n_m)$. Let

$$S_{n,(u)} = \begin{cases} S_{n,m} & \text{if } u = m \in \{0,1,\dots,n\} \\ \text{linear for } u \in [m-1,m] & \text{when } m \in \{1,\dots,n\} \end{cases}$$

Lemma 8.1.9 *If $\tau^n_{[ns]} \to s$ in probability for each $s \in [0,1]$, then*

$$\|S_{n,(n\cdot)} - B(\cdot)\| \to 0 \quad \textit{in probability}$$

To make the connection with the original problem, let $X_{n,m} = X_m/\sqrt{n}$ and define $\tau^n_1, \dots, \tau^n_n$ so that $(S_{n,1}, \dots, S_{n,n}) =_d (B(\tau^n_1), \dots, B(\tau^n_n))$. If $T_1, T_2, \dots$ are the stopping times defined in the proof of Theorem 8.1.3, Brownian scaling implies $\tau^n_m =_d T_m/n$, so the hypothesis of Lemma 8.1.9 is satisfied.

Proof The fact that B has continuous paths (and hence uniformly continuous on [0,1]) implies that if $\epsilon > 0$, then there is a $\delta > 0$ so that $1/\delta$ is an integer and

(a) $$P(|B_t - B_s| < \epsilon \text{ for all } 0 \le s \le 1, |t-s| < 2\delta) > 1 - \epsilon$$

The hypothesis of Lemma 8.1.9 implies that if $n \ge N_\delta$, then

$$P(|\tau^n_{[nk\delta]} - k\delta| < \delta \quad \text{for } k = 1,2,\dots,1/\delta) \ge 1 - \epsilon$$

Since $m \to \tau^n_m$ is increasing, it follows that if $s \in ((k-1)\delta, k\delta)$

$$\tau^n_{[ns]} - s \ge \tau^n_{[n(k-1)\delta]} - k\delta$$
$$\tau^n_{[ns]} - s \le \tau^n_{[nk\delta]} - (k+1)\delta$$

so if $n \ge N_\delta$,

(b) $$P\left(\sup_{0\le s\le 1} |\tau^n_{[ns]} - s| < 2\delta\right) \ge 1 - \epsilon$$

When the events in (a) and (b) occur

(c) $$|S_{n,m} - B_{m/n}| < \epsilon \text{ for all } m \le n$$

To deal with $t = (m+\theta)/n$ with $0 < \theta < 1$, we observe that

$$|S_{n,(nt)} - B_t| \le (1-\theta)|S_{n,m} - B_{m/n}| + \theta|S_{n,m+1} - B_{(m+1)/n}|$$
$$+ (1-\theta)|B_{m/n} - B_t| + \theta|B_{(m+1)/n} - B_t|$$

Using (c) on the first two terms and (a) on the last two, we see that if $n \ge N_\delta$ and $1/n < 2\delta$, then $\|S_{n,(n\cdot)} - B(\cdot)\| < 2\epsilon$ with probability $\ge 1 - 2\epsilon$. Since ϵ is arbitrary, the proof of Lemma 8.1.9 is complete. □

To get Theorem 8.1.4 now, we have to show:

Lemma 8.1.10 *If φ is bounded and continuous, then $E\varphi(S_{n,(n\cdot)}) \to E\varphi(B(\cdot))$.*

Proof For fixed $\epsilon > 0$, let $G_\delta = \{\omega : \text{if } \|\omega - \omega'\| < \delta, \text{ then } |\varphi(\omega) - \varphi(\omega')| < \epsilon\}$. Since φ is continuous, $G_\delta \uparrow C[0,1]$ as $\delta \downarrow 0$. Let $\Delta = \|S_{n,(n\cdot)} - B(\cdot)\|$. The desired result now follows from Lemma 8.1.9 and the trivial inequality

$$|E\varphi(S_{n,(n\cdot)}) - E\varphi(B(\cdot))| \le \epsilon + (2\sup|\varphi(\omega)|)\{P(G_\delta^c) + P(\Delta \ge \delta)\}$$ □

To accommodate our final example, we need a trivial generalization of Theorem 8.1.4. Let $C[0,\infty) = \{\text{continuous } \omega : [0,\infty) \to \mathbf{R}\}$ and let $\mathcal{C}[0,\infty)$ be the σ-field generated by the finite dimensional sets. Given a probability measure μ on $C[0,\infty)$, there is a corresponding measure $\pi_M\mu$ on $C[0,M] = \{\text{continuous } \omega : [0,M] \to \mathbf{R}\}$ (with $\mathcal{C}[0,M]$ the σ-field generated by the finite dimensional sets) obtained by "cutting off the paths at time M." Let $(\psi_M\omega)(t) = \omega(t)$ for $t \in [0,M]$ and let $\pi_M\mu = \mu \circ \psi_M^{-1}$. We say that a sequence of probability measures μ_n on $C[0,\infty)$ converges weakly to μ if for all M, $\pi_M\mu_n$ converges weakly to $\pi_M\mu$ on $C[0,M]$, the last concept being defined by a trivial extension of the definitions for $M = 1$. With these definitions, it is easy to conclude:

Theorem 8.1.11 *$S(n\cdot)/\sqrt{n} \Rightarrow B(\cdot)$, i.e., the associated measures on $C[0,\infty)$ converge weakly.*

Proof By definition, all we have to show is that weak convergence occurs on $C[0,M]$ for all $M < \infty$. The proof of Theorem 8.1.4 works in the same way when 1 is replaced by M. □

Example 8.1.12 Let $N_n = \inf\{m : S_m \ge \sqrt{n}\}$ and $T_1 = \inf\{t : B_t \ge 1\}$. Since $\psi(\omega) = T_1(\omega) \wedge 1$ is continuous P_0 a.s. on $C[0,1]$ and the distribution of T_1 is continuous, it follows from Theorem 8.1.5 that for $0 < t < 1$

$$P(N_n \le nt) \to P(T_1 \le t)$$

Repeating the last argument with 1 replaced by M and using Theorem 8.1.11 shows that the last conclusion holds for all t.

Exercises

8.1.1 Use Exercise 7.5.4 to conclude that $E(T_{U,V}^2) \le 4EX^4$.

8.1.2 Suppose S_n is one-dimensional simple random walk and let

$$R_n = 1 + \max_{m \le n} S_m - \min_{m \le n} S_m$$

be the number of points visited by time n. Show that $R_n/\sqrt{n} \Rightarrow$ a limit.

8.2 CLTs for Martingales

In this section we will prove central limit theorems for martingales. The key is an embedding theorem due to Strassen (1967). See his Theorem 4.3.

Theorem 8.2.1 *If S_n is a square integrable martingale with $S_0 = 0$, and B_t is a Brownian motion, then there is a sequence of stopping times $0 = T_0 \le T_1 \le T_2 \ldots$ for the Brownian motion so that*

$$(S_0, S_1, \ldots, S_k) \stackrel{d}{=} (B(T_0), B(T_1), \ldots, B(T_k)) \quad \textit{for all } k \ge 0$$

Proof We include $S_0 = 0 = B(T_0)$ only for the sake of starting the induction argument. Suppose we have $(S_0, \ldots, S_{k-1}) =_d (B(T_0), \ldots, B(T_{k-1}))$ for some $k \ge 1$. The strong Markov property implies that $\{B(T_{k-1} + t) - B(T_{k-1}) : t \ge 0\}$ is a Brownian motion that is independent of $\mathcal{F}(T_{k-1})$. Let $\mu_k(s_0, \ldots, s_{k-1}; \cdot)$ be a regular conditional distribution of $S_k - S_{k-1}$, given $S_j = s_j, 0 \le j \le k-1$, that is, for each Borel set A

$$P(S_k - S_{k-1} \in A | S_0, \ldots, S_{k-1}) = \mu_k(S_0, \ldots, S_{k-1}; A)$$

By Exercises 4.1.18 and 4.1.16, this exists, and we have

$$0 = E(S_k - S_{k-1} | S_0, \ldots, S_{k-1}) = \int x \mu_k(S_0, \ldots, S_{k-1}; dx)$$

so the mean of the conditional distribution is 0 almost surely. Using Theorem 8.1.1 now, we see that for almost every $\hat{S} \equiv (S_0, \ldots, S_{k-1})$, there is a stopping time $\tau_{\hat{S}}$ (for $\{B(T_{k-1} + t) - B(T_{k-1}) : t \ge 0\}$) so that

$$B(T_{k-1} + \tau_{\hat{S}}) - B(T_{k-1}) \stackrel{d}{=} \mu_k(S_0, \ldots, S_{k-1}; \cdot)$$

If we let $T_k = T_{k-1} + \tau_{\hat{S}}$, then $(S_0, \ldots, S_k) \stackrel{d}{=} (B(T_0), \ldots, B(T_k))$ and the result follows by induction. □

Remark While the details of the proof are fresh in the reader's mind, we would like to observe that if $E(S_k - S_{k-1})^2 < \infty$, then

$$E(\tau_{\hat{S}} | S_0, \ldots, S_{k-1}) = \int x^2 \mu_k(S_0, \ldots, S_{k-1}; dx)$$

since $B_t^2 - t$ is a martingale and $\tau_{\hat{S}}$ is the exit time from a randomly chosen interval $(S_{k-1} + U_k, S_{k-1} + V_k)$.

Before getting involved in the details of the central limit theorem, we show that the existence of the embedding allows us to give an easy proof of Theorem 4.5.2.

Theorem 8.2.2 *Let $\mathcal{F}_m = \sigma(S_0, S_1, \ldots S_m)$. $\lim_{n\to\infty} S_n$ exists and is finite on $\sum_{m=1}^{\infty} E((S_m - S_{m-1})^2|\mathcal{F}_{m-1}) < \infty$.*

Proof Let $\mathcal{B}_t$ be the filtration generated by Brownian motion, and let $t_m = T_m - T_{m-1}$. By construction we have

$$E((S_m - S_{m-1})^2|\mathcal{F}_{m-1}) = E(t_m|\mathcal{B}(T_{m-1}))$$

Let $M = \inf\{n : \sum_{m=1}^{n+1} E(t_m|\mathcal{B}(T_{m-1})) > A\}$. M is a stopping time and by its definition has $\sum_{m=1}^{M} E(t_m|\mathcal{B}(T_{m-1})) \le A$. Now $\{M \ge m\} \in \mathcal{F}_{m-1}$, so

$$\begin{aligned} E\sum_{m=1}^{M} t_m &= \sum_{m=1}^{\infty} E(t_m; M \ge m) \\ &= \sum_{m=1}^{\infty} E(E(t_M|\mathcal{B}(T_{m-1}); M \ge m) = E\sum_{m=1}^{M} E(t_m|\mathcal{B}(T_{m-1})) \le A \end{aligned}$$

From this we see that $\sum_{m=1}^{M} t_m < \infty$. As $A \to \infty$, $P(M < \infty) \downarrow 0$, so $T_\infty = \sum_{m=1}^{\infty} t_m < \infty$. Since $B(T_n) \to B(T_\infty)$ as $n \to \infty$ the desired result follows. □

Our first step toward the promised central limit theorem is to prove a result of Freedman (1971a). We say that $X_{n,m}, \mathcal{F}_{n,m}$, $1 \le m \le n$, is a **martingale difference array** if $X_{n,m} \in \mathcal{F}_{n,m}$ and $E(X_{n,m}|\mathcal{F}_{n,m-1}) = 0$ for $1 \le m \le n$, where $\mathcal{F}_{n,0} = \{\emptyset, \Omega\}$. Let $S_{n,m} = X_{n,1} + \cdots + X_{n,m}$. Let $N = \{0, 1, 2, \ldots\}$. Throughout the section, $S_{n,(n\cdot)}$ denotes the linear interpolation of $S_{n,m}$ defined by

$$S_{n,(u)} = \begin{cases} S_k & \text{if } u = k \in N \\ \text{linear for } u \in [k, k+1] & \text{when } k \in N \end{cases}$$

and $B(\cdot)$ is a Brownian motion with $B_0 = 0$. Let

$$V_{n,k} = \sum_{1 \le m \le k} E(X_{n,m}^2|\mathcal{F}_{n,m-1})$$

Theorem 8.2.3 *Suppose $\{X_{n,m}, \mathcal{F}_{n,m}\}$ is a martingale difference array.*

If (i) for each t, $V_{n,[nt]} \to t$ in probability,

(i) $|X_{n,m}| \le \epsilon_n$ for all m with $\epsilon_n \to 0$, and

then $S_{n,(n\cdot)} \Rightarrow B(\cdot)$.

Proof (i) implies $V_{n,n} \to 1$ in probability. By stopping each sequence at the first time $V_{n,k} > 2$ and setting the later $X_{n,m} = 0$, we can suppose without loss of generality that $V_{n,n} \le 2 + \epsilon_n^2$ for all n. By Theorem 8.2.1, we can find stopping times $T_{n,1}, \ldots, T_{n,n}$ so that $(S_{n,1}, \ldots, S_{n,n}) =_d (B(T_{n,1}), \ldots, B(T_{n,n}))$. By Lemma 8.1.9, it suffices to show that $T_{n,[nt]} \to t$ in probability for each $t \in [0, 1]$. To do this we let $t_{n,m} = T_{n,m} - T_{n,m-1}$ (with

$T_{n,0} = 0$) observe that by the remark after the proof of Theorem 8.2.1, $E(t_{n,m}|\mathcal{F}_{n,m-1}) = E(X_{n,m}^2|\mathcal{F}_{n,m-1})$. The last observation and hypothesis (i) imply

$$\sum_{m=1}^{[nt]} E(t_{n,m}|\mathcal{F}_{n,m-1}) \to t \quad \text{in probability}$$

To get from this to $T_{n,[nt]} \to t$ in probability, we observe

$$E\left(\sum_{m=1}^{[nt]} t_{n,m} - E(t_{n,m}|\mathcal{F}_{n,m-1})\right)^2 = E\sum_{m=1}^{[nt]} \{t_{n,m} - E(t_{n,m}|\mathcal{F}_{n,m-1})\}^2$$

by the orthogonality of martingale increments, Theorem 4.4.7. Now

$$\begin{aligned} E(\{t_{n,m} - E(t_{n,m}|\mathcal{F}_{n,m-1})\}^2|\mathcal{F}_{n,m-1}) &\le E(t_{n,m}^2|\mathcal{F}_{n,m-1}) \\ &\le CE(X_{n,m}^4|\mathcal{F}_{n,m-1}) \le C\epsilon_n^2 E(X_{n,m}^2|\mathcal{F}_{n,m-1}) \end{aligned}$$

by Exercise 7.5.4 and assumption (ii). Summing over n, taking expected values, and recalling we have assumed $V_{n,n} \le 2 + \epsilon_n^2$, it follows that

$$E\left(\sum_{m=1}^{[nt]} t_{n,m} - E(t_{n,m}|\mathcal{F}_{n,m-1})\right)^2 \to C\epsilon_n^2 EV_{n,n} \to 0$$

Unscrambling the definitions, we have shown $E(T_{n,[nt]} - V_{n,[nt]})^2 \to 0$, so Chebyshev's inequality implies $P(|T_{n,[nt]} - V_{n,[nt]}| > \epsilon) \to 0$, and using (ii) now completes the proof. □

With Theorem 8.2.3 established, a truncation argument gives us:

Theorem 8.2.4 (Lindeberg-Feller theorem for martingales) *Suppose $X_{n,m}$, $\mathcal{F}_{n,m}$, $1 \le m \le n$ is a martingale difference array.*

If (i) $V_{n,[nt]} \to t$ in probability for all $t \in [0,1]$ and

(ii) for all $\epsilon > 0$, $\sum_{m\le n} E(X_{n,m}^2 1_{(|X_{n,m}|>\epsilon)}|\mathcal{F}_{n,m-1}) \to 0$ in probability,

then $S_{n,(n\cdot)} \Rightarrow B(\cdot)$.

Remark 8.2.1 If $X_{n,m}$ on each row are independent, this reduces to the martingale central limit theorem.

Proof The first step is to truncate so that we can apply Theorem 8.2.3. Let

$$\hat{V}_n(\epsilon) = \sum_{m=1}^{n} E(X_{n,m}^2 1_{(|X_{n,m}|>\epsilon_n)}|\mathcal{F}_{n,m-1})$$

Lemma 8.2.5 *If $\epsilon_n \to 0$ slowly enough, then $\epsilon_n^{-2}\hat{V}_n(\epsilon_n) \to 0$ in probability.*

Remark The ϵ_n^{-2} in front is so that we can conclude

$$\sum_{m\le n} P(|X_{n,m}| > \epsilon_n|\mathcal{F}_{n,m-1}) \to 0 \text{ in probability}$$

Proof Let N_m be chosen so that $P(m^2\hat{V}_n(1/m) > 1/m) \le 1/m$ for $n \ge N_m$. Let $\epsilon_n = 1/m$ for $n \in [N_m, N_{m+1})$ and $\epsilon_n = 1$ if $n < N_1$. If $\delta > 0$ and $1/m < \delta$, then for $n \in [N_m, N_{m+1})$

$$P(\epsilon_n^{-2}\hat{V}_n(\epsilon_n) > \delta) \le P(m^2\hat{V}_n(1/m) > 1/m) \le 1/m \qquad \square$$

Let $\hat{X}_{n,m} = X_{n,m}1_{(|X_{n,m}|>\epsilon_n)}$, $\bar{X}_{n,m} = X_{n,m}1_{(|X_{n,m}|\le\epsilon_n)}$ and

$$\tilde{X}_{n,m} = \bar{X}_{n,m} - E(\bar{X}_{n,m}|\mathcal{F}_{n,m-1})$$

Our next step is to show:

Lemma 8.2.6 *If we define* $\tilde{S}_{n,(n\cdot)}$ *in the obvious way, then Theorem 8.2.3 implies* $\tilde{S}_{n,(n\cdot)} \Rightarrow B(\cdot)$.

Proof Since $|\tilde{X}_{n,m}| \le 2\epsilon_n$, we only have to check (ii) in Theorem 8.2.3. To do this, we observe that the conditional variance formula, Theorem 4.4.8, implies

$$E(\tilde{X}^2_{n,m}|\mathcal{F}_{n,m-1}) = E(\bar{X}^2_{n,m}|\mathcal{F}_{n,m-1}) - E(\bar{X}_{n,m}|\mathcal{F}_{n,m-1})^2$$

For the first term, we observe

$$E(\bar{X}^2_{n,m}|\mathcal{F}_{n,m-1}) = E(X^2_{n,m}|\mathcal{F}_{n,m-1}) - E(\hat{X}^2_{n,m}|\mathcal{F}_{n,m-1})$$

For the second, we observe that $E(X_{n,m}|\mathcal{F}_{n,m-1}) = 0$ implies

$$E(\bar{X}_{n,m}|\mathcal{F}_{n,m-1})^2 = E(\hat{X}_{n,m}|\mathcal{F}_{n,m-1})^2 \le E(\hat{X}^2_{n,m}|\mathcal{F}_{n,m-1})$$

by Jensen's inequality, so it follows from (a) and (i) that

$$\sum_{m=1}^{[nt]} E(\tilde{X}^2_{n,m}|\mathcal{F}_{n,m-1}) \to t \quad \text{for all } t \in [0,1] \qquad \square$$

Having proved Lemma 8.2.6, it remains to estimate the difference between $S_{n,(n\cdot)}$ and $\tilde{S}_{n,(n\cdot)}$. On $\{|X_{n,m}| \le \epsilon_n \text{ for all } 1 \le m \le n\}$, we have

$$\|S_{n,(n\cdot)} - \tilde{S}_{n,(n\cdot)}\| \le \sum_{m=1}^{n} |E(\bar{X}_{n,m}|\mathcal{F}_{n,m-1})| \tag{8.2.1}$$

To handle the right-hand side, we observe

$$\begin{aligned}\sum_{m=1}^{n} |E(\bar{X}_{n,m}|\mathcal{F}_{n,m-1})| &= \sum_{m=1}^{n} |E(\hat{X}_{n,m}|\mathcal{F}_{n,m-1})| \\ &\le \sum_{m=1}^{n} E(|\hat{X}_{n,m}||\mathcal{F}_{n,m-1}) \qquad (8.2.2) \\ &\le \epsilon_n^{-1}\sum_{m=1}^{n} E(\hat{X}^2_{n,m}|\mathcal{F}_{n,m-1}) \to 0\end{aligned}$$

in probability by Lemma 8.2.5. To complete the proof now, it suffices to show

$$P(|X_{n,m}| > \epsilon_n \text{ for some } m \le n) \to 0 \tag{8.2.3}$$

for with (8.2.2) and (8.2.1) this implies $\|S_{n,(n\cdot)} - \tilde{S}_{n,(n\cdot)}\| \to 0$ in probability. The proof of Theorem 8.2.3 constructs a Brownian motion with $\|\tilde{S}_{n,(n\cdot)} - B(\cdot)\| \to 0$, so the desired result follows from the triangle inequality and Lemma 8.1.10.

To prove (8.2.3), we will use Lemma 3.5 of Dvoretsky (1972).

Lemma 8.2.7 *If A_n is adapted to $\mathcal{G}_n$, then for any nonnegative $\delta \in \mathcal{G}_0$,*

$$P\left(\cup_{m=1}^n A_m|\mathcal{G}_0\right) \le \delta + P\left(\sum_{m=1}^n P(A_m|\mathcal{G}_{m-1}) > \delta \,\middle|\, \mathcal{G}_0\right)$$

Proof We proceed by induction. When $n = 1$, the conclusion says

$$P(A_1|\mathcal{G}_0) \le \delta + P(P(A_1|\mathcal{G}_0) > \delta|\mathcal{G}_0)$$

This is obviously true on $\Omega_- \equiv \{P(A_1|\mathcal{G}_0) \le \delta\}$ and also on $\Omega_+ \equiv \{P(A_1|\mathcal{G}_0) > \delta\} \in \mathcal{G}_0$, since on Ω_+

$$P(P(A_1|\mathcal{G}_0) > \delta|\mathcal{G}_0) = 1 \ge P(A_1|\mathcal{G}_0)$$

To prove the result for n sets, observe that by the last argument the inequality is trivial on Ω_+. Let $B_m = A_m \cap \Omega_-$. Since $\Omega_- \in \mathcal{G}_0 \subset \mathcal{G}_{m-1}$, $P(B_m|\mathcal{G}_{m-1}) = P(A_m|\mathcal{G}_{m-1})$ on Ω_-. (See Exercise 4.1.2.) Applying the result for $n-1$ sets with $\gamma = \delta - P(B_1|\mathcal{G}_0) \ge 0$,

$$P\left(\cup_{m=2}^n B_m|\mathcal{G}_1\right) \le \gamma + P\left(\sum_{m=2}^n P(B_m|\mathcal{G}_{m-1}) > \gamma \,\middle|\, \mathcal{G}_1\right)$$

Taking conditional expectation w.r.t. $\mathcal{G}_0$ and noting $\gamma \in \mathcal{G}_0$,

$$P\left(\cup_{m=2}^n B_m|\mathcal{G}_0\right) \le \gamma + P\left(\sum_{m=1}^n P(B_m|\mathcal{G}_{m-1}) > \delta \,\middle|\, \mathcal{G}_0\right)$$

$\cup_{2\le m\le n} B_m = (\cup_{2\le m\le n} A_m) \cap \Omega_-$ and another use of Exercise 4.1.2 shows

$$\sum_{1\le m\le n} P(B_m|\mathcal{G}_{m-1}) = \sum_{1\le m\le n} P(A_m|\mathcal{G}_{m-1})$$

on Ω_-. So, on Ω_-,

$$P\left(\cup_{m=2}^n A_m|\mathcal{G}_0\right) \le \delta - P(A_1|\mathcal{G}_0) + P\left(\sum_{m=1}^n P(A_m|\mathcal{G}_{m-1}) > \delta|\mathcal{G}_0\right)$$

The result now follows from

$$P\left(\cup_{m=1}^n A_m|\mathcal{G}_0\right) \le P(A_1|\mathcal{G}_0) + P\left(\cup_{m=2}^n A_m|\mathcal{G}_0\right)$$

To see this, let $C = \cup_{2\le m\le n} A_m$, observe that $1_{A_1\cup C} \le 1_{A_1} + 1_C$, and use the monotonicity of conditional expectations. □

Proof of (8.2.3) Let $A_m = \{|X_{n,m}| > \epsilon_n\}$, $\mathcal{G}_m = \mathcal{F}_{n,m}$, and let δ be a positive number. Lemma 8.2.7 implies

$$P(|X_{n,m}| > \epsilon_n \text{ for some } m \le n) \le \delta + P\left(\sum_{m=1}^n P(|X_{n,m}| > \epsilon_n|\mathcal{F}_{n,m-1}) > \delta\right)$$

To estimate the right-hand side, we observe that "Chebyshev's inequality" (Exercise 1.3 in Chapter 4) implies

$$\sum_{m=1}^{n} P(|X_{n,m}| > \epsilon_n | \mathcal{F}_{n,m-1}) \le \epsilon_n^{-2} \sum_{m=1}^{n} E(\hat{X}_{n,m}^2 | \mathcal{F}_{n,m-1}) \to 0$$

so $\limsup_{n\to\infty} P(|X_{n,m}| > \epsilon_n \text{ for some } m \le n) \le \delta$. Since δ is arbitrary, the proof of (e) and hence of Theorem 8.2.4 is complete. □

For applications, it is useful to have a result for a single sequence.

Theorem 8.2.8 (Martingale central limit theorem) *Suppose X_n, $\mathcal{F}_n$, $n \ge 1$, is a martingale difference sequence and let $V_k = \sum_{1\le n\le k} E(X_n^2|\mathcal{F}_{n-1})$.*
If (i) $V_k/k \to \sigma^2 > 0$ in probability and

(ii) $n^{-1}\sum_{m\le n} E(X_m^2 1_{(|X_m|>\epsilon\sqrt{n})}) \to 0$

then $S_{(n\cdot)}/\sqrt{n} \Rightarrow \sigma B(\cdot)$.

Proof Let $X_{n,m} = X_m/\sigma\sqrt{n}$, $\mathcal{F}_{n,m} = \mathcal{F}_m$. Changing notation and letting $k = nt$, our first assumption becomes (i) of Theorem 8.2.4. To check (ii), observe that

$$E\sum_{m=1}^{n} E(X_{n,m}^2 1_{(|X_{n,m}|>\epsilon)}|\mathcal{F}_{n,m-1}) = \sigma^{-2} n^{-1} \sum_{m=1}^{n} E(X_m^2 1_{(|X_m|>\epsilon\sigma\sqrt{n})}) \to 0 \qquad \square$$

8.3 CLTs for Stationary Sequences

We begin by considering the martingale case.

Theorem 8.3.1 *Suppose X_n, $n \in \mathbf{Z}$, is an ergodic stationary sequence of square integrable martingale differences, i.e., $\sigma^2 = EX_n^2 < \infty$ and $E(X_n|\mathcal{F}_{n-1}) = 0$, where $\mathcal{F}_n = \sigma(X_m, m \le n)$. Let $S_n = X_1 + \cdots + X_n$. Then*

$$S_{(n\cdot)}/n^{1/2} \Rightarrow \sigma B(\cdot)$$

Remark 8.3.1 This result was discovered independently by Billingsley (1961) and Ibragimov (1963).

Proof $u_n \equiv E(X_n^2|\mathcal{F}_{n-1})$ can be written as $\varphi(X_{n-1}, X_{n-2}, \ldots)$, so Theorem 7.1.3 implies u_n is stationary and ergodic, and the ergodic theorem implies

$$n^{-1}\sum_{m=1}^{n} u_m \to Eu_0 = EX_0^2 \quad \text{a.s.}$$

The last conclusion shows that (i) of Theorem 8.2.8 holds. To verify (ii), we observe

$$n^{-1}\sum_{m=1}^{n} E(X_m^2 1_{(|X_m|>\epsilon\sqrt{n})}) = E(X_0^2 1_{(|X_0|>\epsilon\sqrt{n})}) \to 0$$

by the dominated convergence theorem. □

We prove our first central limit theorem for stationary sequences by reducing to the martingale case. Our proof is based on Scott (1973), but the key is an idea of Gordin (1969). In some applications of this result (see e.g., Exercise 8.3.1), it is convenient to use σ-fields richer than $\mathcal{F}_m = \sigma(X_n, n \le m)$. So we adopt abstract conditions that satisfy the demands of martingale theory and ergodic theory:

(i) $n \to \mathcal{F}_n$ is increasing, and $X_n \in \mathcal{F}_n$

(ii) $\theta^{-1}\mathcal{F}_m = \mathcal{F}_{m+1}$

Theorem 8.3.2 *Suppose X_n, $n \in \mathbf{Z}$ is an ergodic stationary sequence with $EX_n = 0$. Let $\mathcal{F}_n$ satisfy (i), (ii), and*

$$\sum_{n\ge 1} \|E(X_0|\mathcal{F}_{-n})\|_2 < \infty$$

where $\|Y\|_2 = (EY^2)^{1/2}$. Let $S_n = X_1 + \cdots + X_n$. Then $S_{(n\cdot)}/\sqrt{n} \Rightarrow \sigma B(\cdot)$, where

$$\sigma^2 = EX_0^2 + 2\sum_{n=1}^{\infty} EX_0 X_n$$

and the series in the definition converges absolutely.

Proof Suppose X_n, $n \in \mathbf{Z}$ is defined on sequence space $(\mathbf{R}^{\mathbf{Z}}, \mathcal{R}^{\mathbf{Z}}, P)$ with $X_n(\omega) = \omega_n$ and let $(\theta^n\omega)(m) = \omega(m+n)$. Let

$$H_n = \{Y \in \mathcal{F}_n \text{ with } EY^2 < \infty\}$$
$$K_n = \{Y \in H_n \text{ with } E(YZ) = 0 \text{ for all } Z \in H_{n-1}\}$$

Geometrically, $H_0 \supset H_{-1} \supset H_{-2} \ldots$ is a sequence of subspaces of L^2 and K_n is the orthogonal complement of H_{n-1} in H_n. If Y is a random variable, let $(\theta^n Y)(\omega) = Y(\theta^n\omega)$. Generalizing from the example $Y = f(X_{-j}, \ldots, X_k)$, which has $\theta^n Y = f(X_{n-j}, \ldots, X_{n+k})$, it is easy to see that if $Y \in H_k$, then $\theta^n Y \in H_{k+n}$, and hence if $Y \in K_j$, then $\theta^n Y \in K_{n+j}$.

If X_0 happened to be in K_0, then we would be happy since then $X_n = \theta^n X_0 \in K_n$ for all n, and taking $Z = 1_A \in H_{n-1}$ we would have $E(X_n 1_A) = 0$ for all $A \in \mathcal{F}_{n-1}$ and hence $E(X_n|\mathcal{F}_{n-1}) = 0$. The next best thing to having $X_0 \in K_0$ is to have

$$X_0 = Y_0 + Z_0 - \theta Z_0 \tag{8.3.1}$$

with $Y_0 \in K_0$ and $Z_0 \in L^2$, for then if we let

$$S_n = \sum_{m=1}^{n} X_m = \sum_{m=1}^{n} \theta^m X_0 \quad \text{and} \quad T_n = \sum_{m=1}^{n} \theta^m Y_0$$

then $S_n = T_n + \theta Z_0 - \theta^{n+1} Z_0$. The $\theta^m Y_0$ are a stationary ergodic martingale difference sequence (ergodicity follows from Theorem 7.1.3), so Theorem 8.3.1 implies

$$T_{(n\cdot)}/\sqrt{n} \Rightarrow \sigma B(\cdot) \quad \text{where } \sigma^2 = EY_0^2$$

To get rid of the other term, we observe $\theta Z_0/\sqrt{n} \to 0$ a.s. and

$$P\left(\sup_{1\le m\le n} \theta^{m+1} Z_0 > \epsilon\sqrt{n}\right) \le nP(Z_0 > \epsilon\sqrt{n}) \le \epsilon^{-2} E(Z_0^2; Z_0 > \epsilon\sqrt{n}) \to 0$$

by dominated convergence. To solve (8.3.1) formally, we let

$$Z_0 = \sum_{j=0}^{\infty} E(X_j|\mathcal{F}_{-1})$$

$$\theta Z_0 = \sum_{j=0}^{\infty} E(X_{j+1}|\mathcal{F}_0)$$

$$Y_0 = \sum_{j=0}^{\infty} \{E(X_j|\mathcal{F}_0) - E(X_j|\mathcal{F}_{-1})\}$$

and check that

$$Y_0 + Z_0 - \theta Z_0 = E(X_0|\mathcal{F}_0) = X_0$$

To justify the last calculation, we need to know that the series in the definitions of Z_0 and Y_0 converge. Our assumption and shift invariance imply

$$\sum_{j=0}^{\infty} \|E(X_j|\mathcal{F}_{-1})\|_2 < \infty$$

so the triangle inequality implies that the series for Z_0 converges in L^2. Since

$$\|E(X_j|\mathcal{F}_0) - E(X_j|\mathcal{F}_{-1})\|_2 \le \|E(X_j|\mathcal{F}_0)\|_2$$

(the left-hand side is the projection onto $K_0 \subset H_0$), the series for Y_0 also converges in L^2. Putting the pieces together, we have shown Theorem 8.3.2 with $\sigma^2 = EY_0^2$. To get the indicated formula for σ^2, observe that conditioning and using Cauchy-Schwarz

$$|EX_0X_n| = |E(X_0E(X_n|\mathcal{F}_0))| \le \|X_0\|_2 \|E(X_n|\mathcal{F}_0)\|_2$$

Shift invariance implies $\|E(X_n|\mathcal{F}_0)\|_2 = \|E(X_0|\mathcal{F}_{-n})\|_2$, so the series converges absolutely.

$$ES_n^2 = \sum_{j=1}^{n}\sum_{k=1}^{n} EX_jX_k = n\, EX_0^2 + 2\sum_{m=1}^{n-1}(n-m)EX_0X_m$$

From this, it follows easily that

$$n^{-1}ES_n^2 \to EX_0^2 + 2\sum_{m=1}^{\infty} EX_0X_m$$

To finish the proof, let $T_n = \sum_{m=1}^{n} \theta^m Y_0$, observe $\sigma^2 = EY_0^2$ and

$$n^{-1}E(S_n - T_n)^2 = n^{-1}E(\theta Z_0 - \theta^{n+1} Z_0)^2 \le 4EZ_0^2/n \to 0$$

since $(a-b)^2 \le (2a)^2 + (2b)^2$. □

We turn now to examples. In the first one, it is trivial to check the hypothesis of Theorem 8.3.2.

Example 8.3.3 (M-dependent sequences) Let X_n, $n \in \mathbf{Z}$, be a stationary sequence with $EX_n = 0$, $EX_n^2 < \infty$, and suppose $\{X_j, j \le 0\}$ and $\{X_k, k > M\}$ are independent. In this case, $E(X_0|\mathcal{F}_{-n}) = 0$ for $n > M$, so Theorem 8.3.2 implies $S_{n,(n\cdot)}/\sqrt{n} \Rightarrow \sigma B(\cdot)$, where

$$\sigma^2 = EX_0^2 + 2\sum_{m=1}^{M} EX_0 X_m$$

Remark 8.3.2 For a bare-hands approach to this problem, see theorem 7.3.1 in Chung (1974).

Exercise 8.3.1 Consider the special case of Example 8.3.3 in which ξ_i, $i \in \mathbf{Z}$, are i.i.d. and take values H and T with equal probability. Let $\eta_n = f(\xi_n, \xi_{n+1})$, where $f(H,T) = 1$ and $f(i,j) = 0$ otherwise. $S_n = \eta_1 + \cdots + \eta_n$ counts the number of head runs in $\xi_1, \ldots, \xi_{n+1}$. Apply Theorem 8.3.2 with $\mathcal{F}_n = \sigma(X_m : m \le n+1)$ to show that there are constants μ and σ so that $(S_n - n\mu)/\sigma n^{1/2} \Rightarrow \chi$. What is the random variable Y_0 constructed in the proof of Theorem 8.3.2 in this case?

Example 8.3.4 (Markov chains) Let ζ_n, $n \in \mathbf{Z}$ be an irreducible Markov chain on a countable state space S in which each ζ_n has the stationary distribution π. Let $X_n = f(\zeta_n)$, where $\sum f(x)\pi(x) = 0$ and $\sum f(x)^2\pi(x) < \infty$. Results in Chapter 6 imply that X_n is an ergodic stationary sequence. If we let $\mathcal{F}_{-n} = \sigma(\zeta_m, m \le -n)$, then

$$E(X_0|\mathcal{F}_{-n}) = \sum_y p^n(\zeta_{-n}, y) f(y)$$

where $p^n(x,y)$ is the n step transition probability, so

$$\|E(X_0|\mathcal{F}_{-n})\|_2^2 = \sum_x \pi(x)\Big(\sum_y p^n(x,y) f(y)\Big)^2$$

When f is bounded, we can use $\sum f(x)\pi(x) = 0$ to get the following bound,

$$\|E(X_0|\mathcal{F}_{-n})\|_2^2 \le \|f\|_\infty \sum_x \pi(x)\|p^n(x,\cdot) - \pi(\cdot)\|^2$$

where $\|f\|_\infty = \sup|f(x)|$ and $\|\cdot\|$ is the total variation norm.

When S is finite, all f are bounded. If the chain is aperiodic, Exercise 5.6 in Chapter 5 implies that

$$\sup_x \|p^n(x,\cdot) - \pi(\cdot)\| \le Ce^{-\epsilon n}$$

and the hypothesis of Theorem 8.3.2 is satisfied. To see that the limiting variance σ^2 may be 0 in this case, consider the modification of Exercise 7.1 in which $X_n = (\xi_n, \xi_{n+1})$, $f(H,T) = 1$, $f(T,H) = -1$ and $f(H,H) = f(T,T) = 0$. In this case, $\sum_{m=1}^{n} f(X_m) \in \{-1, 0, 1\}$ so there is no central limit theorem.

Our last example is simple to treat directly, but will help us evaluate the strength of the conditions in Theorem 8.3.2 and in later theorems.

Example 8.3.5 (Moving average process) Suppose

$$X_m = \sum_{k \geq 0} c_k \xi_{m-k} \quad \text{where} \quad \sum_{k \geq 0} c_k^2 < \infty$$

and the ξ_i, $i \in \mathbf{Z}$, are i.i.d. with $E\xi_i = 0$ and $E\xi_i^2 = 1$. If $\mathcal{F}_{-n} = \sigma(\xi_m \,;\, m \leq -n)$, then

$$\|E(X_0|\mathcal{F}_{-n})\|_2 = \left\| \sum_{k \geq n} c_k \xi_{-k} \right\|_2 = \left(\sum_{k \geq n} c_k^2 \right)^{1/2}$$

If, for example, $c_k = (1+k)^{-p}$, $\|E(X_0|\mathcal{F}_{-n})\|_2 \sim n^{(1/2)-p}$, and Theorem 8.3.5 applies if $p > 3/2$.

Remark 8.3.2 Theorem 5.3 in Hall and Heyde (1980) shows that

$$\sum_{j \geq 0} \|E(X_j|\mathcal{F}_0) - E(X_j|\mathcal{F}_{-1})\|_2 < \infty$$

is sufficient for a central limit theorem. Using this result shows that $\sum |c_k| < \infty$ is sufficient for the central limit theorem in Example 7.3.

The condition in the improved result is close to the best possible. Suppose the ξ_i take values 1 and -1 with equal probability and let $c_k = (1+k)^{-p}$, where $1/2 < p < 1$. The Lindeberg-Feller theorem can be used to show that $S_n / n^{3/2-p} \Rightarrow \sigma\chi$. To check the normalization, note that

$$\sum_{m=1}^{n} X_m = \sum_{j \leq n} a_{n,j} \xi_j$$

If $j \geq 0$, then $a_{n,j} = \sum_{i=0}^{n-j} c_k \approx (n-j)^{1-p}$, so using $0 \leq j \leq n/2$ the variance is at least n^{3-2p}. Further details are left to the reader.

8.3.1 Mixing Properties

The last three examples show that in many cases it is easy to verify the hypothesis of Theorem 8.3.2 directly. To connect Theorem 8.3.2 with other results in the literature, we will introduce sufficient conditions phrased in terms of quantitative measures of dependence between σ-fields. Our first is

$$\alpha(\mathcal{G}, \mathcal{H}) = \sup\{|P(A \cap B) - P(A)P(B)| : A \in \mathcal{G}, B \in \mathcal{H}\}$$

If $\alpha = 0$, $\mathcal{G}$ and $\mathcal{H}$ are independent, so α measures the dependence between the σ-fields.

Lemma 8.3.6 *Let $p, q, r \in (1, \infty]$ with $1/p + 1/q + 1/r = 1$, and suppose $X \in \mathcal{G}$, $Y \in \mathcal{H}$ have $E|X|^p$, $E|Y|^q < \infty$. Then*

$$|EXY - EXEY| \leq 8\|X\|_p \|Y\|_q (\alpha(\mathcal{G}, \mathcal{H}))^{1/r}$$

Here, we interpret $x^0 = 1$ for $x > 0$ and $0^0 = 0$.

Proof If $\alpha = 0$, X and Y are independent and the result is true, so we can suppose $\alpha > 0$. We build up to the result in three steps, starting with the case $r = \infty$.

$$|EXY - EXEY| \leq 2\|X\|_p\|Y\|_q \tag{8.3.2}$$

Proof Hölder's inequality implies $|EXY| \leq \|X\|_p\|Y\|_q$, and Jensen's inequality implies

$$\|X\|_p\|Y\|_q \geq |E|X|E|Y|| \geq |EXEY|$$

so the result follows from the triangle inequality. □

$$|EXY - EXEY| \leq 4\|X\|_\infty\|Y\|_\infty\alpha(\mathcal{G},\mathcal{H}) \tag{8.3.3}$$

Proof Let $\eta = \text{sgn}\,\{E(Y|\mathcal{G}) - EY\} \in \mathcal{G}$. $EXY = E(XE(Y|\mathcal{G}))$, so

$$\begin{aligned}
|EXY - EXEY| &= |E(X\{E(Y|\mathcal{G}) - EY\})| \\
&\leq \|X\|_\infty E|E(Y|\mathcal{G}) - EY| \\
&= \|X\|_\infty E(\eta\{E(Y|\mathcal{G}) - EY\}) \\
&= \|X\|_\infty\{E(\eta Y) - E\eta EY\}
\end{aligned}$$

Applying the last result with $X = Y$ and $Y = \eta$ gives

$$|E(Y\eta) - EYE\eta| \leq \|Y\|_\infty|E(\zeta\eta) - E\zeta E\eta|$$

where $\zeta = \text{sgn}\,\{E(\eta|\mathcal{H}) - E\eta\}$. Now $\eta = 1_A - 1_B$ and $\zeta = 1_C - 1_D$, so

$$\begin{aligned}
|E(\zeta\eta) - E\zeta E\eta| &= |P(A \cap C) - P(B \cap C) - P(A \cap D) + P(B \cap D) \\
&\quad - P(A)P(C) + P(B)P(C) + P(A)P(D) - P(B)P(D)| \\
&\leq 4\alpha(\mathcal{G},\mathcal{H})
\end{aligned}$$

Combining the last three displays gives the desired result. □

$$|EXY - EXEY| \leq 6\|X\|_p\|Y\|_\infty\,\alpha(\mathcal{G},\mathcal{H})^{1-1/p} \tag{8.3.4}$$

Proof Let $C = \alpha^{-1/p}\|X\|_p$, $X_1 = X1_{(|X|\leq C)}$ and $X_2 = X - X_1$

$$\begin{aligned}
|EXY - EXEY| &\leq |EX_1Y - EX_1EY| + |EX_2Y - EX_2EY| \\
&\leq 4\alpha C\|Y\|_\infty + 2\|Y\|_\infty E|X_2|
\end{aligned}$$

by (8.3.3) and (8.3.2). Now

$$E|X_2| \leq C^{-(p-1)}E(|X|^p1_{(|X|>C)}) \leq C^{-p+1}E|X|^p$$

Combining the last two inequalities and using the definition of C gives

$$|EXY - EX\,EY| \leq 4\alpha^{1-1/p}\|X\|_p\|Y\|_\infty + 2\|Y\|_\infty\alpha^{1-1/p}\|X\|_p^{-p+1+p}$$

which proves (8.3.4) □

Finally, to prove Lemma 8.3.6, let $C = \alpha^{-1/q}\|Y\|_q$, $Y_1 = Y\, 1_{(|Y|\le C)}$, and $Y_2 = Y - Y_1$.

$$\begin{aligned}|EXY - EXEY| &\le |EXY_1 - EXEY_1| + |EXY_2 - EXEY_2| \\ &\le 6C\|X\|_p \alpha^{1-1/p} + 2\|X\|_p\|Y_2\|_\theta\end{aligned}$$

where $\theta = (1-1/p)^{-1}$ by (8.3.4) and (8.3.2). Now

$$E|Y_2|^\theta \le C^{-q+\theta} E(|Y|^q 1_{(|Y|>C)}) \le C^{-q+\theta} E|Y|^q$$

Taking the $1/\theta$ root of each side and recalling the definition of C

$$\|Y_2\|_\theta \le C^{-(q-\theta)/\theta}\|Y\|_q^{q/\theta} \le \alpha^{(q-\theta)/\theta q}\|Y\|_q$$

so we have

$$|EXY - EXEY| \le 6\alpha^{-1/q}\|Y\|_q\|X\|_p\alpha^{1-1/p} + 2\|X\|_p\alpha^{1/\theta - 1/q}\|Y\|_q^{1/\theta+1/q}$$

proving the desired result. □

Remark 8.3.3 The last proof is from appendix III of Hall and Heyde (1980). They attribute (b) to Ibragimov (1962) and (c) and Lemma 8.3.6 to Davydov (1968).

Theorem 8.3.7 *Suppose X_n, $n \in \mathbf{Z}$ is an ergodic stationary sequence with $EX_n = 0$, $E|X_0|^{2+\delta} < \infty$. Let $\alpha(n) = \alpha(\mathcal{F}_{-n}, \sigma(X_0))$, where $\mathcal{F}_{-n} = \sigma(X_m : m \le -n)$, and suppose*

$$\sum_{n=1}^\infty \alpha(n)^{\delta/2(2+\delta)} < \infty$$

If $S_n = X_1 + \cdots + X_n$, then $S_{(n\cdot)}/\sqrt{n} \Rightarrow \sigma B(\cdot)$, where

$$\sigma^2 = EX_0^2 + 2\sum_{n=1}^\infty EX_0X_n$$

Remark If one strengthens the mixing condition, one can weaken the summability condition. Let $\bar\alpha(n) = \alpha(\mathcal{F}_{-n}, \mathcal{F}_0')$, where $\mathcal{F}_0' = \sigma(X_k, k \ge 0)$. When $\bar\alpha(n) \downarrow 0$, the sequence is called **strong mixing**. Rosenblatt (1956) introduced the concept as a condition under which the central limit theorem for stationary sequences could be obtained. Ibragimov (1962) proved $S_n/\sqrt{n} \Rightarrow \sigma\chi$, where $\sigma^2 = \lim_{n\to\infty} ES_n^2/n$ under the assumption

$$\sum_{n=1}^\infty \bar\alpha(n)^{\delta/(2+\delta)} < \infty$$

See Ibragimov and Linnik (1971), Theorem 18.5.3, or Hall and Heyde (1980), Corollary 5.1, for a proof.

Proof To use Lemma 8.3.6 to estimate the quantity in Theorem 8.3.2, we begin with

$$\|E(X|\mathcal{F})\|_2 = \sup\{E(XY) : Y \in \mathcal{F}, \|Y\|_2 = 1\} \tag{8.3.5}$$

Proof of (8.3.5) If $Y \in \mathcal{F}$ with $\|Y\|_2 = 1$, then using a by now familiar property of conditional expectation and the Cauchy-Schwarz inequality

$$EXY = E(E(XY|\mathcal{F})) = E(YE(X|\mathcal{F})) \le \|E(X|\mathcal{F})\|_2\|Y\|_2$$

Equality holds when $Y = E(X|\mathcal{F})/\|E(X|\mathcal{F})\|_2$. □

Letting $p = 2+\delta$ and $q = 2$ in Lemma 8.3.6, noticing

$$\frac{1}{r} = 1 - \frac{1}{p} - \frac{1}{q} = 1 - \frac{1}{2+\delta} - \frac{1}{2} = \frac{4+2\delta-2-(2+\delta)}{(2+\delta)2} = \frac{\delta}{(2+\delta)2}$$

and recalling $EX_0 = 0$, shows that if $Y \in \mathcal{F}_{-n}$

$$|EX_0Y| \le 8\|X_0\|_{2+\delta}\|Y\|_2\,\alpha(n)^{\delta/2(2+\delta)}$$

Combining this with (8.3.5) gives

$$\|E(X_0|\mathcal{F}_{-n})\|_2 \le 8\|X_0\|_{2+\delta}\,\alpha(n)^{\delta/2(2+\delta)}$$

and it follows that the hypotheses of Theorem 8.3.2 are satisfied. □

In the M dependent case (Example 8.3.3), $\alpha(n) = 0$ for $n > M$, so Theorem 8.3.7 applies. As for Markov chains (Example 8.3.4), in this case

$$\begin{aligned}\alpha(n) &= \sup_{A,B} |P(X_{-n} \in A, X_0 \in B) - \pi(A)\pi(B)| \\ &= \sup_{A,B}\left| \sum_{x\in A,\, y\in B} \pi(x)p^n(x,y) - \pi(x)\pi(y)\right| \\ &\le \sum_x \pi(x)2\|p^n(x,\cdot) - \pi(\cdot)\|\end{aligned}$$

so the hypothesis of Theorem 8.3.7 can be checked if we know enough about the rate of convergence to equilibrium.

Finally, to see how good the conditions in Theorem 8.3.7 are, we consider the special case of the moving average process (Example 8.3.5) in which the ξ_i are i.i.d. standard normals. Let

$$\rho(\mathcal{G},\mathcal{H}) = \sup\{\text{corr}(X,Y) : X \in \mathcal{G}, Y \in \mathcal{H}\}$$

where

$$\text{corr}(X,Y) = \frac{EXY - EXEY}{\|X - EX\|_2\|Y - EY\|_2}$$

Clearly, $\alpha(\mathcal{G},\mathcal{H}) \le \rho(\mathcal{G},\mathcal{H})$. Kolmogorov and Rozanov (1960) have shown that when $\mathcal{G}$ and $\mathcal{H}$ are generated by Gaussian random variables, then $\rho(\mathcal{G},\mathcal{H}) \le 2\pi\alpha(\mathcal{G},\mathcal{H})$. They proved this by showing that $\rho(\mathcal{G},\mathcal{H})$ is the angle between $L^2(\mathcal{G})$ and $L^2(\mathcal{H})$. Using the geometric interpretation, we see that if $|c_k|$ is decreasing, then $\alpha(n) \le \bar{\alpha}(n) \le |c_n|$, so Theorem 8.3.7 requires $\sum |c_n|^{(1/2)-\epsilon} < \infty$, but Ibragimov's result applies if $\sum |c_n|^{1-\epsilon} < \infty$. However, as remarked earlier, the central limit theorem is valid if $\sum |c_n| < \infty$.

8.4 Empirical Distributions, Brownian Bridge

Let $X_1, X_2, \ldots$ be i.i.d. with distribution distribution F. Let

$$\hat{F}_n(x) = \frac{1}{n}|\{m \le n : X_m \le x\}|$$

be the empirical distribution. In Theorem 2.4.9 we showed

Theorem 8.4.1 (The Glivenko-Cantelli theorem) *As* $n \to \infty$,

$$\sup_x |F_n(x) - F(x)| \to 0 \quad a.s.$$

In this section, we will investigate the rate of convergence when F is continuous. We impose this restriction, so we can reduce to the case of a uniform distribution on (0,1) by setting $Y_n = F(X_n)$. Since $x \to F(x)$ is nondecreasing and continuous and no observations land in intervals of constancy of F, it is easy to see that if we let

$$\hat{G}_n(y) = \frac{1}{n}|\{m \le n : Y_m \le y\}|$$

then

$$\sup_x |\hat{F}_n(x) - F(x)| = \sup_{0<y<1} |\hat{G}_n(y) - y|$$

For the rest of the section then, we will assume $Y_1, Y_2, \dots$ is i.i.d. uniform on (0,1). To be able to apply Donsker's theorem, we will transform the problem. Put the observations $Y_1, \dots, Y_n$ in increasing order: $U_1^n < U_2^n < \dots < U_n^n$. I claim that

$$\begin{aligned}\sup_{0<y<1} \hat{G}_n(y) - y &= \sup_{1 \le m \le n} \frac{m}{n} - U_m^n \\ \inf_{0<y<1} \hat{G}_n(y) - y &= \inf_{1 \le m \le n} \frac{m-1}{n} - U_m^n \end{aligned} \tag{8.4.1}$$

since the sup occurs at a jump of $\hat{G}_n$ and the inf right before a jump. For a picture, see Figure 8.1. We will show that

$$D_n \equiv n^{1/2} \sup_{0<y<1} |\hat{G}_n(y) - y|$$

has a limit, so the extra $-1/n$ in the inf does not make any difference.

Our third and final maneuver is to give a special construction of the order statistics $U_1^n < U_2^n \dots < U_n^n$. Let $W_1, W_2, \dots$ be i.i.d. with $P(W_i > t) = e^{-t}$ and let $Z_n = W_1 + \dots + W_n$.

Lemma 8.4.2 $\{U_k^n : 1 \le k \le n\} \stackrel{d}{=} \{Z_k/Z_{n+1} : 1 \le k \le n\}$

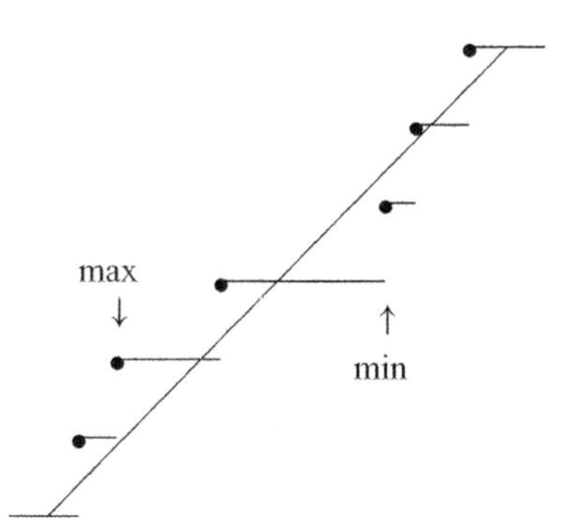

Figure 8.1 Picture proof of formulas in (8.4.1).

Proof We change variables $v = r(t)$, where $v_i = t_i/t_{n+1}$ for $i \le n$, $v_{n+1} = t_{n+1}$. The inverse function is

$$s(v) = (v_1 v_{n+1}, \dots, v_n v_{n+1}, v_{n+1})$$

which has matrix of partial derivatives $\partial s_i/\partial v_j$ given by

$$\begin{pmatrix} v_{n+1} & 0 & \dots & 0 & v_1 \\ 0 & v_{n+1} & \dots & 0 & v_2 \\ \vdots & \vdots & \ddots & \vdots & \vdots \\ 0 & 0 & \dots & v_{n+1} & v_n \\ 0 & 0 & \dots & 0 & 1 \end{pmatrix}$$

The determinant of this matrix is v_{n+1}^n, so if we let $W = (V_1, \dots, V_{n+1}) = r(Z_1, \dots, Z_{n+1})$, the change of variables formula implies W has joint density

$$f_W(v_1, \dots, v_n, v_{n+1}) = \left(\prod_{m=1}^{n} \lambda e^{-\lambda v_{n+1}(v_m - v_{m-1})} \right) \lambda e^{-\lambda v_{n+1}(1-v_n)} v_{n+1}^n$$

To find the joint density of $V = (V_1, \dots, V_n)$, we simplify the preceding formula and integrate out the last coordinate to get

$$f_V(v_1, \dots, v_n) = \int_0^\infty \lambda^{n+1} v_{n+1}^n e^{-\lambda v_{n+1}} \, dv_{n+1} = n!$$

for $0 < v_1 < v_2 \dots < v_n < 1$, which is the desired joint density. □

We turn now to the limit law for D_n. As argued earlier, it suffices to consider

$$\begin{aligned} D_n' &= n^{1/2} \max_{1 \le m \le n} \left| \frac{Z_m}{Z_{n+1}} - \frac{m}{n} \right| \\ &= \frac{n}{Z_{n+1}} \max_{1 \le m \le n} \left| \frac{Z_m}{n^{1/2}} - \frac{m}{n} \cdot \frac{Z_{n+1}}{n^{1/2}} \right| \\ &= \frac{n}{Z_{n+1}} \max_{1 \le m \le n} \left| \frac{Z_m - m}{n^{1/2}} - \frac{m}{n} \cdot \frac{Z_{n+1} - n}{n^{1/2}} \right| \end{aligned} \tag{8.4.2}$$

If we let

$$B_n(t) = \begin{cases} (Z_m - m)/n^{1/2} & \text{if } t = m/n \text{ with } m \in \{0, 1, \dots, n\} \\ \text{linear} & \text{on } [(m-1)/n, m/n] \end{cases}$$

then

$$D_n' = \frac{n}{Z_{n+1}} \max_{0 \le t \le 1} \left| B_n(t) - t \left\{ B_n(1) + \frac{Z_{n+1} - Z_n}{n^{1/2}} \right\} \right|$$

The strong law of large numbers implies $Z_{n+1}/n \to 1$ a.s., so the first factor will disappear in the limit. To find the limit of the second, we observe that Donsker's theorem, Theorem 8.1.4, implies $B_n(\cdot) \Rightarrow B(\cdot)$, a Brownian motion, and computing second moments shows

$$(Z_{n+1} - Z_n)/n^{1/2} \to 0 \text{ in probability}$$

$\psi(\omega) = \max_{0\le t\le 1} |\omega(t) - t\omega(1)|$ is a continuous function from $C[0,1]$ to $\mathbf{R}$, so it follows from Donsker's theorem that:

Theorem 8.4.3 $D_n \Rightarrow \max_{0\le t\le 1} |B_t - tB_1|$, *where* B_t *is a Brownian motion starting at 0.*

Remark Doob (1949) suggested this approach to deriving results of Kolmogorov and Smirnov, which was later justified by Donsker (1952). Our proof follows Breiman (1968).

To identify the distribution of the limit in Theorem 8.4.3, we will first prove

$$\{B_t - tB_1, 0 \le t \le 1\} \stackrel{d}{=} \{B_t, 0 \le t \le 1 | B_1 = 0\} \tag{8.4.3}$$

a process we will denote by B_t^0 and call the **Brownian bridge**. The event $B_1 = 0$ has probability 0, but it is easy to see what the conditional probability should mean. If $0 = t_0 < t_1 < \ldots < t_n < t_{n+1} = 1$, $x_0 = 0$, $x_{n+1} = 0$, and $x_1, \ldots, x_n \in \mathbf{R}$, then

$$\begin{aligned} P(B(t_1) &= x_1, \ldots, B(t_n) = x_n | B(1) = 0) \\ &= \frac{1}{p_1(0,0)} \prod_{m=1}^{n+1} p_{t_m - t_{m-1}}(x_{m-1}, x_m) \end{aligned} \tag{8.4.4}$$

where $p_t(x,y) = (2\pi t)^{-1/2} \exp(-(y-x)^2/2t)$.

Proof of (8.4.3) Formula (8.4.4) shows that the f.d.d.'s of B_t^0 are multivariate normal and have mean 0. Since $B_t - tB_1$ also has this property, it suffices to show that the covariances are equal. We begin with the easier computation. If $s < t$, then

$$E((B_s - sB_1)(B_t - tB_1)) = s - st - st + st = s(1-t) \tag{8.4.5}$$

For the other process, $P(B_s^0 = x, B_t^0 = y)$ is

$$\begin{aligned} &\frac{\exp(-x^2/2s)}{(2\pi s)^{1/2}} \cdot \frac{\exp(-(y-x)^2/2(t-s))}{(2\pi(t-s))^{1/2}} \cdot \frac{\exp(-y^2/2(1-t))}{(2\pi(1-t))^{1/2}} \cdot (2\pi)^{1/2} \\ &= (2\pi)^{-1}(s(t-s)(1-t))^{-1/2} \exp(-(ax^2 + 2bxy + cy^2)/2) \end{aligned} \tag{8.4.6}$$

where

$$a = \frac{1}{s} + \frac{1}{t-s} = \frac{t}{s(t-s)} \qquad b = -\frac{1}{t-s}$$
$$c = \frac{1}{t-s} + \frac{1}{1-t} = \frac{1-s}{(t-s)(1-t)}$$

Recalling the discussion at the end of Section 3.9 and noticing

$$\begin{pmatrix} \frac{t}{s(t-s)} & \frac{-1}{(t-s)} \\ \frac{-1}{(t-s)} & \frac{1-s}{(t-s)(1-t)} \end{pmatrix}^{-1} = \begin{pmatrix} s(1-s) & s(1-t) \\ s(1-t) & t(1-t) \end{pmatrix}$$

(multiply the matrices!) shows (8.4.3) holds. □

Remark To do computations with the joint density, it is useful to know that if (X,Y) have a bivariate normal distribution with mean 0 $EX^2 = \sigma_X^2$, $EY^2 = \sigma_Y$, and correlation ρ, then the joint density is given by

$$\frac{1}{2\pi\sigma_X\sigma_Y\sqrt{1-\rho^2}}\exp\left[-\frac{1}{2(1-\rho^2)}\left(\frac{x^2}{\sigma_X^2}-\frac{2xy}{\sigma_X\sigma_Y}+\frac{y^2}{\sigma_Y^2}\right)\right]$$

In (8.4.6) if $X = B_s^0$ and $Y = B_s^0$, then $\sigma_X = \sqrt{s(1-s)}$, $\sigma_Y = \sqrt{t(1-t)}$ and

$$\rho = \frac{s(1-t)}{\sqrt{s(1-s)t(1-t)}} = \sqrt{\frac{s(1-t)}{(1-s)t}}$$

so we have

$$1-\rho^2 - \frac{(1-s)t - s(1-t)}{(1-s)t} = \frac{t-s}{(1-s)t}.$$

We leave it to the reader to check the well-known fact that the conditional distribution

$$(Y|X = x) = \text{normal}(x\rho\sigma_Y/\sigma_X, (1-\rho^2)\sigma_Y^2)$$

which in our special case implies

$$(B_t^0|B_s^0 = x) = \text{normal}(x(1-t)/(1-s), (1-t)(1-s)/1-s)) \tag{8.4.7}$$

Our final step in investigating the limit distribution of D_n is to compute the distribution of $\max_{0\le t\le 1}|B_t^0|$. To do this, we first prove

Theorem 8.4.4 *The density function of B_t on $\{T_a \wedge T_b > t\}$ is*

$$P_x(T_a \wedge T_b > t, B_t = y) = \sum_{n=-\infty}^{\infty} P_x(B_t = y + 2n(b-a)) \tag{8.4.8}$$
$$- P_x(B_t = 2b - y + 2n(b-a))$$

Proof We begin by observing that if $A \subset (a, b)$

$$P_x(T_a \wedge T_b > t, B_t \in A) = P_x(B_t \in A) - P_x(T_a < T_b, T_a < t, B_t \in A)$$
$$- P_x(T_b < T_a, T_b < t, B_t \in A) \tag{8.4.9}$$

If we let $\rho_a(y) = 2a - y$ be reflection through a and observe that $\{T_a < T_b\}$ is $\mathcal{F}(T_a)$ measurable, then it follows from the proof of (7.4.5) that

$$P_x(T_a < T_b, T_a < t, B_t \in A) = P_x(T_a < T_b, B_t \in \rho_a A)$$

where $\rho_a A = \{\rho_a(y) : y \in A\}$. To get rid of the $T_a < T_b$, we observe that

$$P_x(T_a < T_b, B_t \in \rho_a A) = P_x(B_t \in \rho_a A) - P_x(T_b < T_a, B_t \in \rho_a A)$$

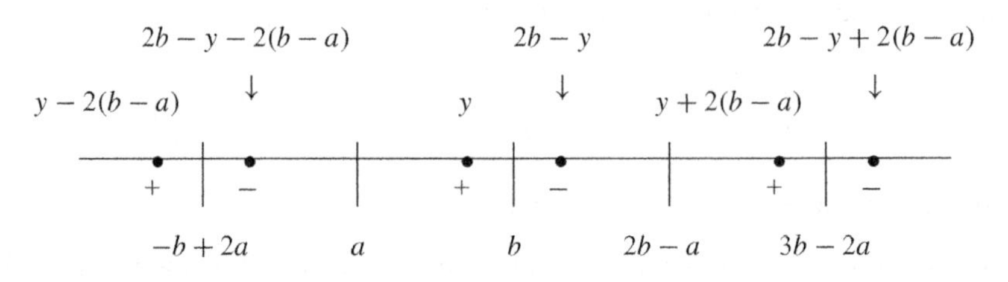

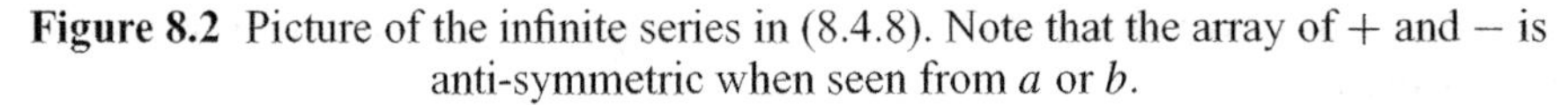

Figure 8.2 Picture of the infinite series in (8.4.8). Note that the array of + and − is anti-symmetric when seen from a or b.

Noticing that $B_t \in \rho_a A$ and $T_b < T_a$ imply $T_b < t$ and using the reflection principle again gives

$$\begin{aligned} P_x(T_b < T_a, B_t \in \rho_a A) &= P_x(T_b < T_a, B_t \in \rho_b \rho_a A) \\ &= P_x(B_t \in \rho_b \rho_a A) - P_x(T_a < T_b, B_t \in \rho_b \rho_a A) \end{aligned}$$

Repeating the last two calculations n more times gives

$$\begin{aligned} P_x(T_a < T_b, B_t \in \rho_a A) &= \sum_{m=0}^{n} P_x(B_t \in \rho_a (\rho_b \rho_a)^m A) - P_x(B_t \in (\rho_b \rho_a)^{m+1} A) \\ &\quad + P_x(T_a < T_b, B_t \in (\rho_b \rho_a)^{n+1} A) \end{aligned}$$

Each pair of reflections pushes A further away from 0, so letting $n \to \infty$ shows

$$P_x(T_a < T_b, B_t \in \rho_a A) = \sum_{m=0}^{\infty} P_x(B_t \in \rho_a (\rho_b \rho_a)^m A) - P_x(B_t \in (\rho_b \rho_a)^{m+1} A)$$

Interchanging the roles of a and b gives

$$P_x(T_b < T_a, B_t \in \rho_b A) = \sum_{m=0}^{\infty} P_x(B_t \in \rho_b (\rho_a \rho_b)^m A) - P_x(B_t \in (\rho_a \rho_b)^{m+1} A)$$

Combining the last two expressions with (8.4.9) and using $\rho_c^{-1} = \rho_c$, $(\rho_a \rho_b)^{-1} = \rho_b^{-1} \rho_a^{-1}$ gives

$$P_x(T_a \wedge T_b > t, B_t \in A) = \sum_{m=-\infty}^{\infty} P_x(B_t \in (\rho_b \rho_a)^n A) - P_x(B_t \in \rho_a (\rho_b \rho_a)^n A)$$

To prepare for applications, let $A = (u, v)$, where $a < u < v < b$, notice that $\rho_b \rho_a(y) = y + 2(b - a)$, and change variables in the second sum to get

$$\begin{aligned} &P_x(T_a \wedge T_b > t, u < B_t < v) = \\ &\sum_{n=-\infty}^{\infty} \{P_x(u + 2n(b-a) < B_t < v + 2n(b-a)) \\ &\qquad - P_x(2b - v + 2n(b-a) < B_t < 2b - u + 2n(b-a))\} \end{aligned} \tag{8.4.10}$$

Letting $u = y - \epsilon$, $v = y + \epsilon$, dividing both sides by 2ϵ, and letting $\epsilon \to 0$ (leaving it to the reader to check that the dominated convergence theorem applies) gives the desired result. □

Setting $x = y = 0$, $t = 1$, and dividing by $(2\pi)^{-1/2} = P_0(B_1 = 0)$, we get a result for the Brownian bridge B_t^0:

$$\begin{aligned} &P_0\left(a < \min_{0 \le t \le 1} B_t^0 < \max_{0 \le t \le 1} B_t^0 < b\right) \\ &\quad = \sum_{n=-\infty}^{\infty} e^{-(2n(b-a))^2/2} - e^{-(2b+2n(b-a))^2/2} \end{aligned} \tag{8.4.11}$$

Taking $a = -b$, we have

$$P_0\left(\max_{0\le t\le 1} |B_t^0| < b\right) = \sum_{m=-\infty}^{\infty} (-1)^m e^{-2m^2b^2} \tag{8.4.12}$$

This formula gives the distribution of the Kolmogorov-Smirnov statistic, which can be used to test if an i.i.d. sequence $X_1, \ldots, X_n$ has distribution F. To do this, we transform the data to $F(X_n)$ and look at the maximum discrepancy between the empirical distribution and the uniform. (8.4.12) tells us the distribution of the error when the X_i have distribution F.

Exercises

8.4.1 Use Exercise 7.4.8 and the reasoning that led to (8.4.11) to conclude

$$P\left(\max_{0\le t\le 1} B_t^0 > b\right) = \exp(-2b^2)$$

8.4.2 Let B_t be a Brownian motion starting at 0. Show that $X_t = (1-t)B(t/1-t)$ is a Brownian bridge.

8.4.3 Show that Brownian bridge is a Markov process by checking that if $s_1 < \ldots s_n < s < t < 1$

$$P(B(t) = y|B(s) = x, B(s_n) = x_n, \ldots B(s_1) = x_1) = P(B(t) = y|B(s) = x)$$

8.4.4 Let B_t^0 be Brownian bridge. Show that $X_t = B_t^0/(1-t)$ is a martingale but is not L^2 bounded.

8.5 Laws of the Iterated Logarithm

Our first goal is to show:

Theorem 8.5.1 (LIL for Brownian motion)

$$\limsup_{t\to\infty} B_t/(2t\log\log t)^{1/2} = 1 \quad a.s.$$

Here LIL is short for "law of the iterated logarithm," a name that refers to the $\log\log t$ in the denominator. Once Theorem 8.5.1 is established, we can use the Skorokhod representation to prove the analogous result for random walks with mean 0 and finite variance.

Proof The key to the proof is (7.4.4).

$$P_0\left(\max_{0\le s\le 1} B_s > a\right) = P_0(T_a \le 1) = 2\, P_0(B_1 \ge a) \tag{8.5.1}$$

To bound the right-hand side, we use the result in Lemma 1.2.6. For $x > 0$,

$$(x^{-1} - x^{-3})\exp(-x^2/2) \le \int_x^\infty \exp(-y^2/2)dy \le x^{-1}\exp(-x^2/2) \tag{8.5.2}$$

and hence

$$\int_x^\infty \exp(-y^2/2)\,dy \sim \frac{1}{x}\exp(-x^2/2) \quad \text{as } x \to \infty \tag{8.5.3}$$

(8.5.3) and Brownian scaling imply that

$$P_0(B_t > (tf(t))^{1/2}) \sim \kappa f(t)^{-1/2} \exp(-f(t)/2)$$

where $\kappa = (2\pi)^{-1/2}$ is a constant that we will try to ignore. The last result implies that if $\epsilon > 0$, then

$$\sum_{n=1}^{\infty} P_0(B_n > (nf(n))^{1/2}) \begin{cases} < \infty & \text{when } f(n) = (2+\epsilon)\log n \\ = \infty & \text{when } f(n) = (2-\epsilon)\log n \end{cases}$$

and hence by the Borel-Cantelli lemma that

$$\limsup_{n\to\infty} B_n/(2n \log n)^{1/2} \le 1 \quad \text{a.s.}$$

To replace $\log n$ by $\log\log n$, we have to look along exponentially growing sequences. Let $t_n = \alpha^n$, where $\alpha > 1$.

$$\begin{aligned} P_0\left(\max_{t_n \le s \le t_{n+1}} B_s > (t_n f(t_n))^{1/2}\right) &\le P_0\left(\max_{0 \le s \le t_{n+1}} B_s/t_{n+1}^{1/2} > \left(\frac{f(t_n)}{\alpha}\right)^{1/2}\right) \\ &\le 2\kappa (f(t_n)/\alpha)^{-1/2} \exp(-f(t_n)/2\alpha) \end{aligned}$$

by (8.5.1) and (8.5.2). If $f(t) = 2\alpha^2 \log\log t$, then

$$\log\log t_n = \log(n \log\alpha) = \log n + \log\log\alpha$$

so $\exp(-f(t_n)/2\alpha) \le C_\alpha n^{-\alpha}$, where C_α is a constant that depends only on α, and hence

$$\sum_{n=1}^{\infty} P_0\left(\max_{t_n \le s \le t_{n+1}} B_s > (t_n f(t_n))^{1/2}\right) < \infty$$

Since $t \to (tf(t))^{1/2}$ is increasing and $\alpha > 1$ is arbitrary, it follows that

$$\limsup B_t/(2t \log\log t)^{1/2} \le 1 \tag{8.5.4}$$

To prove the other half of Theorem 8.5.1, again let $t_n = \alpha^n$, but this time α will be large, since to get independent events, we will we look at

$$P_0\left(B(t_{n+1}) - B(t_n) > (t_{n+1} f(t_{n+1}))^{1/2}\right) = P_0\left(B_1 > (\beta f(t_{n+1}))^{1/2}\right)$$

where $\beta = t_{n+1}/(t_{n+1} - t_n) = \alpha/(\alpha - 1) > 1$. The last quantity is

$$\ge \frac{\kappa}{2} (\beta f(t_{n+1}))^{-1/2} \exp(-\beta f(t_{n+1})/2)$$

if n is large by (8.5.3). If $f(t) = (2/\beta^2)\log\log t$, then $\log\log t_n = \log n + \log\log\alpha$, so

$$\exp(-\beta f(t_{n+1})/2) \ge C_\alpha n^{-1/\beta}$$

where C_α is a constant that depends only on α, and hence

$$\sum_{n=1}^{\infty} P_0\left(B(t_{n+1}) - B(t_n) > (t_{n+1} f(t_{n+1}))^{1/2}\right) = \infty$$

Since the events in question are independent, it follows from the second Borel-Cantelli lemma that

$$B(t_{n+1}) - B(t_n) > ((2/\beta^2)t_{n+1} \log\log t_{n+1})^{1/2} \quad \text{i.o.} \tag{8.5.5}$$

From (8.5.4), we get

$$\limsup_{n\to\infty} B(t_n)/(2t_n \log\log t_n)^{1/2} \le 1 \tag{8.5.6}$$

Since $t_n = t_{n+1}/\alpha$ and $t \to \log\log t$ is increasing, combining (8.5.5) and (8.5.6), and recalling $\beta = \alpha/(\alpha-1)$ gives

$$\limsup_{n\to\infty} B(t_{n+1})/(2t_{n+1} \log\log t_{n+1})^{1/2} \ge \frac{\alpha-1}{\alpha} - \frac{1}{\alpha^{1/2}}$$

Letting $\alpha \to \infty$ now gives the desired lower bound, and the proof of Theorem 8.5.1 is complete. □

Turning to random walk, we will prove a result due to Hartman and Wintner (1941):

Theorem 8.5.2 *If $X_1, X_2, \ldots$ are i.i.d. with $EX_i = 0$ and $EX_i^2 = 1$, then*

$$\limsup_{n\to\infty} S_n/(2n \log\log n)^{1/2} = 1$$

Proof By Theorem 8.1.2, we can write $S_n = B(T_n)$ with $T_n/n \to 1$ a.s. As in the proof of Donsker's theorem, this is all we will use in the following argument. Theorem 8.5.2 will follow from Theorem 8.5.1 once we show

$$(S_{[t]} - B_t)/(t \log\log t)^{1/2} \to 0 \quad \text{a.s.} \tag{8.5.7}$$

To do this, we begin by observing that if $\epsilon > 0$ and $t \ge t_o(\omega)$

$$T_{[t]} \in [t/(1+\epsilon), t(1+\epsilon)] \tag{8.5.8}$$

To estimate $S_{[t]} - B_t$, we let $M(t) = \sup\{|B(s) - B(t)| : t/(1+\epsilon) \le s \le t(1+\epsilon)\}$. To control the last quantity, we let $t_k = (1+\epsilon)^k$ and notice that if $t_k \le t \le t_{k+1}$

$$\begin{aligned} M(t) &\le \sup\{|B(s) - B(t)| : t_{k-1} \le s, t \le t_{k+2}\} \\ &\le 2 \sup\{|B(s) - B(t_{k-1})| : t_{k-1} \le s \le t_{k+2}\} \end{aligned}$$

Noticing $t_{k+2} - t_{k-1} = \delta t_{k-1}$, where $\delta = (1+\epsilon)^3 - 1$, scaling implies

$$\begin{aligned} &P\left(\max_{t_{k-1}\le s\le t_{k+2}} |B(s) - B(t)| > (3\delta t_{k-1} \log\log t_{k-1})^{1/2} \right) \\ &= P\left(\max_{0\le r\le 1} |B(r)| > (3 \log\log t_{k-1})^{1/2} \right) \\ &\le 2\kappa (3 \log\log t_{k-1})^{-1/2} \exp(-3 \log\log t_{k-1}/2) \end{aligned}$$

by a now familiar application of (8.5.1) and (8.5.2). Summing over k and using (b) gives

$$\limsup_{t\to\infty} (S_{[t]} - B_t)/(t \log\log t)^{1/2} \le (3\delta)^{1/2}$$

If we recall $\delta = (1+\epsilon)^3 - 1$ and let $\epsilon \downarrow 0$, (a) follows and the proof is complete. □

Strassen (1965) has shown an exact converse. If Theorem 8.5.2 holds, then $EX_i = 0$ and $EX_i^2 = 1$. Another one of his contributions to this subject is:

Theorem 8.5.3 (Strassen's (1964) invariance principle) *Let $X_1, X_2, \ldots$ be i.i.d. with $EX_i = 0$ and $EX_i^2 = 1$, let $S_n = X_1 + \cdots + X_n$, and let $S_{(n\cdot)}$ be the usual linear interpolation. The limit set (i.e., the collection of limits of convergent subsequences) of*

$$Z_n(\cdot) = (2n \log\log n)^{-1/2} S(n\cdot) \quad \text{for } n \geq 3$$

is $\mathcal{K} = \{f : f(x) = \int_0^x g(y)\, dy \text{ with } \int_0^1 g(y)^2\, dy \leq 1\}$.

Jensen's inequality implies $f(1)^2 \leq \int_0^1 g(y)^2\, dy \leq 1$ with equality if and only if $f(t) = t$, so Theorem 8.5.3 contains Theorem 8.5.2 as a special case and provides some information about how the large value of S_n came about.

Exercises

8.5.1 Show that if $E|X_i|^\alpha = \infty$ for some $\alpha < 2$, then

$$\limsup_{n\to\infty} |X_n|/n^{1/\alpha} = \infty \quad \text{a.s.}$$

so the law of the iterated logarithm fails.

8.5.2 Give a direct proof that, under the hypotheses of Theorem 8.5.3, the limit set of $\{S_n/(2n \log\log n)^{1/2}\}$ is $[-1, 1]$.

9

Multidimensional Brownian Motion

The point of this chapter is to establish the connection between Brownian motion and various parabolic and elliptic partial differential equations. The first section uses martingales to do computations for multidimensional Brownian motion.

9.1 Martingales

Let $\Delta f = \sum_{i=1}^d \partial^2 f/\partial x_i^2$ be the Laplacian of f. Combining Ito's formula (7.6.9) with Theorem 7.6.4:

Theorem 9.1.1 *Suppose $v \in C^2$, i.e., all first- and second-order partial derivatives exist and are continuous, and $E \int_0^t |\nabla v(B_s)|^2\, ds < \infty$ for all t. Then*

$$v(B_t) - \int_0^t \frac{1}{2}\Delta v(B_s)\, ds \quad \textit{is a continuous martingale.}$$

Let $S_r = \inf\{t : |B_t| = r\}$, $r < R$, and $\tau = S_r \wedge S_R$. The first detail is to note that if $|x| < R$, then $P_x(S_R < \infty) = 1$. Once we know this we can conclude

Theorem 9.1.2 *If $|x| < R$, then $E_x S_R = (R^2 - |x|^2)/d$.*

Proof It follows from Theorem 7.5.4 that $|B_t|^2 - dt = \sum_{i=1}^d (B_t^i)^2 - t$ is a martingale. Theorem 7.5.1 implies $|x|^2 = E|B_{S_R \wedge t}|^2 - dE(S_R \wedge t)$. Letting $t \to \infty$ gives the desired result. □

To further investigate Brownian motion, we introduce two new functions:

$$\varphi(x) = \begin{cases} \log|x| & d = 2 \\ |x|^{2-d} & d \geq 3 \end{cases}$$

We leave it to the reader to check that in each case $\Delta\varphi = 0$.

Lemma 9.1.3 $\varphi(x) = E_x\varphi(B_\tau)$.

Proof Define $\psi(x) = g(|x|)$ to be C^2 and have compact support, and have $\psi(x) = \phi(x)$ when $r < |x| < R$. Theorem 9.1.1 implies that $\psi(x) = E_x\psi(B_{t\wedge\tau})$. Letting $t \to \infty$ now gives the desired result. □

Lemma 9.1.3 implies that

$$\varphi(x) = \varphi(r)P_x(S_r < S_R) + \varphi(R)(1 - P_x(S_r < S_R))$$

where $\varphi(r)$ is short for the value of $\varphi(x)$ on $\{x : |x| = r\}$. Solving now gives

$$P_x(S_r < S_R) = \frac{\varphi(R) - \varphi(x)}{\varphi(R) - \varphi(r)} \tag{9.1.1}$$

In $d = 2$, the last formula says

$$P_x(S_r < S_R) = \frac{\log R - \log |x|}{\log R - \log r} \tag{9.1.2}$$

If we fix r and let $R \to \infty$ in (9.1.2), the right-hand side goes to 1. So

$$P_x(S_r < \infty) = 1 \quad \text{for any } x \text{ and any } r > 0$$

It follows that two-dimensional Brownian motion is **recurrent** in the sense that if G is any open set, then $P_x(B_t \in G \text{ i.o.}) \equiv 1$.

If we fix R, let $r \to 0$ in (9.1.2), and let $S_0 = \inf\{t > 0 : B_t = 0\}$, then for $x \neq 0$

$$P_x(S_0 < S_R) \leq \lim_{r \to 0} P_x(S_r < S_R) = 0$$

Since this holds for all R and since the continuity of Brownian paths implies $S_R \uparrow \infty$ as $R \uparrow \infty$, we have $P_x(S_0 < \infty) = 0$ for all $x \neq 0$. To extend the last result to $x = 0$, we note that the Markov property implies

$$P_0(B_t = 0 \text{ for some } t \geq \epsilon) = E_0[P_{B_\epsilon}(T_0 < \infty)] = 0$$

for all $\epsilon > 0$, so $P_0(B_t = 0 \text{ for some } t > 0) = 0$, and thanks to our definition of $S_0 = \inf\{t > 0 : B_t = 0\}$, we have

$$P_x(S_0 < \infty) = 0 \quad \text{for all } x \tag{9.1.3}$$

Thus, in $d \geq 2$ Brownian motion will not hit 0 at a positive time even if it starts there. As a corollary we see that Brownian motion in $d = 3$ will not hit a straight line.

For $d \geq 3$, formula (9.1.1) says

$$P_x(S_r < S_R) = \frac{R^{2-d} - |x|^{2-d}}{R^{2-d} - r^{2-d}} \tag{9.1.4}$$

There is no point in fixing R and letting $r \to 0$, here. The fact that two-dimensional Brownian motion does not hit 0 implies that three-dimensional Brownian motion does not hit 0 and indeed will not hit the line $\{x : x_1 = x_2 = 0\}$. If we fix r and let $R \to \infty$ in (9.1.4), we get

$$P_x(S_r < \infty) = (r/|x|)^{d-2} < 1 \quad \text{if } |x| > r \tag{9.1.5}$$

From the last result it follows easily that for $d \geq 3$, Brownian motion is **transient**, i.e., it does not return infinitely often to any bounded set.

Theorem 9.1.4 *As $t \to \infty$, $|B_t| \to \infty$ a.s.*

Proof Let $A_n = \{|B_t| > n^{1-\epsilon} \text{ for all } t \geq S_n\}$. The strong Markov property implies

$$P_x(A_n^c) = E_x(P_{B(S_n)}(S_{n^{1-\epsilon}} < \infty)) = (n^{1-\epsilon}/n)^{d-2} \to 0$$

as $n \to \infty$. Now $\limsup A_n = \cap_{N=1}^{\infty} \cup_{n=N}^{\infty} A_n$ has

$$P(\limsup A_n) \geq \limsup P(A_n) = 1$$

So infinitely often the Brownian path never returns to $\{x : |x| \le n^{1-\epsilon}\}$ after time S_n and this implies the desired result. □

The scaling relation (7.1.1) implies that $S_n =_d n^2 S_1$, so the proof of Theorem 9.1.4 suggests that

$$|B_t|/t^{(1-\epsilon)/2} \to \infty$$

Dvoretsky and Erdös (1951) have proved the following result about how fast Brownian motion goes to ∞ in $d \ge 3$.

Theorem 9.1.5 *Suppose $g(t)$ is positive and decreasing. Then*

$$P_0(|B_t| \le g(t)\sqrt{t} \text{ i.o. as } t \uparrow \infty) = 1 \text{ or } 0$$

according as $\int^\infty g(t)^{d-2}/t\, dt = \infty$ *or* $< \infty$.

Here the absence of the lower limit implies that we are only concerned with the behavior of the integral "near ∞." A little calculus shows that

$$\int^\infty t^{-1} \log^{-\alpha} t\, dt = \infty \text{ or } < \infty$$

according as $\alpha \le 1$ or $\alpha > 1$, so B_t goes to ∞ faster than

$$t^{1/2}(\log t)^{-\alpha/(d-2)} \qquad \text{for any } \alpha > 1.$$

Exercises

9.1.1 Combine scaling with the 0-1 law to show that

$$\liminf_{t\to\infty} |B_t|/t^{1/2} = 0 \quad \text{a.s.}$$

9.1.2 Suppose $f \in C^2$ is nonnegative and superharmonic. Show that in dimensions 1 and 2, f must be constant. Give an example of the form $f(x) = g(|x|)$ in $d \ge 3$.

9.1.3 Suppose $f \in C^2$ is bounded and harmonic. Then f must be constant (in any dimension).

9.2 Heat Equation

In this section, we will consider the following equation:

(a) $u \in C^{1,2}$ and $u_t = \frac{1}{2}\Delta u$ in $(0,\infty) \times \mathbf{R}^d$.

(b) u is continuous at each point of $\{0\} \times \mathbf{R}^d$ and $u(0,x) = f(x)$, where f is bounded and continuous.

This equation derives its name from the fact that if the units of measurement are chosen suitably, then the solution $u(t,x)$ gives the temperature at the point $x \in R^d$ at time t when the temperature profile at time 0 is given by $f(x)$.

Obviously, (b) cannot hold unless f is continuous. The results below can be generalized to unbounded f. We assume f is bounded to keep things simple. In this and the next five sections the assumptions stated at the beginning are in force throughout the section.

The first step in solving this equation, as it will be six times in the following, is to find a local martingale M_s, $s \in [0,t]$, i.e., a process in which there is an increasing seqeunce of stopping times so that $M(s \wedge T_n)$ is a martingale and $P(T_n < t) \to 0$.

Theorem 9.2.1 *If u satisfies (a), $M_s = u(t-s, B_s)$ is a local martingale on $[0,t]$.*

Proof Applying Itô's formula, (7.6.9) gives

$$u(t-s, B_s) - u(t, B_0) = \int_0^s -u_t(t-r, B_r)\,dr$$
$$+ \int_0^s \nabla u(t-r, B_r) \cdot dB_r + \frac{1}{2}\int_0^s \Delta u(t-r, B_r)\,dr$$

The result follows, since $-u_t + \frac{1}{2}\Delta u = 0$ and the second term on the right-hand side is a local martingale by Theorem 7.6.4. Here we need the stopping times to make sure the integrability condition holds. □

Our next step is to prove a uniqueness theorem.

Theorem 9.2.2 *If there is a solution that is bounded, then it must be*

$$v(t,x) \equiv E_x f(B_t)$$

Here $\equiv$ means that the last equation defines v. We will always use u for a generic solution of the equation and v for our special solution.

Proof If we now assume that u is bounded, then $M_s, 0 \le s < t$, is a bounded martingale. The martingale convergence theorem implies that

$$M_t \equiv \lim_{s \uparrow t} M_s \quad \text{exists a.s.}$$

If u satisfies (b), this limit must be $f(B_t)$. Since M_s is uniformly integrable, it follows that $u(t,x) = E_x M_0 = E_x M_t = v(t,x)$. □

Now that Theorem 9.2.2 has told us what the solution must be, the next logical step is to find conditions under which v is a solution. Here and in the other five examples it is easy to show that if v is smooth enough, then it is a solution.

Theorem 9.2.3 *If $v \in C^{1,2}$, then it satisfies (a).*

Proof If we think of running space time Brownian motion $(t-s, B_s)$ to construct the solution, then we see

$$v(t+h, x) = E_x v(t, B_h)$$

Subtracting $v(t,x)$ from each side and using Taylor series

$$
\begin{aligned}
v(t+h,x) - v(t,x) =& E_x\left(\sum_{i=1}^{d} \frac{\partial v}{\partial x_i}(t,x)(B_i(h)-x_i)\right. \\
&\left. + \frac{1}{2}\sum_{1\le i,j\le d} \frac{\partial^2 v}{\partial x_i \partial x_j}(t,x)(B_i(h)-x_i)(B_j(h)-x_j)\right) + o(h)
\end{aligned}
$$

Since for $i \neq j$, $B_i(t)$, $B_i^2(t) - t$, and $B_i(t)B_j(t)$ are martingales

$$
v(t+h,x) - v(t,x) = \frac{1}{2}\sum_{i=1}^{d} \frac{\partial^2 v}{\partial x_i^2}(t,x)h + o(h)
$$

Dividing by h and letting $h \to 0$ gives the desired result. □

Theorem 9.2.4 *v satisfies (b).*

Proof $(B_t - B_0) \stackrel{d}{=} t^{1/2}\chi$, where χ has a normal distribution with mean 0 and variance 1, so if $t_n \to 0$ and $x_n \to x$, the bounded convergence theorem implies that

$$
v(t_n, x_n) = Ef(x_n + t_n^{1/2}\chi) \to f(x)
$$

which proves the desired result. □

The final step in showing that v is a solution is to find conditions that guarantee that it is smooth.

Theorem 9.2.5 *$v \in C^{1,2}$ and hence satisfies (a).*

Proof Our solution is

$$
v(t,x) = E_x f(B_t) = \int (2\pi t)^{-d/2} e^{-|x-y|^2/2t} f(y)\, dy
$$

Using calculus and some basic facts from analysis one can prove the desired result. The details are not exciting and are left to the reader. □

9.3 Inhomogeneous Heat Equation

In this section, we will consider what happens when we add a bounded continuous function $g(x)$ to the heat equation. That is, we will study

(a) $u \in C^{1,2}$ and $u_t = \frac{1}{2}\Delta u + g$ in $(0,\infty) \times \mathbf{R}^d$

(b) u is continuous at each point of $\{0\} \times \mathbf{R}^d$ and $u(0,x) = f(x)$ is bounded and continuous.

We observed in the previous section that (b) cannot hold unless f is continuous. Here $g = u_t - (1/2)\Delta u$, so the equation in (a) cannot hold with $u \in C^{1,2}$ unless g is continuous. Our first step in treating the new equation is to observe that if u_1 is a solution of the equation with $f = f_0$ and $g = 0$ which we studied in the last section, and u_2 is a solution of the equation with $f = 0$ and $g = g_0$, then $u_1 + u_2$ is a solution of the equation with $f = f_0$ and $g = g_0$, so we can restrict our attention to the case $f \equiv 0$.

Having made this simplification, we will now study the previous equation by following the procedure used in the last section.

Theorem 9.3.1 *If u satisfies (a), then*

$$M_s = u(t-s, B_s) + \int_0^s g(B_r)\,dr$$

is a local martingale on $[0,t]$.

Proof Applying Itô's formula as in the proof of Theorem 9.2.1 gives

$$u(t-s,B_s) - u(t,B_0) = \int_0^s (-u_t + \frac{1}{2}\Delta u)(t-r,B_r)\,dr + \int_0^s \nabla u(t-r,B_r)\cdot dB_r$$

which proves the result, since $-u_t + \frac{1}{2}\Delta u = -g$ and the second term on the right-hand side is a local martingale by Theorem 7.6.4. □

Again the next step is a uniqueness result.

Theorem 9.3.2 *If there is a solution that is bounded on* $[0,T]\times\mathbf{R}^d$ *for any* $T<\infty$, *it must be*

$$v(t,x) \equiv E_x\left(\int_0^t g(B_s)\,ds\right)$$

Proof Under the assumptions on g and u, $M_s, 0 \le s < t$, defined in Theorem 9.3.1 is a bounded martingale and $u(0,x)\equiv 0$, so

$$M_t \equiv \lim_{s\uparrow t} M_s = \int_0^t g(B_s)\,ds$$

Since M_s is uniformly integrable, $u(t,x) = E_x M_0 = E_x M_t = v(t,x)$. □

Again, it is easy to show that if v is smooth enough, it is a solution.

Theorem 9.3.3 *If $v \in C^{1,2}$, then it satisfies (a) in* $(0,\infty)\times\mathbf{R}^d$.

Proof If we think of running space time Brownian motion $(t-s, B_s)$ to construct the solution, then we see

$$v(t+h,x) = E_x\left(v(t,B_h) + \int_0^h g(B_s)\,ds\right)$$

Subtracting $v(t,x)$ from each side and using Taylor series

$$\begin{aligned} v(t+h,x) - v(t,x) = E_x\Bigg(&\sum_{i=1}^d \frac{\partial v}{\partial x_i}(t,x)(B_i(h)-x_i) \\ &+ \frac{1}{2}\sum_{1\le i,j\le d} \frac{\partial^2 v}{\partial x_i \partial x_j}(t,x)(B_i(h)-x_i)(B_j(h)-x_j) \\ &+ \int_0^h g(B_s)\,ds\Bigg) + o(h)\end{aligned}$$

Since for $i \neq j$, $B_i(t)$, $B_i^2(t) - t$, and $B_i(t)B_j(t)$ are martingales

$$v(t+h,x) - v(t,x) = \frac{1}{2}\sum_{i=1}^{d} \frac{\partial^2 v}{\partial x_i^2}(t,x)h + E_x \int_0^h g(B_s)\, ds + o(h)$$

Dividing by h and letting $h \to 0$ gives $v_t = \frac{1}{2}\Delta V + g(x)$, which is the desired result. □

Theorem 9.3.4 *v satisfies (b).*

Proof If $|g| \le M$, then as $t \to 0$

$$|v(t,x)| \le E_x \int_0^t |g(B_s)|\, ds \le Mt \to 0$$

which proves the desired result. □

The final step in showing that v is a solution is to check that $v \in C^{1,2}$. The calculations are messy, so we just state the result. Here and in what follows when we say a function is Hölder continuous, that means $|g(x) - g(y)| \le C|x-y|^\alpha$ for some $\alpha > 0$.

Theorem 9.3.5 *If g is Hölder continuous, then $v \in C^{1,2}$.*

The problem is that $v(t,x) = \int G_t(x,y)g(y)\, dy$, where

$$G_t(x,y) = \int_0^t (2\pi s)^{-d/2} e^{-|y-x|^2/2s}\, ds$$

Looking at the y with $|y-x|^2 \le s$ leads to a lower bound

$$G_t(x,y) \ge C \int_{|y-x|^2}^t s^{-d/2}\, ds = O(|y-x|^{2-d})$$

From this it follows that first derivative blows up like δ^{1-d} and the second like is δ^{-d}, which is not integrable near 0. However, if g is Hölder continuity of g, its smoothness can save the day. For details, see pp 12–13 of Friedman (1964).

9.4 Feynman-Kac Formula

In this section, we will consider what happens when we add cu to the right-hand side of the heat equation, where $c(x)$ is bounded and continuous. That is, we will study

(a) $u \in C^{1,2}$ and $u_t = \frac{1}{2}\Delta u + cu$ in $(0,\infty) \times \mathbf{R}^d$.

(b) u is continuous at each point of $\{0\} \times \mathbf{R}^d$ and $u(0,x) = f(x)$, where f is bounded and continuous.

If $c(x) \le 0$, then this equation describes heat flow with cooling. That is, $u(t,x)$ gives the temperature at the point $x \in \mathbf{R}^d$ at time t, when the heat at x at time t dissipates at the rate $-c(x)$.

The first step, as usual, is to find a local martingale.

Theorem 9.4.1 *Let $c_s = \int_0^s c(B_r)\, dr$. If u satisfies (a), then*

$$M_s = u(t-s, B_s)\exp(c_s)$$

is a local martingale on $[0,t]$

Proof The first step in doing this is to generalize our third version of Itô's formula

$$\begin{aligned} &u(t-s,B_s)\exp(c_s) - u(t,B_0) \qquad (9.4.1)\\ &= \int_0^s -u_t(t-r,B_r)\exp(c_r)\,dr + \int_0^s \exp(c_r)\nabla u(t-r,B_r)\cdot dB_r \\ &\quad + \int_0^s u(t-r,B_r)c(B_r)\exp(c_r)\,dr + \frac{1}{2}\int_0^s \Delta u(t-r,B_r)\exp(c_r)\,dr \end{aligned}$$

Proof of (9.4.1). As in the three previous proofs of Itô's formula, we subdivide the interval $[0,s]$ with $s_i^n = si/2^n$. We write the difference as

$$\begin{aligned} &[u(t-s_{i+1}^n, B(s_{i+1}^n)) - u(t-s_i^n, B(s_i^n))]\exp(c(s_{i+1}^n)) \\ &+ u(t-s_i^n, B(s_i^n))[\exp(c(s_{i+1}^n)) - \exp(c(s_i^n))] \end{aligned}$$

The limit of the first line gives the first, second, and fourth terms. Using

$$\exp(c(s_{i+1}^n)) - \exp(c(s_i^n)) = \exp(c(s_i^n))\left\{\exp\left(\int_{s_i^n}^{s_{i+1}^n} c(B_r)\,dr\right) - 1\right\}$$

we see that the limit of the second line gives the third term. □

Rearranging, the right-hand side of (9.4.1) is

$$\begin{aligned} &= \int_0^s \left(-u_t + cu + \frac{1}{2}\Delta u\right)(t-r,B_r)\exp(c_r)\,dr \\ &\quad + \int_0^s \exp(c_r)\,\nabla u(t-r,B_r)\cdot dB_r \end{aligned}$$

which proves the desired result, since $-u_t + cu + \frac{1}{2}\Delta u = 0$, so using Theorem 7.6.4 the second term is a local martingale. □

The next step, again, is a uniqueness result.

Theorem 9.4.2 *If there is a solution that is bounded on* $[0,T]\times\mathbf{R}^d$ *for any* $T<\infty$, *it must be*

$$v(t,x) \equiv E_x\{f(B_t)\exp(c_t^t)\} \qquad (9.4.2)$$

Proof Under our assumptions on c and u, M_s, $0\le s<t$, is a bounded martingale and $M_t \equiv \lim_{s\to t} M_s = f(B_t)\exp(c_t)$. Since M_s is uniformly integrable, it follows that

$$u(t,x) = E_x M_0 = E_x M_t = v(t,x)$$

which proves the desired result. □

As before, it is easy to show that if v is smooth enough, it is a solution.

Theorem 9.4.3 *If* $v\in C^{1,2}$, *then it satisfies (a).*

Proof When we write the solution in (9.4.2), we think about running space-time Brownian motion $(t-s,B_s)$ from time 0 to t, so we have

$$v(t+h,x) = E_x[v(t,B_h)\exp(c_h)]$$

Subtracting $v(t,x)$ from each side

$$v(t+h,x) - v(t,x) = E_x[v(t,B_h)\exp(c_h) - v(t,x)]$$

Adding and subtracting $v(t,B_h)$ and the right-hand side and using Taylor's theorem

$$v(t,B_h)\exp(c_h) - v(t,x) \approx \sum_{i=1}^{d} \frac{\partial v}{\partial x_i}(t,x)(B_h^i - x)$$

$$+\frac{1}{2}\sum_{1\le i,j\le d} \frac{\partial^2 v}{\partial x_i \partial x_j}(t,x)(B_h^i - x)(B_h^j - x) + v(t,B_h)(\exp(c_h)-1) + o(h)$$

Since for $i \neq j$, $W_i(t)$, $W_i^2(t) = t$ and $W_i(t)W_j(t)$ are martingales.

$$E_x[v(t,B_h)\exp(c_h) - v(t,x)] \approx \sum_{i=1}^{d} \frac{\partial^2 v}{\partial x_i^2} h + E_x[v(t,B_h)(\exp(c_h)-1)] + o(h)$$

Dividing each side by h and letting $h \to 0$, we get

$$\frac{\partial v}{\partial t} = \frac{1}{2}\Delta v(t,x) + c(x)v(t,x)$$

which proves the desired result. □

Theorem 9.4.4 *v satisfies (b).*

Proof If $|c| \le M$, then $e^{-Mt} \le \exp(c_t^t) \le e^{Mt}$, so $\exp(c_t^t) \to 1$ as $t \to 0$. Letting $\|f\|_\infty = \sup_x |f(x)|$ this result implies

$$|E_x \exp(c_t^t) f(B_t) - E_x f(B_t)| \le \|f\|_\infty E_x|\exp(c_t^t) - 1| \to 0$$

Theorem 9.2.4 implies that $(t,x) \to E_x f(B_t)$ is continuous at each point of $\{0\} \times \mathbf{R}^d$ and the desired result follows. □

This brings us to the problem of determining when v is smooth enough to be a solution.

Theorem 9.4.5 *If f and c Hölder continuous, then $v \in C^{1,2}$ and, hence, satisfies (a).*

To solve the problem in this case, we use a trick to reduce our result to the previous case. We begin by observing that

$$c_s = \int_0^s c(B_r)\,dr$$

is continuous and locally of bounded variation. So using the ordinary change of variables formula, if $h \in C^1$, then

$$h(c_t) - h(c_0) = \int_0^t h'(c_s)c(B_s)\,ds$$

Taking $h(x) = e^{-x}$, we have

$$\exp(-c_t) - 1 = -\int_0^t \exp(-c_s)c(B_s)\,ds$$

Multiplying by $-\exp(c_t)$ gives

$$\exp(c_t) - 1 = \int_0^t c(B_s) \exp(c_t - c_s)\, ds$$

Plugging in the definitions of c_t^t and c_s^t, we have

$$\exp\left(\int_0^t c(B_r)\, dr\right) = 1 + \int_0^t c(B_s) \exp\left(\int_s^t c(B_r)\, dr\right) ds$$

Multiplying by $f(B_t)$, taking expected values, and using Fubini's theorem, which is justified since everything is bounded, gives

$$v(t,x) = E_x f(B_t) + \int_0^t E_x \left\{ c(B_s) \exp\left(\int_s^t c(B_r)\, dr\right) f(B_t) \right\} ds$$

Conditioning on $\mathcal{F}_s$, noticing $c(B_s) \in \mathcal{F}_s$, and using the Markov property

$$E_x \left(c(B_s) \exp\left(\int_s^t c(B_r)\, dr\right) f(B_t) \middle| \mathcal{F}_s \right) = c(B_s) v(t-s, B_s)$$

Taking the expected value of the last equation and plugging into the previous one, we have

$$\begin{aligned} v(t,x) &= E_x f(B_t) + \int_0^t E_x\{c(B_s) v(t-s, B_s)\} ds \\ &\equiv v_1(t,x) + v_2(t,x) \end{aligned} \tag{9.4.3}$$

The first term on the right, $v_1(t,x)$, is $C^{1,2}$ by Theorem 9.2.3. The second term, $v_2(t,x)$, is of the form considered in the last section with $g(r,x) = c(x)v(r,x)$. If we start with the trivial observation that if c and f are bounded, then v is bounded on $[0,N] \times \mathbf{R}^d$, and apply (2.6b), we see that

$$|v_2(t,x) - v_2(t,y)| \le C_N |x-y| \quad \text{whenever } t \le N$$

To get a similar estimate for v_1 let $\bar{B}_t$ be a Brownian motion starting at 0 and observe that since f is Hölder continuous

$$\begin{aligned} |v_1(t,x) - v_1(t,y)| &= |E\{f(x+\bar{B}_t) - f(y+\bar{B}_t)\}| \\ &\le E|f(x+\bar{B}_t) - f(y+\bar{B}_t)| \le C|x-y|^\alpha \end{aligned}$$

Combining the last two estimates, we see that $v(r,x)$ is Hölder continuous locally in t. Since c and v are bounded and we have supposed that $c,x)$ is Hölder continuous, the triangle inequality implies $g(r,x)$ is Hölder continuous locally in t. The result then follows from smoothness results for the inhomogeneous heat equation.

9.5 Dirichlet Problem

The developments here and in Section 9.7 follow Port and Stone (1978). In this section, we will consider the Dirichlet problem. That is, given an open set G, we will seek a function that satisfies

(a) $u \in C^2$ and $\Delta u = 0$ in G.

(b) At each point of ∂G, u is continuous and $u = f$, a bounded function.

Physically, u is an equilibrium temperature distribution when ∂G is held at a fixed temperature profile f.

Let $\tau = \inf\{t > 0 : B_t \notin G\}$. We assume throughout this section that

$$P_x(\tau < \infty) = 1 \quad \text{for all } x \in G. \tag{9.5.1}$$

If u satisfies (a), then $M_t = u(B_t)$ is a local martingale on $[0, \tau)$. That is, there is an increasing sequence of stopping times T_n so that $u(B(t \wedge T_n)$ is a martingale. To turn a local martingale into an honest martingale, we introduce a function $\gamma : [0, \infty) \to [0, \tau)$ stays constant for one unit of time when its value reaches T_n. That is, $\gamma(t) = t$ for $0 \le t \le T_1$, $\gamma(t) = T_1$ for $T_1 \le t \le T_1 + 1$, and in general

$$\gamma(t) = t - k \text{ for } T_k + k \le t \le T_{k+1} + k$$

$$\gamma(t) = T_{k+1} \text{ for } T_{k+1} + k \le T_{k+1} + k + 1$$

Since $\gamma(t)$ is a stopping time and $\gamma(n) \le T_n$.

Lemma 9.5.1 *$M_t = u(B(\gamma(t)))$ is a martingale.*

Theorem 9.5.2 *If there is a solution of the Dirichlet problem that is bounded, it must be*

$$v(x) \equiv E_x f(B_\tau)$$

Proof Since u is bounded, M_t is a bounded martingale, and hence $\lim_{t\to\infty} M_t$ exists. The boundary condition implies $M_\infty = f(B_\tau)$. Since the martingale is uniformly integrable, $u(x) = E_x M_\infty = E_x f(B_\tau)$. □

Theorem 9.5.3 *If $v \in C^2$, then it satisfies (a).*

Proof We learned this argument, which avoids the use of stochastic calculus, from Liggett (2010). Let $x \in D$ and $B = B(x,r) \subset D$. If τ_B is the exit time from B, then the strong Markov property implies

$$v(x) = E_x v(B(\tau_B)).$$

Using Taylor's theorem

$$E_x v(B(\tau_B)) - v(x) = E_x \sum_{i=1}^d \frac{\partial v}{\partial x_i}(x)[B_i(\tau_B) - x_i]$$
$$+ \frac{1}{2} E_x \sum_{1\le i,j\le d} \frac{\partial^2 v}{\partial x_i \partial x_j}(x)[B_i(\tau_B) - x_i][B_j(\tau_B) - x_j] + o(r^2)$$

Since for $i \ne j$, $B_i(t)$, $B_i^2(t) - t$ and $B_i(t)B_j(t)$ are martingales, we have

$$E_x[B_i(\tau_B) - x_i] = 0 \qquad E_x[B_i(\tau_B) - x_i][B_j(\tau_B) - x_j] = 0 \quad \text{when } i \ne j$$
$$E_x[B_i(\tau_B) - x_i]^2 = E_0[B_i^2(\tau_B)] = E_x \tau_B$$

Using these results in the previous formula

$$0 = E_x v(B(\tau_B)) - v(x) = \frac{1}{2} \Delta v(x) E_x \tau_B + o(r^2)$$

which proves the desired result. □

Theorem 9.5.4 $v \in C^\infty$.

Proof Let $x \in D$ and $B = B(x,r) \subset D$. If τ_B is the exit time from B, then the strong Markov property implies

$$v(x) = E_x v(B(\tau_B)) = \int_{\partial D(x,r)} \bar{v}(y)\, \pi(dy).$$

where π is surface measure on the boundary $\partial D(x,r)$ normalized to be a probability measure.

The last result is the **averaging property** of harmonic functions. Now a simple analytical argument takes over to show $v \in C^\infty$. To prove this, we let ψ be a nonnegative infinitely differentiable function that vanishes on $[\delta^2, \infty)$ but is not $\equiv 0$. It is routine to show that

$$h(x) = \int_{D(x,\delta)} \psi(|y-x|^2) v(y)\, dy \in C^\infty$$

Changing to spherical coordinates and using the averaging property, it is easy to see that $h(x)$ is a constant multiple of $v(x)$ so $v \in C^\infty$. □

As the next example shows, v does not always satisfy (b).

Example 9.5.5 (Punctured Disc) Let $d = 3$ and let $G = D - K$, where $D = \{x : |x| < 1\}$, and $K = \{x : x_1 = x_2 = 0, x_3 \geq 0\}$. Our uniqueness result and the fact that three-dimensional Brownian motion never hits a line implies that the solution must be $v(x) = E_x f(B_{\tau(D)})$, where $\tau(D)$ is the exit time from D. If we let f be 0 on the boundary of D and $1 - x_3$ on K, then the solution $v \equiv 0$ in G and boundary condition will not be satisfied.

Example 9.5.6 (Lebesgue's Thorn) Let $d \geq 3$ and let

$$G = (-1,1)^d - \cup_{1 \leq n \leq \infty} \{[2^{-n}, 2^{-n+1}] \times [-a_n, a_n]^{d-1}\}$$

where the $n = \infty$ term is the single point $\{0\}$. We will now show that if $a_n \downarrow 0$ sufficiently fast, then $P_0(\tau = 0) = 0$. From this, it will follow that if we define f to be 0 on the boundary of $(-1,1)^d$ and $1 - x_1$ on the rest of the boundary, $E_0 f(B_\tau) < 1$ and the boundary condition is not satisfied.

Proof $P_0((B_t^2, B_t^3) = (0,0)$ for some $t > 0) = 0$, so with probability 1, a Brownian motion B_t starting at 0 will not hit

$$I_n = \{x : x_1 \in [2^{-n}, 2^{-n+1}], x_2 = x_3 = \ldots = x_d = 0\}$$

Since B_t is transient in $d \geq 3$, it follows that for a.e. ω the distance between $\{B_s : 0 \leq s < \infty\}$ and I_n is positive. From the last observation, it follows immediately that if we let $T_n = \inf\{t : B_t \in [2^{-n}, 2^{-n+1}] \times [a_n, a_n]^{d-1}\}$ and pick a_n small enough, then $P_0(T_n < \infty) \leq 3^{-n}$. Now $\sum_{n=1}^\infty 3^{-n} = 3^{-1}(3/2) = 1/2$, so if we let $\tau = \inf\{t > 0 : B_t \notin G\}$ and $\sigma = \inf\{t > 0 : B_t \notin (-1,1)^d\}$, then we have

$$P_0(\tau < \sigma) \leq \sum_{n=1}^{\infty} P_0(T_n < \infty) \leq \frac{1}{2}$$

Thus $P_0(\tau > 0) \geq P_0(\tau = \sigma) \geq 1/2$. Since $\tau = 0$ is measurable with respect to $\mathcal{F}_0^+$, we have $P_0(\tau = 0) = 1$. □

A point $y \in \partial G$ is said to be a **regular point** if $P_y(\tau = 0) = 1$.

Theorem 9.5.7 *Let G be any open set. Suppose f is bounded and continuous and y is a regular point of ∂G. If $x_n \in G$ and $x_n \to y$, then $v(x_n) \to f(y)$.*

The first step is to show:

Lemma 9.5.8 *If $t > 0$, then $x \to P_x(\tau \le t)$ is lower semicontinuous. That is, if $x_n \to x$, then*

$$\liminf_{n\to\infty} P_{x_n}(\tau \le t) \ge P_x(\tau \le t)$$

Proof By the Markov property

$$P_x(B_s \in G^c \text{ for some } s \in (\epsilon, t]) = \int p_\epsilon(x,y) P_y(\tau \le t - \epsilon)\, dy$$

Since $y \to P_y(\tau \le t - \epsilon)$ is bounded and measurable and

$$p_\epsilon(x,y) = (2\pi\epsilon)^{-d/2} e^{-|x-y|^2/2\epsilon}$$

it follows from the dominated convergence theorem that

$$x \to P_x(B_s \in G^c \text{ for some } s \in (\epsilon, t])$$

is continuous for each $\epsilon > 0$. Letting $\epsilon \downarrow 0$ shows that $x \to P_x(\tau \le t)$ is an increasing limit of continuous functions and, hence, lower semicontinuous. □

If y is regular for G and $t > 0$, then $P_y(\tau \le t) = 1$, so it follows from Lemma 9.5.8 that if $x_n \to y$, then

$$\liminf_{n\to\infty} P_{x_n}(\tau \le t) \ge 1 \quad \text{for all } t > 0$$

With this established, it is easy to complete the proof. Since f is bounded and continuous, it suffices to show that if $D(y,\delta) = \{x : |x - y| < \delta\}$, then

Lemma 9.5.9 *If y is regular for G and $x_n \to y$, then for all $\delta > 0$*

$$P_{x_n}(\tau < \infty, B_\tau \in D(y,\delta)) \to 1$$

Proof Let $\epsilon > 0$ and pick t so small that

$$P_0\left(\sup_{0\le s\le t} |B_s| > \frac{\delta}{2}\right) < \epsilon$$

Since $P_{x_n}(\tau \le t) \to 1$ as $x_n \to y$, it follows from the choices above that

$$\liminf_{n\to\infty} P_{x_n}(\tau < \infty, B_\tau \in D(y,\delta)) \ge \liminf_{n\to\infty} P_{x_n}\left(\tau \le t,\ \sup_{0\le s\le t} |B_s - x_n| \le \frac{\delta}{2}\right)$$

$$\ge \liminf_{n\to\infty} P_{x_n}(\tau \le t) - P_0\left(\sup_{0\le s\le t} |B_t| > \frac{\delta}{2}\right) > 1 - \epsilon$$

Since ϵ was arbitrary, this proves the lemma and hence Theorem 9.5.7. □

The next result shows that if G^c is not too small near y, then y is regular.

Theorem 9.5.10 (Cone Condition) *If there is a cone V having vertex y and an $r > 0$ such that $V \cap D(y,r) \subset G^c$, then y is a regular point.*

Proof The first thing to do is to define a **cone with vertex y, pointing in direction v, with opening a** as follows:

$$V(y,v,a) = \{x : x = y + \theta(v+z), \text{ where } \theta \in (0,\infty),\ z \perp v, \text{ and } |z| < a\}$$

Now that we have defined a cone, the rest is easy. Since the normal distribution is spherically symmetric,

$$P_y(B_t \in V(y,v,a)) = \epsilon_a > 0$$

where ϵ_a is a constant that depends only on the opening a. Let $r > 0$ be such that $V(y,v,a) \cap D(y,r) \subset G^c$. The continuity of Brownian paths implies

$$\lim_{t \to 0} P_y\left(\sup_{s \le t} |B_s - y| > r\right) = 0$$

Combining the last two results with a trivial inequality, we have

$$\epsilon_a \le \liminf_{t \downarrow 0} P_y(B_t \in G^c) \le \lim_{t \downarrow 0} P_y(\tau \le t) \le P_y(\tau = 0)$$

and it follows from Blumenthal's zero-one law that $P_y(\tau = 0) = 1$. □

9.5.1 Exit Distributions

Half-space. Let $\tau = \inf\{t : B_t \notin H\}$, where $H = \{(x,y) : x \in \mathbf{R}^{d-1}, y > 0\}$. For $\theta \in \mathbf{R}^{d-1}$ let

$$h_\theta(x,y) = \frac{C_d y}{(|x-\theta|^2 + y^2)^{d/2}} \tag{9.5.2}$$

where C_d is chosen so that $\int d\theta\, h_\theta(0,1) = 1$.

Lemma 9.5.11 $\Delta h_\theta = 0$ *in* H

Proof (a) Ignoring C_d and differentiating gives

$$D_{x_i} h_\theta = -\frac{d}{2} \cdot \frac{2(x_i - \theta_i)y}{(|x-\theta|^2 + y^2)^{(d+2)/2}}$$

$$D_{x_i x_i} h_\theta = -d\, \frac{y}{(|x-\theta|^2 + y^2)^{(d+2)/2}} + \frac{d(d+2)(x_i - \theta)^2 y}{(|x-\theta|^2 + y^2)^{(d+4)/2}}$$

$$D_y h_\theta = \frac{1}{(|x-\theta|^2 + y^2)^{d/2}} - d \cdot \frac{y^2}{(|x-\theta|^2 + y^2)^{(d+2)/2}}$$

$$D_{yy} h_\theta = -d \cdot \frac{y + 2y}{(|x-\theta|^2 + y^2)^{(d+2)/2}} + \frac{d(d+2)y^3}{(|x-\theta|^2 + y^2)^{(d+4)/2}}$$

Adding up, we see that

$$\sum_{i=1}^{d-1} D_{x_i x_i} h_\theta + D_{yy} h_\theta = \frac{\{(-d)(d-1) + (-d)\cdot 3\}y}{(|x-\theta|^2 + y^2)^{(d+2)/2}} + \frac{d(d+2)y(|x-\theta|^2 + y^2)}{(|x-\theta|^2 + y^2)^{(d+4)/2}} = 0$$

which completes the proof. □

Theorem 9.5.12 *Let f is bounded and continuous and*

$$u(x,y) = \int d\theta\, h_\theta(x,y) f(\theta, 0)$$

u satisfies the Dirichlet problem in H with boundary value f. From this it follows that $h_\theta(x,y)$ gives the probability density $P_{x,y}(B(\tau_H) = (\theta,0))$.

Proof Interchanging the derivative and integral, we see that $\Delta u = 0$. The next step is to show that it satisfies the boundary condition.

Clearly, $\int d\theta\, h_\theta(x,y)$ is independent of x. To show that it is independent of y, let $x = 0$ and change variables $\theta_i = y\phi_i$ for $1 \le i \le d-1$ to get

$$\int d\theta\, h_\theta(0,y) = \int \frac{C_d y}{(y^2|\phi|^2 + y^2)^{d/2}} \cdot y^{d-1} d\phi = \int d\phi\, h_\phi(0,1) = 1$$

Changing variables $\theta_i = x_i + r_i y$ and using dominated convergence

$$\int_{D(x,\epsilon)^c} d\theta\, h_\theta(x,y) = \int_{D(0,\epsilon/y)^c} dr\, h_r(0,1) \to 0$$

Since $P_{x,y}(\tau < \infty) = 1$ for all $x \in H$, the last conclusion follows from Theorem 9.5.2. □

When $d = 2$

$$h_\theta(0,y) = \frac{y}{\pi(|\theta|^2 + y^2)}$$

is Cauchy density. To see why this should be true, start the Brownian motion at $(0,0)$ and let $T_y = \inf\{t : B_t^2 = y\}$. The strong Markov property implies that $B^1(T_y)$ has independent increments. Brownian scaling implies $B^1(T_y) =_d yB^1(T_1)$. Since $B^1(T_y) =_d -B^1(T_y)$, the distribution must be the symmetric stable law with $\alpha = 1$.

Exit distributions for the ball. Let $D = \{x : |x| < 1\}$, let $\tau_D = \inf\{t : B_t \notin D\|$, and define the Poisson kernel by

$$k_y(x) = \frac{1 - |x|^2}{|x-y|^d} \tag{9.5.3}$$

Theorem 9.5.13 *If f is bounded and continuous, then*

$$E_x f(B_\tau) = u(x) = \int_{\partial D} k_y(x) f(y) \pi(dy) \tag{9.5.4}$$

where π is surface measure on ∂D normalized to be a probability measure.

The proof technique is the same as for the result in the half-space. We show that u satisfies the Dirichlet problem with boundary condition f. We leave it to the reader to show that for fixed y, $\Delta k_y(x) = 0$ and then complete the proof as previously.

Exercises

9.5.1 Suppose h is bounded and harmonic on $\mathbf{R}^d$. Prove that h is constant.

9.5.2 Suppose $f \in C^2$ is superharmonic, takes values in $[0,\infty]$ and is continuous. Show that in dimensions 1 and 2, f must be constant. Give an example of the form $f(x) = g(|x|)$ in $d \geq 3$.

9.5.3 Let B_t be a d-dimensional Brownian motion and let $\tau = \inf\{t : B_t \notin G\}$. Suppose G is a connected open set G, all boundary points are regular, and $P_x(\tau < \infty) < 1$ in G. Show that all bounded solutions of the Dirichlet problem with boundary value 0 on ∂G have the form $u(x) = cP_x(\tau = \infty)$. Hint: Define $\bar{u} = u + \|u\|_\infty P_x(\tau = \infty)$ in G, $\bar{u} = 0$ on G^c, and show that $u(B_y)$ is a submartingale.

9.5.4 Let $\tau = \inf\{t : B_t \notin G\}$ and consider now the Dirichlet problem in a connected open set G in $d \geq 3$, where all boundary points are regular and which has $P_x(\tau < \infty) < 1$ in G. Suppose that the boundary condition $f \equiv 0$. Generalize the previous result to conclude that all bounded solutions have the form $u(x) = cP_x(\tau = \infty)$.

9.5.5 Define a flat cone $\bar{V}(y,v,a)$ to be the intersection of $V(y,v,a)$ with a $d-1$ dimensional hyperplane that contains the line $\{y+\theta v : \theta \in \mathbf{R}\}$. Show that Theorem 9.5.10 remains true if "cone" is replaced by "flat cone."

9.6 Green's Functions and Potential Kernels

Let $D = B(0,r) = \{y : |y| < r\}$ the ball of radius r centered at 0. B_t will return to D i.o. in $d \leq 2$ but not in $d \geq 3$. In this section we will investigate the occupation time $\int_0^\infty 1_D(B_t)\,dt$. The first step is to show that for any x

$$P_x\left(\int_0^\infty 1_D(B_t)\,dt = \infty\right) = 1 \quad \text{in } d \leq 2, \tag{9.6.1}$$

$$E_x \int_0^\infty 1_D(B_t)\,dt < \infty \quad \text{in } d \geq 3. \tag{9.6.2}$$

Proof Let $T_0 = 0$ and $G = B(0,2r)$. For $k \geq 1$, let

$$S_k = \inf\{t > T_{k-1} : B_t \in D\}$$
$$T_k = \inf\{t > S_k : B_t \in G\}$$

Writing τ for T_1 and using the strong Markov property, we get for $k \geq 1$

$$P_x\left(\int_{S_k}^{T_k} 1_D(B_t)\,dt \geq s \,\middle|\, \mathcal{F}_{S_k}\right) = P_{B(S_k)}\left(\int_0^\tau 1_D(B_t)\,dt \geq s\right) = H(s)$$

From the strong Markov property it follows that

$$\int_{S_k}^{T_k} 1_D(B_t)\,dt \quad \text{are i.i.d.}$$

Since these random variables have positive mean, it follows from the strong law of large numbers that

$$\int_0^\infty 1_D(B_t)\,dt \ge \lim_{n\to\infty} \sum_{k=1}^n \int_{S_k}^{T_k} 1_D(B_t)\,dt = \infty \quad \text{a.s.}$$

proving (9.6.1).

If f is a nonnegative function, then Fubini's theorem implies

$$\begin{aligned} E_x \int_0^\infty f(B_t)\,dt &= \int_0^\infty E_x f(B_t)\,dt = \int_0^\infty \int p_t(x,y) f(y)\,dy\,dt \\ &= \int \int_0^\infty p_t(x,y)\,dt\, f(y)\,dy \end{aligned}$$

where $p_t(x,y) = (2\pi t)^{-d/2} e^{-|x-y|^2/2t}$ is the transition density for Brownian motion. As $t \to \infty$, $p_t(x,y) \sim (2\pi t)^{-d/2}$, so if $d \le 2$, then $\int p_t(x,y) dt = \infty$. When $d \ge 3$, changing variables $t = |x-y|^2/2s$ gives

$$\begin{aligned} \int_0^\infty p_t(x,y)\,dt &= \int_0^\infty \frac{1}{(2\pi t)^{d/2}} e^{-|y-x|^2/2t}\,dt \\ &= \int_\infty^0 \left(\frac{s}{\pi |x-y|^2}\right)^{d/2} e^{-s} \left(-\frac{|x-y|^2}{2s^2}\right) ds \\ &= \frac{|x-y|^{2-d}}{2\pi^{d/2}} \int_0^\infty s^{(d/2)-2} e^{-s}\,ds \\ &= \frac{\Gamma(\frac{d}{2}-1)}{2\pi^{d/2}} |x-y|^{2-d} \end{aligned} \tag{9.6.3}$$

where $\Gamma(\alpha) = \int_0^\infty s^{\alpha-1} e^{-s}\,ds$ is the usual gamma function. If we define

$$G(x,y) = \int_0^\infty p_t(x,y)\,dt$$

then in $d \ge 3$, $G(x,y) < \infty$ for $x \ne y$, and

$$\text{(2.4)} \qquad E_x \int_0^\infty f(B_t)\,dt = \int G(x,y) f(y)\,dy$$

To complete the proof now, we observe that taking $f = 1_D$ with $D = B(0,r)$ and changing to polar coordinates

$$\int_D G(0,y)\,dy = \int_0^r s^{d-1} C_d s^{2-d}\,ds = \frac{C_d}{2} r^2 < \infty$$

To extend the last conclusion to $x \neq 0$, observe that applying the strong Markov property at the exit time τ from $B(0,|x|)$ for a Brownian motion starting at 0 and using the rotational symmetry we have

$$E_x \int_0^\infty 1_D(B_s)\,ds = E_0 \int_\tau^\infty 1_D(B_s)\,ds \leq E_0 \int_0^\infty 1_D(B_s)\,ds$$

which completes the proof. □

We call $G(x,y)$ the **potential kernel**, because $G(\cdot,y)$ is the electrostatic potential of a unit charge at y. See chapter 3 of Port and Stone (1978) for more on this. In $d \leq 2$, $\int_0^\infty p_t(x,y)\,dt \equiv \infty$, so we have to take another approach to define a useful G:

$$G(x,y) = \int_0^\infty (p_t(x,y) - a_t)\,dt \tag{9.6.4}$$

where the a_t are constants we will choose to make the integral converge (at least when $x \neq y$). To see why this modified definition might be useful note that if $\int f(y)\,dy = 0$, then

$$\int G(x,y) f(y)\,dy = \lim_{t\to\infty} \int_0^t E_x f(B_s)\,ds$$

When $d = 1$ we let $a_t = p_t(0,0)$. With this choice,

$$G(x,y) = \frac{1}{\sqrt{2\pi}} \int_0^\infty (e^{-(y-x)^2/2t} - 1) t^{-1/2}\,dt$$

and the integral converges, since the integrand is ≤ 0 and $\sim -(y-x)^2/2t^{3/2}$ as $t \to \infty$. Changing variables $t = (y-x)^2/2u$ gives

$$\begin{aligned}
G(x,y) &= \frac{1}{\sqrt{2\pi}} \int_\infty^0 (e^{-u} - 1) \left(\frac{2u}{(y-x)^2}\right)^{1/2} \frac{-(y-x)^2}{2u^2}\,du \\
&= -\frac{|y-x|}{2\sqrt{\pi}} \int_0^\infty \left(\int_0^u e^{-s}\,ds\right) u^{-3/2}\,du \\
&= -\frac{|y-x|}{\sqrt{\pi}} \int_0^\infty ds\, e^{-s} \int_s^\infty \frac{u^{-3/2}}{2}\,du \\
&= -\frac{|y-x|}{\sqrt{\pi}} \int_0^\infty ds\, e^{-s} s^{-1/2} = -|y-x|
\end{aligned} \tag{9.6.5}$$

since

$$\int_0^\infty ds\, e^{-s} s^{-1/2} = \int_0^\infty dr\, e^{-r^2/2} \sqrt{2} = \frac{1}{2}\sqrt{2} \cdot \sqrt{2\pi} = \sqrt{\pi}$$

The computation is almost the same for $d = 2$. The only thing that changes is the choice of a_t. If we try $a_t = p_t(0,0)$ again, then for $x \neq y$ the integrand $\sim -t^{-1}$ as $t \to 0$ and the integral diverges, so we let $a_t = p_t(0,e_1)$, where $e_1 = (1,0)$. With this choice of a_t, we get

$$\begin{aligned}
G(x,y) &= \frac{1}{2\pi}\int_0^\infty (e^{-|x-y|^2/2t} - e^{-1/2t})\,t^{-1}\,dt \\
&= \frac{1}{2\pi}\int_0^\infty \left(\int_{|x-y|^2/2t}^{1/2t} e^{-s}\,ds\right) t^{-1}\,dt \\
&= \frac{1}{2\pi}\int_0^\infty ds\, e^{-s}\int_{|x-y|^2/2s}^{1/2s} t^{-1}\,dt \\
&= \frac{1}{2\pi}\left(\int_0^\infty ds\, e^{-s}\right)(-\log(|x-y|^2)) \\
&= \frac{-1}{\pi}\log(|x-y|) \qquad (9.6.6)
\end{aligned}$$

To sum up, the potential kernels are given by

$$G(x,y) = \begin{cases} (\Gamma(d/2-1)/2\pi^{d/2})\cdot|x-y|^{2-d} & d\ge 3 \\ (-1/\pi)\cdot\log(|x-y|) & d=2 \\ -1\cdot|x-y| & d=1 \end{cases} \qquad (9.6.7)$$

The reader should note that in each case $G(x,y) = C\varphi(|x-y|)$, where φ is the harmonic function we used in Section 9.1. This is, of course, no accident. $x \to G(x,0)$ is obviously spherically symmetric and, as we will see, satisfies $\Delta G(x,0) = 0$ for $x \neq 0$, so the results above imply $G(x,0) = A + B\varphi(|x|)$.

The formulas above correspond to $A = 0$, which is nice and simple. But what about the weird-looking B's? What is special about them? The answer is simple: They are chosen to make $\frac{1}{2}\Delta G(x,0) = -\delta_0$ (a point mass at 0) in the distributional sense. It is easy to see that this happens in $d = 1$. In that case, $\phi(x) = |x|$ then

$$\varphi'(x) = \begin{cases} 1 & x>0 \\ -1 & x<0 \end{cases}$$

so if $B = -1$, then $B\varphi''(x) = -2\delta_0$. More sophisticated readers can check that this is also true in $d \ge 2$.

9.7 Poisson's Equation

In this section, we study the following equation in a bounded open set G:

(a) $u \in C^2$ and $\frac{1}{2}\Delta u = -g$ in G, where g is bounded and continuous.

(b) At each point of ∂G, u is continuous and $u = 0$.

We can add a solution of the Dirichlet problem to replace $u = 0$ in (b) by $u = f$. It follows from Theorem 9.1.1 that

Theorem 9.7.1 *Let $\tau = \inf\{t > 0 : B_t \notin G\}$. If u satisfies (a), then*

$$M_t = u(B_t) + \int_0^t g(B_s)\,ds$$

is a local martingale on $[0,\tau)$.

The next step is to prove a uniqueness result.

Theorem 9.7.2 *If there is a solution of (a) that is bounded, it must be*

$$v(x) \equiv E_x \left(\int_0^\tau g(B_t)dt \right)$$

Note that if $g \equiv 1$, the solution is $v(x) = E_x\tau$.

Proof If u satisfies (a), then M_t defined is a local martingale on $[0,\tau)$. If G is bounded, then $E_x\tau < \infty$ for all $x \in G$. If u and g are bounded, then for $t < \tau$

$$|M_t| \le \|u\|_\infty + \tau\|g\|_\infty$$

Since M_t is dominated by an integrable random variable, it follows that

$$M_\tau \equiv \lim_{t\uparrow\tau} M_t = \int_0^\tau g(B_t)\,dt$$

and $u(x) = E_x(M_\tau)$. □

As in our treatment of the Dirichlet problem:

Theorem 9.7.3 *If $v \in C^2$, then it satisfies (a).*

Proof Let $x \in D$ and $B(x,r) \subset D$. The strong Markov property implies

$$v(x) = E_x \int_0^{\tau_B} g(B_s)\,ds + E_x v(B(\tau_B))$$

Using Taylor's theorem

$$E_x v(B(\tau_B)) - v(x) = E_x \sum_{i=1}^d \frac{\partial v}{\partial x_i}(x)[B_i(\tau_B) - x_i]$$
$$+ \frac{1}{2} E_x \sum_{1\le i,j\le d} \frac{\partial^2 v}{\partial x_i \partial x_j}(x)[B_i(\tau_D) - x_i][B_j(\tau_D) - x_j] + o(r^2)$$

Since for $i \neq j$, $B_i(t)$, $B_i^2(t) - t$ and $B_i(t)B_j(t)$ are martingales, we have

$$E_x v(B(\tau_B)) - v(x) = \frac{1}{2}\Delta v(x) E_x\tau_B + o(r^2)$$

On the other hand, $E_x \int_0^{\tau_B} g(B_s)\,ds = [g(x) + o(1)]E_x\tau_B$, so letting $r \to 0$ we have $(1/2)\Delta v = -g$ □

After the extensive discussion in the last section on the Dirichlet problem, the conditions needed to guarantee that the boundary conditions hold should come as no surprise.

Theorem 9.7.4 *Suppose that G and g are bounded. Let y be a regular point of ∂G. If $x_n \in G$ and $x_n \to y$, then $v(x_n) \to 0$.*

Proof We begin by observing:

(i) It follows from Lemma 9.5.9 that if $\epsilon > 0$, then $P_{x_n}(\tau > \epsilon) \to 0$.

(ii) If G is bounded, then $C = \sup_x E_x\tau < \infty$ and, hence, $\|v\|_\infty \le C\|g\|_\infty < \infty$.

Let $\epsilon > 0$. Beginning with some elementary inequalities then using the Markov property, we have

$$|v(x_n)| \le E_{x_n}\left(\int_0^{\tau\wedge\epsilon} |g(B_s)|\,ds\right) + E_{x_n}\left(\left|\int_\epsilon^\tau g(B_s)\,ds\right|; \tau > \epsilon\right)$$
$$\le \epsilon\|g\|_\infty + E_{x_n}(|v(B_\epsilon)|; \tau > \epsilon) \le \epsilon\|g\|_\infty + \|v\|_\infty P_{x_n}(\tau > \epsilon)$$

Letting $n \to \infty$, and using (i) and (ii) proves the desired result, since ϵ is arbitrary. □

Last, but not least, we come to the question of proving $v \in C^2$. For these developments we will assume that g is defined on $\mathbf{R}^d$ not just in G and has compact support. Recall we are supposing G is bounded and notice that the values of g on G^c are irrelevant, so there will be no loss of generality if we later want to suppose that $\int g(x)\,dx = 0$. We begin with the case $d \ge 3$, because in this case implies

$$\bar{w}(x) = E_x \int_0^\infty |g(B_t)|\,dt < \infty$$

and, moreover, is a bounded function of x. This means we can define

$$w(x) = E_x \int_0^\infty g(B_t)\,dt,$$

use the strong Markov property to conclude

$$w(x) = E_x \int_0^\tau g(B_t)\,dt + E_x w(B_\tau),$$

and change notation to get

$$v(x) = w(x) - E_x w(B_\tau). \tag{9.7.1}$$

Theorem 9.5.4 tells us that the second term is C^∞ in G, so to verify that $v \in C^2$ we need only prove that w is a task in $d \ge 3$ that is made simpler by the fact that gives the following explicit formula

$$w(x) = C_d \int |x - y|^{2-d} g(y)\,dy$$

The details are tedious, so we will content ourselves to state the result.

Theorem 9.7.5 *In $d \ge 2$ if g is Hölder continuous, then w is C^2. In $d = 1$ it is sufficient to suppose g is continuous.*

The last result settles the question of smoothness in $d \ge 3$. To extend the result to $d \le 2$, we use the potential kernels defined in (9.6.4). Using this interpretation of w, it is easy to see that (9.7.1) holds, so again our problem is reduced to proving that w is C^2, which is a problem in calculus. Once all the computations are done, we find that Theorem 9.7.5 holds in $d = 2$ and that in $d = 1$, it is sufficient to assume that g is continuous. The reader can find proof either in Port and Stone (1978), pp. 116-117, or in Gilbarg and Trudinger (1977), pp. 53-55.

9.7.1 Occupation Times

Half space. To explain why we want to define the potential kernels, we return to Brownian motion in the half space, $H = \{y : y_d > 0\}$. Let $\tau = \inf\{t : B_t \notin H\}$, and let

$$\bar{y} = (y_1, \ldots, y_{d-1}, -y_d)$$

be the reflection of y through the plane $\{y \in \mathbf{R}^d : y_d = 0\}$.

Theorem 9.7.6 *If $x \in H$, $f \geq 0$ has compact support, and $\{x : f(x) > 0\} \subset H$, then*

$$E_x\left(\int_0^\tau f(B_t)dt\right) = \int G(x,y)f(y)dy - \int G(x,\bar{y})f(y)dy$$

Proof Using Fubini's theorem, which is justified since $f \geq 0$, then using the reflection principle and the fact that $\{x : f(x) > 0\} \subset H$, we have

$$\begin{aligned}
E_x \int_0^\tau f(B_t)\, dt &= \int_0^\infty E_x(f(B_t); \tau > t)\, dt \\
&= \int_0^\infty \int_H (p_t(x,y) - p_t(x,\bar{y}))f(y)\, dy\, dt \\
&= \int_0^\infty \int_H (p_t(x,y) - a_t)f(y)\, dy\, dt \\
&\quad - \int_0^\infty \int_H (p_t(x,\bar{y}) - a_t)f(y)\, dy\, dt \\
&= \int G(x,y)f(y)\, dy - \int G(x,\bar{y})f(y)\, dy
\end{aligned}$$

The compact support of f and the formulas for G imply that $\int |G(x,y)f(y)|dy$ and $\int |G(x,\bar{y})f(y)|dy$ are finite, so the last two equalities are valid. □

The proof given here simplifies considerably in the case $d \geq 3$; however, part of the point of this proof is that, with the definition we have chosen for G in the recurrent case, the formulas and proofs can be the same for all d.

Let $G_H(x,y) = G(x,y) - G(x,\bar{y})$. We think of $G_H(x,y)$ as the "expected occupation time (density) at y for a Brownian motion starting at x and killed when it leaves H." The rationale for this interpretation is that (for suitable f)

$$E_x\left(\int_0^\tau f(B_t)dt\right) = \int G_H(x,y)f(y)\, dy$$

The answer is very simple in $d = 1$.

$$G_H(x,y) = 2(x \wedge y) \quad \text{for all } x, y > 0 \tag{9.7.2}$$

It is somewhat surprising that $y \to G_H(x,y)$ is constant $= 2x$ for $y \geq x$, that is, all points $y > x$ have the same expected occupation time!

Proof $G(x,y) = -|x-y|$, so $G_H(x,y) = -|x-y| + |x+y|$. Separating things into cases, we see that

$$G_H(x,y) = \begin{cases} -(x-y) + (x+y) = 2y & \text{when } 0 < y < x \\ -(y-x) + (x+y) = 2x & \text{when } x < y \end{cases}$$

which gives the desired formula. □

Ball. Let $G(x, y)$ be the Green's functions defined in (9.6.7).

Theorem 9.7.7 *If g is bounded and measurable, then*

$$E_x \int_0^{\tau} g(B_t)\,dt = \int G_D(x,y)g(y)\,dy$$

where

$$G_D(x,y) = G(x,y) - \int \frac{1-|x|^2}{|x-z|^d} G(z,y)\pi(dz) \tag{9.7.3}$$

This follows from (9.7.1) and (9.5.4). This formula can be simplified considerably. Turning to page 123 in Folland (1976), we find.

Theorem 9.7.8 *In $d \geq 3$, if $0 < |y| < 1$, then*

$$G_D(x,y) = G(x,y) - |y|^{2-d} G(x, y/|y|^2)$$

Proof By (9.7.1) it is enough to show that (i) the second term is harmonic in D and (ii) $= G(x, y)$ on the boundary. $y \to y/|y|^2$ maps points $y \in D$ to D^c, so (i) holds. To prove (ii), we note

$$\begin{aligned} G(x,y) - |y|^{2-d} G(x, y/|y|^2) &= \frac{C_d}{|x-y|^{d-2}} - \frac{C_d}{|y|^{d-2}} \left| x - \frac{y}{|y|^2} \right|^{-(d-2)} \\ &= \frac{C_d}{|x-y|^{d-2}} - \frac{C_d}{|x|y| - y|y|^{-1}|^{d-2}} = 0 \end{aligned}$$

The last equality follows from a fact useful for the next proof:

Lemma 9.7.9 *If $|x| = 1$, then $|x|y| - y|y|^{-1}| = |x - y|$.*

Using $|z|^2 = z \cdot z$ and then $|x|^2 = 1$, we have

$$\begin{aligned} |x|y| - y|y|^{-1}|^2 &= |x|^2|y|^2 - 2x \cdot y + 1 \\ &= |y|^2 - 2x \cdot y + |x|^2 = |x-y|^2 \end{aligned}$$

which completes the proof. □

Again turning to page 123 of Folland (1976) we "guess"

Theorem 9.7.10 *In $d = 2$ if $0 < |y| < 1$, then*

$$G_D(x,y) = \frac{-1}{\pi}\left(\ln|x-y| - \ln|x|y| - y|y|^{-1}|\right)$$

We leave the proof to the reader.

In $d = 1$, where $D = (-1, 1)$, (9.7.3) tells us that

$$\begin{aligned} G_D(x,y) &= G(x,y) - \frac{x+1}{2} G(1,y) - \frac{1-x}{2} G(-1,y) \\ &= -|x-y| + \frac{x+1}{2}(1-y) + \frac{1-x}{2}(y+1) \\ &= -|x-y| + 1 - xy \end{aligned}$$

Considering the two cases $x \geq y$ and $x \leq y$ leads to

$$G_D(x,y) = \begin{cases} (1-x)(1+y) & -1 \leq y \leq x \leq 1 \\ (1-y)(1+x) & -1 \leq x \leq y \leq 1 \end{cases} \tag{9.7.4}$$

Geometrically, if we fix y, then $x \to G_D(x,y)$ is determined by the conditions that it is linear on $[0,y]$ and on $[y,1]$ with $G_D(0,y) = G_D(1,y) = 0$, and $(1/2)G''_D(y,y) = -\delta_y$, a point mass at y. That is, G'_D is discontinuous at y and $G'_D(y-,y) - G'_D(y+,y) = 2$.

Exercise 9.7.1 Let G be a bounded open set and let $\tau = \inf\{t : B_t \notin G\}$. What equation should we solve to find $w(x) = E_x\tau^2$?

9.8 Schrödinger Equation

The developments here follow Chung and Zhao (1995). In this section we will consider what happens when we add cu where c is a bounded continuous function to the left-hand side of the Dirichlet problem. That is, given a bounded continuous function f on ∂G, we will study

(a) $u \in C^2$ and $\frac{1}{2}\Delta u + cu = 0$ in G.

(b) At each point of ∂G, u is continuous and $u = f$.

As always, the first step in solving the PDE is to find a local martingale.

Theorem 9.8.1 *Let* $\tau = \inf\{t > 0 : B_t \notin G\}$. *If* u *satisfies (a), then*

$$M_t = u(B_t)\exp\left(\int_0^t c(B_s)ds\right)$$

is a local martingale on $[0,\tau)$.

Proof Let $c_t = \int_0^t c(B_s)ds$. Applying the version of Itô's from (9.4.1) gives

$$\begin{aligned} u(B_t)\exp(c_t) - u(B_0) &= \int_0^t \exp(c_s)\nabla u(B_s)\cdot dB_s + \int_0^t u(B_s)\exp(c_s)c(B_s)\,ds \\ &\quad + \frac{1}{2}\int_0^t \Delta u(B_s)\exp(c_s)\,ds \end{aligned}$$

for $t < \tau$. This proves the result, since $\frac{1}{2}\Delta u + cu = 0$, and using theorem 7.6.4 the first term on the right-hand side is a local martingale on $[0,\tau)$. □

At this point, the reader might expect that the next step, as it has been five times before, is to assume that everything is bounded and conclude that if there is a solution that is bounded, it must be

$$v(x) \equiv E_x(f(B_\tau)\exp(c_\tau))$$

We will not do this, however, because the following simple example shows that this result is false.

Example 9.8.2 Let $d = 1$, $G = (-a,a)$, $c \equiv \gamma$, and $f \equiv 1$. The equation we are considering is

$$\frac{1}{2}u'' + \gamma u = 0 \qquad u(a) = u(-a) = 1$$

The general solution is $A \cos bx + B \sin bx$, where $b = \sqrt{2\gamma}$. So if we want the boundary condition to be satisfied, we must have

$$1 = A \cos ba + B \sin ba$$
$$1 = A \cos(-ba) + B \sin(-ba) = A \cos ba - B \sin ba$$

Adding the two equations and then subtracting them it follows that

$$2 = 2A \cos ba \qquad 0 = 2B \sin ba$$

From this we see that $B = 0$ always works and we may or may not be able to solve for A.

If $\cos ba = 0$, then there is no solution. If $\cos ba \neq 0$, then $u(x) = \cos bx / \cos ba$ is a solution. We will see later that if $ab < \pi/2$, then

$$v(x) = \cos bx / \cos ba$$

However, this cannot possibly hold for $ab > \pi/2$ since $v(x) \geq 0$, while the right-hand side is < 0 for some values of x.

We will see that the trouble with the last example is that if $ab > \pi/2$, then $c \equiv \gamma$ is too large, or to be precise, if we let

$$w(x) = E_x \exp\left(\int_0^\tau c(B_s)ds\right)$$

then $w \equiv \infty$ in $(-a,a)$. The rest of this section is devoted to showing that if $w \not\equiv \infty$, then "everything is fine." The development will require several stages. The first step is to show:

Lemma 9.8.3 *Let $\theta > 0$. There is a $\mu > 0$ so that if H is an open set with Lebesgue measure $|H| \leq \mu$ and $\tau_H = \inf\{t > 0 : B_t \notin H\}$, then*

$$\sup_x E_x(\exp(\theta \tau_H)) \leq 2$$

Proof Pick $\gamma > 0$ so that $e^{\theta\gamma} \leq 4/3$. Clearly,

$$P_x(\tau_H > \gamma) \leq \int_H \frac{1}{(2\pi\gamma)^{d/2}} e^{-|x-y|^2/2\gamma}\, dy \leq \frac{|H|}{(2\pi\gamma)^{d/2}} \leq \frac{1}{4}$$

if we pick μ small enough. Using the Markov property, we can conclude that

$$\begin{aligned}P_x(\tau_H > k\gamma) &= E_x(P_{B_\gamma}(\tau_H > (k-1)\gamma); \tau_H > \gamma) \\ &\leq \frac{1}{4} \sup_y P_y(\tau_H > (k-1)\gamma)\end{aligned}$$

So it follows by induction that for all integers $k \geq 0$ we have

$$\sup_x P_x(\tau_H > k\gamma) \leq \frac{1}{4^k}$$

Since e^x is increasing, and $e^{\theta\gamma} \le 4/3$ by assumption, we have

$$E_x \exp(\theta\tau_H) \le \sum_{k=1}^{\infty} \exp(\theta\gamma k) P_x((k-1)\gamma < \tau_H \le k\gamma)$$
$$\le \sum_{k=1}^{\infty} \left(\frac{4}{3}\right)^k \frac{1}{4^{k-1}} = \frac{4}{3} \sum_{k=1}^{\infty} \frac{1}{3^{k-1}} = \frac{4}{3} \cdot \frac{1}{1-\frac{1}{3}} = 2$$

Careful readers may have noticed that we left $\tau_H = 0$ out of the expected value. However, by the Blumenthal 0-1 law either $P_x(\tau_H = 0) = 0$, in which case our computation is valid, or $P_x(\tau_H = 0) = 1$, in which case $E_x \exp(\theta\tau_H) = 1$. □

Let $c^* = \sup_x |c(x)|$. By Lemma 9.8.3, we can pick r_0 so small that if $T_r = \inf\{t : |B_t - B_0| > r\}$ and $r \le r_0$, then $E_x \exp(c^* T_r) \le 2$ for all x.

Lemma 9.8.4 *Let $2\delta \le r_0$. If $D(x, 2\delta) \subset G$ and $y \in D(x,\delta)$, then*

$$w(y) \le 2^{d+2} w(x)$$

The reason for our interest in this is that it shows $w(x) < \infty$ implies $w(y) < \infty$ for $y \in D(x,\delta)$.

Proof If $D(y,r) \subset G$, and $r \le r_0$, then the strong Markov property implies

$$\begin{aligned} w(y) &= E_y[\exp(c_{T_r}) w(B(T_r))] \le E_y[\exp(c^* T_r) w(B(T_r))] \\ &= E_y[\exp(c^* T_r)] \int_{\partial D(y,r)} w(z)\pi(dz) \le 2 \int_{\partial D(y,r)} w(z)\pi(dz) \end{aligned} \tag{9.8.1}$$

where π is surface measure on $\partial D(y,r)$ normalized to be a probability measure, since the exit time, T_r, and exit location, $B(T_r)$, are independent.

If $\delta \le r_0$ and $D(y,\delta) \subset G$, multiplying the last inequality by r^{d-1} and integrating from 0 to δ gives

$$\frac{\delta^d}{d} w(y) \le 2 \cdot \frac{1}{\sigma_d} \int_{D(y,\delta)} w(z)\, dz$$

where σ_d is the surface area of $\{x : |x| = 1\}$. Rearranging we have

$$\int_{D(y,\delta)} w(z)\, dz \ge 2^{-1} \frac{\delta^d}{C_o} w(y) \tag{9.8.2}$$

where $C_o = d/\sigma_d$ is a constant that depends only on d.

Repeating the argument in (9.8.1) with $y = x$ and using the fact that $c_{T_r} \ge -c^* T_r$ gives

$$\begin{aligned} w(x) &= E_x[\exp(c_{T_r}) w(B(T_r))] \ge E_x[\exp(-c^* T_r) w(B(T_r))] \\ &= E_x[\exp(-c^* T_r)] \int_{\partial D(x,r)} w(z)\pi(dz) \end{aligned}$$

Since $1/x$ is convex, Jensen's inequality implies

$$E_x[\exp(-c^* T_r)] \ge 1/E_x[\exp(c^* T_r)] \ge 1/2$$

Combining the last two displays, multiplying by r^{d-1}, and integrating from 0 to 2δ, we get the lower bound

$$\frac{(2\delta)^d}{d} w(x) \geq 2^{-1} \cdot \frac{1}{\sigma_d} \int_{D(x,2\delta)} w(z)\, dz$$

Rearranging and using $D(x,2\delta) \supset D(y,\delta)$, $w \geq 0$, we have

$$w(x) \geq 2^{-1} \frac{C_o}{(2\delta)^d} \int_{D(y,\delta)} w(z)\, dz \tag{9.8.3}$$

where again $C_o = d/\sigma_d$. Combining (9.8.3) and (9.8.2), it follows that

$$\begin{aligned} &\geq 2^{-1} \frac{C_o}{(2\delta)^d} \int_{D(y,\delta)} w(z)\, dz \\ &\geq 2^{-1} \frac{C_o}{(2\delta)^d} \cdot 2^{-1} \frac{\delta^d}{C_o} w(y) = 2^{-(d+2)} w(y) \end{aligned}$$

which gives the desired result. □

Lemma 9.8.4 and a simple covering argument lead to:

Theorem 9.8.5 *Let G be a connected open set. If $w \not\equiv \infty$, then*

$$w(x) < \infty \quad \text{for all } x \in G$$

Proof From Lemma 9.8.4, we see that if $w(x) < \infty$, $2\delta \leq r_0$, and $D(x,2\delta) \subset G$, then $w < \infty$ on $D(x,\delta)$. From this result, it follows that $G_0 = \{x : w(x) < \infty\}$ is an open subset of G. To argue now that G_0 is also closed (when considered as a subset of G), we observe that if $2\delta < r_0$, $D(y,3\delta) \subset G$, and we have $x_n \in G_0$ with $x_n \to y \in G$, then for n sufficiently large, $y \in D(x_n,\delta)$ and $D(x_n,2\delta) \subset G$, so $w(y) < \infty$. □

Before we proceed to the uniqueness result, we want to strengthen the last conclusion.

Theorem 9.8.6 *Let G be a connected open set with finite Lebesgue measure, $|G| < \infty$. If $w \not\equiv \infty$, then*

$$\sup_x w(x) < \infty$$

Proof Let $K \subset G$ be compact, so that $|G - K| < \mu$ the constant in Lemma 9.8.3 for $\theta = c^*$. For each $x \in K$ we can pick a δ_x so that $2\delta_x \leq r_0$ and $D(x,2\delta_x) \subset G$. The open sets $D(x,\delta_x)$ cover K, so there is a finite subcover $D(x_i,\delta_{x_i})$, $1 \leq i \leq I$. Clearly,

$$\sup_{1 \leq i \leq I} w(x_i) < \infty$$

Lemma 9.8.4 implies $w(y) \leq 2^{d+2} w(x_i)$ for $y \in D(x_i,\delta_{x_i})$, so

$$M = \sup_{y \in K} w(y) < \infty$$

If $y \in H = G - K$, then $E_y(\exp(c^* \tau_H)) \leq 2$ by (6.3a), so using the strong Markov property

$$\begin{aligned} w(y) &= E_y\left(\exp(c\tau_H); \tau_H = \tau\right) + E_y\left(\exp(c\tau_H) w(B_{\tau_H}); B_{\tau_H} \in K\right) \\ &\leq 2 + M E_y\left(\exp(c\tau_H); B_{\tau_H} \in K\right) \leq 2 + 2M \end{aligned}$$

we have the desired result. □

With Theorem 9.8.6 established, we are now more than ready to prove our uniqueness result. To simplify the statements of the results that follow, we will now list the assumptions that we will make for the rest of the section.

(A1) G is a bounded connected open set.

(A2) f and c are bounded and continuous.

(A3) $w \not\equiv \infty$.

Theorem 9.8.7 *If there is a solution that is bounded, it must be*

$$v(x) \equiv E_x(f(B_\tau)\exp(c_\tau))$$

Proof Theorem 9.8.1 implies that $M_s = u(B_s)\exp(c_s)$ is a local martingale on $[0, \tau)$. Since f, c, and u are bounded, letting $s \uparrow \tau \wedge t$ and using the bounded convergence theorem gives

$$u(x) = E_x(f(B_\tau)\exp(c_\tau); \tau \le t) + E_x(u(B_t)\exp(c_t); \tau > t)$$

Since f is bounded and $w(x) = E_x \exp(c_\tau) < \infty$, the dominated convergence theorem implies that as $t \to \infty$, the first term converges to $E_x(f(B_\tau)\exp(c_\tau))$. To show that the second term $\to 0$, we begin with the observation that since $\{\tau > t\} \in \mathcal{F}_t$, the definition of conditional expectation and the Markov property imply

$$\begin{aligned} E_x(u(B_t)\exp(c_\tau); \tau > t) &= E_x(E_x(u(B_t)\exp(c_\tau)|\mathcal{F}_t); \tau > t) \\ &= E_x(u(B_t)\exp(c_t)w(B_t); \tau > t) \end{aligned}$$

Now we claim that for all $y \in G$

$$w(y) \ge \exp(-c^*)P_y(\tau \le 1) \ge \epsilon > 0$$

The first inequality is trivial. The last two follow easily from (A1).

Replacing $w(B_t)$ by ϵ,

$$\begin{aligned} E_x(|u(B_t)|\exp(c_t); \tau > t) &\le \epsilon^{-1}E_x(|u(B_t)|\exp(c_\tau); \tau > t) \\ &\le \epsilon^{-1}\|u\|_\infty E_x(\exp(c_\tau); \tau > t) \to 0 \end{aligned}$$

as $t \to \infty$, by the dominated convergence theorem, since $w(x) = E_x \exp(c_\tau) < \infty$ and $P_x(\tau < \infty) = 1$. Going back to the first equation in the proof, we have shown $u(x) = v(x)$ and the proof is complete. □

This completes our consideration of uniqueness. The next stage in our program, fortunately, is as easy as it always has been. Recall that here and in what follows we are assuming (A1)–(A3).

Theorem 9.8.8 *If $v \in C^2$, then it satisfies (a) in G.*

Proof Let $x \in D$ and $B(x,r) \subset D$. Let σ be the exit time from $B(x,r)$. The strong Markov property implies

$$v(x) = E_x v(B_\sigma \exp(c_\sigma)$$

Using Taylor's theorem

$$E_x v(B(\sigma)\exp(c_\sigma)) - v(x) = E_x \sum_{i=1}^{d} \frac{\partial v}{\partial x_i}(x)[B_i(\tau_D) - x_i]$$

$$+ \frac{1}{2} E_x \sum_{1 \le i,j \le d} \frac{\partial^2 v}{\partial x_i \partial x_j}(x)[B_i(\tau_D) - x_i][B_j(\tau_D) - x_j] + o(r^2)$$

Since for $i \neq j$, $B_i(t)$, $B_i^2(t) - t$ and $B_i(t)B_j(t)$ are martingales, we have

$$E_x v(B(\tau_D)) - v(x) = \frac{1}{2}\Delta v(x) E_x \tau_B + o(r^2)$$

On the other hand, $E_x \int_0^{\tau_B} g(B_s)\, ds = [g(x) + o(1)]E_x \tau_B$, so letting $r \to 0$ we have $(1/2)\Delta v = -g$. □

Having proved Theorem 9.8.8, the next step is to consider the boundary condition. As in the last two sections, we need the boundary to be regular.

Theorem 9.8.9 *v satisfies (b) at each regular point of ∂G.*

Proof Let y be a regular point of ∂G. We showed in (4.5a) and (4.5b) that if $x_n \to y$, then $P_{x_n}(\tau \le \delta) \to 1$ and $P_{x_n}(B_\tau \in D(y,\delta)) \to 1$ for all $\delta > 0$. Since c is bounded and f is bounded and continuous, the bounded convergence theorem implies that

$$E_{x_n}(\exp(c_\tau) f(B_\tau); \tau \le 1) \to f(y)$$

To control the contribution from the rest of the space, we observe that if $|c| \le M$, then using the Markov property and the boundedness of w established in Theorem 9.8.6, we have

$$E_{x_n}(\exp(c_\tau) f(B_\tau); \tau > 1) \le e^M \|f\|_\infty E_{x_n}(w(B_1); \tau > 1)$$

$$\le e^M \|f\|_\infty \|w\|_\infty P_{x_n}(\tau > 1) \to 0$$

which completes the proof. □

This brings us finally to the problem of determining when v is smooth enough to be a solution.

Theorem 9.8.10 *If in addition to (A1)–(A3), c is Hölder continuous, then $v \in C^2$ and, hence, satisfies (a).*

Return to Example 9.8.2. Recall $d = 1$, $G = (-a,a)$, $c(x) \equiv \gamma > 0$, and $f \equiv 1$.

Theorem 9.8.11 *Let $\tau = \inf\{t : B_t \notin (-a,a)\}$. If $0 < \gamma < \pi^2/8a^2$, then*

$$E_x e^{\gamma\tau} = \frac{\cos(x\sqrt{2\gamma})}{\cos(a\sqrt{2\gamma})}$$

If $\gamma \ge \pi^2/8a^2$, then $E_x e^{\gamma\tau} \equiv \infty$.

Proof By results in Example 9.8.2

$$u(x) = \cos(x\sqrt{2\gamma})/\cos(a\sqrt{2\gamma})$$

is a nonnegative solution of

$$\frac{1}{2}u'' + \gamma u = 0 \qquad u(-a) = u(a) = 1$$

To check that $w \not\equiv \infty$, we observe that Theorem 9.8.1 implies that $M_t = u(B_t)e^{\gamma t}$ is a local martingale on $[0, \tau)$. If we let T_n be a sequence of stopping times that reduces M_t, then

$$u(x) = E_x\left(u(B_{T_n \wedge n})e^{\gamma(T_n \wedge n)}\right)$$

Letting $n \to \infty$ and noting $T_n \wedge n \uparrow \tau$, it follows from Fatou's lemma that

$$u(x) \geq E_x e^{\gamma\tau}$$

The last equation implies $w \not\equiv \infty$, so Theorem 9.8.8 implies the result for $\gamma < \pi^2/8a^2$.

To see that $w(x) \equiv \infty$ when $\gamma \geq \pi^2/8a^2$, suppose not. In this case, (6.3) implies $v(x) = E_x(f(B_\tau)\exp(c_\tau))$ is the unique solution of our equation but computations in Example 9.8.2 imply that in this case there is no nonnegative solution. □

Appendix A

Measure Theory Details

This appendix proves the results from measure theory that were stated but not proved in the text.

A.1 Carathéodory's Extension Theorem

This section is devoted to the proof of:

Theorem A.1.1 *Let $\mathcal{S}$ be a semialgebra and let μ defined on $\mathcal{S}$ have $\mu(\emptyset) = 0$. Suppose (i) if $S \in \mathcal{S}$ is a finite disjoint union of sets $S_i \in \mathcal{S}$, then $\mu(S) = \sum_i \mu(S_i)$, and (ii) if $S_i, S \in \mathcal{S}$ with $S = +_{i\geq 1} S_i$ then $\mu(S) \leq \sum_i \mu(S_i)$. Then μ has a unique extension $\bar{\mu}$ that is a measure on $\bar{\mathcal{S}}$ the algebra generated by $\mathcal{S}$. If the extension is σ-finite, then there is a unique extension ν that is a measure on $\sigma(\mathcal{S})$.*

Proof Lemma 1.1.7 shows that $\bar{\mathcal{S}}$ is the collection of finite disjoint unions of sets in $\mathcal{S}$. We define $\bar{\mu}$ on $\bar{\mathcal{S}}$ by $\bar{\mu}(A) = \sum_i \mu(S_i)$ whenever $A = +_i S_i$. To check that $\bar{\mu}$ is well defined, suppose that $A = +_j T_j$ and observe $S_i = +_j (S_i \cap T_j)$ and $T_j = +_i (S_i \cap T_j)$, so (i) implies

$$\sum_i \mu(S_i) = \sum_{i,j} \mu(S_i \cap T_j) = \sum_j \mu(T_j)$$

In Section 1.1 we proved:

Lemma A.1.2 *Suppose only that (i) holds.*
(a) If $A, B_i \in \bar{\mathcal{S}}$ with $A = +_{i=1}^n B_i$ then $\bar{\mu}(A) = \sum_i \bar{\mu}(B_i)$.
(b) If $A, B_i \in \bar{\mathcal{S}}$ with $A \subset \cup_{i=1}^n B_i$ then $\bar{\mu}(A) \leq \sum_i \bar{\mu}(B_i)$.

To extend the additivity property to $A \in \bar{\mathcal{S}}$ that are countable disjoint unions $A = +_{i\geq 1} B_i$, where $B_i \in \bar{\mathcal{S}}$, we observe that each $B_i = +_j S_{i,j}$ with $S_{i,j} \in \mathcal{S}$ and $\sum_{i\geq 1} \bar{\mu}(B_i) = \sum_{i\geq 1, j} \mu(S_{i,j})$, so replacing the B_i's by $S_{i,j}$'s we can without loss of generality suppose that the $B_i \in \mathcal{S}$. Now $A \in \bar{\mathcal{S}}$ implies $A = +_j T_j$ (a finite disjoint union) and $T_j = +_{i\geq 1} T_j \cap B_i$, so (ii) implies

$$\mu(T_j) \leq \sum_{i\geq 1} \mu(T_j \cap B_i)$$

Summing over j and observing that nonnegative numbers can be summed in any order,

$$\bar{\mu}(A) = \sum_j \mu(T_j) \leq \sum_{i\geq 1} \sum_j \mu(T_j \cap B_i) = \sum_{i\geq 1} \mu(B_i)$$

the last equality following from (i). To prove the opposite inequality, let $A_n = B_1 + \cdots + B_n$, and $C_n = A \cap A_n^c$. $C_n \in \bar{\mathcal{S}}$, since $\bar{\mathcal{S}}$ is an algebra, so finite additivity of $\bar{\mu}$ implies

$$\bar{\mu}(A) = \bar{\mu}(B_1) + \cdots + \bar{\mu}(B_n) + \bar{\mu}(C_n) \geq \bar{\mu}(B_1) + \cdots + \bar{\mu}(B_n)$$

and letting $n \to \infty$, $\bar{\mu}(A) \geq \sum_{i \geq 1} \bar{\mu}(B_i)$.

Having defined a measure on the algebra $\bar{\mathcal{S}}$, we now complete the proof by establishing:

Theorem A.1.3 (Carathéodory's Extension Theorem) *Let μ be a σ-finite measure on an algebra $\mathcal{A}$. Then μ has a unique extension to $\sigma(\mathcal{A})$ = the smallest σ-algebra containing $\mathcal{A}$.*

Uniqueness We will prove that the extension is unique before tackling the more difficult problem of proving its existence. The key to our uniqueness proof is Dynkin's $\pi - \lambda$ theorem, a result that we will use many times in the book. As usual, we need a few definitions before we can state the result. $\mathcal{P}$ is said to be a **π-system** if it is closed under intersection, i.e., if $A, B \in \mathcal{P}$, then $A \cap B \in \mathcal{P}$. For example, the collection of rectangles $(a_1, b_1] \times \cdots \times (a_d, b_d]$ is a π-system. $\mathcal{L}$ is said to be a **λ-system** if it satisfies: (i) $\Omega \in \mathcal{L}$. (ii) If $A, B \in \mathcal{L}$ and $A \subset B$, then $B - A \in \mathcal{L}$. (iii) If $A_n \in \mathcal{L}$ and $A_n \uparrow A$, then $A \in \mathcal{L}$. The reader will see in a moment that the next result is just what we need to prove uniqueness of the extension.

Theorem A.1.4 ($\pi - \lambda$ Theorem) *If $\mathcal{P}$ is a π-system and $\mathcal{L}$ is a λ-system that contains $\mathcal{P}$, then $\sigma(\mathcal{P}) \subset \mathcal{L}$.*

Proof We will show that

(a) if $\ell(\mathcal{P})$ is the smallest λ-system containing $\mathcal{P}$, then $\ell(\mathcal{P})$ is a σ-field.

The desired result follows from (a). To see this, note that since $\sigma(\mathcal{P})$ is the smallest σ-field and $\ell(\mathcal{P})$ is the smallest λ-system containing $\mathcal{P}$, we have

$$\sigma(\mathcal{P}) \subset \ell(\mathcal{P}) \subset \mathcal{L}$$

To prove (a), we begin by noting that a λ-system that is closed under intersection is a σ-field, since

$$\text{if } A \in \mathcal{L} \text{ then } A^c = \Omega - A \in \mathcal{L}$$
$$A \cup B = (A^c \cap B^c)^c$$
$$\cup_{i=1}^n A_i \uparrow \cup_{i=1}^\infty A_i \text{ as } n \uparrow \infty$$

Thus, it is enough to show

(b) $\ell(\mathcal{P})$ is closed under intersection.

To prove (b), we let $\mathcal{G}_A = \{B : A \cap B \in \ell(\mathcal{P})\}$ and prove

(c) if $A \in \ell(\mathcal{P})$, then $\mathcal{G}_A$ is a λ-system.

To check this, we note: (i) $\Omega \in \mathcal{G}_A$, since $A \in \ell(\mathcal{P})$.

(ii) if $B, C \in \mathcal{G}_A$ and $B \supset C$, then $A \cap (B - C) = (A \cap B) - (A \cap C) \in \ell(\mathcal{P})$, since $A \cap B, A \cap C \in \ell(\mathcal{P})$ and $\ell(\mathcal{P})$ is a λ-system.

(iii) if $B_n \in \mathcal{G}_A$ and $B_n \uparrow B$, then $A \cap B_n \uparrow A \cap B \in \ell(\mathcal{P})$, since $A \cap B_n \in \ell(\mathcal{P})$ and $\ell(\mathcal{P})$ is a λ-system.

To get from (c) to (b), we note that since $\mathcal{P}$ is a π-system,

$$\text{if } A \in \mathcal{P}, \text{ then } \mathcal{G}_A \supset \mathcal{P} \text{ and so (c) implies } \mathcal{G}_A \supset \ell(\mathcal{P})$$

i.e., if $A \in \mathcal{P}$ and $B \in \ell(\mathcal{P})$, then $A \cap B \in \ell(\mathcal{P})$. Interchanging A and B in the last sentence: if $A \in \ell(\mathcal{P})$ and $B \in \mathcal{P}$, then $A \cap B \in \ell(\mathcal{P})$ but this implies

$$\text{if } A \in \ell(\mathcal{P}), \text{ then } \mathcal{G}_A \supset \mathcal{P}, \text{ and so (c) implies } \mathcal{G}_A \supset \ell(\mathcal{P}).$$

This conclusion implies that if $A, B \in \ell(\mathcal{P})$, then $A \cap B \in \ell(\mathcal{P})$, which proves (b) and completes the proof. □

To prove that the extension in Theorem A.1.3 is unique, we will show:

Theorem A.1.5 *Let $\mathcal{P}$ be a π-system. If ν_1 and ν_2 are measures (on σ-fields $\mathcal{F}_1$ and $\mathcal{F}_2$) that agree on $\mathcal{P}$ and there is a sequence $A_n \in \mathcal{P}$ with $A_n \uparrow \Omega$ and $\nu_i(A_n) < \infty$, then ν_1 and ν_2 agree on $\sigma(\mathcal{P})$.*

Proof Let $A \in \mathcal{P}$ have $\nu_1(A) = \nu_2(A) < \infty$. Let

$$\mathcal{L} = \{B \in \sigma(\mathcal{P}) : \nu_1(A \cap B) = \nu_2(A \cap B)\}$$

We will now show that $\mathcal{L}$ is a λ-system. Since $A \in \mathcal{P}$, $\nu_1(A) = \nu_2(A)$ and $\Omega \in \mathcal{L}$. If $B, C \in \mathcal{L}$ with $C \subset B$, then

$$\begin{aligned}\nu_1(A \cap (B - C)) &= \nu_1(A \cap B) - \nu_1(A \cap C)\\ &= \nu_2(A \cap B) - \nu_2(A \cap C) = \nu_2(A \cap (B - C))\end{aligned}$$

Here we use the fact that $\nu_i(A) < \infty$ to justify the subtraction. Finally, if $B_n \in \mathcal{L}$ and $B_n \uparrow B$, then part (iii) of Theorem 1.1.1 implies

$$\nu_1(A \cap B) = \lim_{n\to\infty} \nu_1(A \cap B_n) = \lim_{n\to\infty} \nu_2(A \cap B_n) = \nu_2(A \cap B)$$

Since $\mathcal{P}$ is closed under intersection by assumption, the $\pi - \lambda$ theorem implies $\mathcal{L} \supset \sigma(\mathcal{P})$, i.e., if $A \in \mathcal{P}$ with $\nu_1(A) = \nu_2(A) < \infty$ and $B \in \sigma(\mathcal{P})$, then $\nu_1(A \cap B) = \nu_2(A \cap B)$. Letting $A_n \in \mathcal{P}$ with $A_n \uparrow \Omega$, $\nu_1(A_n) = \nu_2(A_n) < \infty$, and using the last result and part (iii) of Theorem 1.1.1, we have the desired conclusion. □

Exercise A.1.1 Give an example of two probability measures $\mu \neq \nu$ on $\mathcal{F} =$ all subsets of $\{1, 2, 3, 4\}$ that agree on a collection of sets $\mathcal{C}$ with $\sigma(\mathcal{C}) = \mathcal{F}$, i.e., the smallest σ-algebra containing $\mathcal{C}$ is $\mathcal{F}$.

Existence Our next step is to show that a measure (not necessarily σ-finite) defined on an algebra $\mathcal{A}$ has an extension to the σ-algebra generated by $\mathcal{A}$. If $E \subset \Omega$, we let $\mu^*(E) = \inf \sum_i \mu(A_i)$, where the infimum is taken over all sequences from $\mathcal{A}$ so that $E \subset \cup_i A_i$. Intuitively, if ν is a measure that agrees with μ on $\mathcal{A}$, then it follows from part (ii) of Theorem 1.1.1 that

$$\nu(E) \leq \nu(\cup_i A_i) \leq \sum_i \nu(A_i) = \sum_i \mu(A_i)$$

so $\mu^*(E)$ is an upper bound on the measure of E. Intuitively, the measurable sets are the ones for which the upper bound is tight. Formally, we say that E is **measurable** if

$$\mu^*(F) = \mu^*(F \cap E) + \mu^*(F \cap E^c) \quad \text{for all sets } F \subset \Omega \tag{A.1.1}$$

The last definition is not very intuitive, but we will see in the following proofs that it works very well.

It is immediate from the definition that μ^* has the following properties:

(i) **monotonicity**. If $E \subset F$, then $\mu^*(E) \le \mu^*(F)$.

(ii) **subadditivity**. If $F \subset \cup_i F_i$, a countable union, then $\mu^*(F) \le \sum_i \mu^*(F_i)$.

Any set function with $\mu^*(\emptyset) = 0$ that satisfies (i) and (ii) is called an **outer measure**. Using (ii) with $F_1 = F \cap E$ and $F_2 = F \cap E^c$ (and $F_i = \emptyset$ otherwise), we see that to prove a set is measurable, it is enough to show

$$\mu^*(F) \ge \mu^*(F \cap E) + \mu^*(F \cap E^c) \tag{A.1.2}$$

We begin by showing that our new definition extends the old one.

Lemma A.1.6 *If $A \in \mathcal{A}$, then $\mu^*(A) = \mu(A)$ and A is measurable.*

Proof Part (ii) of Theorem 1.1.1 implies that if $A \subset \cup_i A_i$, then

$$\mu(A) \le \sum_i \mu(A_i)$$

so $\mu(A) \le \mu^*(A)$. Of course, we can always take $A_1 = A$ and the other $A_i = \emptyset$, so $\mu^*(A) \le \mu(A)$.

To prove that any $A \in \mathcal{A}$ is measurable, we begin by noting that the inequality is (A.1.2) trivial when $\mu^*(F) = \infty$, so we can without loss of generality assume $\mu^*(F) < \infty$. To prove that (A.1.2) holds when $E = A$, we observe that since $\mu^*(F) < \infty$ there is a sequence $B_i \in \mathcal{A}$, so that $\cup_i B_i \supset F$ and

$$\sum_i \mu(B_i) \le \mu^*(F) + \epsilon$$

Since μ is additive on $\mathcal{A}$, and $\mu = \mu^*$ on $\mathcal{A}$, we have

$$\mu(B_i) = \mu^*(B_i \cap A) + \mu^*(B_i \cap A^c)$$

Summing over i and using the subadditivity of μ^* gives

$$\mu^*(F) + \epsilon \ge \sum_i \mu^*(B_i \cap A) + \sum_i \mu^*(B_i \cap A^c) \ge \mu^*(F \cap A) + \mu^*(F^c \cap A)$$

which proves the desired result, since ϵ is arbitrary. □

Lemma A.1.7 *The class $\mathcal{A}^*$ of measurable sets is a σ-field, and the restriction of μ^* to $\mathcal{A}^*$ is a measure.*

Remark This result is true for any outer measure.

Proof It is clear from the definition that:

(a) If E is measurable, then E^c is.

Our first nontrivial task is to prove:

(b) If E_1 and E_2 are measurable, then $E_1 \cup E_2$ and $E_1 \cap E_2$ are.

Proof of (b) To prove the first conclusion, let G be any subset of Ω. Using subadditivity, the measurability of E_2 (let $F = G \cap E_1^c$ in (A.1.1), and the measurability of E_1, we get

$$\begin{aligned}\mu^*(G \cap (E_1 \cup E_2)) &+ \mu^*(G \cap (E_1^c \cap E_2^c)) \\ &\le \mu^*(G \cap E_1) + \mu^*(G \cap E_1^c \cap E_2) + \mu^*(G \cap E_1^c \cap E_2^c) \\ &= \mu^*(G \cap E_1) + \mu^*(G \cap E_1^c) = \mu^*(G)\end{aligned}$$

To prove that $E_1 \cap E_2$ is measurable, we observe $E_1 \cap E_2 = (E_1^c \cup E_2^c)^c$ and use (a). □

(c) Let $G \subset \Omega$ and $E_1, \ldots, E_n$ be disjoint measurable sets. Then

$$\mu^*\left(G \cap \cup_{i=1}^n E_i\right) = \sum_{i=1}^n \mu^*(G \cap E_i)$$

Proof of (c). Let $F_m = \cup_{i \le m} E_i$. E_n is measurable, $F_n \supset E_n$, and $F_{n-1} \cap E_n = \emptyset$, so

$$\begin{aligned}\mu^*(G \cap F_n) &= \mu^*(G \cap F_n \cap E_n) + \mu^*(G \cap F_n \cap E_n^c) \\ &= \mu^*(G \cap E_n) + \mu^*(G \cap F_{n-1})\end{aligned}$$

The desired result follows from this by induction. □

(d) If the sets E_i are measurable, then $E = \cup_{i=1}^\infty E_i$ is measurable.

Proof of (d). Let $E_i' = E_i \cap \left(\cap_{j<i} E_j^c\right)$. (a) and (b) imply E_i' is measurable, so we can suppose without loss of generality that the E_i are pairwise disjoint. Let $F_n = E_1 \cup \ldots \cup E_n$. F_n is measurable by (b), so using monotonicity and (c) we have

$$\begin{aligned}\mu^*(G) &= \mu^*(G \cap F_n) + \mu^*(G \cap F_n^c) \ge \mu^*(G \cap F_n) + \mu^*(G \cap E^c) \\ &= \sum_{i=1}^n \mu^*(G \cap E_i) + \mu^*(G \cap E^c)\end{aligned}$$

Letting $n \to \infty$ and using subadditivity

$$\mu^*(G) \ge \sum_{i=1}^\infty \mu^*(G \cap E_i) + \mu^*(G \cap E^c) \ge \mu^*(G \cap E) + \mu^*(G \cap E^c)$$

which is (A.1.2). □

The last step in the proof of Theorem A.1.7 is

(e) If $E = \cup_i E_i$, where $E_1, E_2, \ldots$ are disjoint and measurable, then

$$\mu^*(E) = \sum_{i=1}^{\infty} \mu^*(E_i)$$

Proof of (e). Let $F_n = E_1 \cup \ldots \cup E_n$. By monotonicity and (c)

$$\mu^*(E) \geq \mu^*(F_n) = \sum_{i=1}^{n} \mu^*(E_i)$$

Letting $n \to \infty$ now and using subadditivity gives the desired conclusion. □

A.2 Which Sets Are Measurable?

The proof of Theorem A.1.3 given in the last section defines an extension to $\mathcal{A}^* \supset \sigma(\mathcal{A})$. Our next goal is to describe the relationship between these two σ-algebras. Let $\mathcal{A}_\sigma$ denote the collection of countable unions of sets in $\mathcal{A}$, and let $\mathcal{A}_{\sigma\delta}$ denote the collection of countable intersections of sets in $\mathcal{A}_\sigma$. Our first goal is to show that every measurable set is almost a set in $\mathcal{A}_{\sigma\delta}$.

Define the symetric difference by $A \Delta B = (A - B) \cup (B - A)$.

Lemma A.2.1 *Let E be any set with $\mu^*(E) < \infty$.*
(i) For any $\epsilon > 0$, there is an $A \in \mathcal{A}_\sigma$ with $A \supset E$ and $\mu^*(A) \leq \mu^*(E) + \epsilon$.
(ii) For any $\epsilon > 0$, there is a $B \in \mathcal{A}$ with $\mu(B \Delta E) \leq 2\epsilon$, where
(ii) There is a $C \in \mathcal{A}_{\sigma\delta}$ with $C \supset E$ and $\mu^*(C) = \mu^*(E)$.

Proof By the definition of μ^*, there is a sequence A_i so that $A \equiv \cup_i A_i \supset E$ and $\sum_i \mu(A_i) \leq \mu^*(E) + \epsilon$. The definition of μ^* implies $\mu^*(A) \leq \sum_i \mu(A_i)$, establishing (i).

For (ii) we note that there is a finite union $B = \cup i = 1^n A_i$ so that $\mu(A - B) \leq \epsilon$, and hence $\mu(E - B) \leq \epsilon$. Since $\mu(B - E) \leq \mu(A - E) \leq \epsilon$, the desired result follows.

For (iii), let $A_n \in \mathcal{A}_\sigma$ with $A_n \supset E$ and $\mu^*(A_n) \leq \mu^*(E) + 1/n$, and let $C = \cap_n A_n$. Clearly, $C \in \mathcal{A}_{\sigma\delta}$, $B \supset E$, and hence by monotonicity, $\mu^*(C) \geq \mu^*(E)$. To prove the other inequality, notice that $B \subset A_n$ and hence $\mu^*(C) \leq \mu^*(A_n) \leq \mu^*(E) + 1/n$ for any n. □

Theorem A.2.2 *Suppose μ is σ-finite on $\mathcal{A}$. $B \in \mathcal{A}^*$ if and only if there is an $A \in \mathcal{A}_{\sigma\delta}$ and a set N with $\mu^*(N) = 0$, so that $B = A - N (= A \cap N^c)$.*

Proof It follows from Lemma A.1.6 and A.1.7 that if $A \in \mathcal{A}_{\sigma\delta}$, then $A \in \mathcal{A}^*$. A.1.2 in Section A.1 and monotonicity imply sets with $\mu^*(N) = 0$ are measurable, so using Lemma A.1.7 again it follows that $A \cap N^c \in \mathcal{A}^*$. To prove the other direction, let Ω_i be a disjoint collection of sets with $\mu(\Omega_i) < \infty$ and $\Omega = \cup_i \Omega_i$. Let $B_i = B \cap \Omega_i$ and use Lemma A.2.1 to find $A_i^n \in \mathcal{A}_\sigma$, so that $A_i^n \supset B_i$ and $\mu(A_i^n) \leq \mu^*(E_i) + 1/n2^i$. Let $A_n = \cup_i A_i^n$. $B \subset A_n$ and

$$A_n - B \subset \sum_{i=1}^{\infty} (A_i^n - B_i)$$

so, by subadditivity,

$$\mu^*(A_n - B) \le \sum_{i=1}^{\infty} \mu^*(A_i^n - B_i) \le 1/n$$

Since $A_n \in \mathcal{A}_\sigma$, the set $A = \cap_n A_n \in \mathcal{A}_{\sigma\delta}$. Clearly, $A \supset B$. Since $N \equiv A - B \subset A_n - B$ for all n, monotonicity implies $\mu^*(N) = 0$, and the proof is complete. □

A measure space $(\Omega, \mathcal{F}, \mu)$ is said to be **complete** if $\mathcal{F}$ contains all subsets of sets of measure 0. In the proof of Theorem A.2.2, we showed that $(\Omega, \mathcal{A}^*, \mu^*)$ is complete. Our next result shows that $(\Omega, \mathcal{A}^*, \mu^*)$ is the completion of $(\Omega, \sigma(\mathcal{A}), \mu)$.

Theorem A.2.3 *If $(\Omega, \mathcal{F}, \mu)$ is a measure space, then there is a complete measure space $(\Omega, \bar{\mathcal{F}}, \bar{\mu})$ called the* **completion** *of $(\Omega, \mathcal{F}, \mu)$, so that: (i) $E \in \bar{\mathcal{F}}$ if and only if $E = A \cup B$, where $A \in \mathcal{F}$ and $B \subset N \in \mathcal{F}$ with $\mu(N) = 0$, (ii) $\bar{\mu}$ agrees with μ on $\mathcal{F}$.*

Proof The first step is to check that $\bar{\mathcal{F}}$ is a σ-algebra. If $E_i = A_i \cup B_i$ where $A_i \in \mathcal{F}$ and $B_i \subset N_i$ where $\mu(N_i) = 0$, then $\cup_i A_i \in \mathcal{F}$ and subadditivity implies $\mu(\cup_i N_i) \le \sum_i \mu(N_i) = 0$, so $\cup_i E_i \in \bar{\mathcal{F}}$. As for complements, if $E = A \cup B$ and $B \subset N$, then $B^c \supset N^c$, so

$$E^c = A^c \cap B^c = (A^c \cap N^c) \cup (A^c \cap B^c \cap N)$$

$A^c \cap N^c$ is in $\mathcal{F}$ and $A^c \cap B^c \cap N \subset N$, so $E^c \in \bar{\mathcal{F}}$.

We define $\bar{\mu}$ in the obvious way: If $E = A \cup B$ where $A \in \mathcal{F}$ and $B \subset N$ where $\mu(N) = 0$, then we let $\bar{\mu}(E) = \mu(A)$. The first thing to show is that $\bar{\mu}$ is well defined, i.e., if $E = A_i \cup B_i$, $i = 1, 2$, are two decompositions, then $\mu(A_1) = \mu(A_2)$. Let $A_0 = A_1 \cap A_2$ and $B_0 = B_1 \cup B_2$. $E = A_0 \cup B_0$ is a third decomposition with $A_0 \in \mathcal{F}$ and $B_0 \subset N_1 \cup N_2$, and has the pleasant property that if $i = 1$ or 2

$$\mu(A_0) \le \mu(A_i) \le \mu(A_0) + \mu(N_1 \cup N_2) = \mu(A_0)$$

The last detail is to check that $\bar{\mu}$ is measure, but that is easy. If $E_i = A_i \cup B_i$ are disjoint, then $\cup_i E_i$ can be decomposed as $\cup_i A_i \cup (\cup_i B_i)$, and the $A_i \subset E_i$ are disjoint, so

$$\bar{\mu}(\cup_i E_i) = \mu(\cup_i A_i) = \sum_i \mu(A_i) = \sum_i \bar{\mu}(E_i)$$ □

Theorem 1.1.11 allows us to construct Lebesgue measure λ on $(\mathbf{R}^d, \mathcal{R}^d)$. Using Theorem A.2.3, we can extend λ to be a measure on $(\mathbf{R}, \bar{\mathcal{R}}^d)$, where $\bar{\mathcal{R}}^d$ is the completion of $\mathcal{R}^d$. Having done this, it is natural (if somewhat optimistic) to ask: Are there any sets that are not in $\bar{\mathcal{R}}^d$? The answer is "Yes" and we will now give an example of a nonmeasurable B in $\mathbf{R}$.

A Nonmeasurable Subset of [0,1)

The key to our construction is the observation that λ is translation invariant: i.e., if $A \in \bar{\mathcal{R}}$ and $x + A = \{x + y : y \in A\}$, then $x + A \in \bar{\mathcal{R}}$ and $\lambda(A) = \lambda(x + A)$. We say that $x, y \in [0, 1)$ are equivalent and write $x \sim y$ if $x - y$ is a rational number. By the axiom of choice, there is a set B that contains exactly one element from each equivalence class. B is our nonmeasurable set, that is,

Theorem A.2.4 $B \notin \bar{\mathcal{R}}$.

Proof The key is the following:

Lemma A.2.5 *If* $E \subset [0, 1)$ *is in* $\bar{\mathcal{R}}$, $x \in (0, 1)$, *and* $x +' E = \{(x + y) \bmod 1 : y \in E\}$, *then* $\lambda(E) = \lambda(x +' E)$.

Proof Let $A = E \cap [0, 1 - x)$ and $B = E \cap [1 - x, 1)$. Let $A' = x + A = \{x + y : y \in A\}$ and $B' = x - 1 + B$. $A, B \in \bar{\mathcal{R}}$, so by translation invariance $A', B' \in \bar{\mathcal{R}}$ and $\lambda(A) = \lambda(A')$, $\lambda(B) = \lambda(B')$. Since $A' \subset [x, 1)$ and $B' \subset [0, x)$ are disjoint,

$$\lambda(E) = \lambda(A) + \lambda(B) = \lambda(A') + \lambda(B') = \lambda(x +' E)$$

□

From Lemma A.2.5, it follows easily that B is not measurable; if it were, then $q +' B$, $q \in \mathbf{Q} \cap [0, 1)$ would be a countable disjoint collection of measurable subsets of [0,1), all with the same measure α and having

$$\cup_{q \in \mathbf{Q} \cap [0,1)} (q +' B) = [0, 1)$$

If $\alpha > 0$, then $\lambda([0, 1)) = \infty$, and if $\alpha = 0$, then $\lambda([0, 1)) = 0$. Neither conclusion is compatible with the fact that $\lambda([0, 1)) = 1$, so $B \notin \bar{\mathcal{R}}$. □

Exercise A.2.1 Let B be the nonmeasurable set constructed in Theorem A.2.4. (i) Let $B_q = q +' B$ and show that if $D_q \subset B_q$ is measurable, then $\lambda(D_q) = 0$. (ii) Use (i) to conclude that if $A \subset \mathbf{R}$ has $\lambda(A) > 0$, there is a nonmeasurable $S \subset A$.

Letting $B' = B \times [0, 1]^{d-1}$, where B is our nonmeasurable subset of (0,1), we get a nonmeasurable set in $d > 1$. In $d = 3$, there is a much more interesting example, but we need the reader to do some preliminary work. In Euclidean geometry, two subsets of $\mathbf{R}^d$ are said to be **congruent** if one set can be mapped onto the other by translations and rotations.

Claim Two congruent measurable sets must have the same Lebesgue measure.

Exercise A.2.2 Prove the claim in $d = 2$ by showing (i) if B is a rotation of a rectangle A, then $\lambda^*(B) = \lambda(A)$. (ii) If C is congruent to D, then $\lambda^*(C) = \lambda^*(D)$.

Banach-Tarski Theorem

Banach and Tarski (1924) used the axiom of choice to show that it is possible to partition the sphere $\{x : |x| \le 1\}$ in $\mathbf{R}^3$ into a finite number of sets $A_1, \dots, A_n$ and find congruent sets $B_1, \dots, B_n$ whose union is two disjoint spheres of radius 1! Since congruent sets have the same Lebesgue measure, at least one of the sets A_i must be nonmeasurable. The construction relies on the fact that the group generated by rotations in $\mathbf{R}^3$ is not Abelian. Lindenbaum (1926) showed that this cannot be done with any bounded set in $\mathbf{R}^2$. For a popular account of the Banach-Tarski theorem, see French (1988).

Solovay's Theorem

The axiom of choice played an important role in the last two constructions of nonmeasurable sets. Solovay (1970) proved that its use is unavoidable. In his own words, "We show that the existence of a non-Lebesgue measurable set cannot be proved in Zermelo-Frankel set theory if the use of the axiom of choice is disallowed." This should convince the reader that all subsets of $\mathbf{R}^d$ that arise "in practice" are in $\bar{\mathcal{R}}^d$.

A.3 Kolmogorov's Extension Theorem

To construct some of the basic objects of study in probability theory, we will need an existence theorem for measures on infinite product spaces. Let $\mathbf{N} = \{1,2,\ldots\}$ and

$$\mathbf{R}^{\mathbf{N}} = \{(\omega_1,\omega_2,\ldots) : \omega_i \in \mathbf{R}\}$$

We equip $\mathbf{R}^{\mathbf{N}}$ with the product σ-algebra $\mathcal{R}^{\mathbf{N}}$, which is generated by the **finite dimensional rectangles** = sets of the form $\{\omega : \omega_i \in (a_i,b_i] \text{ for } i = 1,\ldots,n\}$, where $-\infty \le a_i < b_i \le \infty$.

Theorem A.3.1 (Kolmogorov's extension theorem) *Suppose we are given probability measures μ_n on $(\mathbf{R}^n,\mathcal{R}^n)$ that are consistent, that is,*

$$\mu_{n+1}((a_1,b_1] \times \ldots \times (a_n,b_n] \times \mathbf{R}) = \mu_n((a_1,b_1] \times \ldots \times (a_n,b_n])$$

Then there is a unique probability measure P on $(\mathbf{R}^{\mathbf{N}},\mathcal{R}^{\mathbf{N}})$ with

$$(*) \qquad P(\omega : \omega_i \in (a_i,b_i], 1 \le i \le n) = \mu_n((a_1,b_1] \times \ldots \times (a_n,b_n])$$

An important example of a consistent sequence of measures is

Example A.3.2 Let $F_1,F_2,\ldots$ be distribution functions and let μ_n be the measure on $\mathbf{R}^n$ with

$$\mu_n((a_1,b_1] \times \ldots \times (a_n,b_n]) = \prod_{m=1}^{n} (F_m(b_m) - F_m(a_m))$$

In this case, if we let $X_n(\omega) = \omega_n$, then the X_n are independent and X_n has distribution F_n.

Proof of Theorem A.3.1. Let $\mathcal{S}$ be the sets of the form $\{\omega : \omega_i \in (a_i,b_i], 1 \le i \le n\}$, and use $(*)$ to define P on $\mathcal{S}$. $\mathcal{S}$ is a semialgebra, so by Theorem A.1.1 it is enough to show that if $A \in \mathcal{S}$ is a disjoint union of $A_i \in \mathcal{S}$, then $P(A) \le \sum_i P(A_i)$. If the union is finite, then all the A_i are determined by the values of a finite number of coordinates and the conclusion follows from the proof of Theorem 1.1.11.

Suppose now that the union is infinite. Let $\mathcal{A} = \{$finite disjoint unions of sets in $\mathcal{S}\}$ be the algebra generated by $\mathcal{S}$. Since $\mathcal{A}$ is an algebra (by Lemma 1.1.7)

$$B_n \equiv A - \cup_{i=1}^{n} A_i$$

is a finite disjoint union of rectangles, and by the result for finite unions,

$$P(A) = \sum_{i=1}^{n} P(A_i) + P(B_n)$$

It suffices then to show:

Lemma A.3.3 *If $B_n \in \mathcal{A}$ and $B_n \downarrow \emptyset$, then $P(B_n) \downarrow 0$.*

Proof Suppose $P(B_n) \downarrow \delta > 0$. By repeating sets in the sequence, we can suppose

$$B_n = \cup_{k=1}^{K_n} \{\omega : \omega_i \in (a_i^k, b_i^k], 1 \le i \le n\} \quad \text{where } -\infty \le a_i^k < b_i^k \le \infty$$

The strategy of the proof is to approximate the B_n from within by compact rectangles with almost the same probability and then use a diagonal argument to show that $\cap_n B_n \ne \emptyset$. There is a set $C_n \subset B_n$ of the form

$$C_n = \cup_{k=1}^{K_n} \{\omega : \omega_i \in [\bar{a}_i^k, \bar{b}_i^k], 1 \le i \le n\} \quad \text{with } -\infty < \bar{a}_k^i < \bar{b}_k^i < \infty$$

that has $P(B_n - C_n) \le \delta/2^{n+1}$. Let $D_n = \cap_{m=1}^n C_m$.

$$P(B_n - D_n) \le \sum_{m=1}^{n} P(B_m - C_m) \le \delta/2$$

so $P(D_n) \downarrow$ a limit $\ge \delta/2$. Now there are sets $C_n^*, D_n^* \subset \mathbf{R}^n$, so that

$$C_n = \{\omega : (\omega_1, \ldots, \omega_n) \in C_n^*\} \quad \text{and} \quad D_n = \{\omega : (\omega_1, \ldots, \omega_n) \in D_n^*\}$$

Note that

$$C_n = C_n^* \times \mathbf{R} \times \mathbf{R} \times \ldots \quad \text{and} \quad D_n = D_n^* \times \mathbf{R} \times \mathbf{R} \times \ldots$$

so C_n and C_n^* (and D_n and D_n^*) are closely related but $C_n \subset \Omega$ and $C_n^* \subset \mathbf{R}^n$.

C_n^* is a finite union of closed rectangles, so

$$D_n^* = C_n^* \cap_{m=1}^{n-1} (C_m^* \times \mathbf{R}^{n-m})$$

is a compact set. For each m, let $\omega_m \in D_m$. $D_m \subset D_1$, so $\omega_{m,1}$ (i.e., the first coordinate of ω_m) is in D_1^* Since D_1^* is compact, we can pick a subsequence $m(1, j) \ge j$, so that as $j \to \infty$,

$$\omega_{m(1,j),1} \to \text{ a limit } \theta_1$$

For $m \ge 2$, $D_m \subset D_2$ and hence $(\omega_{m,1}, \omega_{m,2}) \in D_2^*$. Since D_2^* is compact, we can pick a subsequence of the previous subsequence (i.e., $m(2, j) = m(1, i_j)$ with $i_j \ge j$), so that as $j \to \infty$

$$\omega_{m(2,j),2} \to \text{ a limit } \theta_2$$

Continuing in this way, we define $m(k, j)$ a subsequence of $m(k-1, j)$, so that as $j \to \infty$,

$$\omega_{m(k,j),k} \to \text{ a limit } \theta_k$$

Let $\omega_i' = \omega_{m(i,i)}$. ω_i' is a subsequence of all the subsequences, so $\omega_{i,k}' \to \theta_k$ for all k. Now $\omega_{i,1}' \in D_1^*$ for all $i \ge 1$ and D_1^* is closed, so $\theta_1 \in D_1^*$. Turning to the second set, $(\omega_{i,1}', \omega_{i,2}') \in D_2^*$ for $i \ge 2$ and D_2^* is closed, so $(\theta_1, \theta_2) \in D_2^*$. Repeating the last argument, we conclude that $(\theta_1, \ldots, \theta_k) \in D_k^*$ for all k, so $\omega = (\theta_1, \theta_2, \ldots) \in D_k$ (no star here, since we are now talking about subsets of Ω) for all k and

$$\emptyset \ne \cap_k D_k \subset \cap_k B_k$$

a contradiction that proves the desired result. □

A.4 Radon-Nikodym Theorem

In this section, we prove the Radon-Nikodym theorem. To develop that result, we begin with a topic that at first may appear to be unrelated. Let $(\Omega, \mathcal{F})$ be a measurable space. α is said to be a **signed measure** on $(\Omega, \mathcal{F})$ if (i) α takes values in $(-\infty, \infty]$, (ii) $\alpha(\emptyset) = 0$, and (iii) if $E = +_i E_i$ is a disjoint union then $\alpha(E) = \sum_i \alpha(E_i)$, in the following sense:

If $\alpha(E) < \infty$, the sum converges absolutely and $= \alpha(E)$.

If $\alpha(E) = \infty$, then $\sum_i \alpha(E_i)^- < \infty$ and $\sum_i \alpha(E_i)^+ = \infty$.

Clearly, a signed measure cannot be allowed to take both the values ∞ and $-\infty$, since $\alpha(A) + \alpha(B)$ might not make sense. In most formulations, a signed measure is allowed to take values in either $(-\infty, \infty]$ or $[-\infty, \infty)$. We will ignore the second possibility to simplify statements later. As usual, we turn to examples to help explain the definition.

Example A.4.1 Let μ be a measure, f be a function with $\int f^- \, d\mu < \infty$, and let $\alpha(A) = \int_A f \, d\mu$. Exercise 5.8 implies that α is a signed measure.

Example A.4.2 Let μ_1 and μ_2 be measures with $\mu_2(\Omega) < \infty$, and let $\alpha(A) = \mu_1(A) - \mu_2(A)$.

The Jordan decomposition, Theorem A.4.6, will show that Example A.4.2 is the general case. To derive that result, we begin with two definitions. A set A is **positive** if every measurable $B \subset A$ has $\alpha(B) \geq 0$. A set A is **negative** if every measurable $B \subset A$ has $\alpha(B) \leq 0$.

Exercise A.4.1 In Example A.4.1, A is positive if and only if $\mu(A \cap \{x : f(x) < 0\}) = 0$.

Lemma A.4.3 *(i) Every measurable subset of a positive set is positive. (ii) If the sets A_n are positive, then $A = \cup_n A_n$ is also positive.*

Proof (i) is trivial. To prove (ii), observe that

$$B_n = A_n \cap \left(\cap_{m=1}^{n-1} A_m^c \right) \subset A_n$$

are positive, disjoint, and $\cup_n B_n = \cup_n A_n$. Let $E \subset A$ be measurable, and let $E_n = E \cap B_n$. $\alpha(E_n) \geq 0$, since B_n is positive, so $\alpha(E) = \sum_n \alpha(E_n) \geq 0$. □

The conclusions in Lemma A.4.3 remain valid if positive is replaced by negative. The next result is the key to the proof of Theorem A.4.5.

Lemma A.4.4 *Let E be a measurable set with $\alpha(E) < 0$. Then there is a negative set $F \subset E$ with $\alpha(F) < 0$.*

Proof If E is negative, this is true. If not, let n_1 be the smallest positive integer, so that there is an $E_1 \subset E$ with $\alpha(E_1) \geq 1/n_1$. Let $k \geq 2$. If $F_k = E - (E_1 \cup \ldots \cup E_{k-1})$ is negative, we are done. If not, we continue the construction letting n_k be the smallest positive integer, so that there is an $E_k \subset F_k$ with $\alpha(E_k) \geq 1/n_k$. If the construction does not stop for any $k < \infty$, let

$$F = \cap_k F_k = E - (\cup_k E_k)$$

Since $0 > \alpha(E) > -\infty$ and $\alpha(E_k) \geq 0$, it follows from the definition of signed measure that

$$\alpha(E) = \alpha(F) + \sum_{k=1}^{\infty} \alpha(E_k)$$

$\alpha(F) \leq \alpha(E) < 0$, and the sum is finite. From the last observation and the construction, it follows that F can have no subset G with $\alpha(G) > 0$, for then $\alpha(G) \geq 1/N$ for some N and we would have a contradiction. □

Theorem A.4.5 (Hahn decompositon) *Let α be a signed measure. Then there is a positive set A and a negative set B, so that $\Omega = A \cup B$ and $A \cap B = \emptyset$.*

Proof Let $c = \inf\{\alpha(B) : B \text{ is negative}\} \le 0$. Let B_i be negative sets with $\alpha(B_i) \downarrow c$. Let $B = \cup_i B_i$. By Lemma A.4.3, B is negative, so by the definition of c, $\alpha(B) \ge c$. To prove $\alpha(B) \le c$, we observe that $\alpha(B) = \alpha(B_i) + \alpha(B - B_i) \le \alpha(B_i)$, since B is negative, and let $i \to \infty$. The last two inequalities show that $\alpha(B) = c$, and it follows from our definition of a signed measure that $c > -\infty$. Let $A = B^c$. To show A is positive, observe that if A contains a set with $\alpha(E) < 0$, then by Lemma A.4.4, it contains a negative set F with $\alpha(F) < 0$, but then $B \cup F$ would be a negative set that has $\alpha(B \cup F) = \alpha(B) + \alpha(F) < c$, a contradiction. □

The Hahn decomposition is not unique. In Example A.4.1, A can be any set with

$$\{x : f(x) > 0\} \subset A \subset \{x : f(x) \ge 0\} \quad \text{a.e.}$$

where $B \subset C$ a.e. means $\mu(B \cap C^c) = 0$. The last example is typical of the general situation. Suppose $\Omega = A_1 \cup B_1 = A_2 \cup B_2$ are two Hahn decompositions. $A_2 \cap B_1$ is positive and negative, so it is a **null set**: All its subsets have measure 0. Similarly, $A_1 \cap B_2$ is a null set.

Two measures μ_1 and μ_2 are said to be **mutually singular** if there is a set A with $\mu_1(A) = 0$ and $\mu_2(A^c) = 0$. In this case, we also say μ_1 is **singular with respect to** μ_2 and write $\mu_1 \perp \mu_2$.

Exercise A.4.2 Show that the uniform distribution on the Cantor set (Example 1.2.7) is singular with respect to Lebesgue measure.

Theorem A.4.6 (Jordan decomposition) *Let α be a signed measure. There are mutually singular measures α_+ and α_- so that $\alpha = \alpha_+ - \alpha_-$. Moreover, there is only one such pair.*

Proof Let $\Omega = A \cup B$ be a Hahn decomposition. Let

$$\alpha_+(E) = \alpha(E \cap A) \quad \text{and} \quad \alpha_-(E) = -\alpha(E \cap B)$$

Since A is positive and B is negative, α_+ and α_- are measures. $\alpha_+(A^c) = 0$ and $\alpha_-(A) = 0$, so they are mutually singular. To prove uniqueness, suppose $\alpha = \nu_1 - \nu_2$ and D is a set with $\nu_1(D) = 0$ and $\nu_2(D^c) = 0$. If we set $C = D^c$, then $\Omega = C \cup D$ is a Hahn decomposition, and it follows from the choice of D that

$$\nu_1(E) = \alpha(C \cap E) \quad \text{and} \quad \nu_2(E) = -\alpha(D \cap E)$$

Our uniqueness result for the Hahn decomposition shows that $A \cap D = A \cap C^c$ and $B \cap C = A^c \cap C$ are null sets, so $\alpha(E \cap C) = \alpha(E \cap (A \cup C)) = \alpha(E \cap A)$ and $\nu_1 = \alpha_+$. □

Exercise A.4.3 Show that $\alpha_+(E) = \sup\{\alpha(F) : F \subset E\}$.

Remark Let α be a **finite signed measure** (i.e., one that does not take the value ∞ or $-\infty$) on $(\mathbf{R}, \mathcal{R})$. Let $\alpha = \alpha_+ - \alpha_-$ be its Jordan decomposition. Let $A(x) = \alpha((-\infty, x])$, $F(x) = \alpha_+((-\infty, x])$, and $G(x) = \alpha_-((-\infty, x])$. $A(x) = F(x) - G(x)$, so the distribution function for a finite signed measure can be written as a difference of two bounded increasing functions. It follows from Example A.4.2 that the converse is also true. Let $|\alpha| = \alpha^+ + \alpha^-$. $|\alpha|$ is called the **total variation of** α, since in this example $|\alpha|((a, b])$ is the total variation of A over $(a, b]$ as defined in analysis textbooks. See, for example, Royden (1988), p. 103. We exclude the left endpoint of the interval, since a jump there makes no contribution to the total variation on $[a, b]$, but it does appear in $|\alpha|$.

Our third and final decomposition is:

Theorem A.4.7 (Lebesgue decomposition) *Let μ, ν be σ-finite measures. ν can be written as $\nu_r + \nu_s$, where ν_s is singular with respect to μ and*

$$\nu_r(E) = \int_E g\, d\mu$$

Proof By decomposing $\Omega = +_i \Omega_i$, we can suppose without loss of generality that μ and ν are finite measures. Let $\mathcal{G}$ be the set of $g \geq 0$, so that $\int_E g\, d\mu \leq \nu(E)$ for all E.

(a) If $g, h \in \mathcal{G}$, then $g \vee h \in \mathcal{G}$.

Proof of (a). Let $A = \{g > h\}$, $B = \{g \leq h\}$.

$$\int_E g \vee h\, d\mu = \int_{E \cap A} g\, d\mu + \int_{E \cap B} h\, d\mu \leq \nu(E \cap A) + \nu(E \cap B) = \nu(E)$$

Let $\kappa = \sup\{\int g\, d\mu : g \in \mathcal{G}\} \leq \nu(\Omega) < \infty$. Pick g_n, so that $\int g_n\, d\mu > \kappa - 1/n$ and let $h_n = g_1 \vee \ldots \vee g_n$. By (a), $h_n \in \mathcal{G}$. As $n \uparrow \infty$, $h_n \uparrow h$. The definition of κ, the monotone convergence theorem, and the choice of g_n imply that

$$\kappa \geq \int h\, d\mu = \lim_{n \to \infty} \int h_n\, d\mu \geq \lim_{n \to \infty} \int g_n\, d\mu = \kappa$$

Let $\nu_r(E) = \int_E h\, d\mu$ and $\nu_s(E) = \nu(E) - \nu_r(E)$. The last detail is to show:

(b) ν_s is singular with respect to μ.

Proof of (b). Let $\epsilon > 0$ and let $\Omega = A_\epsilon \cup B_\epsilon$ be a Hahn decomposition for $\nu_s - \epsilon\mu$. Using the definition of ν_r and then the fact that A_ϵ is positive for $\nu_s - \epsilon\mu$ (so $\epsilon\mu(A_\epsilon \cap E) \leq \nu_s(A_\epsilon \cap E)$),

$$\int_E (h + \epsilon 1_{A_\epsilon})\, d\mu = \nu_r(E) + \epsilon\mu(A_\epsilon \cap E) \leq \nu(E)$$

This holds for all E, so $k = h + \epsilon 1_{A_\epsilon} \in \mathcal{G}$. It follows that $\mu(A_\epsilon) = 0$ for if not, then $\int k\, d\mu > \kappa$ a contradiction. Letting $A = \cup_n A_{1/n}$, we have $\mu(A) = 0$. To see that $\nu_s(A^c) = 0$, observe that if $\nu_s(A^c) > 0$, then $(\nu_s - \epsilon\mu)(A^c) > 0$ for small ϵ, a contradiction, since $A^c \subset B_\epsilon$, a negative set. □

Exercise A.4.4 Prove that the Lebesgue decomposition is unique. Note that you can suppose without loss of generality that μ and ν are finite.

We are finally ready for the main business of the section. We say a measure ν is **absolutely continuous with respect to** μ (and write $\nu \ll \mu$) if $\mu(A) = 0$ implies that $\nu(A) = 0$.

Exercise A.4.5 If $\mu_1 \ll \mu_2$ and $\mu_2 \perp \nu$, then $\mu_1 \perp \nu$.

Theorem A.4.8 (Radon-Nikodym theorem) *If μ, ν are σ-finite measures and ν is absolutely continuous with respect to μ, then there is a $g \geq 0$, so that $\nu(E) = \int_E g\, d\mu$. If h is another such function, then $g = h$ μ a.e.*

Proof Let $\nu = \nu_r + \nu_s$ be any Lebesgue decomposition. Let A be chosen, so that $\nu_s(A^c) = 0$ and $\mu(A) = 0$. Since $\nu \ll \mu$, $0 = \nu(A) \geq \nu_s(A)$ and $\nu_s \equiv 0$. To prove uniqueness, observe that if $\int_E g\, d\mu = \int_E h\, d\mu$ for all E, then letting $E \subset \{g > h, g \leq n\}$ be any subset of finite measure, we conclude $\mu(g > h, g \leq n) = 0$ for all n, so $\mu(g > h) = 0$, and, similarly, $\mu(g < h) = 0$. □

Example A.4.9 Theorem A.4.8 may fail if μ is not σ-finite. Let $(\Omega, \mathcal{F}) = (\mathbf{R}, \mathcal{R})$, $\mu =$ counting measure and $\nu =$ Lebesgue measure.

The function g whose existence is proved in Theorem A.4.8 is often denoted $d\nu/d\mu$. This notation suggests the following properties, whose proofs are left to the reader.

Exercise A.4.6 If $\nu_1, \nu_2 \ll \mu$, then $\nu_1 + \nu_2 \ll \mu$

$$d(\nu_1 + \nu_2)/d\mu = d\nu_1/d\mu + d\nu_2/d\mu.$$

Exercise A.4.7 If $\nu \ll \mu$ and $f \geq 0$, then $\int f \, d\nu = \int f \frac{d\nu}{d\mu} \, d\mu$.

Exercise A.4.8 If $\pi \ll \nu \ll \mu$, then $d\pi/d\mu = (d\pi/d\nu) \cdot (d\nu/d\mu)$.

Exercise A.4.9 If $\nu \ll \mu$ and $\mu \ll \nu$, then $d\mu/d\nu = (d\nu/d\mu)^{-1}$.

A.5 Differentiating under the Integral

At several places in the text, we need to interchange differentiate inside a sum or an integral. This section is devoted to results that can be used to justify those computations.

Theorem A.5.1 *Let $(S, \mathcal{S}, \mu)$ be a measure space. Let f be a complex valued function defined on $\mathbf{R} \times S$. Let $\delta > 0$, and suppose that for $x \in (y - \delta, y + \delta)$ we have*

(i) $u(x) = \int_S f(x,s) \, \mu(ds)$ with $\int_S |f(x,s)| \, \mu(ds) < \infty$

(ii) for fixed s, $\partial f / \partial x(x,s)$ exists and is a continuous function of x,

(iii) $v(x) = \int_S \frac{\partial f}{\partial x}(x,s) \, \mu(ds)$ is continuous at $x = y$,

and (iv) $\int_S \int_{-\delta}^{\delta} \left| \frac{\partial f}{\partial x}(y + \theta, s) \right| \, d\theta \, \mu(ds) < \infty$

then $u'(y) = v(y)$.

Proof Letting $|h| \leq \delta$ and using (i), (ii), (iv), and Fubini's theorem in the form given in Exercise 1.7.4, we have

$$\begin{aligned}
u(y+h) - u(y) &= \int_S f(y+h,s) - f(y,s) \, \mu(ds) \\
&= \int_S \int_0^h \frac{\partial f}{\partial x}(y+\theta,s) \, d\theta \, \mu(ds) \\
&= \int_0^h \int_S \frac{\partial f}{\partial x}(y+\theta,s) \, \mu(ds) \, d\theta
\end{aligned}$$

The last equation implies

$$\frac{u(y+h) - u(y)}{h} = \frac{1}{h} \int_0^h v(y+\theta) \, d\theta$$

Since v is continuous at y by (iii), letting $h \to 0$ gives the desired result. □

Example A.5.2 For a result in Section 3.3, we need to know that we can differentiate under the integral sign in

$$u(x) = \int \cos(xs) e^{-s^2/2} \, ds$$

For convenience, we have dropped a factor $(2\pi)^{-1/2}$ and changed variables to match Theorem A.5.1. Clearly, (i) and (ii) hold. The dominated convergence theorem implies (iii)

$$x \to \int -s\sin(sx)e^{-s^2/2}\,ds$$

is a continuous. For (iv), we note

$$\int \left|\frac{\partial f}{\partial x}(x,s)\right|\,ds = \int |s|e^{-s^2/2}\,ds < \infty$$

and the value does not depend on x, so (iv) holds.

For some examples the following form is more convenient:

Theorem A.5.3 *Let $(S,\mathcal{S},\mu)$ be a measure space. Let f be a complex valued function defined on $\mathbf{R}\times S$. Let $\delta > 0$, and suppose that for $x \in (y-\delta, y+\delta)$ we have*

(i) $u(x) = \int_S f(x,s)\,\mu(ds)$ with $\int_S |f(x,s)|\,\mu(ds) < \infty$

(ii) for fixed s, $\partial f/\partial x(x,s)$ exists and is a continuous function of x,

(iii′)
$$\int_S \sup_{\theta\in[-\delta,\delta]} \left|\frac{\partial f}{\partial x}(y+\theta,s)\right|\,\mu(ds) < \infty$$

then $u'(y) = v(y)$.

Proof In view of Theorem A.5.1 it is enough to show that (iii) and (iv) of that result hold. Since

$$\int_{-\delta}^{\delta} \left|\frac{\partial f}{\partial x}(y+\theta,s)\right|\,d\theta \le 2\delta \sup_{\theta\in[-\delta,\delta]} \left|\frac{\partial f}{\partial x}(y+\theta,s)\right|$$

it is clear that (iv) holds. To check (iii), we note that

$$|v(x)-v(y)| \le \int_S \left|\frac{\partial f}{\partial x}(x,s) - \frac{\partial f}{\partial x}(y,s)\right|\,\mu(ds)$$

(ii) implies that the integrand $\to 0$ as $x \to y$. The desired result follows from (iii′) and the dominated convergence theorem. □

To indicate the usefulness of the new result, we prove:

Example A.5.4 If $\phi(\theta) = Ee^{\theta Z} < \infty$ for $\theta \in [-\epsilon,\epsilon]$, then $\phi'(0) = EZ$.

Proof Here θ plays the role of x, and we take μ to be the distribution of Z. Let $\delta = \epsilon/2$. $f(x,s) = e^{xs} \ge 0$, so (i) holds by assumption. $\partial f/\partial x = se^{xs}$ is clearly a continuous function, so (ii) holds. To check (iii′), we note that there is a constant C, so that if $x \in (-\delta,\delta)$, then $|s|e^{xs} \le C\left(e^{-\epsilon s} + e^{\epsilon s}\right)$. □

Taking $S = \mathbf{Z}$ with $\mathcal{S}$ = all subsets of S and μ = counting measure in Theorem A.5.3 gives the following:

Theorem A.5.5 *Let $\delta > 0$. Suppose that for $x \in (y-\delta, y+\delta)$ we have*

(i) $u(x) = \sum_{n=1}^{\infty} f_n(x)$ with $\sum_{n=1}^{\infty} |f_n(x)| < \infty$

(ii) for each n, $f_n'(x)$ *exists and is a continuous function of* x,

and (iii) $\sum_{n=1}^{\infty} \sup_{\theta \in (-\delta,\delta)} |f_n'(y+\theta)| < \infty$

then $u'(x) = v(x)$.

Example A.5.6 In Section 2.6 we want to show that if $p \in (0,1)$, then

$$\left(\sum_{n=1}^{\infty}(1-p)^n\right)' = -\sum_{n=1}^{\infty} n(1-p)^{n-1}$$

Proof Let $f_n(x) = (1-x)^n$, $y = p$, and pick δ so that $[y-\delta, y+\delta] \subset (0,1)$. Clearly, (i) $\sum_{n=1}^{\infty} |(1-x)^n| < \infty$ and (ii) $f_n'(x) = n(1-x)^{n-1}$ is continuous for x in $[y-\delta, y+\delta]$. To check (iii), we note that if we let $2\eta = y - \delta$, then there is a constant C, so that if $x \in [y-\delta, y+\delta]$ and $n \geq 1$, then

$$n(1-x)^{n-1} = \frac{n(1-x)^{n-1}}{(1-\eta)^{n-1}} \cdot (1-\eta)^{n-1} \leq C(1-\eta)^{n-1}$$ □

Table of the Normal Distribution

$$\Phi(x) = \int_{-\infty}^{x} \frac{1}{\sqrt{2\pi}} e^{-y^2/2}\, dy$$

To illustrate the use of the table: $\Phi(0.36) = 0.6406$, $\Phi(1.34) = 0.9099$

	0	1	2	3	4	5	6	7	8	9
0.0	0.5000	0.5040	0.5080	0.5120	0.5160	0.5199	0.5239	0.5279	0.5319	0.5359
0.1	0.5398	0.5438	0.5478	0.5517	0.5557	0.5596	0.5636	0.5675	0.5714	0.5753
0.2	0.5793	0.5832	0.5871	0.5910	0.5948	0.5987	0.6026	0.6064	0.6103	0.6141
0.3	0.6179	0.6217	0.6255	0.6293	0.6331	0.6368	0.6406	0.6443	0.6480	0.6517
0.4	0.6554	0.6591	0.6628	0.6664	0.6700	0.6736	0.6772	0.6808	0.6844	0.6879
0.5	0.6915	0.6950	0.6985	0.7019	0.7054	0.7088	0.7123	0.7157	0.7190	0.7224
0.6	0.7257	0.7291	0.7324	0.7357	0.7389	0.7422	0.7454	0.7486	0.7517	0.7549
0.7	0.7580	0.7611	0.7642	0.7673	0.7703	0.7734	0.7764	0.7793	0.7823	0.7852
0.8	0.7881	0.7910	0.7939	0.7967	0.7995	0.8023	0.8051	0.8078	0.8106	0.8133
0.9	0.8159	0.8186	0.8212	0.8238	0.8264	0.8289	0.8315	0.8340	0.8365	0.8389
1.0	0.8413	0.8438	0.8461	0.8485	0.8508	0.8531	0.8554	0.8577	0.8599	0.8621
1.1	0.8643	0.8665	0.8686	0.8708	0.8729	0.8749	0.8770	0.8790	0.8810	0.8830
1.2	0.8849	0.8869	0.8888	0.8907	0.8925	0.8943	0.8962	0.8980	0.8997	0.9015
1.3	0.9032	0.9049	0.9066	0.9082	0.9099	0.9115	0.9131	0.9147	0.9162	0.9177
1.4	0.9192	0.9207	0.9222	0.9236	0.9251	0.9265	0.9279	0.9292	0.9306	0.9319
1.5	0.9332	0.9345	0.9357	0.9370	0.9382	0.9394	0.9406	0.9418	0.9429	0.9441
1.6	0.9452	0.9463	0.9474	0.9484	0.9495	0.9505	0.9515	0.9525	0.9535	0.9545
1.7	0.9554	0.9564	0.9573	0.9582	0.9591	0.9599	0.9608	0.9616	0.9625	0.9633
1.8	0.9641	0.9649	0.9656	0.9664	0.9671	0.9678	0.9686	0.9693	0.9699	0.9706
1.9	0.9713	0.9719	0.9726	0.9732	0.9738	0.9744	0.9750	0.9756	0.9761	0.9767
2.0	0.9772	0.9778	0.9783	0.9788	0.9793	0.9798	0.9803	0.9808	0.9812	0.9817
2.1	0.9821	0.9826	0.9830	0.9834	0.9838	0.9842	0.9846	0.9850	0.9854	0.9857
2.2	0.9861	0.9864	0.9868	0.9871	0.9875	0.9878	0.9881	0.9884	0.9887	0.9890
2.3	0.9893	0.9896	0.9898	0.9901	0.9904	0.9906	0.9909	0.9911	0.9913	0.9916
2.4	0.9918	0.9920	0.9922	0.9924	0.9927	0.9929	0.9931	0.9932	0.9934	0.9936
2.5	0.9938	0.9940	0.9941	0.9943	0.9945	0.9946	0.9948	0.9949	0.9951	0.9952
2.6	0.9953	0.9955	0.9956	0.9957	0.9959	0.9960	0.9961	0.9962	0.9963	0.9964
2.7	0.9965	0.9966	0.9967	0.9968	0.9969	0.9970	0.9971	0.9972	0.9973	0.9974
2.8	0.9974	0.9975	0.9976	0.9977	0.9977	0.9978	0.9979	0.9979	0.9980	0.9981
2.9	0.9981	0.9982	0.9982	0.9983	0.9984	0.9984	0.9985	0.9985	0.9986	0.9986
3.0	0.9986	0.9987	0.9987	0.9988	0.9988	0.9989	0.9989	0.9989	0.9990	0.9990

References

D. Aldous and P. Diaconis (1986) Shuffling cards and stopping times. *Amer. Math. Monthly.* 93, 333–348

E. S. Andersen and B. Jessen (1984) On the introduction of measures in infinite product spaces. *Danske Vid. Selsk. Mat.-Fys. Medd.* 25, No. 4

K. Athreya and P. Ney (1972) *Branching Processes.* Springer-Verlag, New York

K. Athreya and P. Ney (1978) A new approach to the limit theory of recurrent Markov chains. *Trans. AMS.* 245, 493–501

K. Athreya, D. McDonald, and P. Ney (1978) Coupling and the renewal theorem. *Amer. Math. Monthly.* 85, 809–814

L. Bachelier (1900) Théorie de la spéculation. *Ann. Sci. École Norm. Sup.* 17, 21–86

S. Banach and A. Tarski (1924) Sur la décomposition des ensembles de points en parties respectivements congruent. *Fund. Math.* 6, 244–277

L. E. Baum and P. Billingsley (1966) Asymptotic distributions for the coupon collector's problem. *Ann. Math. Statist.* 36, 1835–1839

F. Benford (1938) The law of anomalous numbers. *Proc. Amer. Phil. Soc.* 78, 552–572

J. D. Biggins (1977) Chernoff's theorem in branching random walk. *J. Appl. Probab.* 14, 630–636

J. D. Biggins (1978) The asymptotic shape of branching random walk. *Adv. Appl. Probab.* 10, 62–84

J. D. Biggins (1979) Growth rates in branching random walk. *Z. Warsch. verw. Gebiete.* 48, 17–34

P. Billingsley (1961) The Lindeberg-Lévy theorem for martingales. *Proc. AMS.* 12, 788–792

P. Billingsley (1979) *Probability and Measure.* John Wiley & Sons, New York

G. D. Birkhoff (1931) Proof of the ergodic theorem. *Proc. Nat. Acad. Sci.* 17, 656–660

D. Blackwell and D. Freedman (1964) The tail σ-field of a Markov chain and a theorem of Orey. *Ann. Math. Statist.* 35, 1291–1295

R. M. Blumenthal and R. K. Getoor (1968) *Markov Processes and Their Potential Theory.* Academic Press, New York

E. Borel (1909) Les probabilités dénombrables et leur applications arithmét- iques. *Rend. Circ. Mat. Palermo.* 27, 247–271

L. Breiman (1968) *Probability.* Addison-Wesley, Reading, MA

K. L. Chung (1974) *A Course in Probability Theory*, second edition. Academic Press, New York.

K. L. Chung, P. Erdös, and T. Sirao (1959) On the Lipschitz's condition for Brownian motion. *J. Math. Soc. Japan.* 11, 263–274

K. L. Chung and W. H. J. Fuchs (1951) On the distribution of values of sums of independent random variables. *Memoirs of the AMS*, No. 6

K. L. Chung and Z. Zhao (1995) *From Brownian Motion to Schrödinger's Equation.* Spinger, New York

V. Chvátal and D. Sankoff (1975) Longest common subsequences of two random sequences. *J. Appl. Probab.* 12, 306–315

J. Cohen, H. Kesten, and C. Newman (1985) *Random Matrices and Their Applications.* AMS Contemporary Math. 50, Providence, RI

J. T. Cox and R. Durrett (1981) Limit theorems for percolation processes with necessary and sufficient conditions. *Ann. Probab.* 9, 583–603

B. Davis (1983) On Brownian slow points. *Z. Warsch. verw. Gebiete.* 64, 359–367

P. Diaconis and D. Freedman (1980) Finite exchangeable sequences. *Ann. Prob.* 8, 745–764

J. Dieudonné (1948) Sur la théorème de Lebesgue-Nikodym, II. *Ann. Univ. Grenoble.* 23, 25–53

M. Donsker (1951) An invariance principle for certain probability limit theorems. *Memoirs of the AMS*, No. 6

M. Donsker (1952) Justification and extension of Doob's heurisitc approach to the Kolmogorov-Smirnov theorems. *Ann. Math. Statist.* 23, 277–281

J. L. Doob (1949) A heuristic approach to the Kolmogorov-Smirnov theorems. *Ann. Math. Statist.* 20, 393–403

J. L. Doob (1953) *Stochastic Processes.* John Wiley & Sons, New York

L. E. Dubins (1968) On a theorem of Skorkhod. *Ann. Math. Statist.* 39, 2094–2097

L. E. Dubins and D. A. Freedman (1965) A sharper form of the Borel-Cantelli lemma and the strong law. *Ann. Math. Statist.* 36, 800–807

L. E. Dubins and D. A. Freedman (1979) Exchangeable processes need not be distributed mixtures of independent and identically distributed random variables. *Z. Warsch. verw. Gebiete.* 48, 115–132

R. M. Dudley (1989) *Real Analysis and Probability.* Wadsworth Pub. Co., Pacific Grove, CA

R. Durrett and S. Resnick (1978) Functional limit theorems for dependent random variables. *Ann. Probab.* 6, 829–846

A. Dvoretsky (1972) Asymptotic normality for sums of dependent random variables. *Proc. 6th Berkeley Symp.*, Vol. II, 513–535

A. Dvoretsky and P. Erdös (1951) Some problems on random walk in space. *Proc. 2nd Berkeley Symp.* 353–367

A. Dvoretsky, P. Erdös, and S. Kakutani (1961) Nonincrease everywhere of the Brownian motion process. *Proc. 4th Berkeley Symp.*, Vol. II, 103–116

E. B. Dynkin (1965) *Markov processes.* Springer-Verlag, New York

E. B. Dynkin and A. A. Yushkevich (1956) Strong Markov processes. *Theor. Probab. Appl.* 1, 134–139.

P. Erdös and M. Kac (1946) On certain limit theorems of the theory of probability. *Bull. AMS.* 52, 292–302

P. Erdös and M. Kac (1947) On the number of positive sums of independent random variables. *Bull. AMS.* 53, 1011–1020

N. Etemadi (1981) An elementary proof of the strong law of large numbers. *Z. Warsch. verw. Gebiete.* 55, 119–122

S. N. Ethier and T. G. Kurtz (1986) *Markov Processes: Characterization and Convergence.* John Wiley & Sons, New York.

A. M. Faden (1985) The existence of regular conditional probabilities: necessary and sufficient conditions. *Ann. Probab.* 13, 288–298

W. Feller (1946) A limit theorem for random variables with infinite moments. *Amer. J. Math.* 68, 257–262

W. Feller (1961) A simple proof of renewal theorems. *Comm. Pure Appl. Math.* 14, 285–293

W. Feller (1968) *An Introduction to Probability Theory and Its Applications, Vol. I,* third edition. John Wiley & Sons, New York

W. Feller (1971) *An Introduction to Probability Theory and Its Applications, Vol. II,* second edition. John Wiley & Sons, New York

D. Freedman (1965) Bernard Friedman's urn. *Ann. Math. Statist.* 36, 956–970

D. Freedman (1971a) *Brownian Motion and Diffusion.* Originally published by Holden Day, San Francisco, CA. Second edition by Springer-Verlag, New York

D. Freedman (1971b) *Markov chains.* Originally published by Holden Day, San Francisco, CA. Second edition by Springer-Verlag, New York

D. Freedman (1980) A mixture of independent and identically distributed random variables need not admit a regular conditional probability given the exchangeable σ-field. *Z. Warsch. verw. Gebiete.* 51, 239–248

R. M. French (1988) The Banach-Tarski theorem. *Math. Intelligencer.* 10, No. 4, 21–28

B. Friedman (1949) A simple urn model. *Comm. Pure Appl. Math.* 2, 59–70

H. Furstenburg (1970) Random walks in discrete subgroups of Lie Groups. In *Advances in Probability* edited by P. E. Ney, Marcel Dekker, New York

H. Furstenburg and H. Kesten (1960) Products of random matrices. *Ann. Math. Statist.* 31, 451–469

A. Garsia (1965) A simple proof of E. Hopf's maximal ergodic theorem. *J. Math. Mech.* 14, 381–382

M. L. Glasser and I. J. Zucker (1977) Extended Watson integrals for the cubic lattice. *Proc. Nat. Acad. Sci.* 74, 1800–1801

B. V. Gnedenko (1943) Sur la distribution limité du terme maximum d'une série aléatoire. *Ann. Math.* 44, 423–453

B. V. Gnedenko and A. V. Kolmogorov (1954) *Limit Distributions for Sums of Independent Random Variables.* Addison-Wesley, Reading, MA

M. I. Gordin (1969) The central limit theorem for stationary processes. *Soviet Math. Doklady.* 10, 1174–1176

P. Hall (1982) *Rates of Convergence in the Central Limit Theorem.* Pitman Pub. Co., Boston, MA

P. Hall and C. C. Heyde (1976) On a unified approach to the law of the iterated logarithm for martingales. *Bull. Austral. Math. Soc.* 14, 435–447

P. Hall and C. C. Heyde (1980) *Martingale Limit Theory and Its Application.* Academic Press, New York

P. R. Halmos (1950) *Measure Theory.* Van Nostrand, New York

J. M. Hammersley (1970) A few seedlings of research. *Proc. 6th Berkeley Symp.*, Vol. I, 345–394

G. H. Hardy and J. E. Littlewood (1914) Some problems of Diophantine approximation. *Acta Math.* 37, 155–239

G. H. Hardy and E. M. Wright (1959) *An Introduction to the Theory of Numbers*, fourth edition. Oxford University Press, London

T. E. Harris (1956) The existence of stationary measures for certain Markov processes. *Proc. 3rd Berkeley Symp.*, Vol. II, 113–124

P. Hartman and A. Wintner (1941) On the law of the iterated logarithm. *Amer. J. Math.* 63, 169–176

E. Hewitt and L. J. Savage (1956) Symmetric measures on Cartesian products. *Trans. AMS.* 80, 470–501

E. Hewitt and K. Stromberg (1965) *Real and Abstract Analysis.* Springer-Verlag, New York

C. C. Heyde (1963) On a property of the lognormal distribution. *J. Royal. Stat. Soc. B.* 29, 392–393

C. C. Heyde (1967) On the influence of moments on the rate of convergence to the normal distribution. *Z. Warsch. verw. Gebiete.* 8, 12–18

J. L. Hodges, Jr. and L. Le Cam (1960) The Poisson approximation to the binomial distribution. *Ann. Math. Statist.* 31, 737–740

G. Hunt (1956) Some theorems concerning Brownian motion. *Trans. AMS.* 81, 294–319

I. A. Ibragimov (1962) Some limit theorems for stationary processes. *Theor. Probab. Appl.* 7, 349–382

I. A. Ibragimov (1963) A central limit theorem for a class of dependent random variables. *Theor. Probab. Appl.* 8, 83–89

I. A. Ibragimov and Y. V. Linnik (1971) *Independent and Stationary Sequences of Random Variables.* Wolters-Noordhoff, Groningen

H. Ishitani (1977) A central limit theorem for the subadditive process and its application to products of random matrices. *RIMS, Kyoto.* 12, 565–575

J. Jacod and A. N. Shiryaev, (2003) *Limit Theorems for Stochastic Processes* Second edition. Springer-Verlag, Berlin

K. Itô and H. P. McKean (1965). *Diffusion Processes and Their Sample Paths.* Springer-Verlag, New York

J. Jacod and A. N. Shiryaev (1987) *Limit Theorems for Stochastic Processes.* Springer, New York

M. Kac (1947a) Random walk and the theory of Brownian motion. *Amer. Math. Monthly.* 54, 369–391

M. Kac (1947b) On the notion of recurrence in discrete stochastic processes. *Bull. AMS.* 53, 1002–1010

M. Kac (1959) *Statistical Independence in Probability, Analysis, and Number Theory.* Carus Monographs, Math. Assoc. of America

Y. Katznelson and B. Weiss (1982) A simple proof of some ergodic theorems. *Israel J. Math.* 42, 291–296

E. Keeler and J. Spencer (1975) Optimal doubling in backgammon. *Operations Research.* 23, 1063–1071

H. Kesten (1986) Aspects of first passage percolation. In *École d'été de probabilités de Saint-Flour XIV.* Lecture Notes in Math 1180, Springer-Verlag, New York

H. Kesten (1987) Percolation theory and first passage percolation. *Ann. Probab.* 15, 1231–1271

J. F. C. Kingman (1968) The ergodic theory of subadditive processes. *J. Roy. Stat. Soc. B* 30, 499–510

J. F. C. Kingman (1973) Subadditive ergodic theory. *Ann. Probab.* 1, 883–909

J. F. C. Kingman (1975) The first birth problem for age dependent branching processes. *Ann. Probab.* 3, 790–801

A. N. Kolmogorov and Y. A. Rozanov (1964) On strong mixing conditions for stationary Gaussian processes. *Theor. Probab. Appl.* 5, 204–208

K. Kondo and T. Hara (1987) Critical exponent of susceptibility for a general class of ferromagnets in $d > 4$ dimensions. *J. Math. Phys.* 28, 1206–1208

U. Krengel (1985) *Ergodic Theorems.* deGruyter, New York

S. Leventhal (1988) A proof of Liggett's version of the subadditive ergodic theorem. *Proc. AMS.* 102, 169–173

P. Lévy (1937) *Théorie de l'addition des variables aléatoires.* Gauthier -Villars, Paris

T. M. Liggett (1985) An improved subadditive ergodic theorem. *Ann. Probab.* 13, 1279–1285

T. M. Liggett (2010) *Continuous Tiem Markov Processes. An Introduction.* American Mathematical Society, Providence, RI

A. Lindenbaum (1926) Contributions à l'étude de l'espace metrique. *Fund. Math.* 8, 209–222

T. Lindvall (1977) A probabilistic proof of Blackwell's renewal theorem. *Ann. Probab.* 5, 482–485

B. F. Logan and L. A. Shepp (1977) A variational problem for random Young tableaux. *Adv. in Math.* 26, 206–222

H. P. McKean (1969) *Stochastic Integrals.* Academic Press, New York

M. Motoo (1959) Proof of the law of the iterated logarithm through diffusion equation. *Ann. Inst. Stat. Math.* 10, 21–28

J. Neveu (1965) *Mathematical Foundations of the Calculus of Probabilities.* Holden-Day, San Francisco, CA

J. Neveu (1975) *Discrete Parameter Martingales.* North Holland, Amsterdam

S. Newcomb (1881) Note on the frequency of use of the different digits in natural numbers. *Amer. J. Math.* 4, 39–40

D. Ornstein (1969) Random walks. *Trans. AMS.* 138, 1–60

V. I. Oseleděc (1968) A multiplicative ergodic theorem. Lyapunov characteristic numbers for synmaical systems. *Trans. Moscow Math. Soc.* 19, 197–231

R. E. A. C. Paley, N. Wiener, and A. Zygmund (1933) Notes on random functions. *Math. Z.* 37, 647–668

E. J. G. Pitman (1956) On derivatives of characteristic functions at the origin. *Ann. Math. Statist.* 27, 1156–1160

E. Perkins (1983) On the Hausdorff dimension of the Brownian slow points. *Z. Warsch. verw. Gebiete.* 64, 369–399

S. C. Port and C. J. Stone (1969) Potential theory of random walks on abelian groups. *Acta Math.* 122, 19–114

S. C. Port and C. J. Stone (1978) *Brownian Motion and Classical Potential Theory.* Academic Press, New York

M. S. Ragunathan (1979) A proof of Oseledĕc's multiplicative ergodic theorem. *Israel J. Math.* 32, 356–362

R. Raimi (1976) The first digit problem. *Amer. Math. Monthly.* 83, 521–538

S. Resnick (1987) *Extreme Values, Regular Variation, and Point Processes.* Springer-Verlag, New York

D. Revuz (1984) *Markov Chains*, second edition. North Holland, Amsterdam

D. H. Root (1969) The existence of certain stopping times on Brownian motion. *Ann. Math. Statist.* 40, 715–718

M. Rosenblatt (1956) A central limit theorem and a strong mixing condition. *Proc. Nat. Acad. Sci.* 42, 43–47

H. Royden (1988) *Real Analysis*, third edition. McMillan, New York

D. Ruelle (1979) Ergodic theory of differentiable dynamical systems. *IHES Pub. Math.* 50, 275–306

C. Ryll-Nardzewski (1951) On the ergodic theorems, II. *Studia Math.* 12, 74–79

L. J. Savage (1972) *The Foundations of Statistics*, second edition. Dover, New York

D. J. Scott (1973) Central limit theorems for martingales and for processes with stationary independent increments using a Skorokhod representation approach. *Adv. Appl. Probab.* 5, 119–137

L. A. Shepp (1964) Recurrent random walks may take arbitrarily large steps. *Bull. AMS.* 70, 540–542

S. Sheu (1986) Representing a distribution by stopping Brownian motion: Root's construction. *Bull. Austral. Math. Soc.* 34, 427–431

N. V. Smirnov (1949) Limit distributions for the terms of a variational series. *AMS Transl. Series.* 1, No. 67

R. Smythe and J. C. Wierman (1978) *First Passage Percolation on the Square Lattice.* Lecture Notes in Math 671, Springer-Verlag, New York

R. M. Solovay (1970) A model of set theory in which every set of reals is Lebesgue measurable. *Ann. Math.* 92, 1–56

F. Spitzer (1964) *Principles of Random Walk.* Van Nostrand, Princeton, NJ

C. Stein (1987) *Approximate Computation of Expectations.* IMS Lecture Notes Vol. 7

C. J. Stone (1969) On the potential operator for one dimensional recurrent random walks. *Trans. AMS.* 136, 427–445

J. Stoyanov (1987) *Counterexamples in Probability.* John Wiley & Sons, New York

V. Strassen (1964) An invariance principle for the law of the iterated logarithm. *Z. Warsch. verw. Gebiete.* 3, 211–226

V. Strassen (1965) A converse to the law of the iterated logarithm. *Z. Warsch. verw. Gebiete.* 4, 265–268

V. Strassen (1967) Almost sure behavior of the sums of independent random variables and martingales. *Proc. 5th Berkeley Symp.*, Vol. II, 315–343

H. Thorisson (1987) A complete coupling proof of Blackwell's renewal theorem. *Stoch. Proc. Appl.* 26, 87–97

P. van Beek (1972) An application of Fourier methods to the problem of sharpening the Berry-Esseen inequality. *Z. Warsch. verw. Gebiete.* 23, 187–196

A. M. Vershik and S. V. Kerov (1977) Asymptotic behavior of the Plancherel measure of the symmetric group and the limit form of random Young tableau. *Dokl. Akad. Nauk SSR* 233, 1024–1027

H. Wegner (1973) On consistency of probability measures. *Z. Warsch. verw. Gebiete.* 27, 335–338

L. Weiss (1955) The stochastic convergence of a function of sample successive differences. *Ann. Math. Statist.* 26, 532–536

N. Wiener (1923) Differential space. *J. Math. Phys.* 2, 131–174

K. Yosida and S. Kakutani (1939) Birkhoff's ergodic theorem and the maximal ergodic theorem. *Proc. Imp. Acad. Tokyo* 15, 165–168

Index